PUBLICATIONS OF THE NEWTON INSTITUTE

Worldwide Asset and Liability Modeling

Publications of the Newton Institute

Edited by H.K. Moffatt

Director, Isaac Newton Institute for Mathematical Sciences

The Isaac Newton Institute of Mathematical Sciences of the University of Cambridge exists to stimulate research in all branches of the mathematical sciences, including pure mathematics, statistics, applied mathematics, theoretical physics, theoretical computer science, mathematical biology and economics. The four six-month long research programmes it runs each year bring together leading mathematical scientists from all over the world to exchange ideas through seminars, teaching and informal interaction.

WORLDWIDE ASSET AND LIABILITY MODELING

edited by

William T. Ziemba

University of British Columbia

and

John M. Mulvey

Princeton University

PUBLISHED BY THE PRESS SYNDICATE OF THE UNIVERSITY OF CAMBRIDGE
The Pitt Building, Trumpington Street, Cambridge CB2 1RP, United Kingdom

CAMBRIDGE UNIVERSITY PRESS
The Edinburgh Building, Cambridge CB2 2RU, United Kingdom
40 West 20th Street, New York, NY 10011-4211, USA
10 Stamford Road, Oakleigh, Melbourne 3166, Australia

First published 1998

Printed in the United Kingdom at the University Press, Cambridge

Typeset in 12pt Computer Modern

A catalogue record for this book is available from the British Library

ISBN 0 521 57187 1 hardback

CONTENTS

ACKNOWLEDGEMENTS

We thank the following publishers and authors for allowing us to reproduce the articles listed below:

'The importance of the asset allocation decision,' Chris R. Hensel, D. Don Ezra and John H. Ilkiw. Reprinted from *Financial Analysts Journal*, 1991 (July-August): 65–72.

'The effect of errors in means, variances, and covariances on optimal portfolio choice,' Vijay K. Chopra and William T. Ziemba. Reprinted from *Journal of Portfolio Management*, 1993 (Winter): 6–11.

'The comparative performance of global vs international models of equity market risk,' Stan Beckers, Gregory Connor and Ross Curds. Reprinted from *Financial Analysts Journal* **52** (2) 1996: 31–39.

'A global stock and bond model,' Lucie Chaumeton, Gregory Connor and Ross Curds. Reprint from *Financial Analysts Journal* **52** (6) 1996: 65–74.

'Optimal investment strategies for university endowment funds,' Robert C. Merton. Reprinted from *Studies of Supply and Demand in Higher Education*, Charles T. Clotfelter and Michael Rothschild, eds., University of Chicago Press, 1993: 211–236.

'The CALM stochastic programming model for dynamic asset-liability management,' Georgio Consigli and Michael A.H. Dempster. Reprinted from *Annals of Operations Research*, 1998 (in press).

'Asset and liability management under uncertainty for fixed income securities,' Stavros A. Zenios. Reprinted from *Annals of Operations Research* **59** 1995: 77–97.

'The Russell–Yasuda Kasai model: an asset/liability model for a Japanese insurance company using multistage stochastic programming,' David R. Cariño, Terry Kent, David H. Myers, Celine Stacy, Michael Sylvanus, Andrew Turner, Kanji Watanabe and William T. Ziemba. Reprinted from *Interfaces* **24** (1) 1994: 24–49.

CONTRIBUTORS

Stan Beckers, BARRA International Ltd, 1 Whittington Ave, London EC3V 1LE, England.

Adam J. Berger, Lattice Financial LLC, 55 Princeton–Hightstown Road, Princeton Junction, NJ 08550, USA

Marida Bertocchi, University of Bergamo, Department of Mathematics, Piazza Rosate 2, I-24129 Bergamo, Italy.

C.G.E. Boender, ORTEC Consultants bv, Groningenweg 6–33, 2803 PV Gouda, The Netherlands; and Free University Amsterdam, De Boelelaan 1107, 1081 HV Amsterdam, The Netherlands; and Erasmus University Rotterdam, P.O. Box 1738, 3000 DR Rotterdam, The Netherlands.

Michael J. Brennan, John E. Anderson Graduate School of Management, UCLA, 110 Westwood Plaza, Box 951481, Los Angeles, CA 90095–1481, USA; and London Business School, Sussex Place, Regent's Park, London NW 1 4SA, UK.

David R. Cariño, Frank Russell Company Pty Ltd, GPO Box 5291, Sydney NSW 2001, Australia.

Lucie Chaumeton, BARRA International Ltd, 1 Whittington Ave, London EC3V 1LE, England.

Vijay Kumar Chopra, Bankers Trust Company, 130 Liberty Street, MS 2355 New York, NY 10006, USA.

Gregory Connor, BARRA International Ltd, 1 Whittington Ave, London EC3V 1LE, England.

Georgio Consigli, Judge Institute of Management Studies, University of Cambridge, Cambridge CB2 1AG, England.

Sal Correnti, Falcon Asset Management Inc., Harborplace Tower, 18th Floor, 111 South Charles Street, Baltimore, MD 21202, USA.

Ross Curds, BARRA International Ltd, 1 Whittington Ave, London EC3V 1LE, England.

Michael A.H. Dempster, Judge Institute of Management Studies, University of Cambridge, Cambridge CB2 1AG, England.

Cees Dert, Vrije Universiteit Amsterdam, Faculteit der Economische Wetenschappen en Econometrie, Vakgroep BFS, De Boelelaan 1105, NL 1081 HV Amsterdam, The Netherlands; and ABN AMRO Asset Management, PAC AA 3260, PO Box 283, NL 1000 AE Amsterdam, The Netherlands.

Jitka Dupačovà, Charles University Prague, Department of Probability and Mathematical Statistics, Sokolovská 83, CZ-186 00 Prague, Czech Republic.

Kelly A. Easton, BZW Barclays Global Investors, 45 Fremont St, San Francisco, CA 94105, USA.

D. Don Ezra, Frank Russell Company, 909 A Street Tacoma, Washington 98402, USA.

Karl Frauendorfer, Institute of Operations Research, University of St. Gallen, Holzstrasse 15, CH-9010 St. Gallen, Switzerland.

Robert R. Grauer, Department of Economics and Faculty of Business, Simon Fraser University, Burnaby, British Columbia, Canada V5A 1S6.

Richard C. Grinold, BZW Barclays Global Investors, 45 Fremont St, San Francisco, CA 94105, USA.

Nils Hakansson, Haas School of Business, University of California, Berkeley, 350 Barrows Hall, Berkeley, CA 94720, USA.

Fred Heemskerk, ORTEC Consultants bv Groningenweg 6-33, 2803 PV Gouda, The Netherlands.

Chris R. Hensel, Frank Russell Company, 909 A Street Tacoma, Washington 98402, USA.

Martin Holmer, Policy Simulation Group, 1314 Kearney Street, NE Washington, DC 20017, USA

John H. Ilkiw, Frank Russell Company, 909 A Street Tacoma, Washington 98402, USA.

Terry Kent, US Olympic Society, Lake Placid, NY, USA.

Pieter Klaassen, Vrije Universiteit, Financial Sector Management, Department of Economics and Econometrics, De Boelelaan 1105, 1081 HV Amsterdam, The Netherlands; and Rabobank International, Postbus 17100, 3500 HG Utrecht, The Netherlands.

Robert C. Merton, Harvard Business School, Harvard University, Morgan 397, Soldiers Field, Boston, MA 02163, USA.

Vittorio Moriggia, University of Bergamo, Department of Mathematics, Piazza Rosate 2, I-24129 Bergamo, Italy.

John M. Mulvey, School of Engineering and Applied Science, Princeton University, Princeton, NJ 08544, USA

David H. Myers, School of Business, University of Washington, Seattle, WA 98185, USA.

Markus Rudolf, University of St. Gallen, Merkurstr. 1, CH-9000 St. Gallen, Switzerland.

Michael Schürle, Institute of Operations Research, University of St. Gallen, Holzstrasse 15, CH-9010 St. Gallen, Switzerland.

Edwardo S. Schwartz, John E. Anderson Graduate School of Management, UCLA, 110 Westwood Plaza, Box 951481, Los Angeles, CA 90095–1481, USA.

Suresh Sethi, Faculty of Management, Joseph L. Rotman Centre for Management, University of Toronto, 105 St George St., Toronto, Canada M5S 3E6.

Steve Sonlin, Falcon Asset Management Inc., Harborplace Tower, 18th Floor, 111 South Charles Street, Baltimore, MD 21202, USA.

Celine Stacy, Frank Russell Company, 909 A Street Tacoma, Washington 98402, USA.

John C. Sweeney, Falcon Asset Management Inc., Harborplace Tower, 18th Floor, 111 South Charles Street, Baltimore, MD 21202, USA.

Michael Sylvanus, Frank Russell Company, 909 A Street Tacoma, Washington 98402, USA.

A. Eric Thorlacius, Falcon Asset Management Inc., Harborplace Tower, 18th Floor, 111 South Charles Street, Baltimore, MD 21202, USA.

Andrew L. Turner, Frank Russell Company, 909 A Street Tacoma, Washington 98402, USA.

Paul van Aalst, Erasmus University Rotterdam, P.O. Box 1738, 3000 DR Rotterdam, The Netherlands 4; *and* KPMG Brans and Co.

Kanji Watanabe, The Yasuda Fire and Marine Insurance Co., Ltd. Shinjuku-ku, Tokyo 160, Japan.

Amy P. Williams, Falcon Asset Management Inc., Harborplace Tower, 18th Floor, 111 South Charles Street, Baltimore, MD 21202, USA.

Stavros A. Zenios, University of Cyprus; *and* HERMES Laboratory for Financial Modeling and Simulation, Decision Sciences Department, The Wharton School, University of Pennsylvania, Philadelphia PA 19104, USA.

William T. Ziemba, Faculty of Commerce and Business Administration, The University of British Columbia, Vancouver, British Columbia V6T 1Y8, Canada.

Heinz Zimmermann, Swiss Institute of Banking, University of St. Gallen, Merkurstr. 1, CH-9000 St. Gallen, Switzerland.

Preface

Few problems are as important and complex to institutions and individuals as the management of their assets in such a way that their liabilities can be covered and their goals achieved. The assets must be invested over time to achieve favorable returns subject to various uncertainties, policy and legal constraints, taxes and other requirements, and liability commitments. Most investors, be they individuals or institutions, do not diversify properly across markets or across time, particularly in relation to their liability commitments. There are many motivations for studying asset liability management, including:

(a) the results may be useful to set guidelines for institutions and individual investors concerning their asset allocation mixes; the models integrate various decisions over time with the constraints, preferences and uncertainties inherent in the investment problem; and

(b) the models consider temporal dependence of asset returns and liability commitments, path dependent preferences, short and long term tradeoffs and provide for realistic measurement of risks and their tradeoff with investment returns considering the effects of taxes, transaction costs and other problem features.

To study this area, I organized a week-long set of research seminars under the general theme 'Worldwide asset and liability modeling,' on May 15–20, 1995 at the Isaac Newton Institute for Mathematical Science on the campus of the University of Cambridge. This research program was followed by an institutional investor workshop on Saturday May 20th. This week's activities formed part of the six month Financial Mathematics Seminar held at the Newton Institute from January to June 1995. I organized this part of the program under the general direction of the financial mathematics seminar organizers Mark Davis, Stewart Hodges, Ioannis Karatzas and Chris Rogers. This volume consists of twenty-five papers arising from this program. Most of the papers appearing here were presented in Cambridge with a few added to round out the volume.

The research papers in this volume utilize several approaches and integrate a number of techniques such as single period mean-variance, multi-period models using stochastic programming with and without specific decision rules, dynamic stochastic control, stochastic dynamic programming and simulation. These papers discuss a variety of models that have been implemented, are close to being implemented, or represent new innovative approaches that may lead to future novel applications.

The volume also discusses issues concerned with the future of asset-liability management modeling. This includes models for individuals and various financial institutions such as banks and insurance companies. This will lead to custom financial engineering products. These models hold much promise for the future to provide users with organized, diversified systems to help manage their financial affairs in an increasingly complex financial world. The models force diversification and attention to extreme events and hence help minimize the possibility of financial disasters while at the same time providing good advice in ordinary circumstances balancing the various complex elements of the investor's situation.

The seminar in Cambridge took place in the efficient and most pleasant facilities of the Isaac Newton Institute on the campus of the University of Cambridge. The staff of the Institute, particularly Anne Cartwright, Florence Leroy, the Associate Director John Wright and Director Michael Atiyah were most helpful before and during our pleasant stay in Cambridge. Financial mathematics seminar chairman Chris Rogers was most supportive and helpful throughout this activity. My work in the practical use of asset and liability allocation models has been supported, encouraged and improved by my consulting association since 1989 with the Frank Russell company. Special thanks go to my Russell colleagues Chris Hensel and Andy Turner for their encouragement and our joint work. The Natural Sciences and Engineering Research Council of Canada has supported my research in stochastic programming including financial theory and asset-liability applications at the University of British Columbia since 1969. This support was very helpful with this project as well.

I was pleased to have John Mulvey join me as a co-editor of this volume. Besides co-authoring the introduction with me, John adds his special insights gained from years of outstanding research and consulting to improve the papers in this volume as well as contributing several outstanding co-authored papers based on his own pioneering work. Our editor David Tranah has been most helpful and patient in the preparation of this volume. Finally special thanks go to my wife Sandra Schwartz for much encouragement and help on the seminar in Cambridge and in the preparation of this volume.

William T. Ziemba
Vancouver

Part I
Introduction

Asset and Liability Management Systems for Long-Term Investors: Discussion of the Issues

John M. Mulvey and William T. Ziemba

1   Introduction

This book surveys optimal investment policies for long term investors – especially those who wish to achieve goals and meet future obligations. This domain is called asset and liability management (ALM). Allocating assets lies at the heart of a strategic risk management system. In addition, liability streams and their uncertainty, institutional constraints and policies, taxes, transaction costs and the like are important features in real financial planning. Application areas described here include pension plans, insurance companies, investment conglomerates, banks, university endowments, and wealthy and ordinary individuals. These investors possess future liabilities and goals. They must make investment decisions while considering the use of their funds, that is, investing for a purpose. Risks must be measured in the context of the entire organization's or individual's financial situation.

Many investors do not manage their strategic asset mix despite much evidence that asset allocation decisions are critical for investors with diversified portfolios. Why do these investors ignore strategic planning? First, there are few computer systems for evaluating asset allocation decisions while taking into account an investor's particular temporal circumstances, liabilities and goals. The analysis is made more difficult since each investor has a unique set of qualifying factors. The adage 'one size fits all' is clearly wrong when it comes to long-term financial planning. The difficult actuarial task of evaluating the soundness of a pension plan illustrates the problem. Each U.S. company with a defined-benefit pension plan must conduct a comprehensive analysis of its plan in the context of its long-term liabilities.

As a second barrier, evaluating long-term investment strategies requires several components that are not now commonly available. There must be a way to generate scenarios that is logically consistent and based on sound economic principles. Parameters of the scenario generator must fit past data and trends. Yet the stochastic model must take into account changing economic conditions, for example, the deregulation of interest rates and currencies. The behavior of the investor must also be accounted for. We must include risk aversion in the context of the temporal nature of the decision problems. The optimization module must integrate asset decisions, liability decisions, and goal payments over time. Combining these elements into an easily computable stochastic optimization system is technically challenging.

For large investors, asset allocation typically consists of two steps. First, there is the setting of target values for the major asset categories (large cap stock, cash, bonds, international stocks, etc.) Once these targets are set, the investor (or an investment committee) hires managers who attempt to beat the associated risk adjusted indices. Alternatively they might purchase index funds to match the target proportions. Typically about one quarter of active managers beat their benchmarks, and sometimes like 1996 to 1998, much less for the widely followed S&P500 index. Reasons for this and sample data appear in the financial press and in investment research papers and books; see e.g. Ziemba & Schwartz (1991). Active managers must meet or beat their index, or face being fired for lack of acceptable performance. A portfolio manager is generally given several years to sort out the performance issue, in order to reduce the chances that luck enters into the manager choice decision. Periodically, the strategic issue is revisited. Typically, the allocation decision is evaluated at least annually, for example, at the company's board of directors meeting.

This is not to minimize the strong results from passive management strategies. Hensel, Ezra and Ilkiw's paper in this volume studies this. They considered seven representative US clients of the Frank Russell Company who were using professional money managers whose goal was to 'beat their benchmarks with lower risk' for sixteen quarters from January 1985 to December 1988. The specified fix mix was 50% US equity, 5% nonUS equity, 30% fixed income, 5% real estate and 10% cash equivalents, which was rebalanced quarterly. The results detailed in Table 1 indicate that most of the volatility is explained by the naive fixed-mix policy allocation. T-bills and the naive portfolio allocation explain most of the return and market timing security selection, etc. provided little benefit.

The papers by Chopra and Ziemba, Hensel and Turner, and Grinold and King deal with static portfolio problems in a mean variance context. Chopra and Ziemba show that errors in mean estimation are crucial to obtaining accurate portfolio weights. Errors in variances are about twice as damaging as errors in co-variances with the means being about ten times more important than the variances. Hensel and Turner explore this further, focusing on ways to massage the inputs and constrain the outputs to obtain superior investment decisions. Grinold and Kelly develop a model that allows one to track over time the impact of various sources to the performance of the portfolio. Specifically they decompose the expected returns used to select the optimal portfolio into those from an ex-ante efficient portfolio, those from the manager's special

Table 1. Average Return and Return Variation Explained (Quarterly)
by Seven Clients of the Frank Russell Company

Decision Level	Average Contribution	Additional Variation Explained by this Level (Volatility)
Minimum Risk Portfolio (T-Bills)	1.62%	2.66%
Naive Policy Allocation (what other pension funds are doing; fixed-mix)	2.13	94.35
Specific Policy Allocation	0.49	0.50
Market Timing	(0.10)	0.14
Security Selection	(0.23)	0.40
Interaction and Activity	(0.005)	1.95
Total	3.86%	100.00%

insights and those due to legal, policy, diversification and other constraints.

Multiperiod stochastic asset allocation improves upon static strategies. Instead of passive management between meetings of the investment committee who are responsible for setting asset allocation strategy, a dynamic asset allocation adjusts the mix as conditions change. The Cariño–Turner paper demonstrates how the strategy changes as time and uncertainty unfold themselves with a simple three period five year example. This and other papers in the volume show the superiority of stochastic dynamic models over, for example, simple fixed-mix or buy-and-hold strategies. Dynamic strategies pinpoint the relationship between asset risks, liability risks, and goal achievement which ultimately maximizes the investor's wealth net of liabilities and penalty costs and goals.

Liabilities and goals alter the investment process. First, we differentiate *intrinsic* from *contextual* risks. Intrinsic risk refers to uncertainty surrounding a single security, e.g. Microsoft stock. The price of the stock may increase or decrease in tandem with the market – this risk cannot be easily eliminated through diversification strategies by asset-only investors. In contrast, risks which are unrelated to market movements, namely non-systematic risks, can be mitigated through diversification. For investors with long-term liabilities, market risks can be reduced since the financial well being for long-term investors is a function not only of assets, but also among other elements liabilities, interest rates (through discounting), goals, and possibly inflation. A standard formula for determining financial well being is: Wealth = assets − PV (liabilities). For example, see Sharpe & Tint (1990) and Peskin (1997). We define an alternative measure for financial soundness as: 'surplus wealth', indicating the investor's financial position relative to both liabilities and goals: Surplus wealth = assets − PV(liabilities) − PV (goals). A positive surplus indicates that the investor will be likely to meet future financial obligations along with his goals. A deficit portends the opposite − the investor should then re-evaluate his situation.

To calculate surplus wealth, we must expand the traditional asset and liability framework. Goals must be modeled. The aim is to move up the risk ladder (Figure 1) so that the asset and liability management system includes greater details, and thus is more representative of the investor's financial condition. At the top of the risk ladder, we estimate surplus wealth through the concepts of 'Total integrated risk management' TIRM (Mulvey, Armstrong & Rothberg 1995).

Rung 5:   Total integrated risk management
Rung 4:   Dynamic asset and liability management
Rung 3:   Dynamic asset-only
Rung 2:   Static asset-only portfolios
Rung 1:   Pricing single securities

Figure 1. The Risk Ladder

At every level on the ladder, there are numerous applications of dynamic investment strategies. Most relevant for our purposes are investors who possess long-term liabilities and/or goals and are keen to build a system for evaluating the long-term consequences of today's actions. An integrated approach assists in setting and evaluating financial expectations. It helps investors plan ahead in a consistent fashion. It also helps measure past performance. Noteworthy applications include the following areas:

A. Pension plans

Actuaries evaluate the long-term viability of pension plans with respect to future contributions, anticipated pay-outs to beneficiaries, and other future uncertainties. Uncertainty has been addressed traditionally via a smoothing approach. Legislation and changes in regulation, however, have pushed the analysis towards measuring economic viability and including risks in their studies. ALM examples include implemented models such as: Frank Russell's Mitsubishi Trust Model, see Cariño *et al.* (1995), and Swiss Bank Corporation models, the Towers Perrin CAP:Link System (Mulvey 1996b), and ORTEC's system in the Netherlands (Boender 1995). Elements of these systems are described in this volume: Cariño *et al.* 1994 provides the framework used for the Mitsubishi and Swiss Bank models, Mulvey & Thorlacius describes CAP-Link and Boender, van Aalst & Heemskerk describe ORTEC's model.

B. Insurance companies

Similar to pension plans, insurance companies are highly regulated and therefore their approach for analyzing a company's economic soundness is dictated by past regulations (some of which can be dated). The market is another evaluator of performance. Examples of ALM in insurance are the Russell–Yasuda Kasai Model (Cariño *et al.* 1994, Cariño & Ziemba 1998, Cariño, Myers & Ziemba 1998), Falcon Asset Management (Mulvey, Correnti & Lummis 1997), Renaissance Re-insurance (Lowe & Stanard 1996) and the CALM system described in this volume by Consigli & Dempster.

C. Banks

Banks have been slow to implement integrated risk management systems at the strategic level, despite the severe problems that grew out of the US Savings and Loan crisis in the 1980s and the Japanese banking crisis in the 1990s. The latter was caused by the severe decline in the land and stock markets because of high valuations and high interest rates. The former was caused by regulatory aspects associated with fixed versus variable interest rates. See Pyle (1995), Shaw, Thorp & Ziemba (1995) and Stone & Ziemba (1993) for analyses of these crisis situations, respectively. Funds are often allocated based on a short-term notion of risk – called value at risk (VAR). Recently, however, a thrust has begun to employ tactical risk management systems that tend to be one or two period models such as Algorithmics's riskwatch; see Dembo (1995). For discussions, see e.g. Davidson (1996), Jorion (1996a,b) and Linsmeier & Pearson (1996). Banks rarely employ traditional asset allocation strategies since their portfolios consist of fixed income, real estate and other illiquid asset categories. J.P. Morgan and others are active in this area. Zenios's paper discusses stochastic programming models for fixed-income securities using interest rate contingencies. Portfolios of mortgage backed securities provide the setting for the empirical analyses.

D. Portfolio and mutual fund managers

Many fund managers aim to beat a specified index such as the S&P 500 or the Russell 2000 (small cap stocks), and they are evaluated based on their risk-adjusted performance compared to the benchmark index. An interesting example is Keynes' management of the trading assets of King's College, Cambridge from 1928–45, see Chua & Woodward (1983). In this context, the index equates to the liabilities – the investment goal is to compute risks relative to return on the index. Portfolio managers place constraints on the investments. Asset categories can vary widely, such as broad categories, or sub-indices such as industry sectors, or even individual securities. The number of decision variables increases as investment details are included.

E. Individuals

Individuals can benefit by implementing dynamic asset and management strategies. They can evaluate the level of savings and investment strategies appropriate for meeting future financial goals, such as college education and retirement. Berger and Mulvey's paper describes a multi-stage asset and liability system for individuals that optimizes over decision rules. Fan, Murray & Turner (1997) describe an implemented system for individual customers of the large Italian bank, Banca Fideuram in Rome, using multiperiod stochastic programming.

F. University Endowments

By their nature, universities must consider the long-run when managing their endowment assets. Nevertheless, goals and future liabilities influence their investment risks. This is discussed in Merton's paper in this volume. The basic idea is to locate investment opportunities that are closely correlated with liabilities and goals. All else being equal, risks are reduced by finding asset categories which display co-movements with the present value of liabilities and goals. An example is to invest in real estate for faculty housing in the neighborhood surrounding the university. This investment serves two aims: (1) it maintains the integrity of the area – a worthy goal; and (2) it assists in faculty compensation – an important liability – by supplementing salaries by means of subsidizing housing expenses. This also reduces contextual risks to surplus wealth. The Rudolf & Ziemba (1997) paper is an extension of the Merton (1969, 1990) continuous time model to include liabilities via mutual fund representations.

Another application domain involves insurance for industrial and other large corporations possessing catastrophic property loss exposure. In this case, asset categories are the company's major operating components. Liabilities refer to borrowing decisions. Goals can be dividends, purchases of other companies, etc. The result is an enterprise wide risk management system, built around the scenario generator and the multi-period optimization model. Through integrated risk management, critical insurance decisions are made at the highest corporate level as they affect the overall probability

decision nodes and periods

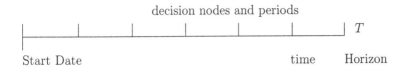

Figure 2. The Planning Period $(t = 1, 2, 3, \ldots, T)$

distribution of shareholders' surplus, rather than at lower levels of the enterprise. Insurance modifies the surplus distribution in a predictable fashion – an example of financial engineering, see Correnti *et al.* (1996). The next section highlights the primary components of long-term asset and liability management systems.

2   Model Structure

The investment process consists of $t = \{1, 2, 3, \ldots, T\}$ time stages. The first stage represents the current date. The end of the planning period, T, is called the planning horizon. Typically, it depicts a point at which the investor has some critical planning purpose, such as the repayment date of a substantial liability. In some models a separate *end effects* period represents periods $T + 1, \ldots;$ – see Grinold (1980, 1983) for general technique and Cariño, Myers & Ziemba (1998) for application to the Russell–Yasuda Kasai model discussed in the Cariño *et al.* paper in this volume. This technique, which assumes that the dual prices in the periods $T + 1, \ldots,$ past the horizon increase in relation to the rate of interest, yields one more steady state period in the model with accompanying variables for that period.

At the beginning of each period, the investor makes decisions regarding the asset mix, the liabilities, and the financial goals. There are uncertainties between time periods. For example, the stock market and bond returns are correlated. The analysis can utilize a system of stochastic differential equations for modeling the stochastic parameters over time of asset pricing models. These relate a set of key economic factors to remaining components, such as asset and liability returns. See, for an example, the Towers–Perrin CAP:Link system discussed by Mulvey (1996b) and in Mulvey & Thorlacius' paper in this volume. Alternative modeling approaches address the integration of the stochastic and the optimization models in a different manner.

The primary decision variables designate asset proportions, liability related decisions, and goal payments, namely:

$x_{j,t}^s$ investment in asset j
$y_{k,t}^s$ liability decision k
$u_{l,t}^s$ goal payment l,
        for time t and scenario s.

In each time period t, the model maximizes its objective function, $f(x)$, by moving resources between asset categories, adjusting liabilities, and achieving goals. There are several candidates for the objective function; see subsection 2.1. In addition, we impose constraints on the process such as limiting borrowing to certain ratios, addressing transactions costs whenever assets are bought or sold, or taking advantage of investment opportunities. There are several modeling approaches for including constraints. These constraints tend to provide preference relations in addition to the objective function. For example, there is utility by satisfying a constraint on a liability. These constraints also provide more curvature of the objective function which, in many cases such as the Frank Russell models, is simply the maximization of expected terminal wealth net of penalty costs on cash flows, goals, etc. Our goal is to find a feasible point, which maximizes a temporal objective function. Since we are dealing with uncertainty in a temporal setting, the optimal solution, like all points, will encompass a set of paths – trajectories – for the investor's wealth (or other measures such as surplus wealth). Ranking these paths is discussed in the next subsection.

There are two basic equations for the flow of funds:

For the jth asset category:

$$x_{j,t+1}^s = (x_{j,t}^s + r_{j,s}^s) - p_{j,t}^s(1+t_j) + q_{j,t}^s(1-t_j^+) \quad \text{for asset } j, \text{ time } t, \text{ scenario } s,$$

where

$$
\begin{aligned}
r_{j,s}^s &= \text{return for asset } j, \\
p_{j,t}^s &= \text{sales of asset } j, \\
q_{j,t}^s &= \text{purchase of asset } j \\
t_j &= \text{transaction costs for asset } j \text{ for time } t \text{ and scenario } s.
\end{aligned}
$$

For the cash flows:

$$x_{l,t+1}^s = (x_{l,t}^s + r_{l,s}^s) - \sum_j q_{j,t}^s + \sum_j p_{j,t}^s(1-t_j^-) + w_t^s - \sum_k y_{k,t}^s - \sum_l u_{l,t}^s$$

where w_t^s = cash inflows at time t, scenario s, cash is asset category l.

The multi-stage investment model avoids looking into the future in an inappropriate fashion. The model cannot optimize over scenarios that do not represent a range of plausible outcomes for the future. To prevent this occurrence, non-anticipatory constraints are added to the model which have the form:

$$x_{j,t}^{s_1} = x_{j,t}^{s_2}$$

for all scenarios s_1 and s_2 inheriting a common past up to time t, that is these prior decisions must be the same; see Rockafellar & Wets (1991).

The financial planning system addresses these non-anticipatory conditions, either explicitly or implicitly, and special purpose algorithms are available for solving the stochastic optimization model.

2.1   Objective Functions

A major element of asset and liability management involves trading off risks and rewards. The standard assets-only theory based on capital asset or arbitrage pricing theory, see e.g. the papers in this volume by Beckers, Connor & Curds, and Chaumeton, Connor & Curds. Connor & Korajczk (1995), survey asset pricing models. Chaumeton *et al.* argue that six fundamental risk factors – four for stocks and two for bonds – explain most of the common volatility of individual international stocks and bonds. The cross-national component of the risk factors is stronger within the European Union than it is elsewhere. Their model is useful for worldwide asset selection and allocation decisions. Ferson (1995) and articles in the companion volume Keim & Ziemba (1998) assume that investments possessing more volatility generate greater expected returns over time than assets with lower levels of volatility. For a factor model that argues the opposite for several countries, see Haugen & Baker (1996). Early multiple fundamental anomaly factor models appear in Jacobs & Levy (1988) for the US and Ziemba & Schwartz (1991) for Japan. The latter is discussed by Ziemba & Schwartz in Keim & Ziemba. The articles by Beckers *et al.* and Chaumeton *et al.* have an international setting in the factor model tradition of their firm BARRA. The temporal issue complicates the decision since longer term horizons dictate a longer time span to recoup losses, thus the more volatile assets may be, in fact, safer in terms of contextual risks. An example is the stock/cash comparison: stocks provide higher expected returns but are more volatile than cash. Indeed, the longer the horizon, the safer are the high return high variance assets. We must consider the time horizon in measuring contextual risks. The paper by Brennan and Schwartz in this volume analyzes this issue among other questions using a Merton-type continuous time instantaneous model where the asset returns depend upon fundamental factors such as interest rates, dividend yields, price–earnings ratios and the like. They show the effect of the time horizon and demonstrate that higher return riskier assets do seem to be safer the longer the horizon.

There are numerous ways to evaluate financial risks, just as there are alternative measures of profitability. We might consider the chance of a loss over the next year to be 15%. Or, we might set a profitability target and evaluate the probability of missing the target. In both cases, risk increases as a function of probability. An improved alternative for evaluating risks is to

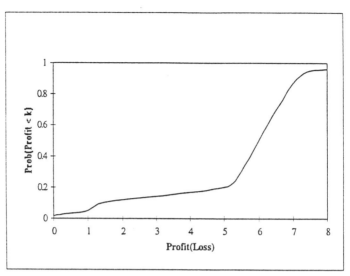

Figure 3. Cumulative Distribution Function for Shareholders Equity

estimate the probability distribution of shareholders equity (Figure 3), along with other measures of financial well being for the company.

Calculating these curves links the major uncertainties in a financial organization. Given a distribution, we can evaluate not only risks but also compare it against reward potential. Typically, we equate reward with expected value. The profit or loss in the next year is

$$\text{Expected profit} = \sum_{s \in S} p_s z^s$$

where p_s is the probability of scenario s, z^s is the profit or loss under scenario s, and S is a set of representative scenarios.

Comparing alternative distributions on a direct basis can be difficult for most decision-makers. To aid in the process, we can employ the concepts of stochastic dominance. For example, if two cumulative distribution functions cross only once, a decision maker who is risk averse, i.e., has a concave utility function, will take the curve with the highest mean return if its variance is less than the alternative, see Hanoch & Levy (1969). Other dominance tests are possible, but these tests are unlikely to apply in a wide set of circumstances; see Ziemba & Vickson (1975) for more on this issue.

There are two primary theories for an objective function. First, we can apply the classical von Neumann–Morgenstern theory and transform random variables to deterministic values, called certainty equivalents. Second, we can fit a classical utility function to the characteristics of the model's output. An example is to define risk as the volatility of a portfolio's return. There are numerous variants of each theory, as discussed in the following two subsections.

2.1.1   Von Neumann–Morgenstern Theory

After more than a half century, the von Neumann–Morgenstern (1944) expected utility maximizing theory remains the pre-eminent approach for making decisions under uncertainty. The optimization model is:

$$\text{Max } E(u(w)) = \sum_s p_s u(w^s) \tag{VM}$$

where $u(w^s)$ is the VM preference function, w^s is the investor's wealth under scenario s, and p_s is the probability of scenario s.

Once the solution, w^*, of VM is found we determine its certainty equivalent value by computing the inverse function at the recommended solution, namely, $\text{CE} = u^{-1}(w^*)$. This CE represents the amount in cash that we would take in order to sell (or buy) the random variable w.

Technically it is convenient in many cases to assume v is exponential and concave since for normally distributed scenarios Max $u(w)$ is equivalent, using Freund's (1957) result, to maximizing the mean–variance model

$$\bar{\mu} - R_A \mu \sigma^2$$

where R_A is the investor's risk Arrow–Pratt aversion index (see Arrow (1971) and Pratt (1964)), namely $-u''/u'$, where primes denote derivatives. For an early application of this to portfolio theory with log normally distributed assets, see Dexter, Yu & Ziemba (1980).

While the VM theory provides a systematic basis for making consistent decisions, it does not address the temporal aspects of decisions over a planning horizon. There are several ways to extend the theory:

1. Select a future target date, for example, the end of the planning horizon. Conduct the CE analysis with respect to the investor's wealth at the target date. Possibly constrain the probability of wealth dropping below some pre-assigned values at intermediate periods ($t = 1, 2, \ldots, T - 1$); a theory for doing this in the context of capital growth theory, i.e. log utility is developed in MacLean & Ziemba (1997). It is discussed in subsection 2.2.3.

2. Transform the random variable z from wealth at period T to a random variable which measures time to achieve a stated objective. For instance, an investor may be interested in the date in which enough has been saved so that retirement with 75% of annual income is assured. Intermediate values of wealth may be constrained.

3. Develop a multi-attribute version of the VM model (see Keeney & Raiffa, 1976, 1994). This approach generally requires simplifying assumptions in order to be practical.

The first two approaches address final wealth as the primary attribute. However, even long-term investors rarely avoid the pain or joy that occurs when their investments underperform or overperform expectations. Investors typically will have a more volatile wealth history to achieve greater returns. Stocks are more volatile than fixed income investments; however they have had higher returns over long time horizons. They are safer the longer the horizon. Thus, there is tension between security and portfolio growth; see subsection 2.2.3.

How does one select a u–function? The theory is straightforward in a static setting. The first step is to find risk premiums for reference gambles:

Risk Premium (RP) = Expected Values − Certainty Equivalence.

For example, suppose that you are faced with the following 50/50 gamble:

20% increase in your surplus wealth over next year (prob. = .5)

−5% increase in your surplus wealth over next year (prob. = .5)

Would you take this gamble over a 1 year Treasury bill – paying 5%? Questions of this type help the investor determine his CE for a wide range of outcomes over wealth (or other suitable dimension) as shown in Figure 4. This example is for a colleague of the second author and was estimated by the certainty equivalent and gain and loss methods, see Keeney & Raifa (1976, 1994).

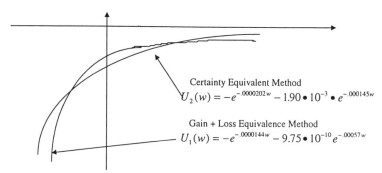

Certainty Equivalent Method
$$U_2(w) = -e^{-.0000202w} - 1.90 \bullet 10^{-3} \bullet e^{-.000145w}$$

Gain + Loss Equivalence Method
$$U_1(w) = -e^{-.0000144w} - 9.75 \bullet 10^{-10} e^{-.00057w}$$

Figure 4. Example of a Utility Function $u(w) = -e^{-bw} - ce^{-dw}$
which is Strictly Increasing and Strictly Concave with
Risk Aversion R_A Decreasing or Constant

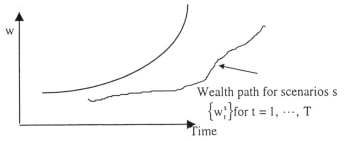

Figure 5. A Target Wealth Path for the Planning Horizon

2.1.2   Classical Utility Functions

In most people's minds, utility functions have little to do with risk premiums and certainty equivalence as advocated by von Neumann and Morgenstern. Instead, a (cardinal) utility function sets a numerical value dictating the relative importance for some characteristic of a model's performance. It is simply a desirability index; it readily applies to the multi-period context. Since our dynamic investment model consists of discrete paths over the planning period with numerous variables of interest, we must be careful to summarize the model's behavior into a small group of summary statistics.

The range of plausible utility functions is large. For example, deviations could be penalized from selected wealth targets at selected time periods – downside risks – as shown in Figure 5; see also Kusy & Ziemba (1986) and the Cariño–Turner and Cariño *et al.* papers in this volume for use in practice. Or we could create a risk measure that takes into account asymmetries in the wealth function as proposed by Bell (1995).

Penalties can be given substantial weights to reflect their importance: goals and liabilities which are time sensitive can be assigned higher priorities than goals which are less critical:

$$\text{Maximize} f(x) = \lambda_1 g_1(x) + \lambda_2 g_2(x) + \cdots + \lambda_k g_k(x)$$

where $g_i(x) =$ the ith goal, $\lambda_i =$ relative importance for goal i. Selecting goals and priorities is a subject of current research, including issues pertaining to regret, habits, changing tastes, and motivational considerations in a temporal setting.

2.1.3   Integrating VM Theory and Utility Functions

Several researchers have brought together the rigor of the VM theory with the intuitive appeal of classical utility functions. For example, Kroll, Levy & Markowitz (1984) show that the two concepts are equivalent for single-period portfolio selection when the investor possesses a quadratic *u*-function; see also

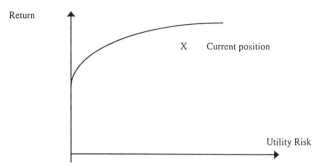

Figure 6. Utility Risk-Modified Efficient Frontier

the earlier discussions in Hanoch & Levy (1969) and in papers and problems in Ziemba & Vickson (1975). Herein, the efficient frontier for returns and risk trace out the optimal solutions to the VM model.

Bell (1995) has shown that the u-function $u(w) = w - be^{-cw}$ for constants $b > 0$, and $c > 0$ is equivalent to a modified risk measure. For a given investor, we can draw efficient frontiers in which risk is based on outcome distributions that are asymmetric in shape. We can analyze the investor's current position to make evaluations regarding the direction of movement – taking on greater risks for greater potential rewards, or moving down in risk. Using these ideas yields a utility risk-modified efficient frontier as shown in Figure 6.

The utility risk measurement is defined as follows:

$$\text{Risk(wealth)} = k \times \left(\log(\text{Expected value } (e^{(\text{wealth}-w^*)})) \right)$$

where $k = 2/(c^2)$, $w^* =$ expected wealth and $c =$ coefficient in VM preference function.

This approach reduces the effort of specifying the risk aversion coefficient – a distinct advantage. If one takes the approach to specify a risk aversion index then the functional form of the utility function does not matter much, at least in the case of normally distributed returns. Then, as shown in calculations and theory in Kallberg & Ziemba (1983), the certainty equivalents and portfolio weights are similar for different utility functions if the average Arrow–Pratt risk aversions are similar.

2.2   Alternative Model Structures

This section describes four frameworks for finding 'optimal' decisions over the planning horizon. The approaches share several key ideas. Three of them start with a system for generating the stochastic parameters, for example, using stochastic differential equations. They employ an objective function that spans the planning horizon. They attempt to optimize an objective function while maintaining a set of constraints or relationships. They enforce non-anticipatory conditions. They make simplifying assumptions in order to carry out the calculations in a reasonable manner. The fourth approach employs single period models to approximate the multi-period asset and liability management problem. Under certain assumptions, see Hakanson (1972) and Hakanson & Ziemba (1995), the single-period model is optimal for the long term. For example, the capital growth model optimizes the growth of the investor's portfolio in the absence of transaction costs, assuming that the asset choices do not affect the stochastic parameters. A fifth approach through stochastic simulation is also discussed in this book in the papers by Holmer and by Boender, van Aalst and Heemskerk. A sixth approach might be characterized as the static mean variance methods discussed by Chopra and Ziemba, Grinold and Kelly, and Hensel and Turner. But here we focus on dynamic asset liability optimization models

2.2.1 Stochastic Programming

A stochastic program determines the optimal investment at each time period as a function of the outlook for the assets in comparison with the investor's circumstances using a large multi-stage stochastic program. A key idea is the generation of scenarios via a tree structure as shown in Figure 7.

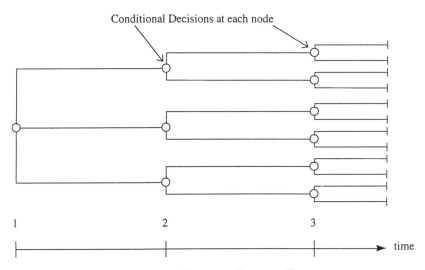

Figure 7. A Scenario Decision Tree

Stochastic programming models are based on the expansion of the decision space, taking into account the conditional nature of the scenario tree. Conditional decisions are made at each node subject to the modeling constraints. We take advantage of every opportunity with the sampled scenarios.

An example of the stochastic programming approach was developed by the second author and staff at the Frank Russell Company. Their models use the following elements in various ways:

- Multiple time periods; end effects – steady state after decision horizon

- Consistency with economic and financial theory

- Discrete scenarios for random elements - returns, liability costs, currency movements

- Alternative forecasting models, handle fat tails

- Institutional, legal and policy constraints

- Penalize for shortfalls in objective function

- Tradeoff short, intermediate and long term goals

- Model derivatives and illiquid assets

- Model transactions costs, taxes, etc

- Multiple expressions of risk in terms understandable to decision makers

- Maximize expected utility of final wealth net of penalty costs

- Can now solve very realistic multiperiod problems on modern workstations using large-scale linear programming and stochastic programming algorithms.

The Frank Russell Company has been active in developing such models. The general model ideas were developed in Kallberg, White & Ziemba (1982), and especially Kusy & Ziemba (1986). More recent models as described in the Cariño *et al.* paper in this volume with more details in Cariño & Ziemba (1998) and Cariño, Myers & Ziemba (1998). A number of implementations have occurred, including in 1997, an assets only model for individual customers of the Banca Fidenram in Rome. Each of the models possesses five periods including initialization and only the Russell–Yasuda model used end effects. The implementations are compared in Table 2.

The success of the models for institutional application demonstrates the practicality of the approach, that it is possible to successfully build and implement large-scale stochastic programming asset and asset-liability models. The application to individual ALM problems is the next frontier and it remains to be seen how successful the models will be in this multi-trillion dollar market; see Berger and Mulvey's paper in this volume for one such model.

Table 2. Large Scale Stochastic Programming Models Developed
and Implemented by the Frank Russell Company

Model	Type of Application	Year Delivered	Number of Scenarios Used	Computer Hardware Implementation
Russell–Yasuda (Tokyo)	Property and Casualty Insurance	1991	256	IBM RISC 6000
Mitsubishi Trust (Tokyo)	Pension Consulting	1994	2000	IBM RISC 6000 with Parallel Processers
Swiss Bank Corporation (Basle)	Pension Consulting	1996	8000	IBM UNIX2
Daido Life Insurance Company (Tokyo)	Life Insurance	1997	25,600	IBM PC
Banca Fideuram (Rome)	Assets Only Personal	1997	10,000	IBM UNIX2 and PC

2.2.2 Decision Rules

A driving decision rule is a function for calculating the values of the investment and other business decisions at each time period. It takes as inputs the state of the world at time t:

$$x_{j,t} = h(a_{j,t}^s, b_{j,t}^s, \ldots)$$

where the $a, b, \ldots$ parameters depict the driving factors.

The decision variables are indexed over time and the state of the organization, but not necessarily over scenarios. A simple example is the fixed-mix strategy. At the end of each time-period, the investor sells overperforming assets and purchases underperforming assets to maintain a target level of the asset categories, e.g. 60% stock and 40% bonds. The fixed-mix rule is as follows:

$$e_j = x_{j,t}^s / \sum_j x_{j,t}^s$$

where e_j = fixed ratio for asset j.

The amount of assets bought and sold can be readily computed by comparing the current portfolio and the ideal portfolio – as set by the e-variables. The ratio for asset j to total assets is fixed across all time periods and scenarios. The investor's surplus wealth may or may not factor into the evaluation. Even in this example the investor must rebalance the portfolio at each stage. Mulvey & Chen (1996) shows that fixed-mix reduces risks and improves returns over a passive buy and hold strategy. Also, from a theoretical standpoint, variants of fixed-mix are optimal for many long-term investors even in the presence of transaction costs (Davis & Norman 1990, and Merton 1969, 1990). Fixed-mix gives rise to small transaction costs compared to other actively managed strategies since the portfolio rarely makes dramatic modifications.

A criticism of fixed-mix pertains to the independence of the investor's risk aversion vis à vis his wealth over the planning period. For instance, most investors would be willing to take on greater risks when their wealth increases with respect to their investment goals. Thus, a pension plan might accept a considerably different mix when they are under-funded as compared with a plan possessing a substantial (e.g. 150%) surplus.

It is relatively easy to develop decision rules to encompass investor's wealth. For example, we can build upon a common variant of portfolio insurance, called constant proportional portfolio insurance CPPI (Perold & Sharpe 1988), which modifies the ratio of risky assets to risk-free assets in the following manner:

$$\% \text{ Risky Assets} = \text{Min}\{D \times (\text{Surplus Wealth} - \text{Floor}) + \text{MinRisk}, \text{MaxRisk}\}$$

where D = risk aversion parameter, Floor = minimum acceptable wealth level, Surplus wealth = estimate of current position with respect to goals and liabilities, MinRisk = minimum level of risky asset and MaxRisk = maximum level of risk asset.

As the investor's wealth increases, the proportion of the risky asset increases with D. Aggressive investors will choose a large value (e.g. $D = 3$) whereas conservative investors will choose a smaller value (e.g. $D = 0.3$). We also impose a limit on the percentage of the risky asset in the portfolio.

The time-step for implementing more aggressive CPPI strategies should be shorter than for the traditional fixed-mix strategy since it renders relatively large changes in the asset mix as the markets move. There may be practical difficulties carrying out this strategy during periods of high volatility and imbalances in supply and demand. The 1987 crash is an example when CPPI would have been inefficient since prices did not clear during substantial periods of time when the market experienced a sudden and sharp decline.

For CPPI, the definition of the risky asset and the risk-free asset are completely general. For instance, the risk-free asset could be an immunized bond portfolio for some projected fixed liabilities. We could purchase a portfolio of US inflation-linked government bonds and money market cash, when the liability cash flows grow with future inflation. Thus, as wealth increases, the investor will hold a higher percentage of riskier assets. The risk-free asset should provide protection against adverse moves in the securities markets.

The D parameter measures risk aversion. A large value of D will cause major swings in the investor's mix. When markets are rising, the investor will switch to the appreciating asset – a momentum strategy. Conversely, when markets are dropping, the investor will sell aggressively. The CPPI strategy performs best when markets are moving either up or down in a persistent fashion. Fixed-mix performs best when markets are volatile but generally moving sideways.

A third approach for incorporating the investor's wealth into the decision rules is to define a target wealth path over the planning horizon. The target wealth path should refer to the surplus wealth for the investor:

$$\text{Surplus wealth} = \text{total assets} - \text{PV(liabilities)} - \text{PV(goals)},$$

where PV is the present value of future liability cashflows. The investor may simply be interested in maintaining a positive surplus wealth over time. In this dynamic environment, we develop an investment strategy and accompanying decision rules which control the relationship of the actual wealth to the target. For instance, we might become more conservative when the surplus wealth drives above the target path by a substantial level. We can decrease or increase cash inflows as part of the strategy – such as those from pension plan contributions. Alternatively, we can select the fixed-mix that matches the target wealth as closely as possible, given the investor's tolerance for risks. Alternatively we could maximize the probability that the investor will reach a target wealth at a specified time.

Given one or more decision rules, we can build a multi-period ALM model for optimizing the setting of decision rules. For example, the best fix-mix proportions can be determined over the set of S scenarios. These optimization problems are relatively small, but they often result in non-convex models and it is difficult to identify the global optimal solution. Examples of optimizing decision rules are Falcon Asset Liability Management (Mulvey, Correnti & Lummis 1997) and Towers Perrin's Opt:Link system, see Mulvey & Thorlacius' paper in this volume.

2.2.3 Capital Growth

Given a set of risky assets how should one invest in each to maximize the long run growth of assets (without borrowing or liabilities)? Kelly (1956) showed that under certain assumptions this is done by maximizing the expected log of asset wealth, that is, by using a logarithmic utility function. Breiman (1961) supplied rigorous mathematical proofs that this strategy did indeed asymptotically maximize long-run asset wealth and minimize the time to achieve a particular goal for large asymptotic goals. Hakansson (1972) showed that the Kelly or capital growth strategy was myopic, that is period by period optimization is optimal for general asset distributions. Advanced proofs of the Breiman and other results in the intertemporally independent and weakly dependent cases appear in Algoet & Cover (1988). Rotando & Thorp (1992) apply the Kelly strategy to long-term investment in the US stock market and demonstrate some of the benefits and liabilities of that strategy. The pros and cons of the capital growth criteria are summarized in Table 1 of MacLean, Ziemba & Blazenko (1992); see also Rubinstein (1977). A major drawback is the size of the investment wagers on the most favorable assets leading to the highest growth rate. The fraction of wealth invested may be unacceptably high because the Arrow–Pratt risk aversion index, see Arrow (1971) and Pratt (1964), is the reciprocal of wealth and hence is essentially zero for reasonable levels of wealth. Indeed the log utility asymptotically provides infinitely more terminal wealth than any other essentially different, i.e. one that differs infinitely often, strategy. However, this increase in wealth is accompanied by large volatility swings in wealth. This motivated Maclean, Ziemba & Blazenko (1992) to develop a theory of growth versus security using fractional Kelly strategies, which are convex combinations of cash, and the Kelly fraction; see also Maclean & Ziemba (1991). Maclean & Ziemba (1997) extend the ideas by adding draw down constraints. They apply this to the dynamic allocation of wealth in stocks, bonds and cash equivalents across international markets. This is done by adjusting the Kelly fraction much in the same way that futures based portfolio insurance strategies blend cash and the market index. See Grossman & Zhou (1996) for an analysis along these lines.

Grauer and Hakansson in their paper in this volume show the typical behavior of the capital growth criterion with the Kelly fraction with investments in cash, stocks and bonds. They use an inflation adapter plus discrete past data scenarios to generate rather good long-term investment results using data for the US from 1934 to 1988. There are large investments in the best current assets and a bumpy ride to high final wealth. Mulvey (1989) extended the single period model such as in Grauer & Hakanson's paper in this volume to address assets and liability for the Pacific Mutual Insurance Company.

Hakansson & Ziemba (1995) review the capital growth literature and various applications. The Kelly (and fraction Kelly) strategies are not for all or even most investors. However, it has helped create many wealthy individuals who are willing to accept considerable risks.

Stochastic Control

Dynamic stochastic control (SC) offers an alternative to stochastic programming for setting dynamic investment strategies. The approach dates to the work of Samuelson (1969), Merton (1969, 1990), and others. A key idea is to form a state space for the driving variables at each time period. Rather than discretizing the scenarios, SC forms a mesh over the state space. Either dynamic programming algorithms or finite element algorithms are available for solving the problem. Brennan, Schwartz & Lagnado (1997) and Brennan & Schwartz (this volume) applied SC for asset allocation.

Short term government interest rates
Long term government interest rates
Dividend yield
Time remaining in planning period

Figure 8. Driving Variables for Brennan–Schwartz–Lagnado SC Model

In order to incorporate liabilities in a dynamic stochastic control model, we must relate the present value of the liability cashflows to the driving economic variables. (See Merton's paper in this volume for university endowment management). As an important example, a pension plan must pay beneficiaries over their retirement years. In many cases, the pension plan increases payoffs as a function of inflation. In addition, the current workforce receives wage increases that must be factored in the model. Thus, the SC model must address at least wage-inflation and the discount rates (the yield curve) for the long-term liabilities. Actuaries estimate the cashflows for pension plans and insurance companies in a systematic fashion. These estimates form the basis for the company's contribution in the future.

2.2.4   Pros and Cons of the Four Approaches

Each of the four approaches to dynamic investing has something to offer. Decision rules are far easier to implement and can be optimized without resorting to large scale linear or non-linear programs. They can be readily tested with out-of-sample scenarios and provide confidence limits on the recommendations. They are intuitive for most professional investors. However, they can lead to non-convex optimization models, requiring extensive searching to find a global optimal solution. Also the rules may lead to suboptimal behavior.

Stochastic programming provides a general purpose-modeling framework that can address real-world features such as turnover constraints, transaction costs, risk aversion, taxes, limits on groups of assets and other considerations.

It requires highly efficient solution algorithms due to the enormous number of decision variables, especially for the multi-stage problems with 4 or more stages. Typical model applications from the Frank Russell research group are 5 stages. Its recommendations can be tested out-of-sample, but the computational costs are so high as to be impractical for many users. A limiting feature is the sampling of scenarios from the stochastic model.

Capital growth models lead to high growth of assets but with considerable risk. When controlled through fractional Kelly strategies or drawn down constraints, the policies provide superior growth security trade-offs. The models' asset generation process however must be simple with simple liability considerations. They are relatively easy to solve once the data structures are in place. The policies tend to concentrate on a small number of superior assets and hence can be poorly diversified; also as with stochastic control, the asset proportions are very sensitive to the input parameters.

Stochastic control is another general-purpose framework: it applies to problems in which the state space can be kept manageable, i.e. with at most 3 or 4 driving variables. Of course, other parameters and variables can be a function of the driving variables. As with stochastic programming, the generation of confidence limits is difficult to calculate. Also, modeling errors may arise due to the state space approximation. The difficulty in specifying general constraints on the process limits SC applications. However, it has a conceptual edge over stochastic programming when it can be implemented since there is no need to sample scenarios. Much needs to be done to implement asset only models with the existing theory, particularly with asset weight constraints, and the development of the theory and applications for asset-liability applications.

In summary, there is no clear-cut winner among the four candidates. We suggest that investors start out with several candidate decision models and rules. They can be readily implemented and optimized. The selected decision rules can serve as benchmarks for the more complex MSP and SC approaches. Also one can link MSP and decision rules to estimate confidence limits on the model's recommendations. Models that combine elements of the four approaches in sections of their analysis may also be desirable.

3   Generating Scenarios

A critical aspect of ALM systems involves modeling the driving stochastic parameters, such as interest rates, inflation, and stock returns. A scenario depicts a single, coherent set of parameter values over the planning horizon $t = \{1, 2, 3, \ldots, T\}$. Coefficients must be internally consistent within a single scenario. Bond returns must correspond to changes in interest rates.; see the paper by Frauendorfer and Schürle in this volume for a one factor interest rate

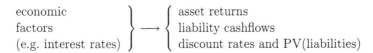

Figure 9. Driving Economic Factors

model applied to a multiperiod stochastic programming model. Barycentric approximation of this process generates the scenario trees which take the various term structure movements into account. Empirical results are given for six and eight period models; see also Murray & Cariño (1994) for approaches to generate discrete scenarios from multivariate lognormal distributions. Currency returns must be generated by currency movements, see e.g. Sweeney, Sonlin & Correnti and Rudolf & Zimmerman in this volume, Gardner & Stone (1995) and Sorensen, Mezrich & Thadani (1993). Maintaining consistency requires that a small set of economic factors drive the overall results as shown in Figure 9. The Towers Perrin CAP:Link system follows these dictums (Mulvey 1996b and Mulvey & Thorlacius' paper in this volume). See also Wilkie (1986, 1995).

Since surplus wealth requires the simultaneous calculation of asset values minus the present value of liabilities for a given scenario; we must ensure that interest rates movements are directly related to asset returns – including government and corporate bonds. At the same time, interest rate movements display certain recurring features, including mean reversion. Generally, we assume that interest rates are monitored and controlled by the central banks, at least for the G7 countries. Thus, we consider as remote the possibility of hyperinflation in the US, Germany, or Japan. See Mulvey and Thorlacius' paper in this volume.

Another issue regarding scenario selection is the need for constructing a scenario tree when employing stochastic programming. Decision rules are not restricted to a tree format. For example, King & Warden (1994) developed an approach for converting isolated scenarios into a tree structure for the Allstate Insurance project.

Estimation of the fat tails which are typical in asset markets can be done several ways. Longin (1996), using 105 years of stock market data, found that the tails were best fit with a Frechet distribution. Jackwert & Rubinstein (1997) devised a method for estimating the market's belief about the tails using put and call option prices with different strike prices at the same moment in time. They found for the S&P500 that these tails are very fat. These tail probabilities for several standard deviations below the mean are much fatter after the 1987 crash. Indeed they are ten to one hundred or more times what a log normal distribution of stock prices would suggest. The stable model of asset prices, while much more complex than the normal or lognormal is frequently convenient for modeling fat tailed asset prices. Mandelbrot (1963) and Fama (1965) introduced these infinite variance models to finance. Under suitable assumptions stable distributed assets are closed under addition so portfolios of stable assets have stable distributions. The extent of the fat tailed behavior is related to the characteristic exponent which is 2 in the case of normal distributions. Moments of this order exist. In examining Japanese stock and land prices (proxied by golf course membership prices which are homogeneous products and are actively traded), Rachev & Ziemba (1990) found characteristic exponents in the range 1.60 and 1.15, respectively. While fat tails are important, their impact on asset allocations remains to be determined. Also, temporal issues are important for long-term investors.

Scenarios should be chosen so that they represent a range of results across the objective function values. We must include, along with the likely scenarios, those rare events that contribute to the optimal value of the expected utility function z^*. If the decision-maker displays severe risk aversion, for instance, we must pre-select a number of scenarios that give rise to poor performance. Of course, the model does its best to avoid or at least minimize the impact of bad scenarios through dynamic asset allocation strategies, see e.g. Mulvey *et al.* (1997).

Aggregation and scenario representation are crucial parts of model development. Cariño, Myers & Ziemba (1998) and Hoyland & Wallace (1996) discuss methods to do this using moments. While crude and weak with respect to tail behavior, this approach which preserves moments like the mean, variance, etc. has been useful in several models. The paper by Dupačovà, Bertocchi and Moriggia in this volume investigates the post optimality and robustness of the optimal value of scenario based stochastic programs by addressing additional non-considered scenarios. Bounds on the optimal value of using the 'old' and 'new' scenarios are obtained. The analysis is applied to Italian bond portfolio management. Dupačovà (1995) surveys the general area of approximation via scenarios.

Correlations and co-moments are crucial to obtain diversified portfolios. Estimation of such relationships is usually done using past data. A problem arises during extreme events since correlations increase during stressful periods. For example, in the seven years up to the crash of October 1987, a sample of twenty three major countries never had all the returns positively related in any one month but they were in October 1987; see Roll (1988). Das & Uppal (1996), Louis (1997) and Mulvey & Zenios (1994) discuss methods to estimate these corrrelations. Klassen's paper in this volume presents a theory of aggregation in multiperiod stochastic programming models useful for asset-liability management models. His formulation addresses the problem of how to obtain approximate descriptions of the true uncertainty. In particular, the methods are free of arbitrage opportunities and are consistent with current market prices. Klassen argues that violation of these properties can lead to poor solutions.

4   Solution Algorithms

In this section, we describe algorithms for solving asset and liability management models. Each algorithm takes advantage of the problem's special structure, resulting in efficient implementations. The survey has four parts corresponding to the underlying ALM framework.

4.1   Stochastic Programs

Computational difficulties arise due to the nature of the scenario tree within the stochastic programming framework. The number of decision variables grows exponentially. In most cases, we can prune the tree by reducing the number of branches emanating out of the nodes, especially for nodes that lie towards the end of the planning horizon. Also, we can apply variance reduction procedures and other statistical procedures, such as importance sampling (Infanger 1994), and the expected value of perfect information (Dempster 1997).

The primary algorithms for solving stochastic programs fall into three categories: direct solvers, especially interior point methods (Berger *et al.* 1995, Birge 1997, Carpenter *et al.* 1991, Gassmann 1990, Yang & Zenios 1995); decomposition methods based on Bender's decomposition (Van Slyke & Wets 1969, Dantzig & Infanger 1993, Infanger 1994, Consigli & Dempster in this volume, Fan & Cariño 1994); and decomposition methods based on augmented Lagrangians (Rockafellar & Wets 1991, Mulvey & Ruszczynski 1995). These algorithms are highly effective in taking advantage of the scenario tree structure. We are now able to solve nonlinear stochastic programs with over 10,000 scenarios. More importantly, run time is a linear function of the number of scenarios (Berger *et al.* 1995). Thus, as computers become 40–50% faster per year, we can grow the size of the stochastic program in a similar fashion. There is a definite trade-off between model realism and ease of use.

4.2   Optimal Decision Rules

The main computational difficulty with solving an optimization model based on decision rules is caused by non-convexity. We cannot directly employ standard nonlinear programming algorithms, since they identify only local optimal points. It is typical to re-start the optimization algorithm from a set of random points or to use a global search algorithm such as Tabu search (see

Glover 1989). However, there is no assurance of reaching the global optimal point. As an alternative, we can apply global optimization methods, such as Adjiman *et al.* (1997). Maranas *et al.* (1997) solved the fixed-mix problem with a global method for a fixed number of scenarios. These methods are limited to solving problem with a modest number of decision variables.

4.3   Capital Growth

These models depict static one period asset-only representations that rely on the myopic property of log utility established by Mossin (1968) for independently distributed assets and by Hakansson (1972) for fairly general dependent assets whose distributions are independent of the asset choices.

In many cases the optimization involves a concave program over a polyhedral or more general convex set and hence can proceed via standard nonlinear programming codes. Grauer and Hakansson use this approach in their paper in this volume. Specific algorithms for such expected log problems appear in Cover (1984) and Ziemba (1972a,b).

Implementation of models with drawdown constraints is developed in Maclean & Ziemba (1997); see also Grossman & Zhou (1996). These models along with other methods can be complicated by integer variables. For example, applications to the investment choices affect the asset price distributions leading to non-convex programming problems.

Capital growth models are generally straightforward to solve. These concepts have been successful in a variety of problems including proprietary investment systems in futures, options and favorable 'gambling' situations such as horseracing. For some published material, see e.g. Clark & Ziemba (1987), MacLean, Ziemba & Blazenko (1992), Hausch, Lo & Ziemba (1994) and Ziemba (1994).

4.4   Stochastic Control

Optimal stochastic control algorithms are practical as long as the state space remains relatively small. Typically, this restricts the domain to models in which the number of driving variables in the stochastic model is at most four. For instance, the models in Brennan, Schwartz & Lagnano (1997) and Brennan & Schwartz (this volume) possess four variables (see Figure 8). Once the state space is defined, the continuous problem is approximated by standard approaches, such as finite elements or dynamic programming. The investment policies derived from such models have asset weights that change dramatically in time and are very sensitive to the mean estimates which the model is trying to predict. Hence errors in these estimates can greatly affect investment policy. This can be seen in the Brennan & Schwartz paper in this volume;

see also similar behavior in the continuous time surplus management model of Rudolf & Ziemba (1997). While many investors would not like such large weighting shifts, measures such as the certainty equivalent as used in the Brennan and Schwartz paper provide promising results. Stochastic control models may provide a motivation for certain classes of decision rules.

Sethi's paper in this volume surveys continuous time investment-consumption models of the Merton type when bankruptcy is allowed.

5   Future Directions

This book surveys the field of dynamic stochastic asset and liability management. Four alternative modeling approaches are proposed: multi-stage decision rules, stochastic programming, capital growth, and stochastic control. Each has advantages over the others; there is no clearly best approach. In addition, simulation and mean-variance models are widely used in practice. There is a great deal at stake when making strategic plans for investors with portfolios worth billions of dollars. A small percentage gain compounded over a number of years results in large relative gains. Thus, for many institutions, the gains that are possible through systematic investing should outweigh the challenges of implementing a dynamic investment strategy.

What are some directions for future research? First, there is a consequential need for robustness measures, such as confidence limits, on the ALM model's recommendations. Both stochastic programming and stochastic control are lacking in this regard. A promising direction involves combining decision rules and stochastic programming (Mulvey *et al.* 1997) so that variance reduction techniques can be employed. Other hybrid approaches may be possible.

Second, a commonly accepted definition of risk incorporating temporal issues is critical for the success of long-term planning. There is a tradeoff between short term pain and long-term gains. It is difficult to make this decision without reference to viable future positions. The area of multi-objective optimization holds a number of promising techniques for assisting investors in analyzing their future opportunities.

Third, there is an occasion to customize securities to fit an investor's environment. For instance, the investor's surplus might be susceptible to an upswing in interest rates along with a drop in the US dollar. While this combination of events might be relatively rare since interest rates and currencies are positively correlated, the institutional investor should not entirely ignore this scenario. The asset and liability management system provides critical information regarding the relative desirability of any customized security. The ALM system also can form the basis for pricing the customized security, e.g. as dual variables from the optimal nonlinear program. In the future, investors will be able to design securities with specified return patterns (within limits,

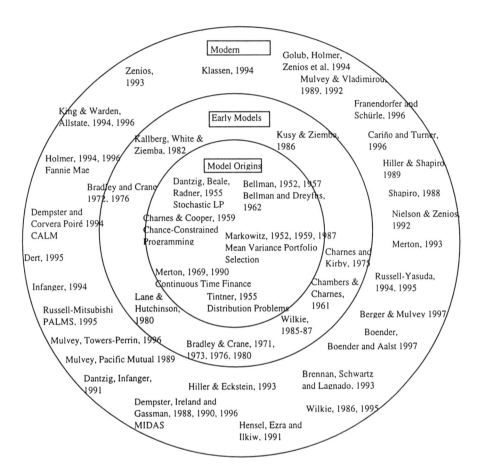

of course). Financial engineering provides greater opportunities for risk management – with an emphasis on the temporal dimension (Greenspan 1995). Thus, we could buy a security that carries out a particular dynamic investment strategy. Dynamic asset and liability management systems are ideal for evaluating these new, complex financial opportunities.

The most sophisticated ALM models have been developed by North American and British researchers who are well represented in this book. But perhaps a more advanced use of such models is by the Dutch, largely because of tradition but also due to their regulatory environment. Three papers by Dert, Boender, van Aalst & Heemskerk and Klassen describe some of this work.

The three concentric circles above show part of the development from the foundations of stochastic optimization by Dantzig, Bellman, Markowitz, Merton and others in the 1950s and 60s to the early models by Bradley & Crane, Charnes and his students, and Ziemba and his students in the 1970s and 80s to the full development of such models in the 1990s by many researchers aided by superior computing power and the vast sums of resources to be managed properly in a very competitive environment. This field is expanding rapidly, but applications are still in their infancy. The potential of stochastic programming models to improve performance and avoid financial disasters will no doubt be greatly increased in the future. Major conferences such as the August 1998 ALM meeting in Vancouver, will discuss and document progress in model theory and practice.

References

Adjiman, C.S., I.P. Androulakis, C.D. Maranas and C.A. Floudas (1997). The αBB algorithm: A general approach for global optimization, Princeton University Department of Chemical Engineering Report

Algoet, P. and T. Cover (1988). Asymptotic optimality and asymptotic equipartition properties of log-optimum investments. *Annals of Probability* **16** 876—98.

Arrow, K.J. (1971). *Essays on the Theory of Risk Bearing*. Markham, Chicago.

Aucamp, D. (1993). On the extensive number of plays to achieve superior performance with the geometric mean strategy. *Management Science* **39** 1163–72.

Bell, David (1995). Risk, return, and utility. *Management Science* **40** 23–30.

Berger, A.J., J.M. Mulvey, E. Rothberg and R. Vanderbei (1995). Solving multistage stochastic programs using tree dissection. Princeton University, Stochastic and Operations Research Report 95-7.

Berger, A.J., J.M. Mulvey and R. Rush (1997). Target-matching in financial scenario generation. Princeton University Report SOR–97–15.

Birge, J.R. (1997). Stochastic programming computation and applications. *Informs Journal on Computing* **9** 111–133.

Black, F. and R. Litterman (1992). Global portfolio optimization. *Financial Analysts Journal* (September-October) 28–43.

Boender, G.C.E. (1995). A hybrid simulation/optimization scenario model for asset/liability management. Report 9513/A, Erasmus University, Rotterdam. To appear in *European Journal of Operations Research*.

Breiman, L. (1961). Optimal gambling system for favorable games. *Proceedings of the 4th Berkeley Symposium on Mathematical Statistics and Probability* **1** 63–68.

Brennan, M.J. and E.S. Schwartz (1982). An equilibrium model of bond pricing and a test of market efficiency. *Journal of Financial and Quantitative Analysis* **17** 75–100.

Brennan, M.J., E.S. Schwartz and R. Lagnado (1998). Strategic asset allocation. *Journal of Economic Dynamics and Control*, in press.

Cariño, D.R. *et al.* 1995). *MTB pension asset/liability management model*. Mimeographed notes, Frank Russell Company, Tacoma, Washington.

Cariño, D.R., T. Kent, D.H. Myers, C. Stacy, M. Sylvanus, A. Turner, K. Watanabe and W.T. Ziemba (1994). The Russell–Yasuda Kasai model: an asset liability model for a Japanese insurance company using multi-stage stochastic programming. *Interfaces* **24** 29–49.

Cariño, D.R. and W.T. Ziemba (1998). *Formulation of the Russell–Yasuda Kasai financial planning*. Report, Frank Russell Company, January, forthcoming *Operations Research*.

Cariño, D.R., D.H. Myers and W.T. Ziemba (1998). *Concepts, technical issues and uses of the Russell–Yasuda Kasai financial planning model*. Report, Frank Russell Company, January, forthcoming *Operations Research*.

Carpenter, T., I. Lustig and J. Mulvey (1991). Formulating stochastic programs for interior point methods. *Operations Research* **39** 757–770.

Chua, J.H. and R.S. Woodward (1983). J.M. Keynes's investment performance: a note. *Journal of Finance* **38** 232–235.

Clark, R. and W.T. Ziemba (1988). Playing the turn-of-the-year effect with index futures. *Operations Research* **35** 799–813.

Connor, G. and R.A. Korajczyk (1995). The arbitrage pricing theory and multi-factor models of asset returns. In *Finance*, R.A. Jarrow, V. Maksimovich and W.T. Ziemba, eds., North Holland, 87–144.

Correnti, S., P. Nealon and S. Sonlin (1996). Decomposing risk to enhance ALM and business decision making for insurance companies. In *Transactions of the 6th AFIR International Colloquium* 443–472.

Cover, T.M. (1984). An algorithm for maximizing expected log investment return. *IEEE Transactions on Information Theory*, Vol. II–30, No. 2, March 369–373.

Dantzig, G.B. (1955). Linear programming under uncertainty. *Management Science* **1** 197–206.

Dantzig, G.B. and G. Infanger (1993). Multi-stage stochastic linear programs for portfolio optimization. *Annals of Operations Research* **45** 59–76.

Das, S.R. and R. Uppal (1994). *International portfolio choice with stochastic correlations*. Report, Harvard Business School.

Davidson, C. (1996). Wider horizons: the technology implications of the drive towards firm-wide risk management. *Risk* **9(4)** 51–53.

Davis, M.H.A. and A.R. Norman (1990). Portfolio selection with transaction costs. *Mathematics of Operations Research* **15** 676–713.

Dembo, R.S. (1995). *Optimal portfolio replication*. Report, Algorithmics Inc., Toronto.

Dempster, M.A.H. (1997). *Parallel Solution of Large Scale Dynamic Stochastic Programmes*. University of Cambridge Report.

Dempster, M.A.H. and A.M. Ireland (1988). A financial expert decision support system. In *Mathematical Models for Decision Support*, NATO ASI Series Vol. F48, B. Mitra, ed., Springer-Verlag.

Dempster, M.A.H., ed. (1980). *Stochastic Programming: Proceedings of the 1st International Conference on Stochastic Programming.* Academic Press.

Dexter, A.S., J. Yu and W.T. Ziemba (1980). Portfolio selection in a lognormal market when the investor has a power utility function: computational results. In *Stochastic Programming*, M.A.H. Dempster, ed., Academic Press, 507–523.

Dupačovà, J. (1995). *Stochastic programming: approximation via scenarios.* Report, Charles University, Prague

Fama, E. (1965). The behaviour of stock market prices. *Journal of Business* **38** 34–105.

Fan, Y. and D.R. Cariño (1994). Nested versus non-nested decomposition for asset allocation problems. Presented at the 15th International Symposium on Mathematical Programming, Ann Arbor, August.

Fan, Y., S. Murray and A. Turner (1997). A retail level stochastic programming asset-liability management model for Italian investors. Report, Frank Russell Company.

Ferson, W.E. (1995). Theory and empirical testing of asset pricing models. In R.A. Jarrow, V. Maksimovich and W.T. Ziemba, eds., *Finance*, North Holland, 145–200.

Frauendorfer, K. (1992). *Stochastic Two-Stage Programming.* Springer-Verlag.

Freund, J. (1956). The introduction of risk into a programming model. *Econometrica* **24** 253–263.

Gardner, G.W. and D. Stone (1995). Estimating currency hedge ratios for international portfolios. *Financial Analysts Journal* (November/December) 58–64.

Gassmann, H.I. (1990). MSLiP: a computer code for the multi-stage stochastic linear programming problem. *Math. Programming* **47** 407–423.

Gassmann, H.I. and A.M. Ireland (1995). Scenario formulation in an algebraic modeling language. *Annals of Operations Research* **59** 45–75.

Glover, F. (1989). Tabu search – Part 1. *ORSA Journal on Computing* **1** 190–206.

Greenspan, A. (1995). Financial innovations and the supervision of financial institutions. *Journal of Financial Engineering* **4** 299–306.

Grinold, R.C. (1980). Time horizons in energy planning models. In *Energy Policy Modeling: United States and Canadian Experiences, Vol. II*, W. T. Ziemba and S.L. Schwartz, eds. Boston, Martinus Nijhoff, 216–237.

Grinold, R.C. (1983). Model building techniques for the correction of end effects in multistage convex programs. *Operations. Research* **31** 407–431.

Grossman, S. and Z. Zhou (1986). Equilibrium analysis of portfolio insurance. *Journal of Finance* **51** 1379–1403.

Hakansson, N.H. (1972). On optimal myopic portfolio policies with and without serial correlation. *Journal of Business* **44** 324–334.

Hakansson, N.H. and W.T. Ziemba (1995). Capital growth theory. In *Finance*, R.A. Jarrow, V. Maksimovic and W.T. Ziemba, eds., 123–144.

Hanoch, G. and H. Levy (1969). The efficiency analysis of choices involving risk. *Review of economic studies* **36** 335–346.

Haugen, R. and N. Baker (1996). Commonality in the determinants of expected stock returns. *Journal of Financial Economics* **41**(3) 401–439.

Hausch, D.B., V. Lo and W.T. Ziemba, eds. (1994). *The Efficiency of Racetrack Betting Markets*. Academic Press.

Higle, J.L. and S. Sen (1991). Stochastic decomposition: an algorithm for two-stage linear programs with recourse. *Math of OR* **16** 650–669.

Holmer, M. (1994). The asset/liability management system at Fannie Mae. *Interfaces* **24** 3–21.

Hoyland, K. and S.W. Wallace (1996). *Generating scenario trees for multistage problems*. Report, Norwegian University of Science and Technology.

Infanger, G. (1994). *Planning Under Uncertainty: Solving Large-Scale Stochastic Linear Programs*. The Scientific Press, Danvers, Massachusetts.

Jackwerth, J.C. and M. Rubinstein (1997). Recovering probability distributions from option prices. *Journal of Finance* **51** 1611–1631.

Jacobs, B.I. and K.N. Levy (1988). Disentangling equity return regularities: new insights and investment opportunities. *Financial Analysts Journal* **44** (May-June) 18–43.

Jarrow, R.A., V. Maksimovic and W.T. Ziemba, eds. (1995). *Finance*, North Holland.

Jorion, P. (1996a). Risk^2: measuring the risk in value at risk. *Financial Analysts Journal* (November/December) 47–56.

Jorion, P. (1996b). *Value at Risk: the New Benchmark for Controlling Market Risk*. Irwin.

Kallberg, J.G., R.W. White and W.T. Ziemba (1982). Short term financial planning under uncertainty. *Management Science* **28** 670–682.

Kallberg, J.G. and W.T. Ziemba (1981). An algorithm for portfolio revision: theory, computational algorithm and empirical results. In *Applications of Management Science*, R.L. Schultz, ed. JAI Press, 267–291.

Kallberg, J.G. and W.T. Ziemba (1983). Comparison of alternative utility functions in portfolio selection problems. *Management Science* **29** 1257–1276.

Kang, P. and S.A. Zenios (1992). Complete prepayment models for mortgage backed securities. *Management Science* **38** 1665–1685.

Keeney, R. and H. Raiffa (1976). *Decisions with Multiple Objectives*, John Wiley and Sons. Reprinted by Cambridge University Press, 1994.

D.M. Keim and W.T Ziemba, eds. (1998). *Security Market Imperfections in World Wide Equity Markets*. Cambridge University Press.

Kelly, J. (1956). A new interpretation of information rate. *Bell System Technology Journal* **35** 917–26.

King, A. and T. Warden (1994). Stochastic programming for strategic portfolio management. *15th International Programming Symposium*, Ann Arbor, August.

Klassen, P. (1994). *Stochastic programming models for interest-rate management* Ph.D. Thesis, MIT.

Konno, H., D.G. Luenberger and J.M. Mulvey, eds. (1993). *Financial Engineering.* J.C. Baltzer AG.

Kroll, Y., H. Levy and H. Markowitz (1984). Mean variance versus direct utility maximization. *Journal of Finance* **39** 47–61.

Kusy, M.I. and W.T. Ziemba (1986). A bank asset and liability management model. *Operations Research* **34** 356–376.

Li, Y. and W.T. Ziemba (1989). Characterizations of optimal portfolios by univariate and multivariate risk aversion. *Management Science* **35** 259–269.

Linsmeier, T.J. and N.D. Pearson (1996). *Risk measurement: an introduction to value at risk.* Report, University of Illinois.

Longin, F.M. (1996). The asymptotic distribution of extreme stock market returns. *Journal of Business* **69** 383–408.

Louis, J.C. (1997). Worrying about correlation derivatives strategy. *Global Finance* March, 53–55.

Lowe, S.P. and J. Stanard (1996). An integrated dynamic financial analysis and decision support system for a property catastrophe reinsurer. CAS Seminar on Dynamic Financial Analysis, *CAS Forum.*

MacLean, L.C. and W.T. Ziemba (1991). Growth-security profiles in capital accumulation under risk. *Annals of Operations Research* **31** 501–10.

MacLean, L.C. and W.T. Ziemba (1997). *Capital growth with security.* Report, University of British Columbia.

MacLean, L.C., W.T. Ziemba and G. Blazenko (1992). Growth versus security in dynamic investment analysis. *Management Science* **38** 1562–85.

Mandelbrot, B. (1963). The variation of certain speculative prices. *Journal of Business* **39** 394–419.

Maranas, C., I. Androulakis, C.A. Floudas, A. Berger and J.M. Mulvey (1997). Solving long-term financial planning problems via global optimization. *Journal of Economics Dynamics and Control* **21** 1405–1425.

Markowitz, H.M. (1952). Portfolio selection. *Journal of Finance* **7** 77–91.

Markowitz, H.M. (1959). Portfolio selection: efficient diversification of investments. In *Cowles Foundation Monograph* **16**. Yale University Press.

Markowitz, H.M. (1976). Investment for the long run: new evidence for an old rule. *Journal of Finance* **31** 1273–86.

Markowitz, H.M. (1991). *Portfolio Choice: Efficient Diversification of Investments*, 2nd edition, Blackwell Publications.

Markowitz, H.M., S. Schaible and W.T. Ziemba (1992). An algorithm for portfolio selection in a lognormal market. *International Journal of Financial Analysis* **1** 109–113.

Merton, R.C. (1969). Lifetime portfolio selection under uncertainty: the continuous time case. *Review of Economics and Statistics* **3** 373–413.

Merton, R.C. (1990). *Continuous-Time Finance.* Blackwell Publishers.

Mossin, J. (1968). Optimal multi-period portfolio policies. *Journal of Finance* **41** 215–229.

Mulvey, J.M. (1989). A surplus optimization perspective. *Investment Management Review* **3** 31–39.

Mulvey, J.M. (1996a). It always pays to look ahead. *Balance Sheet* **4** 23–27.

Mulvey, J.M. (1996b). Generating scenarios for the Towers Perrin investment system. *Interfaces* **26** 1–15.

Mulvey, J.M., J. Armstrong and E. Rothberg (1995). Total integrative risk management. *Risk Special Supplement* June 28–30.

Mulvey, J.M., S. Correnti and J. Lummis (1997). *Total integrated risk management: insurance elements.* Princeton University Report, SOR–97–2.

Mulvey, J.M., R. Rush, J.E. Mitchell and T.R. Willemain (1997). *Stratified filtered sampling in stochastic optimization.* Princeton University Report SOR–97–7. To appear in *European Journal of Operations Research.*

Mulvey, J.M. and A. Ruszczynski (1995). A new scenario decomposition method for large-scale stochastic optimization. *Operations Research* **43** 477–490

Mulvey, J.M. and S.A. Zenios (1994). Capturing the correlations of fixed-income instruments. *Management Science* **40** 1329–1342.

Mulvey, J.M. and Z. Chen (1996). An empirical evaluation of the fixed-mix investment strategy. Princeton University Report SOR–96–21.

Mulvey, J.M. and W.T. Ziemba (1995). Asset and liability allocation in a global environment. In *Finance*, R.A. Jarrow, V. Maksimovic and W.T. Ziemba, eds., North Holland, 435–463.

Murray, S.M. and W.T. Ziemba (1997). *The effect of fat tails in dynamic optimization.* Report in progress, Frank Russell Company and University of British Columbia

Ortec Consultants bv. (1996). *Asset-liability management problem, approach and decision support system.* Report, Gouda, The Netherlands.

Perold, A.F. and W.F. Sharpe (1988). Dynamic strategies for asset allocation. *Financial Analysts Journal* January 16–27.

Peskin, M.W. (1997). Asset allocation and funding policy for corporate-sponsored defined-benefit pension plans. *Journal of Portfolio Management* (Winter) 66–73.

Poterba, J.M. and L.H. Summers (1988). Mean reversion in stock prices: evidence and implications. *Journal of Financial Economics* **22** 27–59.

Pratt, J.W. (1964). Risk aversion in the small and in the large. *Econometrica* **32** 122–136.

Pyle, D. (1995). The US Savings and Loan crisis. In *Finance*, R.A. Jarrow, V. Maksimovich and W.T. Ziemba eds., North Holland, 1105–1125.

Rockafellar, R.T. and R.J.-B. Wets (1991). Scenarios and policy aggregation in optimization under uncertainty. *Mathematics of Operations. Research* **16** 119–147.

S. Rachev and W.T. Ziemba (1990). The distribution of golf course membership in Japan. Presentation to the Berkeley Program in Finance.

Roll, R.W. (1988). The international crash of 1987. In *Black Monday and the Future of Financial Markets*, R.W. Kamphuis, R.C. Kormendi and J.W.H. Watson eds., Dow-Jones Irwin.

Rotando, L.M. and E.O. Thorp (1992). The Kelly criterion and the stock market. *American Mathematical Monthly* (December) 992–1032.

Rubinstein, M.E.. (1977). The strong case for log as the premier model for financial modeling. In *Financial Decisions Under Uncertainty*, H. Levy and M. Sarnet, eds., Academic Press.

Rudolf, M. and W.T. Ziemba (1997). *Intertemporal surplus management.* Report, Swiss Institute of Banking and Finance, St. Gallen.

Samuelson, P. (1969). Lifetime portfolio selection by dynamic stochastic programming. *Review of Economics and Statistics* (August) 239–246.

Sharpe, W.F. and L.G. Tint (1990). Liabilities: a new approach. *Journal of Portfolio Management* (Winter) 5–10.

Shaw, J., E.O. Thorp and W.T. Ziemba (1995). Risk arbitrage in the Nikkei put warrant market of 1989–90). *Applied Mathematical Finance* **2** 243–271.

Sorensen E., E. Mezrich and D. Thadani (1993). Currency hedging through portfolio optimization. *Journal of Portfolio Management* Spring 78–85.

Stone, D. and W.T. Ziemba (1993). Land and stock prices in Japan. *Journal of Economic Perspectives* **7** 149–165.

Van Slyke, R. and R.J.-B. Wets (1969). L-shaped linear programs with applications to optimal control and stochastic programming. *SIAM Journal of Applied Mathematics* **17** 638–663.

von Neumann, J. and O. Morgenstern (1944). *Theory of Games and Economic Behavior.* Princeton University Press.

Wilkie, A.D. (1986). A stochastic investment model for actuarial use. *Transactions of the Faculty of Actuaries* **39** 391–403.

Wilkie, A.D. (1987). Stochastic investment models: theory and applications. *Insurance: Mathematics and Economics* **6** 65–83.

Wilkie, A.D. (1995). *More on a stochastic asset model for actuarial use.* Presented to the Institute of Actuaries and Faculty of Actuaries, London.

Yang, D. and S.A. Zenios (1996). *A scaleable parallel interior algorithm for stochastic linear programming and robust optimization. Computational Optimization and its Applications* **7** 143–158.

Zenios, S.A., ed. (1993a). *Financial Optimization.* Cambridge University Press.

Zenios, S.A. (1993b). A model for portfolio management with mortgage-backed securities. *Annals of Operations Research* **43** 337–356.

Ziemba, W.T. (1972a). Note on 'Optimal growth portfolios when returns are serially correlated.'. *Journal of Financial and Quantitative Analysis* **7** 1995–2000.

Ziemba, W.T. (1972b). Solving nonlinear programming problems with stochastic objective functions. *Journal of Financial and Quantitative Analysis* **7** 1809–1827.

Ziemba, W.T. (1994). Investing in the turn-of-the-year effect in the US futures markets. *INTERFACES* (May–June) **24**(3) 46–61.

Ziemba, W.T. and S.L. Schwartz (1991). *Invest Japan.* Probus Publishers.

Ziemba, W.T. and R.G. Vickson, eds. (1975). *Stochastic Optimization Models in Finance.* Academic Press.

Part II
Static Portfolio Analysis for Asset Allocation

The Importance of the Asset Allocation Decision*

Chris R. Hensel, D. Don Ezra and John H. Ilkiw

Summary

Several different decisions, including asset allocation, security selection and market timing, affect the return to a pension fund (or any investor). The impact of each type of decision can be measured by comparing a portfolio's actual return with the return on a hypothetical portfolio that does not reflect a particular decision that went into the real portfolio. The critical element in the comparison is defining the naive alternative to the decision.

When asset allocation is the decision being evaluated, the naive alternative is not obvious. If Treasury bills are the appropriate naive alternative, then asset allocation is, as commonly thought, the single decision with the greatest impact on a typical pension fund's return. But if a diversified mix (such as the average asset mix across large pension funds) is the alternative, then the impact of departing from this naive allocation may be no greater than the impact of other decisions, including security selection.

Introduction

It is widely believed that the **asset allocation** policy decision of an investor is far more important than decisions such as market timing or security selection.[1] Of course, this belief reflects the practices of the *average* investor – more specifically, the average pension plan sponsor. But it is conceivable that some investors take such large bets on, say, **security** selection (or large departures from the composition of market indexes) that security selection

---

*Reprinted, with permission, from *Financial Analysts Journal*, July/August 1991. Copyright 1991, Association for Investment Management and Research Charlottesville, VA. All rights reserved.

[1] See W.F. Sharpe, 'Asset Allocation,' in Maginn and Tuttle, eds., *Managing Investment Portfolios*, 2nd ed. (Boston: Warren Gorham & Lamont 1990), who accurately quotes the prevailing sentiment: "It is generally agreed by theoreticians and practitioners alike that the asset allocation decision is by far the most important one made by an investor." The definitive quantitative study on the subject is G. Brinson, L.R. Hood & G.L. Beebower, 'Determinants of Portfolio Performance,' *Financial Analysts Journal*, July/August 1986.

has a greater impact on the achieved return of these investors than asset allocation policy. Sponsors might thus like to know how to estimate the relative impacts of different types of investment decisions on their funds' returns.

This article describes a method investors can use to analyze their returns and determine the impacts that several different risk-taking decisions have on these returns. In the course of constructing a suitable model and applying it to a very small sample of actual results, we discovered that the massive influence currently attributed to asset allocation policy depends crucially on one factor – that is, the naive alternative from which the sponsor's policy represents a departure. If the naive alternative is a reasonably diversified portfolio, then asset allocation policy may, for many sponsors, be only as important as (or not much more important than) other types of investment decisions.

Performance Attribution

The impact of any investment decision can be measured by comparing its outcome with the outcome of some alternative decision. This notion is frequently used to **attribute** pension fund investment performance to each of a number of decisions. For example, some funds compare the return on an actively managed portfolio with the return that would have been earned had funds been invested in the market portfolio instead. The difference represents the value added (or perhaps lost) by the investment judgments, which represent departures from the market portfolio.

While the principle is easy to understand, its application is often difficult. It is difficult to define the alternative portfolio that would be held by an investor who is devoid of investment judgment. Conventional wisdom defines the naive alternative as one that represents all available opportunities proportionately; this is usually the same as the average of what everybody else is holding[2]

Furthermore, investment portfolios often reflect judgments of many kinds, each a departure from some naive alternative. A multilevel decision model for a pension plan sponsor (or indeed any investor) might involve three sets of asset allocations:

- the market mix, X,

---

[2]Conventional models of economic equilibrium require that all markets clear so that there is no excess supply in any market. Thus the portfolio positions held in aggregate are the market with no ownable assets excluded. The market so defined represents a benchmark against which average performance can be measured. See, for example, K.J. Arrow, 'The Role of Securities in the Optimal Allocation of Risk Bearing,' *Review of Economic Studies* **31** (1963–64), 91–96, or R. Radner, 'Existence of Equilibrium of Plans, Prices, and Price Expectations in a Sequence of Markets,' *Econometrica*, March 1972.

GLOSSARY

Asset Allocation: The decision of how a fund should be invested across each of several asset classes, assuming neutral capital market conditions exist. This condition implies that asset class return expectations are roughly proportional to the asset classes' assessed riskiness; no class is considered to be underpriced or overpriced.

Security Selection: The decision of how an asset class portfolio should be invested in each of the available securities making up the asset class.

Attribution: The mathematical process of explaining an investment return by relating it to the different risk-taking decisions implicit in the portfolio, and the extent to which each of those risks was rewarded or penalized in the capital markets.

Minimum-Risk Portfolio: That combination of securities or asset classes that reduces the uncertainty of future portfolio returns to a minimum A liability-driven minimum-risk portfolio Is one that reduces the uncertainty of future surplus size (assets minus liabilities) to a minimum.

Coefficients of Determination: Measures of the extent to which any variable is statistically explained by another variable. Such a coefficient has a maximum value of one and a minimum value of zero.

Mean Absolute Deviation: A statistical measure used to show the average extent to which a series of numbers differs in size from a given number. In this calculation, it is immaterial whether the differences are positive or negative; differences of $+2$ and -2 are both considered as differences of 2 in calculating the average.

- the sponsor's customized policy mix y

  and

- the actual mix held from day to day, Z.

The corresponding trio of security weights would be:

- market weights, 1,

- the sponsor's customized 'normal' weights, 2, and

- the actual weights of securities held from day to day, 3

Allocation X could be the average weights of different asset classes held by all US pension funds. Allocation Y could be the customized 60 per cent US equities/40 per cent US bonds selected as the policy for a particular sponsor's pension fund. Allocation Z on a particular day might be 55 per cent US equities, 15 per cent US bonds and 30 per cent cash assuming that mix were deemed by the fund's manager to be temporarily superior to the policy mix.

Weights 1 might be the weights assigned to securities in a well known index such as the S&P 500, attempting to reflect the market's opportunity set. Weights 2 might reflect specific tilts that the sponsor elects from a long-term policy perspective, such as a tilt toward long bonds rather than the intermediate length that reflects all bonds available in the US. Weights 3 would consist of the weights of the actual securities held in the portfolio on a particular day.

The basic naive portfolio consists of asset class weights X and security weights 1 (see Figure 1). It is conveniently labeled Portfolio X_1. It earns no-judgment returns. All other portfolios reflect a choice of some kind, whether made consciously or unconsciously by representatives of the sponsor or by the investment manager (who may also be the sponsor). Portfolio Y_2 is the quintessential representation of the sponsor's decisions, while Z_3 represents the investment manager's decisions.

Differences between the different portfolios' returns measure the impact of various kinds of investment judgments. Thus $(Y_1 - X_1)$ measures the impact of the sponsor's customized asset allocation policy as the two hypothetical portfolios differ only insofar as the sponsor's policy allocation differs from the neutral market allocation.[3]

In summary, Portfolio X_1 represents the results from neutral participation in the markets. Differences between portfolio returns represent the impacts of investment decisions.

Naive Asset Allocations

We mentioned earlier the difficulty of defining the naive asset allocation. While we used the market mix to illustrate Allocation X in the model above, it is by no means obvious that this is what a naive pension plan sponsor would do.

Some candidates for Allocation X frequently found in practice include the following.

A: 100 per cent in T-bills. This could be the minimum-risk portfolio for:
(1) a sponsor concerned with asset growth rather than surplus growth;

---

[3]Actually, $(Y_2 - X_2)$ is an equally valid measure of the impact of the sponsor's policy allocation. This illustrates the principle that there may not always be a unique 'right' way to measure impacts, merely ways with different degrees of usefulness. In fact, however, Weights 2 are rarely found in practice, so $(Y_1 - X_1)$ would usually be used.

Security Weights

		1. Market Security Weights	2: Sponsor's Customized 'Normal' Weights	3: Actual Weights of Securities Held from Day to Day
Asset	X: Market Mix	X_1: Basic Naive Portfolio		
Allocations	Y: Sponsor's Customized Policy Mix		Y_2: Reflects Sponsor's Choices	
	Z: Asset Mix Held from Day to Day			Z_3: Reflects Investment Manager's Choices

Figure 1: Composition of Various Portfolios

or (2) a sponsor concerned with surplus growth, which believes that pension liabilities are real in nature and that real interest rates tend to remain constant over time.[4]

B: 100 per cent in bonds. This could be the minimum-risk portfolio for a sponsor concerned with surplus growth and focused on pension liabilities that are either fixed in nominal terms or inflation-sensitive up to each member's retirement but not beyond retirement.

C: The average asset allocation held by large pension funds. We assumed this to be 50 per cent US stocks, 5 per cent international stocks, 30 per cent US fixed income, 5 per cent real estate and 10 per cent cash.[5] This allocation might be the naive selection of a sponsor who asks, "What is everyone else doing?"[6]

D: The asset allocation representing the 'market mix.' This would include all available investment opportunities. But what is 'all'? Everything available to US. investors? Everything available anywhere in the world? Because of the difficulty in defining 'all,' we did not take this candidate any further.

---

[4]While a discussion of the characteristics of pension liabilities and surplus is beyond the scope of this paper, an explanation is provided in D.D. Ezra, 'Asset Allocation by Surplus Optimization,' *Financial Analysts Journal*, January/February 1991.

[5]See the 1989 Money Market Directory for Pension Funds, p. xvii (corporate pension funds with total assets exceeding \$500 million). In different countries, of course, this allocation will be different, reflecting local preferences and opportunities.

[6]A simple benchmark frequently quoted consists of 60 per cent stocks and 40 per cent bonds. Results using this allocation are virtually indistinguishable from the results using mix C.

Security Weights

		1. Market Security Weights	3: Actual Weights of Securities Held from Day to Day
Asset	A: 100% in T Bills	Minimum-Risk Portfolio	
Allocations	C: Average Mix Held by Sponsors ('Naive Policy Allocation')	Average-Risk Portfolio	
	Y: Sponsor's Customized Policy Mix	Policy Portfolio	Security-Allocated Portfolio
	Z: Asset Mix Held from Day to Day	Timing-Allocated Portfolio	Actual Portfolio

Figure 2: Portfolios Used in Study

Conceptually, we believe the appropriate hierarchy of investment policy portfolios is the one shown in Figure 2. We define the minimum-risk portfolio as liability-driven, in the sense that it minimizes the extent of future surplus uncertainty. For the balance of this article, we will use T-bills as the minimum-risk portfolio, without attempting to defend or convert others to this viewpoint. (Results are virtually identical if bonds are used as the minimum risk portfolio.)

The average-risk portfolio is Naive Allocation C (i.e., "what other large pension funds are doing"), combined with market security weights. For the remainder of this article, we will refer to this allocation as the 'Naive Policy Allocation.' Comparing the return on this portfolio C_1 with the return on T-bills reveals the reward received by the average fund for accepting risk. The sponsor's policy portfolio depends on the specific sponsor. Sponsors usually conduct periodic analyses of the characteristics of their pension assets and liabilities, as well as their own comfort levels (frequently called 'risk tolerance' in the finance literature, although in practice psychological aspects of risk tend to be at least as important as financial aspects). As a result of these analyses, they adopt their own asset allocation policies, which may differ from the average-risk portfolio. Comparing the return on this portfolio Y_1 to the return on the average-risk portfolio C_1 reveals the impact of the sponsor's decision to depart from the implicit average risk tolerance of other sponsors

Once this hierarchy of investment policy portfolios is established, one can proceed as follows

- Ignore the existence of security weights 2. In effect, assume that sponsors invoke no deliberate policy tilts away from standard indexes for the relevant asset classes.

- For the impact of *market timing*, compare the returns on the timing-allocated portfolio (Z_1) to the returns on the sponsor's policy portfolio (Y_1), that is, market timing equals the impact $(Z_1 - Y_1)$.

- For the impact of *security selection*, compare the returns on the security-allocated portfolio (Y_3) with the returns on the policy portfolio (Y_1). That is, security selection equals the impact $(Y_3 Y_1)$.

- The residual part of the actual return earned by each sponsor's fund is attributed to the *interaction* of market timing and security selection, as well as portfolio *activity*; all our hypothetical portfolios were rebalanced quarterly and remained unchanged throughout the quarter. Thus interaction plus activity equals the impact $(Z_3 - Y_1 - \text{timing} - \text{selection})$, which equals the impact $(Z_3 + Y_1 - Z_1 - Y_3)$.

Return Volatility

We examined the volatility of returns for seven Russell US sponsors, using quarterly data over the 1985–88 period.[7] We regressed historical portfolio returns on five different portfolios – the naive alternatives A and C, each sponsor's policy portfolio, the timing-allocated portfolio and the security-allocated portfolio.[8] The results indicate the amount of the variability of each sponsor's returns explained by each decision level. The individual sponsors' squares (**coefficients of determination**) were then averaged over the seven sponsors.

For these seven sponsors over the four-year period, 97.5 per cent of the variation in total plan returns was explained by their policy portfolios.[9] The

---

[7]This study is based on a small sample and a short period of time. Our purpose was to outline an approach rather than to produce definitive numerical results.

[8]The asset class benchmarks were the Russell 3000 for US equity; the MSCI EAFE Index for non-US equity; the Shearson Lehman Hutton Aggregate Bond Index for US fixed income; the Russell NCREIF Property Index (formerly FRC Property Index) for real estate; and the Salomon Brothers 3–Month Treasury Bill Index for cash. Sponsors' actual policy weights were used when available. For sponsors lacking policy portfolios or for periods prior to the adoption of a policy portfolio, policy weights were inferred from the client's average actual allocations to managers for those periods. For these seven sponsors, two had policy portfolios for all quarters and two lacked any policy portfolios. The average length of time these sponsors had policy portfolios was 10 quarters. Thus, on average, the policy portfolios for the initial six quarters were inferred.

[9]This result is reasonably close to the 93.6 per cent calculated for a larger sample, and over a longer time, by Brinson, Hood and Beebower, 'Determinants of Portfolio Performance,' *op. cit.*

Decision Level	Additional Variation Explained by this Level	Cumulative Variation Explained through this Level
Minimum Risk	2.66%	2.66%
Naive Policy Allocation	94.35	97.01
Specific Policy Allocation	0 50	97.51
Market Timing*	0.14	–
Security Selection*	0.40	–
Interaction and Activity*	1.95	–
Total	100.00%	100.00%

* The additional return variation explained by market timing, security selection and interaction and activity need not be sequentially cumulative. Therefore, rather than imply a specific order, they are left blank.

Table 1. Return Variation Explained

diversified naive allocation (Alternative C: average large pension fund exposures to the asset classes) explained 97 per cent of the variation in total plan returns. Table 1 gives the amount of the variation in plan returns explained by different decision levels.

Our results highlight the importance of the naive allocation. If T-bills, rather than a diversified mix, had been selected as the naive alternative, then the sponsor's choice of policy allocation would explain most of the variation in plan returns. The use of a diversified mix as the naive alternative makes the choice among different diversified mixes relatively unimportant; that choice explains only an additional 0.50 per cent of return variation.

Return Impact of Different Decisions

We believe the most significant analysis concerns the potential impact of different types of decisions on returns themselves, rather than their variability. Table 2 shows the results of our return analysis.

The average quarterly return on T-bills over the period was 1.62 per cent.[10] By contrast, the Naive Policy Allocation would have resulted in an average

---

[10] All quarterly return numbers can be annualized by taking the fourth power of the return relatives and subtracting one.

Source of Return	1985–1988 Average Quarterly Return*	1985–1988 Average Quarterly Contribution[†]	Potential Quarterly Impact as Measured by Average MAD
Minimum Risk (T-bills)	1.62%	1.62%	–
Naive Policy Allocation	3.75	2.13	4.43%
Specific Policy Allocation (average of 7 sponsors)	4.24	0.49	0.79
Market Timing (average of 7 sponsors)	–	(0.10)	0.57
Security Selection (average of 7 sponsors)	–	(0.23)	0.85
Interaction and Activity (average of 7 sponsors)	–	(0.05)	0.15
Quarterly Total Fund Return (average of 7 sponsors)	3 86%	3.86%	5.63%

* All returns are before all fees and expenses.

[†] Returns and mean absolute deviations were computed for each sponsor and then averaged across sponsors.

Table 2. Quarterly Return Impacts of Different Types of Decisions

quarterly return of 3.75 per cent – 2.13 per cent higher than the T-bill return. We interpret the 2.13 per cent as the average quarterly reward over the period for the average pension fund's decision to take investment risk as a matter of policy. The reward, of course, is highly dependent on the time period under consideration. There must be periods when the reward is negative that is the essence of risk.

The absolute size of each quarterly reward (or cost) is important. Sponsors will accept negative results If they expect that, in the long term, positive results will predominate.

The size of a decision's impact (regardless of sign) is an indication of its potential effect. It is certainly an *ex post* measure of its effect. We therefore measured the dispersions of the impacts around zero and calculated the average of the absolute values of the dispersions. This measure is the '**mean absolute deviation**' (MAD).[11]

For the Naive Policy Allocation, the potential impact on the fund's return, as measured by the MAD, amounted to 4.43 per cent per quarter. The specific asset allocation policies adopted by the sponsors in the sample added, on average, 0.49 per cent per quarter to the Naive Policy Allocation return; the policy portfolios earned an average quarterly return of 4.24 per cent. The impact of a specific policy allocation relative to the Naive Policy Allocation was occasionally negative. The potential impact of the specific policy allocations, as measured by their average absolute size, was 0.79 per cent per quarter.

The average quarterly return of the timing-allocated portfolios was 4.14 per cent. That is, the average quarterly impact of market-timing decisions by sponsors was −0.10 per cent. When the quarterly impacts were averaged in absolute terms (without regard to sign), the average was 0.57 per cent. This is the potential quarterly impact of managers' market-timing decisions.

Security selection had an average impact of −0.23 per cent per quarter, and a potential impact of 0.85 per cent per quarter. Interaction and activity had an average impact of −0.05 per cent per quarter and a potential impact of 0.15 percent per quarter

Table 2 indicates that T-bills and the risk premium associated with the Naive Policy Allocation accounted for most of the return over the period studied. Market timing security selection and the effect of interactions and activity all on average, reduced returns. Specific asset allocation policy decisions added slightly to returns

The relative magnitudes of the potential *impacts* are particularly interesting

- The potential impact of the Naive Policy Allocation easily surpasses all other decisions.

- The potential impact of interaction and activity is the smallest, when measured over one quarter.

---

[11]Had we used standard deviation instead, we would have measured the variation around the mean and accorded greater weight to extreme observations. We believe measuring the deviations from zero (MAD) provides a better sense of potential future outcomes and is therefore a good measure of the potential impact of different types of decisions.

Source of Potential Quarterly Impact	Lowest Quarterly MAD for any 1 Sponsor	Average Quarterly MAD Over 7 Sponsors	Highest Quarterly for any 1 Sponsor
Specific Policy Allocation	0.51%	0.79%	1.53%
Market Timing	0.23	0.57	0.73
Security Selection	0.53	0.85	1.50
Interaction and Activity	0.07	0.15	0.20

Table 3. Extreme MAD Results in Study

- The potential impacts of specific policy allocation and of security selection are roughly equal in magnitude and somewhat greater than the potential impact of market timing.

The order of average magnitudes in Table 2 may not hold for every sponsor. Table 3 shows, for example, the highest and lowest potential impacts for different sponsors in the sample. One sponsor departed significantly from the Naive Policy Allocation – so much so that the mean potential impact of the specific policy allocation was 1.53 per cent per quarter. Another sponsor departed relatively little from the Naive Policy Allocation; the mean potential impact of the difference was only 0.51 per cent per quarter. Averaged over all seven sponsors, the mean potential quarterly impact of specific policy allocation was 0.79 per cent per quarter, the same number shown in Table 2.

The potential impact of the specific policy allocation will be high for any sponsor whose attitude toward risk differs significantly from the attitude of the average sponsor. This is a decision only the sponsor can make. Similarly, the potential impacts of market timing and security selection will depend on the specific sponsor.

In our sample, the potential impacts of specific policy allocation and security selection are of roughly equal magnitude in all columns and larger than the potential impact of market timing. But there could conceivably be a sponsor with a specific policy allocation identical to the Naive Policy Allocation, which believes in holding indexed portfolios and adding value through market timing. For such a sponsor, the potential impacts of specific policy allocation and of security selection will be zero; market timing will produce the entire impact. This demonstrates the importance of each sponsor deciding individually how much significance each type of decision should be given.

Finally, it is legitimate for the expected reward from a specific policy allocation to be negative for a sponsor that consciously decides to take less investment risk than the average sponsor. However, the expected reward from active management decisions should not be negative; if it is, the exposure to active management should be severely curtailed.

Existing Literature

The Brinson, Hood and Beebower (BHB) study is the seminal work on performance attribution, and deservedly so.[12] Our study follows its methods to a large extent But we made two conceptual changes.

First, in performance attribution, there is usually a base return (representing the naive portfolio and a series of effects (representing the impacts of judgments). If the base return is itself added to one of the effects, it exaggerates the impact of the corresponding judgment. Essentially, this is equivalent to assuming that the naive portfolio always has a zero return. The naive portfolio thus implies no investment whatsoever; this is clearly unrealistic.

We were careful not to measure the impact of the sponsor's policy decision in our numerical results, as the difference from a zero return. This reduces the explanatory power of the sponsor's policy decision as a source of return variability, relative to the results of BHB The difference is small if the naive alternative is T-bills. The reduction is quite large if the naive alternative is a diversified portfolio.

Second, we asked ourselves what the naive alternative to a judgment on asset allocation policy would be. Experience with sponsors indicated that when they study the subject, they almost always consider the average mix held by other sponsors. Hence our decision to consider Naive Allocation C as the point of departure. The same experience underlies our concept of the policy allocation as itself representing a departure from a naive allocation, rather than from no investment whatsoever.

For most sponsors, the decision to depart from a risk-minimizing investment policy is likely to have a greater impact on total plan returns and return variability than any other single decision. Relative to a naive diversified mix, any specific asset allocation policy may have a sizable impact on total return, but nothing like the dominance frequently (and erroneously) attributed to it. Decisions regarding active management (market timing and security selection) can be as worthy of a sponsor's attention as the asset allocation decision.

---

[12]Brinson, Hood and Beebower, 'Determinants of Portfolio Performance,' *op. cit.*

The Effect of Errors in Means, Variances, and Covariances on Optimal Portfolio Choice*

Vijay K. Chopra and William T. Ziemba

Good mean forecasts are critical to the mean-variance framework

There is considerable literature on the strengths and limitations of mean-variance analysis. The basic theory and extensions of MV analysis are discussed in Markowitz [1987] and Ziemba & Vickson [1975]. Bawa, Brown & Klein [1979] and Michaud [1989] review some of its problems.

MV optimization is very sensitive to errors in the estimates of the inputs. Chopra [1993] shows that small changes in the input parameters can result in large changes in composition of the optimal portfolio. Best & Grauer [1991] present some empirical and theoretical results on the sensitivity of optimal portfolios to changes in means. This article examines the relative impact of estimation errors in means, variances, and covariances.

Kallberg & Ziemba [1984] examine the question of mis-specification in normally distributed portfolio selection problems. They discuss three areas of misspecification: the investor's utility function, the mean vector, and the covariance matrix of the return distribution.

They find that utility functions with similar levels of Arrow–Pratt absolute risk aversion result in similar optimal portfolios irrespective of the functional form of the utility[1]; Thus, mis-specification of the utility function is not a major concern because several different utility functions (quadratic, negative exponential, logarithmic, power) result in similar portfolio allocations for similar levels of risk aversion.

Misspecification of the parameters of the return distribution, however, does make a significant difference. Specifically, errors in means are at least ten times as important as errors in variances and covariances.

We show that it is important to distinguish between errors in variances and covariances. The relative impact of errors in means, variances, and covariances also depends on the investor's risk tolerance. For a risk tolerance of

---

*Reprinted, with permission, from *Journal of Portfolio Management*, 1993. Copyright 1993 Institutional Investor Journals.

[1]For an investor with utility function U and wealth W, the Arrow–Pratt absolute risk aversion is $ARA = -U''(W)/U'(W)$. Friend and Blume [1975] show that investor behavior is consistent with decreasing ARA; that is, as investors' wealth increases, their aversion to a given risk decreases.

50, errors in means are about eleven times as important as errors in variances, a result similar to that of Kallberg & Ziemba.[2] Errors in variances are about twice as important as errors in covariances.

At higher risk tolerances, errors in means are even more important relative to errors in variances and covariances. At lower risk tolerances, the relative impact of errors in means, variances, and covariances is closer. Even though errors in means are more important than those in variances and covariances, the difference in importance diminishes with a decline in risk tolerance.

These results have an implication for allocation of resources according to the MV framework. The primary emphasis should be on obtaining superior estimates of means, followed by good estimates of variances. Estimates of covariances are the least important in terms of their influence on the optimal portfolio.

Theory

For a utility function U and gross returns $r - i$ (or return relatives) for assets $i = 1, 2, \ldots, N$, an investor's optimal portfolio is the solution to:

$$\text{maximize } Z(x) = E[U(W_0 \sum_{i=1}^{N} (r_i)x_i)]$$

$$\text{such that } x_i > 0, \quad \sum_{i=1}^{N} = 1,$$

where $Z(x)$ is the investor's expected utility of wealth, W_0 is the investor's initial wealth, the returns r_i have a distribution $F(r)$, and x_i are the portfolio weights that sum to one.

Assuming a negative exponential utility function $U(W) = -\exp(-aW)$ and a joint normal distribution of returns, the expected utility maximization problem is equivalent to the MV-optimization problem:

$$\text{maximize } Z(x) = \sum_{i=1}^{N} E[r_i]x_j - \frac{1}{t} \sum_{i=1}^{N} \sum_{j=1}^{N} x_i x_j E[\sigma_{ij}]$$

$$\text{such that } x_i > 0, \quad \sum_{i=1}^{N} x_i = 1,$$

---

[2]The risk tolerance reflects the investor's desired trade-off between extra return and extra risk (variance). It is the inverse slope of the investor's indifference curve in mean–variance space. The greater the risk tolerance, the more risk an investor is willing to take for a little extra return. Under fairly general input assumptions, a risk tolerance of 50 describes the typical portfolio allocations of large US pensions funds and other institutional investors. Risk tolerances of 25 and 75 characterize extremely conservative and aggressive investors, respectively.

EXHIBIT 1:

List of Ten Randomly Chosen DJIA Securities

1. Aluminum Co. of America
2. American Express Co.
3. Boeing Co.
4. Chevron Co.
5. Coca Cola Co.
6. E.I. Du Pont De Nemours & Co.
7. Minnesota Mining and Manufacturing Co.
8. Procter & Gamble Co.
9. Sears, Roebuck & Co.
10. United Technologies Co.

where $E[r_i]$ is the expected return for asset i, t is the risk tolerance of the investor, and $E[\sigma_{ij}]$ is the covariance between the returns on assets i and j.[3]

A natural question arises: How much worse off is the investor if the distribution of returns is estimated with an error? This is an important consideration because the future distribution of returns is unknown. Investors rely on limited data to estimate the parameters of the distribution, and estimation errors are unavoidable. Our investigation assumes that the distribution of returns is stationary over the sample period. If it is time-varying or non-stationary, the estimated parameters will be erroneous.

To measure how close one portfolio is to another, we compare the cash equivalent (CE) values of the two portfolios. The cash equivalent of a risky portfolio is the certain amount of cash that provides the same utility as the risky portfolio, that is, $U(\text{CE}) = Z(x)$ or $\text{CE} = U^{-1}[Z(x)]$ where, as defined before, $Z(x)$ is the expected utility of the risky portfolio.[4] The cash equivalent is an appropriate measure because it takes into account the investor's risk tolerance and the inherent uncertainty in returns, and it is independent of utility units. For a risk-free portfolio, the cash equivalent is equal to the certain return.

Given a set of asset parameters and the investors risk tolerance, a MV-optimal portfolio has the largest CE value of any portfolio of those assets. The

---

[3]Although the exponential utility function is convenient for deriving the MV problem with normally distributed returns, the MV framework is consistent with expected utility maximization for any concave utility function, assuming normality.

[4]For negative exponential utility, Freund [1956] shows that the expected utility of portfolio x is $Z(x) = 1 - \exp(-aE[x] + (a^2/2)\text{Var}[x])$, where $E[X]$ and $\text{Var}[x]$ are the expected return and variance of the portfolio. The cash equivalent is $\text{CE}_x = (1/a)\log(1 - Z(x))$. If returns are assumed to have a multivariate normal distribution, this is also the cash equivalent of an MV-optimal portfolio. See Dexter, Yu & Ziemba [1980] for more details.

percentage cash equivalent loss (CEL) from holding an arbitrary portfolio, x instead of an optimal portfolio o is

$$CEL = \frac{CE_o - CE_x}{CE_o}$$

where CE_o and CE_x are the cash equivalents of portfolio o and portfolio x respectively.

Data and Methodology

The data consist of monthly observations from January 1980 through December 1989 on ten randomly selected Dow Jones Industrial Average (DJIA) securities. We use the Center for Research in Security Prices (CRSP) database, having deleted one security (Allied–Signal, Inc.) because of lack of data prior to 1985. Each of the remaining twenty–nine securities had an equal probability of being chosen. The securities are listed in Exhibit 1.

MV optimization requires as inputs forecasts for: mean returns, variances, and covariances. We computed historical means $(\bar{r}_i)$, variances (σ_{ii}), and covariances (σ_{ij}), and assumed that these are the 'true' values of these parameters. Thus, we assumed that $E[r_i] = \bar{r}_i$, $E[\sigma_{ii}] = \sigma_{ii}$, and $E[\sigma_{ij}] = \sigma_{ij}$. A base optimal portfolio allocation is computed on the basis of these parameters for a risk tolerance of 50 (equivalent to the parameter $a = 0.04$).

Our results are independent of the source of the inputs. Whether we use historical inputs or those based on a complete forecasting scheme, the results continue to hold as long as the inputs have errors.

Exhibit 2 gives the input parameters and the optimal base portfolio resulting from these inputs. To examine the influence of errors in parameter estimates, we change the true parameters slightly and compute the resulting optimal portfolio. This portfolio will be suboptimal for the investor because it is not based on the true input parameters.

Next we compute the cash equivalent values of the base portfolio and the new optimal portfolio. The percentage cash equivalent loss from holding the suboptimal portfolio instead of the true optimal portfolio measures the impact of errors in input parameters on investor utility.

To evaluate the impact of errors in means, we replaced the assumed true mean $\bar{r}_i$ for asset i by the approximation $\bar{r}_i(1 + kz_i)$ where z_i has a standard normal distribution. The parameter k is varied from 0.05 through 0.20 in steps of 0.05 to examine the impact of errors of different sizes. Larger values of k represent larger errors in the estimates. The variances and covariances are left unchanged in this case to isolate the influence of errors in means.

The percentage cash equivalent loss from holding a portfolio that is optimal for approximate means $\bar{r}_i(1 + kz_i)$ but is suboptimal for the true means r, is

EXHIBIT 2:
Inputs to the Optimization and the Resulting Optimal Portfolio for a
Risk Tolerance of 50 (January 1980–December 1989)

	Alcoa	Amex	Boeing	Chev.	Coke	Du Pont	MMM	P&G	Sears	U Tech
Means (% per month)	1.5617	1.9477	1.907	1.5801	2.1643	1.6010	1.4892	1.6248	1.4075	1.1537
Std. Dev. (% per month)	8.8308	8.4585	10.040	8.6215	5.988	6.8767	5.8162	5.6385	8.0047	8.212

Correlations

Alcoa	1.0000									
Amex	0.3660	1.0000								
Boeing	0.3457	0.5379	1.0000							
Chev.	0.1606	0.2165	0.2218	1.0000						
Coke	0.2279	0.4986	0.4283	0.0569	1.0000					
Du Pont	0.5133	0.5823	0.4051	0.3609	0.3619	1.0000				
MMM	0.5203	0.5569	0.4492	0.2325	0.4811	0.6167	1.0000			
P&G	0.2176	0.4760	0.3867	0.2289	0.5952	0.4996	0.6037	1.0000		
Sears	0.3267	0.6517	0.4883	0.1726	0.4378	0.5811	0.5671	0.5012	1.0000	
U Tech	0.5101	0.5853	0.6569	0.3814	0.4368	0.5644	0.6032	0.4772	0.6039	1.0000

Optimal Port.

	Alcoa	Amex	Boeing	Chev.	Coke	Du Pont	MMM	P&G	Sears	U Tech
Weights	0.0350	0.0082	0.0	0.1626	0.7940	0.0	0.0	0.0	0.00	0.00

then computed. This procedure is repeated with a new set of z values for a total of 100 iterations for each value of k.

To investigate the impact of errors in variances each variance forecast σ_{ii} was replaced by $\sigma_{ii}(1 + kZ_j)$. To isolate the influence of variance errors, the means and covariances are left unchanged.

Finally, the influence of errors in covariances is examined by replacing each covariance σ_{ij} $(i \neq j)$ by $\sigma_{ij} + kz_{ij}$ where z_{ij} has a standard normal distribution, while retaining the original means and variances. The procedure is repeated 100 times for each value of k, each time with a new set of z values, and the cash equivalent loss computed. The entire procedure is repeated for risk tolerances of 25 and 75 to examine how the results vary with investors' risk tolerance.

Results

Exhibit 3 shows the mean, minimum, and maximum cash equivalent loss over the 100 iterations for a risk tolerance of 50. Exhibit 4 plots the average CEL

EXHIBIT 3:
Cash Equivalent Loss (CEL) for Errors of Different Sizes

k (size of error)	Parameter with Error	Mean CEL	Min. CEL	Max. CEL
0.05	Means	0.66	0.01	5.05
0.05	Variances	0.05	0.00	0.34
0.05	Covariances	0.02	0.00	0.25
0.10	Means	2.45	0.01	15.61
0.05	Variances	0.22	0.00	1.39
0.10	Covariances	0.11	0.00	0.66
0.15	Means	5.12	0.15	24.35
0.15	Variances	0.55	0.00	3.35
0.15	Covariances	0.27	0.00	1.11
0.20	Means	10.16	0.17	36 09
0.20	Variances	0.90	0.01	4.16
0.20	Covariances	0.47	0.00	1.94

as a function of k. The CEL for errors in means is approximately eleven times that for errors in variances and over twenty times that for errors in covariances. Thus, it is important to distinguish between errors in variances and errors in covariances.[5] For example, for $k = 0.10$, the CEL is 2.45 for errors in means, 0.22 for errors in variances, and 0.11 for errors in covariances.

Our results on the relative importance of errors in means and variances are similar to those of Kallberg & Ziemba [1984]. They find that errors in means are approximately ten times as important as errors in variances and covariances considered together (they do not distinguish between variances and covariances).

Our results show that for a risk tolerance of 50 the importance of errors in covariances is only half as much as previously believed. Furthermore, the relative importance of errors in means, variances, and covariances depends upon the investor's risk tolerance.

Exhibit 5 shows the average ratio (averaged over errors of different sizes, k) of the CELs for errors in means, variances, and covariances. An investor with a high risk tolerance focuses on raising the expected return of the portfolio

---

[5]The result for covariances also applies to correlation coefficients, as the correlations differ from the covariances only by a scale factor equal to the product of two standard deviations.

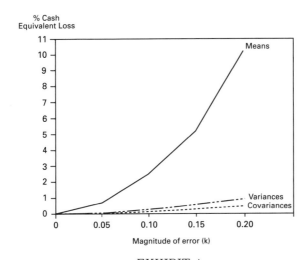

EXHIBIT 4

Mean percentage cash equivalent loss due to errors in inputs

EXHIBIT 5

Average Ratio of CELs for Errors in Means, Variances, and Covariances

Risk Tolerance	Errors in Means versus Variances	Errors in Means versus Covariances	Errors in Variances versus Covariances
25	3.22	5.38	1.67
50	1.98	22.50	2.05
75	21.42	56.84	2.68

and discounts the variance more relative to the expected return. To this investor, errors in expected returns are considerably more important than errors in variances and covariances. For an investor with a risk tolerance of 75, the average CEL for errors in means is over twenty–one times that for errors in variances and over fifty–six times that for errors in covariances.

Minimizing the variance of the portfolio is more important to an investor with a low risk tolerance than raising the expected return. To this investor, errors in means are somewhat less important than errors in variances and covariances. For an investor with a risk tolerance of 25, the average CEL for errors in expected returns is about three times that for errors in variances and about five times that for errors in covariances.

Most large institutional investors have a risk tolerance in the 40 to 60 range. Over that range, there is considerable difference in the relative importance of errors in means, variances, and covariances. Irrespective of the level of risk tolerance, errors in means are the most important, followed by errors in variances. Errors in covariances are the least important in terms of their influence on portfolio optimality.

Implications and Conclusions

Investors have limited resources available to spend on obtaining estimates of necessarily unknowable future parameters of risk and reward. This analysis indicates that the bulk of these resources should be spent on obtaining the best estimates of expected returns of the asset classes under consideration.

Sometimes, investors using the MV framework to allocate wealth among individual stocks set all the expected returns to zero (or a non-zero constant). This can lead to a better portfolio allocation because it is often very difficult to obtain good forecasts for expected returns. Using forecasts that do not accurately reflect the relative expected returns of different securities can substantially degrade MV performance.

In some cases it may be preferable to set all forecasts equal.[6] The optimization then focuses on minimizing portfolio variance and does not suffer from the error-in-means problem. In such cases it is important to have good estimates of variances and covariances for the securities, as MV optimizes only with respect to these characteristics.

Of course, if investors truly believe that they have superior estimates of the means, they should use them. In this case it may be acceptable to use historical values for variances and covariances.

For investors with moderate to high risk tolerance, the cash equivalent loss for errors in means is an order of magnitude greater than that for errors in variances or covariances. As variances and covariances do not much influence the optimal MV allocation (relative to the means), investors with moderate-to-high risk tolerance need not expend considerable resources to obtain better estimates of these parameters.

References

Bawa, Vijay S., Stephen J. Brown and Roger W. Klein (1979). 'Estimation Risk and Optimal Portfolio Choice.' *Studies in Bayesian Econometrics*, Bell Laboratories Series. North Holland.

---

[6]This approach is in the spirit of Stein estimation and is discussed in Chopra, Hensel, and Turner [1993]. As a practical matter, it should be used for assets that belong to the same asset class. e.g., equity indexes of different countries or stocks within a country. It would be inappropriate to apply it to financial instruments with very different characteristics; for example, stocks and T-bills.

Best, Michael J. and Robert R. Grauer (1991). 'On the Sensitivity of Means-Variance-Efficient Portfolios to Changes in Asset Means: Some Analytical and Computational Results.' *Review of Financial Studies* **4**, No. 2, 315–342.

Chopra, Vijay K. (1991). 'Mean-Variance Revisited: Near-Optimal Portfolios and Sensitivity to Input Variations.' *Russell Research Commentary*.

Chopra, Vijay K., Chris R. Hensel and Andrew L. Turner (1993). 'Massaging Mean-Variance Inputs: Returns from Alternative Global Investment Strategies in the 1980s.' *Management Science*, (July): 845–855.

Dexter, Albert S., Johnny N.W. Yu and William T. Ziemba (1980). 'Portfolio Selection in a Lognormal Market when the Investor has a Power Utility Function: Computational Results.' In *Proceedings of the International Conference on Stochastic Programming*, M.A.H. Dempster (ed.), Academic Press, 507–523.

Freund, Robert A. (1956). 'The Introduction of Risk into a Programming Model.' *Econometrica* **24** 253–263.

Freund, L. and M. Blume (1975). 'The Demand for Risky Assets.' *The American Economic Review*, December, 900–922.

Kallberg, Jarl G. and William T. Ziemba (1984). 'Mis-specification in Portfolio Selection Problems'. In *Risk and Capital*, G. Bamberg and K. Spremann (eds.), Lecture Notes in Econometrics and Mathematical Systems. Springer-Verlag.

Klein, Roger W, and Vijay S. Bawa (1976). 'The Effect of Estimation Risk on Optimal Portfolio Choice. *J. of Financial Economics* **3** (June), 215–231.

Markowitz, Harry M. (1987). *Mean-Variance Analysis in Portfolio Choice and Capital Markets*. Basil Blackwell.

Michaud. Richard O. (1989). 'The Markowitz Optimization Enigma: is 'Optimized' Optimal?' *Financial Analysts Journal* **45** (January–February), 31–42.

Ziemba, William T. and Raymond G. Vickson, eds. (1975) *Stochastic Optimization Models in Finance*. Academic Press.

Making Superior Asset Allocation Decisions: A Practitioner's Guide

Chris R. Hensel and Andrew L. Turner

> A man who seeks advice about his actions will not be grateful for the suggestion that he maximize expected utility. *A.D. Roy*

> Normative theories tell us how a rational agent should behave. Descriptive theories tell us how agents do behave. Prescriptive theories offer advice on how to behave when faced with our own cognitive or other limitations.with our own cognitive or other limitations. *Thaler (1992) paraphrasing Bell, Raiffa & Tversky (1988)*

1 Introduction

Since the publication of Markowitz's (1959) monograph on portfolio selection, mean-variance optimization has dominated financial theory as the choice criterion for investments. Many financial theories, such as the Capital Asset Pricing Model (see Sharpe 1964), rest squarely on the assumption that investors are mean-variance utility maximizers or that the market portfolio is mean-variance efficient. And, by and large, most practitioners use mean-variance optimization for asset allocation purposes.

However, many users of mean-variance techniques have discovered that mean-variance optimization has an alarming tendency to act wildly for seemingly minor changes in initial conditions. Concern about the usefulness of mean-variance analysis has appeared in the finance literature only more recently, mostly in the last decade. Exceptions are the work by Bawa, Brown & Klein (1979) and Kallberg & Ziemba (1981, 1984). Studies published by Turner & Hensel (1993), Chopra (1993), Chopra, Hensel & Turner (1993) and Chopra & Ziemba (1993) indicate that the verdict regarding naively applied mean-variance analysis, while not altogether unexpected, is not favorable.

This paper has two purposes. The first purpose is to provide a single, unified interpretation of the four papers cited immediately above and the related finance literature. The second purpose is to outline a guide to the implementation and use of mean-variance asset allocation models. We see the issues in the use of mean-variance analysis as being a set of trade-offs between theory, optimization sensitivity, and forecast reliability. The proposals in this paper can be thought of as an asset allocation tool kit for negotiating tradeoffs. However, it is not a blueprint that can be applied in all situations without modification.

We believe that mean-variance optimization has an important role in asset allocation strategy. The key to successfully using mean-variance analysis is to understand the process and, thereby, realize its full potential. Of the techniques suggested below, many are not new. They have been used on an ad hoc basis for years. Taken together, however, these proposals offer a systematic approach to mean-variance optimization that should produce significant benefits.

The remainder of this paper is organized in three sections. Section 2 reviews the existing literature. Section 3 opens with a brief statement of the problem and shows the effects on hypothetical asset allocation decisions of implementing new procedures and concepts for a representative investor. Section 4 concludes the paper with a statement of opinion on how to achieve superior mean-variance results.

Before proceeding, we caution you against looking for the Holy Grail in these pages. No grand unification theory will be offered for making asset allocation decisions. If our research has demonstrated anything, it is that the more we understand the problem and the data, the less confidence we have in the exact numerical solutions. Given our current knowledge, and in the spirit of the second quote above, we offer a prescriptive theory. What does one do today, knowing there are aspects of the problem that are poorly understood? In the framework we present, we seek to take full advantage of our knowledge of mean-variance analysis while avoiding what we believe to be some serious decision traps.

2 Literature Review

2.1 Model Sensitivity

Michaud (1989) discussed the benefits and limitations of mean-variance optimization. As he states, "The unintuitive character of many 'optimized' portfolios can be traced to the fact that mean-variance optimizers are, in a fundamental sense, 'estimation-error maximizers.' ... mean-variance optimization significantly overweights (underweights) those securities that have large (small) estimated returns, negative (positive) correlations and small (large) variances. These securities are, of course, the ones most likely to have large estimation errors."

Another limitation of mean-variance optimization is the instability of the optimal solution. Small changes in the inputs, especially means, can cause large changes in the optimal asset weightings (see Kallberg & Ziemba 1981, 1984). Chopra & Ziemba (1993) found that errors in the forecasted means are about 11 times more damaging than errors in the forecasted variances at a typical investor risk tolerance level and about double that for the co-variances. Moreover, there are significant movements in the optimal weights over time, and usually a small number of the available assets are included in the optimal mix. These results are consistent with Adler (1987), who states that the output weights of a mean-variance optimizer "are excessively unstable as portfolios are reoptimized and rebalanced over time." The final limitation of interest is the nonuniqueness of optimal solutions (Michaud 1989). The nonunique nature arises because the uniqueness of the solution is dependent upon the erroneous assumption that the mean-variance inputs are without statistical estimation error.

Thus, statistical methods that allow us to reduce the impact of input estimation errors and result in optimal portfolio weights that are more diversified and change slowly over time are intensely interesting. If we clearly had superior forecasting ability, then we might want the portfolio weights to change dramatically and quickly over time to take advantage of these forecasts. We know from the literature on market timing (e.g., Merton & Henriksson 1981 or Ferson & Schadt 1996) that significant bets must be taken to realize measurable benefits from superior forecasts. However, lacking a high degree of confidence in our forecasts, weights that change slowly over time reduce transaction costs and reduce the detrimental impact of estimation errors.

There are two basic approaches to achieving these goals: methodically massaging the inputs and/or constraining the optimal weights. Frost & Savarino (1988) examined the effects of constraining the optimal weights for portfolios of individual securities. Security weight constraints forced the solutions to be more highly diversified than unconstrained solutions. They found that imposing upper bounds on portfolio weights improved portfolio performance by reducing the deleterious effects of estimation errors on the portfolio weights. Chopra (1993) tested this procedure with three asset classes through time and found similar benefits.

2.2   Stein Estimation

Michaud (1989) suggests Stein estimation (shrinkage toward global means) as an approach to reducing estimation error in the inputs. The philosophy behind Stein estimation is that similar entities behave alike on average. For example, if stock indexes of different countries are alike in some fundamental sense, they should exhibit similar returns over time. Jobson & Korkie (1981) examined Stein estimation as a method of reducing estimation risk in individual security expected returns. They found that the practical implementation of mean-variance optimization is greatly enhanced by assuming that the grand mean of return for all securities represents the best estimator of the individual expected returns.

They suggested the use of a global mean as the expected return input. This is equivalent to using a Stein estimation process (Efron & Morris 1973, 1975, 1977), where the weight for the individual mean is zero and the weight for the global mean is one. Stein estimates adjust the parameters of an individual country index to reflect parameters of a global index. Thus, using a weighted average of an individual country index return history and a global index return history would provide a better forecast of the individual country index characteristics.

Stein estimation can be used to prevent the means and covariances from varying substantially both cross-sectionally and intertemporally. Jorion (1985) states that this extreme amount of shrinkage toward the global mean "is difficult to reconcile with the generally accepted trade-off between risk and expected return, unless all stocks fall within the same risk class."

One implication of using the same expected return inputs for all equity markets, and likewise for all fixed income markets, is that it reduces the number of forecasts needed by the mean-variance model. Jobson & Korkie (1981) found that the benefits of just using the grand mean as the expected return inputs for individual securities greatly enhanced the practical implementation of a mean-variance approach. Additionally, they found that "[the] often-suggested five years of monthly data seem quite reasonable providing this new procedure (shrinkage to the global mean) is used."

2.3   Similarity of Country Returns

Turner & Hensel (1993), analyzed total stock and bond returns by country to see if the returns of stocks and bonds were statistically different across markets during the 1980s. They used monthly observations from January 1980 through December 1989 and examined returns based in local currency and hedged and unhedged US dollars. They examined total equity returns of indexes from Australia, Canada, Germany, Japan, the UK, and the US and total fixed income returns of indexes from the same countries except Australia.

Turner & Hensel found that sample mean stock and bond returns during the 1980s were statistically indistinguishable across countries. They found evidence of variance differences, which they speculate may be explainable by other economic factors. They also found evidence that intercountry stock and bond correlations were not significantly different. Thus, they confirmed the results of other researchers, such as Jobson & Korkie (1981), but in a broader global context using more asset types and different statistical tests. Turner & Hensel's work suggests reducing the number of input estimates, but not necessarily the number of asset classes, in a mean-variance global asset allocation problem.

In addition to reducing the number of required expected return estimates, the research of Turner & Hensel suggests further reduction in the number of input estimates by using the average of the correlations within asset class as the correlation inputs for equity and fixed income markets. This reduction in the number of inputs decreases sensitivity of a mean-variance optimizer to forecast errors. By reducing the number of parameter estimates, we should reduce the impact of the estimation errors and improve stability of the optimal weights over time. The effects of reducing the number of estimated inputs on the stability of portfolio weights was examined by Chopra, Hensel & Turner (1993) and is summarized below.

2.4 Effects of Shrinkage

Chopra, Hensel & Turner (1993) explored the impact of adjustments to the inputs on total returns, terminal wealth, and portfolio turnover, in an unconstrained, monthly mean-variance asset allocation over time.

For their study, Chopra, Hensel & Turner used monthly hedged return data from January 1980 through December 1990 for six individual country stock indexes, five country bond indexes, and five cash indexes and examined three levels of Stein estimates. The first involves setting the expected returns of each stock and bond index equal to that of a corresponding global index. The second level involves additionally adjusting stock and bond correlations by setting them equal to their respective averages. Variances are not changed. The third level extends the first level by additionally setting all intercountry stock variances and covariances equal to their averages and by making similar adjustments to intercountry bonds and cash. This last level is a three-asset view of the world: a stock is a stock, a bond is a bond, and cash is cash irrespective of country.

Using the first five years of data, Chopra, Hensel & Turner formed mean-variance optimal portfolios for two risk tolerance levels. These portfolios were held for one month. This 60-month estimation window was rolled forward one month, and the procedure was repeated. This procedure provided 72 months of out-of-sample returns. The returns created by this procedure are not entirely free of look-ahead bias since Turner & Hensel (1993) already had prior knowledge of part of the return series. Four different unconstrained mean-variance approaches and a naive benchmark were tested over time. The naive benchmark was rebalanced monthly to a global 60/40 stock/bond portfolio with equal weighting across countries within asset type.

CHT reached two major conclusions: (1) Any of the suggested adjustments to inputs dominate the results of an unadjusted-input mean-variance optimization. Adjusted-input portfolios had higher mean return, lower variance, and greater terminal wealth than unadjusted-input portfolios. (2) These improvements become even greater with transaction costs. These conclusions were unaffected by October 1987 in the sample, and the results strongly suggest that users of unconstrained mean-variance analysis should use Stein estimation or some other form of shrinkage to adjust the inputs as an alternative to accepting historical data. As a caveat to their work, Chopra, Hensel & Turner state that the presence of October 1987 strongly and adversely affects the comparison of the adjusted-input mean-variance portfolios with the naive benchmark portfolio. The mean returns of the adjusted portfolios tended to be larger than the naive benchmark, though not uniformly larger. With October 1987 removed from the data series, adjust-input mean-variance portfolios had higher mean returns than the naive benchmark for every level of transaction costs and both risk tolerances. However, adjusted-input mean-variance portfolios did uniformly exhibit more variation than the naive benchmark, although this seemed to be substantially linked to October 1987.

2.5 Currency Hedging

Jorian's (1989) paper on currency hedging from a US perspective is the final relevant paper to review. He found hedged foreign assets (especially hedge foreign bonds) were preferred over unhedged assets at risk tolerances consistent with a broad spectrum of professional investors such as pension funds at all but extreme hedging costs (greater than 75 basis points per year).

The benefit of hedging is in the reduction of risk by removing most foreign currency surprise. The impact, at the total portfolio level, of this risk reduction in foreign asset risk depends upon the size (percentage and dollar) of the foreign allocation. Many pension funds who have less than 20% of total assets in foreign investments and for whom foreign segment volatility is not a major issue have chosen not to hedge.

The preference for hedged foreign assets over unhedged is also found when perspectives other than that of US investors are examined. Unpublished research at the Frank Russell Company has confirmed the risk reduction benefits of hedged foreign assets when viewed from a UK perspective and from a Japanese perspective. However, the case for hedged foreign equities appears not as strong when examined from an Australian perspective and from a Canadian perspective in the last 15 years. For additional information, see Appendix B.

3 Implication for Asset Allocation

One of the chief advantages of mean-variance analysis has always been its intuitive simplicity. After all, the notion is simply that one wants as much return for as little risk as one can get. Its second strength, in a world of fast personal computers, has been the relative ease of computations. However, these apparent tremendous advantages have not yielded real practical gains to many investors. The recommendations of mean-variance analysis have often been ignored because of their extremity. And, had they been implemented, they would have often been quite detrimental to performance. What is proposed here is a systematic and extensively tested approach to using mean-variance models; it is not a post hoc adjustment stimulated by the users' distrust or dislike of the results. We believe the proposals contained in this paper offer a practical means of getting more relevant and economically viable asset allocation results from the mean-variance framework.

3.1   Problem Statement

The problem is how to realize the potential of mean-variance optimization. Because some investors have become conditioned to ignore extreme or apparently nonsensical results, one could think of the problem as how to make sense out of nonsense.

3.2   Concepts

This section will outline the key concepts flowing from the papers reviewed in the previous section and briefly describe the implications of those concepts for asset allocation.

Number of asset classes: Depending upon your interpretation of the implications of Turner & Hensel (1993) and Chopra, Hensel & Turner (1993), the number of separate asset classes used in mean-variance optimization either increases or decreases. Taken literally, the research suggests increasing the number of asset classes, treating each single country stock and bond index as a separate asset class. This is based on the Turner & Hensel findings that although the returns and correlations are not significantly different, there are variance differences that should be exploitable and thus result in an increase in the number of asset classes.

The other interpretation essentially assumes that the variance differences are not expected in the future and therefore the variance inputs should be set equal. Treated in this way, expected returns, correlation, and variances are each equal within asset type. This is equivalent to allocating among one global stock portfolio, one global bond portfolio, and one global cash portfolio, with equal weighting within asset type. This approach essentially shrinks the effective number of asset classes to three, although in the allocations we present later we will show the allocations for all of the 16 assets we used.

Composition of asset classes: As the available asset choices, we used hedged returns for stock indexes of Australia, Canada, Germany, Japan, the UK and the US; bond indexes of all but Australia; and cash indexes of all but Germany. At the end of 1990, these countries represented over 85% of the market capitalization of the Morgan Stanley Capital International (MSCI) World Equity Index and over 86% of the Salomon Brothers World Bond Index. Hedged returns from January 1980 through December 1990 were used to estimate the asset allocation inputs for these 16 indexes.

A basic finding – more global: Before we examine representative asset allocation results from these two approaches, note that the unconstrained allocations will probably contain a higher foreign allocation than investors currently find acceptable. The increase of foreign investments is a direct result of including more countries and treating each country's stock and bond indexes as individual asset classes. However, the benefits of increasing investments in foreign markets have been discussed by Jorion (1987, 1989), Solnik (1995), and Solnik & Noetzlin (1982). Unless regulatory constraints or unusual tax schemes prevent it, we expect this conclusion to be robust across markets and investor domiciles.

However, recent work by Sinquefield (1996) contradicts this finding. Using the realizations of the last 20 years as forecasts, Sinquefield argues that the gains from international diversification are too small to justify large international positions. He further asserts there is no reason to believe that the risk factors driving market returns "have *geographically different expected returns* [emphasis added]". As is obvious from our use of shrinkage, we agree. Several recent papers (e.g., Solnik, Boucrelle & Fur(1996)) suggest that correlations between markets may be changing over time. Furthermore, the correlations may be depend on volatility levels. If true, such behavior poses severe problems for appropriate treatment by a mean-variance model. The different findings highlight and reinforce our position that whatever methods are chosen, they can not be applied by rote.

Reasonable expectations: We believe that individual asset return expectations are best estimated by using shrinkage toward a global mean. This belief is based on the research discussed above showing an inability to distinguish mean return differences across countries and the benefits of using global means as inputs for individual returns. The effects of this approach on asset allocations and asset class diversification will be shown.

Currency hedging: In the following analysis, only hedged assets are considered, and the explicit cost of currency hedging is neglected. This may seem to be at odds with common sense. There are two reasons for this. First, unless hedging costs are large, well over 75 basis points, hedged assets are preferred for most risk tolerances relevant to investors. Second, at magnitudes of hedging costs experienced by pension funds for hedging major currencies (about 25 to 35 basis points), the hedging costs are simply swamped by variation in the underlying stock index's expected return. Chopra & Ziemba (1993)

	Historical (US$)	
	Mean Return	Standard Deviation
Australian Stocks	12.9%	23.4%
Canadian stocks	9.5%	18.2%
German stocks	18.8%	21.3%
Japanese stocks	19.0%	18.6%
UK stocks	18.3%	18.9%
US stocks	15.2%	16.8%
Canadian bonds	11.0%	11.9%
German bonds	10.6%	6.1%
Japanese bonds	11.8%	6.8%
UK bonds	11.4%	10.3%
US bonds	12.4%	13.5%
Australian cash	8.6%	1.5%
Canadian cash	9.7%	0.9%
Japanese cash	10.1%	1.1%
UK cash	9.8%	1.0%
US cash	8.8%	0.8%

Table 1: Asset Allocation Optimization Inputs, part 1. For full caption see part 3

showed that variations in expected returns have about 11 times the effect of variations in variances at a typical investor risk tolerance. In sum, subtracting hedging costs encourages misallocation on the basis of such small (albeit known) differences. We present in Appendix A the asset allocation results when foreign asset expected returns are decreased by a 30 basis-point hedging cost. However, if hedging costs are deducted, the conclusions remain unchanged.

3.3 Asset Allocation Inputs

We developed three sets of asset allocation inputs from our 11 years of monthly hedged returns for six country stock indexes, five country bond indexes, and five country cash indexes. The asset allocation results based upon these three sets of inputs are not sensitive to input variation caused by employing varying time horizons precisely because the inputs have been massaged. The expected return and standard deviation inputs are shown in Table 1 (correlation tables are available from the authors upon request).

The first set of inputs is the historical mean returns, standard deviations,

| | Equal Mean Returns | |
	Equal Return	Standard Deviation
Australian Stocks	15.6%	23.4%
Canadian stocks	15.6%	18.2%
German stocks	15.6%	21.3%
Japanese stocks	15.6%	18.6%
UK stocks	15.6%	18.9%
US stocks	15.6%	16.8%
Canadian bonds	11.5%	11.9%
German bonds	11.5%	6.1%
Japanese bonds	11.5%	6.8%
UK bonds	11.5%	10.3%
US bonds	11.5%	13.5%
Australian cash	9.4%	1.5%
Canadian cash	9.4%	0.9%
Japanese cash	9.4%	1.1%
UK cash	9.4%	1.0%
US cash	9.4%	0.8%

Table 1: Asset Allocation Optimization Inputs, part 2. For full caption see part 3

and correlations. The second set changes only the returns, substituting the average of the historical returns within each asset type as the new expected return. The standard deviations and correlations are left unadjusted. The final set of inputs includes the return changes shown in the second set and additionally adjusts the standard deviations and correlations (i.e. covariances). The standard deviation inputs are the square root of the average of the historical variances within each asset type. The correlation inputs are also the average of each like group. For example, the correlation for each stock-stock correlation is the average of all stock-stock correlations; the correlation for each stock-bond correlation is the average of all stock-bond correlations; and the correlation for each stock-cash correlation is the average of all stock-cash correlations.

3.4 Allocation Examples

Table 2 shows the optimal portfolio weights for a mean-variance optimization using Periold's (1984) algorithm with risk tolerances of 50 and 75. The risk tolerance of many US pension funds ranges from about 45 to 85, with 50 to 55 as the average. Therefore, a risk tolerance of 50 is fairly typical

| | Equal Returns and Covariances | |
	Equal Return	Root Mean Variance
Australian Stocks	15.6%	19.7%
Canadian stocks	15.6%	19.7%
German stocks	15.6%	19.7%
Japanese stocks	15.6%	19.7%
UK stocks	15.6%	19.7%
US stocks	15.6%	19.7%
Canadian bonds	11.5%	10.1%
German bonds	11.5%	10.1%
Japanese bonds	11.5%	10.1%
UK bonds	11.5%	10.1%
US bonds	11.5%	10.1%
Australian cash	9.4%	1.1%
Canadian cash	9.4%	1.1%
Japanese cash	9.4%	1.1%
UK cash	9.4%	1.1%
US cash	9.4%	1.1%

Table 1: Asset Allocation Optimization Inputs*, part 3. Annualized: January 1980 Through December 1990**

*Arithmetic average monthly returns were converted to annual equivalents by multiplying by 12. Likewise, monthly standard deviations were converted to annual equivalents by multiplying by the square root of 12. Correlation tables are available from the authors upon request.

** Data sources: Data Resources Inc., Morgan Stanley Capital International, and individual country exchanges.

and 75 represents the more aggressive investor. It is difficult to do a direct comparison of risk tolerances across countries because of the relationship of risk tolerance and the mean-variance inputs. However, the typical risk tolerance of UK pension funds appears to be higher than for US pension funds based on their total equity exposure (domestic plus foreign). For UK funds, the results for a risk tolerance of 75, as used here, would probably be more representative. The total equity exposure of Australian clients is about 57% and is below that of typical US funds — 62% (with real estate excluded from both), which is consistent with a risk tolerance of 50 as used here.

The allocations are unconstrained except for disallowing short sales, and the type of inputs used are listed at the top of each column. The amount of across-asset class diversification increases dramatically as we move away from

		Risk Tolerance = 50	
	Historical	Equal Returns	Eq. Returns and Covariances
Australian Stocks		5%	10%
Canadian stocks		7%	10%
German stocks	22%	8%	10%
Japanese stocks	40%	21%	10%
UK stocks	24%	2%	10%
US stocks		19%	10%
Canadian bonds			8%
German bonds		35%	8%
Japanese bonds	7%	4%	8%
UK bonds			8%
US bonds	7%		8%
Australian cash			
Canadian cash			
Japanese cash			
UK cash			
US cash			

Table 2: Optimal Allocations, part 1. Numbers may not sum to 100% due to rounding.

historical inputs at both risk tolerances.

The foreign allocations in Table 2 range from about 73% to 100%. This is dramatically higher than US investors have been investing outside the US. The Money Market Directory of Pension Funds for 1996 lists the average allocation for corporate pension funds with assets exceeding $500 million, after excluding assets not included above (e.g., real estate), as roughly 54% US stocks, 10% non-US stocks, 31% US bonds, 1% non-US bonds, and 4% cash. Thus, the allocation from Table 2 with the least non-US exposure (73.23%) is more than seven times the average non-US exposure of large US pension plans. Foreign investments of typical UK funds range between 20% and 45%. Although their foreign allocations are much higher than US sponsors', this approach suggests still higher foreign allocations. The Australian Bureau of Statistics lists the average foreign allocation of Australian Superannuation and Approved Deposit Funds as about 15% (after excluding property), still far below the allocations suggested in Table 2, although this may be a result of tax effects.

In broad terms, the shrinkage of estimates toward global means for asset

		Risk Tolerance = 75	
	Historical	Equal Returns	Eq. Returns and Covariances
Australian Stocks		5%	14%
Canadian stocks		12%	14%
German stocks	26%	13%	14%
Japanese stocks	48%	30%	14%
UK stocks	26%	1%	14%
US stocks		27%	14%
Canadian bonds			3%
German bonds		12%	3%
Japanese bonds			3%
UK bonds			3%
US bonds			3%
Australian cash			
Canadian cash			
Japanese cash			
UK cash			
US cash			

Table 2: Optimal Allocations, part 2. Numbers may not sum to 100% due to rounding.

allocation inputs produces a single general effect: an optimal portfolio is much more diversified by country than portfolios using unadjusted historical averages. This general conclusion holds regardless of base currency or country perspective! It means, for example, that one ought to have a sizable position (i.e., greater than 5% to 20%) outside the home country. This is true for both bonds and equities and, hence, hedged foreign bonds ought to be included in the list of available assets.

3.5   Caveats

The markets used in this study were chosen because of their relative size and liquidity. Liquidity is always a concern in placing large positions, foreign or domestic. Nevertheless, foreign positions much larger than those typically in place now among large pension funds can be accommodated in these markets. Following October 19, 1987, US investors should recognize the US is susceptible to a liquidity crunch, as are other major markets. Just as obviously, foreign liquidity is not an issue for investors who are both non-yen and non-dollar based. Clearly, there are also restrictions on foreign ownership of securities in many countries and some sizable differences in regulation.

If the ideas presented here are carried to an extreme, such as equal weighting across 50 equity markets, then liquidity and market size are relevant issues. A possible approach to the size problem would be to treat a group of smaller markets, in total, as one of the larger markets and weight equally with the large markets. For example, the global equity market could be thought of as consisting of Australia, Canada, Germany, Japan, the UK, the US and everything else. If 20 additional countries make up "everything else," then your allocations would equal weight across the seven major markets with 5% of the one seventh allocated to each of the 20 smaller markets that constitute "everything else."

4 Opinion

The strong conclusion of this work is that the typical investor portfolio is grossly underdiversified by country. This is true of investors in the US and Australia (taking due account of tax effects) and, to a lesser extent, the UK. While we do not expect many investors to move to 70% foreign assets, even over a few years, it is the direction in which holdings should go. Moving in this direction would suggest a near-term allocation target of about 40% foreign for an investor with an average risk tolerance — quite far, for example, from the 10% to 20% policy many US sponsors currently target. After all, the US is only 40% to 45% of global equity capitalization — which means that 55% to 60% is foreign. This argument does not rest on superior returns: in fact, there are no superior returns assumed for any country! Nor does it not rest on lower foreign risk. It rests solely on diversification benefits! This position applies equally to hedged foreign equities and hedged foreign bonds. Whether the world is viewed as having more asset classes or fewer, the conclusion remains; increased foreign investments.

For investors domiciled in most markets, the default choice for foreign assets should be to hedge back to the domestic currency. Investors may have valid reasons why they should be unhedged or partially unhedged but, in the absence of reliable currency forecasts, a fully hedged position represents the natural benchmark for asset allocation. This position is based on both published and unpublished research. The case for this default is extremely strong. (Eun & Resnick (1988) or Jorion (1989)). Hedging becomes more critical as risk tolerance falls and/or the proportion of foreign assets grows. However, hedging costs should not be explicitly deducted from expected returns because of the extreme sensitivity of mean-variance asset allocation models to small disturbances in means.

We believe Stein estimation techniques (shrinkage toward global means) should be used to estimate mean-variance inputs instead of relying on unadjusted historical data.

References

M. Adler, 1987. "Global Asset Allocation: Some Uneasy Questions." *Investment Management Review* (September–October): 13–18.

V. Bawa, S.J. Brown, and R. Klein, 1979. *Estimation Risk and Optimal Portfolio Choice.* North-Holland.

D. Bell, H. Raiffa, and A. Tversky, 1988. "Descriptive, Normative, and Prescriptive Interactions in Decision Making." In *Decision Making: Descriptive, Normative, and Prescriptive Interactions,* ed. David Bell.

V.K. Chopra, 1993. "Improving Optimization." *Journal of Investing* (Fall): 51–59.

V.K. Chopra, C.R. Hensel, and A.L. Turner, 1993. "Massaging Mean-Variance Inputs: Returns from Alternative Global Investment Strategies in the 1980s." *Management Science* (July): 845–855.

V.K. Chopra and W.T. Ziemba, 1993. "The Effect of Errors in Means, Variances, and Covariances on Optimal Portfolio Choice." *Journal of Portfolio Management* (Winter):6–11.

B. Efron, and C. Morris, 1973. "Stein's Estimation Rule and Its Competitors: An Empirical Bayes Approach." *Journal of the American Statistical Association* **68**: 117–130.

B. Efron, and C. Morris, 1975. "Data Analysis Using Stein's Estimator and Its Generalizations." *Journal of the American Statistical Association* **70**: 311–319.

B. Efron, and C. Morris, 1977. "Stein's Paradox in Statistics." *Scientific American* (May): 119–127.

C.S. Eun and B.G. Resnick, 1988. "Exchange Rate Uncertainty, Forward Contracts, and International Portfolio Selection." *Journal of Finance* **43** (1): 197–215.

M.D.D. Evans and K. Lewis, 1995 "Do Expected Shifts in Inflation Affect Estimates of the Long-Run Fisher Relation?" *Journal of Finance* **50** (1): 225–253.

D.D. Ezra, 1991. "Asset Allocation by Surplus Optimization." *Financial Analysts Journal* (January–February): 51–57.

W. Ferson and R. Schadt, 1996. "Measuring Fund Strategy and Performance in Changing Economic Conditions." *Journal of Finance* (forthcoming).

K.A. Froot and R.H. Thaler, 1990. "Foreign Exchange." *Journal of Economic Perspectives* **4** (3): 179–192.

P.A. Frost and J.E. Savarino, 1988. "For Better Performance: Constrain Portfolio Weights." *The Journal of Portfolio Management* (Fall): 29–34.

J.D. Jobson and R. Korkie. 1981. "Putting Markowitz Theory to Work." *The Journal of Portfolio Management* (Summer): 70–74.

P. Jorion, 1985. "International Portfolio Diversification with Estimation Risk." *The Journal of Business* **58** (3): 259–278.

P. Jorion, 1987. "Why Buy International Bonds?" *Investment Management Review* (September–October): 19–28.

P. Jorion, 1989. "Asset Allocation with Hedged and Unhedged Foreign Stocks and Bonds." *The Journal of Portfolio Management* (Summer): 49–54.

J.G. Kallberg, and W.T. Ziemba, 1981. "Remarks on Optimal Portfolio Selection." In *Methods of Operations Research*, G. Bamberg and O. Opitz eds. 507–520. Oelgeschlager, Gunn and Hain.

J.G. Kallberg, and W.T. Ziemba, 1984. "Mis-specifications in Portfolio Selection Problems." In *Risk and Capital*, G. Bamberg and A. Spremann eds. 74–87, Springer-Verlag.

H. Markowitz, 1959. *Portfolio Selection: Efficient Diversification of Investments*, Wiley.

R.C. Merton, 1985. "On the Current State of the Stock Market Rationality Hypothesis." In *Macroeconomics and Finance: Essays in Honor of Franco Modigliani* (September).

R.O. Michaud, 1989. "The Markowitz Optimization Enigma: Is 'Optimized' Optimal?" *Financial Analysts Journal* (January–February): 31–42.

A.F. Perold, 1984. "Large-scale Portfolio Optimization." *Management Science* **30**: 1143–1160.

R. Roll, 1992. "Industrial Structure and the Comparative Behavior of International Stock Market Indexes." *The Journal of Finance* **47** (1): 3–42.

W.F. Sharpe, 1964. "Capital Asset Prices: A Theory of Market Equilibrium Under Conditions of Risk." *The Journal of Finance* **19**: 425–442.

R.A. Sinquefield, 1996. "Where Are the Gains from International Diversification?" *Financial Analysts Journal* (January–February): 8-14.

B. Solnik, 1995. *International Investments.*, 3rd edition. Addison-Wesley.

B. Solnik, C. Boucrelle, and Y. le Fur, 1996, "International Market Correlation and Volatility." *Financial Analysts Journal* (September–October): 17-34.

B. Solnik, and B. Noetzlin. 1982. "Optimal International Asset Allocation." *The Journal of Portfolio Management* (Fall): 11–21.

D. Stone, and C.R. Hensel, 1989. "Strategic Currency Hedging Non-US Investments for US-Based Investors." Russell White Paper (August).

R.H. Thaler, 1992. *The Winners Curse: Paradoxes and Anomalies of Economic Life*. The Free Press.

A.L. Turner and C.R. Hensel. 1993. "Were the Returns from Stocks and Bonds of Different Countries Really Different in the 1980s?" *Management Science* (July): 835–844.

| | | Risk Tolerance = 50 | |
	Historical	Equal Returns	Eq. Returns and Covariances
Australian Stocks		5%	10%
Canadian stocks		1%	10%
German stocks	20%	8%	10%
Japanese stocks	40%	20%	10%
UK stocks	26%		10%
US stocks		28%	13%
Canadian bonds			5%
German bonds		34%	5%
Japanese bonds	1%	4%	5%
UK bonds			
US bonds	12%		21%
Australian cash			
Canadian cash			
Japanese cash			
UK cash			
US cash			
Turnover	8%	10%	16%

Table A-1: Optimal Allocations When Hedging Costs Are Explicitly Included, part 1. Numbers may not sum to 100% due to rounding.

Appendix A

Table A-1 presents the allocations that result when all foreign asset returns were decreased 30 basis points to cover the cost of hedging back to the domestic currency. The inclusion of a 30-basis-point hedging cost (foreign expected returns were decreased) caused the least impact on the portfolio generated from historical returns, with increasing impact as shrinkage to global means was used. This is seen by looking at the turnover presented at the bottom of Table A-1. The turnover is calculated from the positions in Table 2, which exclude hedging costs. At a risk tolerance of 50, the turnover almost doubles as we move from historical inputs to inputs with equal returns and covariances. At a risk tolerance of 75, the turnover, while at about the same absolute levels, increases over four times moving from historical inputs to inputs with equal returns and covariances.

While the hedging costs change the allocations, these costs do not alter the general conclusions of the analysis. For example, at a risk tolerance of 50 for the equal return inputs, 61.2% of the portfolio is invested in stocks in Table 2 while 60.9% of the portfolio is invested in stocks in Table A1. Moreover, the overwhelming majority of both portfolios' stock investments is in foreign stocks.

We do not believe hedging costs should be explicitly included because a mean-variance optimizer spuriously allocates on the basis of these small known differences

		Risk Tolerance = 75	
	Historical	Equal Returns	Eq. Returns and Covariances
Australian Stocks		6%	13%
Canadian stocks		3%	13%
German stocks	24%	12%	13%
Japanese stocks	46%	29%	13%
UK stocks	30%		13%
US stocks		40%	19%
Canadian bonds			
German bonds		10%	
Japanese bonds			
UK bonds			
US bonds			17%
Australian cash			
Canadian cash			
Japanese cash			
UK cash			
US cash			
Turnover	4%	13%	18%

Table A1: Optimal Allocations When Hedging Costs Are Explicitly Included, part 2. Numbers may not sum to 100% due to rounding.

while the forecast errors (variance) around these expected return inputs greatly exceed the hedging costs. For the inputs where shrinkage was used, including hedging costs, caused the optimizer to allocate more to domestic assets because their returns have not decreased. The allocations and turnover presented in Table A1 support the findings in Chopra and Ziemba (1993) that mean-variance optimizers are very sensitive to changes in expected return inputs.

Appendix B

Some of the conclusions in this paper are the result of a synthesis of the arguments in Jorion (1989) as well as other works published in academic journals. While it is beyond the scope of this paper to review that literature, the purpose of this appendix is to briefly outline those arguments. The appendix opens with a brief discussion of the relevant concepts, cites some empirical work, reviews the importance of the global context, examines the relevance of individual country experiences, and concludes with a prescription for approaching currency hedging problems in the absence of forecasting skills.

Concepts

Purchasing power parity (PPP) asserts that the foreign exchange rate between two countries depends on the relative price level between the countries. While always thought to apply to flexible exchange rates, it also applies to fixed rates when there is adequate, low-cost trade between the two countries. The empirical evidence in favor of parity is weak except at quite long horizons such as 20 to 30 years, and as long as 75 years may be necessary for parity to hold. Nevertheless, the theory implies that the expected change in parity is related to the expected inflation differential between the two countries. Since one country could print money at a faster rate than the others, it is theoretically possible that a set of exchange rates could rise toward infinity and that a set of expected changes in parity could be positive at every moment.

Covered interest arbitrage requires that the forward prices for foreign exchange must have a parity relationship to interest rates in the respective countries. In theory, interest rate parity (IRP) holds across all maturities on the yield curve. However, in practice it holds best in the short rates, say one to nine months. A consequence of IRP is that the forward premium is related to the current interest differential. Assuming that in the long run the real rate of return to capital is identical across countries (or else real capital would flow between the countries), the interest differential itself is related to the expected inflation differential between the countries (implying that the Fisher equation holds in each country, see for example Evans and Lewis(1995)). Since the expected future spot exchange rate is equal to the current rate adjusted for inflation expectations, it follows that the forward rate is an unbiased estimate of the future spot exchange rate. These two concepts of parity are clearly not country dependent!

The mathematics of currency hedging are straightforward and contained in any textbook on international finance. (See e.g. Solnik (1995)). In brief, it is easy to show that the hedged rate of return differs from the unhedged rate of return by the currency surprise – that proportional change in the exchange rates not predicted by the forward premium. From the above argument, the forward rate is an unbiased estimate of the future spot rate, and so the currency surprise must have an expected value of zero. Then, since the expected hedged and unhedged rates of return are equal, it follows that currency exposures can be thought of as assets having an expected rate of return of zero and positive variance. In a mean-variance asset allocation model, unless zero-return, positive-variance assets are negatively correlated with other assets, they are not desirable.

If local market investment returns and currency surprise are expected to be uncorrelated or positively correlated, then hedging is unambiguously better; that is, currency exposure adds nothing but volatility. However, if the relationship between local market investment returns and currency surprise is expected to be strongly negative, then hedging might not be desirable. Unhedged returns in this case would have a lower standard deviation than hedged returns because currency returns would tend to move in opposite direction from local market returns. Such a relationship might hold for a small, export-driven economy whose economic fortunes are closely tied to its exchange rates. How strongly does the expected negative correlation of local market investment returns and currency surprise need to be expressed for unhedged investments to be preferred? Based on the relative volatility of local market returns and currency surprise, we estimate that the expected negative correlation would need to be at least -0.3 or -0.4. Not only is this a very large (algebraically) correlation, but in order for hedging to be suboptimal, this must be the expected correlation between local market investment returns and currency surprise looking forward.

Evidence

While the forward rate is often found to be an unbiased forecast of the future spot exchange rate, it is almost always found to have high variance and so is not a good forecast. (See e.g. Froot and Thaler (1991)). Recent studies have found that the current spot rate itself is a better predictor of future spot. Also, the evidence on the Fisher effect is mixed. In the long run, higher inflation does lead to higher rates, but it does not always do so in the short run, and real rates of interest do react to inflation shocks. And, while the empirical evidence does vary by country, we must look at the global evidence as a whole, expecting variations in realizations by country and concentrating on what generalizations about the future we can make from the data.

For any country and for a long period of time, the currency surprise can be consistently of a single sign. But, on balance around the globe to a first order approximation, the average surprise over time is zero. It is insufficient to argue that the surprise has not been zero on average for one's own country if one could not have predicted the direction of the surprise! In asset allocation, past observations of anomalies are only relevant to the extent that future anomalies can be predicted from them. This is like a run of heads in the tossing of a fair coin. Unless it enables you to better predict the outcome of the next flip, the observation of a run is useless.

The Global Context

Imagine for a moment that PPP and IRP held at every instant and that we examined a set of realizations of exchange rates and capital market returns for a decade. Imagine further that the currency surprises are uncorrelated with the local market returns. So, in effect, we would be sampling from the joint distribution of the exchange rates and capital markets rates and returns across all the relevant countries and some number of dates.

We would naturally expect that some of the countries would show surprises that trended, while others would not. Even if the surprises were uncorrelated with the capital market returns, we would expect the returns of some of the countries to exhibit positive correlation with the surprises and others to exhibit negative correlation. Virtually none, except in a very large sample, would show zero correlation. But we would expect that across the countries the average correlation would be zero. By placing the data for a particular country in the global context (by accounting for the expected spread of cross-sectional relationships), the fact that some of the countries showed positive average surprises or correlations, for example, would be entirely unsurprising and not at all in conflict with the theory.

This is why in their examination of global stock and bond returns Turner and Hensel (1993) used multivariate statistical tests rather than a series of pairwise statistical tests. Pairwise tests ignore the added context given by the other variables and too often reject the null hypotheses compared to multivariate tests. When the full spread of global relationships is considered, variation that seemed surprising when considered individually seems natural in the broader global context. It would be unthinkable in this context to conclude that, because some subset of the countries did show positive average surprise, the theory had been violated or that, somehow, that particular country was special.

Individual Country Experiences

The 1980s were a time when hedging would have been disadvantageous for some investors and advantageous for others. Some investors (e.g., Australians) would have lost money on hedging against many currencies and not surprisingly would have not experienced material ex post risk reductions. By ignoring the global context that some countries would have had positive experiences and some negative by chance, one could easily conclude that an individual country was different from the rest of the countries.

This type of error is common when evaluating time-series data and the cross-sectional context of the time series is ignored. Merton (1985) cites the case of a researcher who hears of a person who flipped 12 heads in a row with a fair coin. Correctly computing that the probability of that time series (12 heads out of 12 flips) occurring by chance was tremendously low, the researcher asserts that the subject must be different from the general population (i.e., have flipping skill). The conclusion is (narrowly) correct, yet the researcher was unaware of the global context that Merton holds a coin-flipping contest among all his students. In the context of Merton's 3,000 coin-flipping students, the likelihood of at least one student flipping 12 heads was about 50%.

Similarly, in the global context, experiences like the Australian one are not surprising. In fact, it would be surprising if there was not some country whose experience was like that. This makes the use of historical data potentially misleading. When one form of parity has been seriously violated during a sample period, one should not only worry that the mean returns are off but also that the variances and covariances have been misestimated. And while one can calculate the ex post correlations between assets and currencies, one should check whether or not, in the global context, the computed covariances are within the natural variation one would expect in observing exchange rates and capital market rates for a fairly short time period. In short, if the grand mean of the correlations is about what one would expect, the observed variation in the computed correlations is most likely sampling error. Moreover, the use of the historical data when parity has been violated is equivalent to a forecast that parity will continue to be violated with respect to that country.

A Prescription

In contrast, the basic position in the body of this paper is that, as a default (i.e., without a specific reliable forecast), we should expect the various forms of parity to roughly hold; and, in any event, we cannot accurately forecast where and when parity will be violated! As a direct consequence, the risk-reducing strategy is to hedge foreign asset exposures back to the base currency. As long as hedging costs are not too large, most investors will find it beneficial to hedge currency exposure. If an investor believes a particular set of currency forecasts or a particular forecasting methodology, the optimal currency strategy will not be to hedge at all but to speculate in currencies to best exploit the forecasts.

Part III
Performance Measurement Models

Attribution of Performance and Holdings

Richard C. Grinold and Kelly K. Easton

Abstract

The most important output from a portfolio optimization is the composition of the optimal portfolio. A wealth of additional information is available. This information, combined with investment insight, will reveal a great deal about the nature of the process that is generating the expected returns and the interaction of that expected return process with the methods used for portfolio selection. In this paper we show how to capture and analyze this information in two ways.

First, we decompose the expected returns used to select the optimal portfolio into three categories:

(i) the expected returns implied by an ex-ante presumed efficient portfolio;

(ii) the changes in expected return due to the special insights available to the investment manager and;

(iii) the changes in expected return that are, in effect, due to the imposition of constraints and bounds on the portfolio choice.

These constraints and bounds tend to be due to legal restrictions on the investment mandate such as a restriction on short sales or they are diversification constraints. If the constraints are negating the changes in expectations provided by the research, then rather than treating the symptom (wildly unrealistic portfolios) we suggest treating the cause and try to produce forecasts that will result in more realistic and investable portfolios.

Second, we use the notion of a characteristic portfolio to allocate the holdings in our optimal portfolios to various sources: the holdings in the ex-ante presumed efficient portfolio, the holdings implied by our research, and holdings implied by constraints and transactions costs. This procedure allows us to track, over time, the impact of all sources to the final performance of the portfolio.

1   Introduction

The traditional asset allocation manager has it easy. The asset allocation is produced by intuition and, at least, the allusion of genius. Decisions are defended as deep insights, usually framed by current news stories. Positive outcomes are expected and greeted with bows while negative outcomes are defended creatively on an ad hoc basis.

The quantitative or systematic asset allocation manager has none of this luxury. Rather than selling the hope of heroic management, the systematic investor sells a reliable process that is founded on clearly stated ideas and is implemented in a way that is free of caprice. This explicit process has an implicit cost. It requires that the manager can answer questions such as:

- Why am I holding 3.47% of the portfolio in French bonds and 7.06% in Japanese stocks?

- What are the influences that produced these numbers?

- Have these influences contributed to performance in any significant way?

- How can I make this process better?

In this paper we will demonstrate a novel way to address these questions. The procedure is a melding of two ideas: the first order conditions[1] of a portfolio optimization and the notion of a characteristic portfolio.[2]

The plan of the paper is as follows:

- preliminary definitions and notational conventions

- the notion of a characteristic portfolio

- portfolio optimization

- an expected return interpretation of the first order conditions

- a portfolio holdings interpretation of the first order conditions

- performance attribution to the separate elements with an example

- benchmark linked portfolio selection schemes

- portfolio selection in terms of purchase and sell activities.

- a summary

---

[1]These are called the Kuhn–Tucker conditions or often the Karush–Kuhn–Tucker or KKT conditions. We will use the latter term.

[2]See the appendix to chapter II in Grinold and Kahn, (1995).

1.1 Definitions and notation

We have a universe of N risky assets The *excess return* on an asset is the return, say 2.74%, less the return on T-bills, say 0.48%, over the same time period, namely $2.26\% = 2.74\% - 0.48\%$. The excess return is a random vector and is denoted as $\tilde{\mathbf{r}} = \{\tilde{r}_1, \tilde{r}_2, \ldots, \tilde{r}_N\}$. We will make a distinction between unconditional and conditional expectations. The *unconditional expected excess returns* are denoted $\mu = E\langle\tilde{\mathbf{r}}\rangle$. The investment manager has information, represented by $\mathfrak{J}$ that is presumed to be valuable. The *conditional expected excess returns* are given by $\mathbf{f} = E\langle\tilde{\mathbf{r}}|\mathfrak{J}\rangle$. We call the difference between the conditional and the unconditional expected excess returns the expected exceptional returns or alphas.

$$\alpha = E\langle\tilde{\mathbf{r}}|\mathfrak{J}\rangle - E\langle\tilde{\mathbf{r}}\rangle. \tag{1.1}$$

The N by N matrix $\mathbf{V}$ represents the covariances[3] between the risky assets.

$$V_{n,m} = \text{Cov}\langle\tilde{r}_n, \tilde{r}_m\rangle = E\left\langle\{\tilde{r}_n - E\langle\tilde{r}_n\rangle\} \cdot \{\tilde{r}_m - E\langle\tilde{r}_m\rangle\}\right\rangle. \tag{1.2}$$

2 Characteristic Portfolios

Assets have a multitude of attributes: returns, expected returns, earnings price ratios (earnings yield), volumes, yields (coupon or dividend), ROE, *etc.* The notion of a characteristic portfolio allows one to take these dissimilar and perhaps unfamiliar attributes and represent them in a consistent and familiar fashion as portfolios. The key element in this linkage is the covariance structure $\mathbf{V}$.

Let $\mathbf{c} = \{c_1, c_2, \ldots, c_N\}$ be a non-zero vector of asset *characteristics* and let $\mathbf{h}^P = \{h_1^P, h_2^P, \ldots, h_N^P\}$ be the holdings in a portfolio called P. Define portfolio P's *exposure* to characteristic $\mathbf{c} = \{c_1, c_2, \ldots, c_N\}$ as

$$c_p = \sum_{n=1,N} c_n \cdot h_n^P = \mathbf{c}'\mathbf{h}^P. \tag{2.1}$$

For any characteristic $\mathbf{c} \neq 0$ the *characteristic portfolio* of $\mathbf{c}$ is the portfolio with minimum variance among all portfolios that have $c_p = 1$. To find the characteristic portfolio we solve

$$\text{min.}\mathbf{h}' \cdot \mathbf{V} \cdot \mathbf{h} \quad \text{s.t.} \quad \mathbf{c}' \cdot \mathbf{h} = 1. \tag{2.2}$$

---

[3]We will not distinguish between the unconditional and conditional covariance matrices. For the degree of insight one normally associates with forecasting asset returns, there will be little difference. See Grinold and Kahn, (1995) 228–231.

The unique optimal solution of (2.2) is

$$\mathbf{h}^c = \frac{\mathbf{V}^{-1} \cdot \mathbf{c}}{\mathbf{c}' \cdot \mathbf{V}^{-1} \cdot \mathbf{c}} \tag{2.3}$$

One of the easily derived properties of the characteristic portfolio is its variance

$$\sigma_c^2 = \mathbf{h}^{c'} \cdot \mathbf{V} \cdot \mathbf{h}^c = \frac{1}{\mathbf{c}' \cdot v^{-1} \cdot \mathbf{c}}. \tag{2.4}$$

If we put portfolio $\mathbf{h}^c$ in a role similar to that of the market portfolio then, the 'beta' of the assets will be the characteristic $\mathbf{c}$.

$$\mathbf{c} = \frac{\mathbf{V} \cdot \mathbf{h}^c}{\mathbf{h}^c \cdot \mathbf{V} \cdot \mathbf{h}^c} \tag{2.5}$$

We see from (2.3), (2.4), (2.5) that the process works both ways. We can start with a characteristic and use (2.3) to produce a characteristic portfolio. Similarly we can start with any non-zero portfolio and use (2.5) to produce a set of characteristics.

Two different characteristics are linked through the covariance relationship. Thus for two characteristics $\mathbf{a}$ and $\mathbf{b}$

$$\mathbf{h}^a \cdot \mathbf{V} \cdot \mathbf{h}^b = \sigma_{a,b} = \{\mathbf{b}' \cdot \mathbf{h}^a\} \cdot \sigma_b^2 = b_a \cdot \sigma_b^2 = \left\{\mathbf{a}' \cdot \mathbf{h}^b\right\} \cdot \sigma_a^2 = a_b \cdot \sigma_a^2 \tag{2.6}$$

Four characteristics will be used.

1. The summation vector $\mathbf{e}' = \{1, 1, \ldots, 1\}$.

2. The unconditional expected excess returns $\mu = E\langle \tilde{\mathbf{r}} \rangle$.

3. The conditional expected excess returns $\mathbf{f} = E\langle \tilde{\mathbf{r}} | \mathfrak{z} \rangle$.

4. The exceptional forecast $\alpha = E\langle \tilde{\mathbf{r}} | \mathfrak{z} \rangle - E\langle \tilde{\mathbf{r}} \rangle$.

We examine them in turn.

The summation vector: $\mathbf{e}' = \{1, 1, \ldots, 1\}$

A portfolio P is *fully invested* if $e_p = \sum_{n=1,N} h_n^P = \mathbf{e}' \cdot \mathbf{h}^P = 1$. Portfolio mv is the minimum variance fully invested portfolio and the characteristic portfolio of $\mathbf{e}$.

$$\mathbf{h}^{\mathrm{mv}} = \frac{\mathbf{V}^{-1} \cdot \mathbf{e}}{\mathbf{e}' \cdot \mathbf{V}^{-1} \cdot \mathbf{e}} \quad \text{and} \quad \mathbf{e} = \frac{\mathbf{V} \cdot \mathbf{h}^{\mathrm{mv}}}{\mathbf{h}^{\mathrm{mv}} \cdot \mathbf{V} \cdot \mathbf{h}^{\mathrm{mv}}} = \frac{\mathbf{V} \cdot \mathbf{h}^{\mathrm{mv}}}{\sigma_{\mathrm{mv}}^2} \tag{2.7}$$

To illustrate this and other points we will carry an example with $N = 8$ assets through the paper. The assets are stocks and bonds in four countries: United States, Japan, United Kingdom and Germany. The correlations and

	USA-S	JPN-S	UKD-S	GER-S	USA-B	JPN-B	UKD-B	GER-B
USA-S	1	0.4	0.5	0.5	0.4	0.1	0.1	0.1
JPN-S	0.4	1	0.3	0.3	0.1	0.4	0.1	0.1
UKD-S	0.5	0.3	1	0.7	0.1	0.1	0.5	0.1
GER-S	0.5	0.3	0.7	1	0.1	0.1	0.1	0.5
USA-B	0.4	0.1	0.1	0.1	1	0	0	0
JPN-B	0.1	0.4	0.1	0.1	0	1	0	0
UKD-B	0.1	0.1	0.5	0.1	0	0	1	0.2
GER-B	0.1	0.1	0.1	0.5	0	0	0.2	1
Std. Dev.	**17.00%**	**21.00%**	**22.00%**	**20.00%**	**8.00%**	**8.00%**	**8.00%**	**8.00%**

Table 1: Correlations and Annualized Standard Deviations

Country – Asset	mv	eff	$\hat{q}$	a*0.01
United States Stocks	1%	24%	18%	−6%
Japanese Stocks	−3%	18%	5%	−10%
United Kingdom Stocks	−4%	12%	29%	8%
German Stocks	0%	6%	29%	12%
United States Bonds	27%	16%	−11%	−17%
Japanese Bonds	30%	12%	−4%	−10%
United Kingdom Bonds	27%	8%	40%	17%
German Bonds	22%	4%	−7%	−7%
Unconditional Expected Return	0.59%	3.50%	4.39%	0.00%
Conditional Expected Return	0.47%	3.51%	6.19%	1.00%
Alpha	−0.12%	0.01%	1.79%	1.00%
Forecast Standard Deviation	4.06%	9.90%	14.74%	4.43%
Unconditional Sharpe Ratio	0.145	0.353	0.298	0.001
Conditional Sharpe Ratio	0.116	0.354	0.420	0.227

Table 2: Characteristic portfolios and their properties

annualized standard deviations that determine **V** are shown in Table 1. These are the standard deviations for the unhedged returns that a US dollar based investor would receive. The numbers, although reasonable, were conjured up for the sole purpose of this example.

Table 2 shows the four characteristic portfolios that we'll discuss in this section. The minimum variance portfolio is in the first column labeled *mv*. Portfolio *mv* has roughly 0.25% in each of the four bond markets with a small positive or negative position in each of the four stock markets.

The unconditional expected excess returns: $\mu = E\langle \tilde{\mathbf{r}} \rangle$

To get the unconditional expected excess return μ we start with a fully invested portfolio, $\mathbf{h}^{\text{eff}}$, that we assume is efficient[4]. The weightings used in our example are shown in Table 2, column 2, labeled eff. Assume that the[5] portfolio holds a mix of 60% stock and 40% bonds in each country, with country weightings 40% USA, 30% Japan, 20% United Kingdom and 10% Germany.

The assumption that eff is efficient and the specification[6] of a positive expected excess return on $\mathbf{h}^{\text{eff}}$, call it μ_{eff}, yields the expected excess returns on all assets

$$\mu = \mu_{\text{eff}} \cdot \frac{\mathbf{V} \cdot \mathbf{h}^{\text{eff}}}{\mathbf{h}^{\text{eff}} \cdot \mathbf{V} \cdot \mathbf{h}^{\text{eff}}} \text{ and } \mathbf{h}^{\text{eff}} = \mu_{\text{eff}} \cdot \frac{\mathbf{V}^{-1} \cdot \mu}{\mu' \cdot \mathbf{V}^{-1} \cdot \mu}. \tag{2.8}$$

In our example, $\mu_{\text{eff}} = 3.50\%$, $\sigma_{\text{eff}} = 9.90\%$.

Thus $\{\mathbf{h}^{\text{eff}}/\mu_{\text{eff}}\}$ is the characteristic portfolio of $\mu = E\langle \tilde{\mathbf{r}} \rangle$ and the characteristic associated with portfolio $\mathbf{h}^{\text{eff}}$ is μ/μ_{eff} which is also known as β:

$$\mu = \mu_{\text{eff}} \cdot \beta, \text{ where } \beta = \frac{\mathbf{V} \cdot \mathbf{h}^{\text{eff}}}{\mathbf{h}^{\text{eff}} \cdot \mathbf{V} \cdot \mathbf{h}^{\text{eff}}} \text{ and } \mathbf{h}^{\text{eff}} = \frac{\mathbf{V}^{-1} \cdot \beta}{\beta' \cdot \mathbf{V}^{-1} \cdot \beta} \tag{2.9}$$

The unconditional expected excess returns μ and β for our example are in Table 3, in the columns marked μ and β.

The *Sharpe ratio*[7] of a portfolio P is the ratio of the portfolio's expected excess return to portfolio's risk;

$$SR\langle \tilde{r_P} \rangle = \frac{E\langle \tilde{r_P} \rangle}{Std\langle \tilde{r_P} \rangle} = \frac{\mu_P}{\sigma_P} \tag{2.10}$$

The quantity $\sqrt{\mu \cdot \mathbf{V}^{-1} \cdot \mu} = (\mu_{\text{eff}})/(\sigma_{\text{eff}})$ is the Sharpe ratio of portfolio $\mathbf{h}^{\text{eff}}$. Since $\mathbf{h}^{\text{eff}}$ is efficient, thus for any portfolio P

$$SR\langle \tilde{r_P} \rangle \leq SR\langle \tilde{r_{\text{eff}}} \rangle = \frac{\mu_{\text{eff}}}{\sigma_{\text{eff}}} = \sqrt{\mu \cdot \mathbf{V}^{-1} \cdot \mu} \tag{2.11}$$

The curved line in Figure 1 represents the achievable results expected excess return and risk combinations for fully invested portfolios (portfolios with $e_P = 1$). The straight line has a slope equal to $SR\langle \tilde{r_{\text{eff}}} \rangle$. The straight line is tangent to the curved line at the point $(\mu_{\text{eff}}, \sigma_{\text{eff}})$.

The conditional expected excess returns: $\mathbf{f} = E\langle \tilde{\mathbf{r}} | \mathfrak{z} \rangle$

---

[4]For any portfolio P, $\sigma_P \leq \sigma_{\text{eff}}$ implies $\mu_P \leq \mu_{\text{eff}}$.

[5]The example is meant to be instructive. In practice one would use capitalization weights or an easily identified portfolio.

[6]An informed guess is needed; a reasonable value is $\mu_{\text{eff}} \approx 0.3 \cdot \sigma_{\text{eff}}$ where σ_{eff} is the risk of the presumed efficient portfolio.

[7]The Sharpe ratio is a ratio of excess return to volatility. If we are looking forward, these are expectations. If we are looking backwards these are realizations. See Sharpe, 1994 for details.

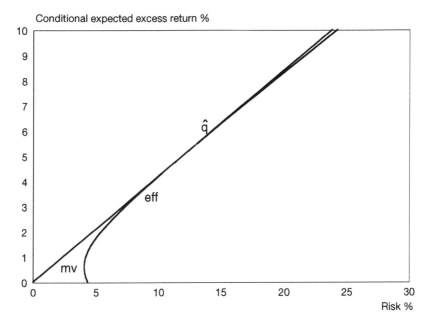

Figure 1: The Unconditional Efficient Frontier

The conditional expected excess returns for our example are in Table 3 in the column marked forecast. $\mathbf{h}^q$ is the characteristic portfolio of the conditional expected excess return:

$$\mathbf{h}^q = \frac{\mathbf{V}^{-1} \cdot \mathbf{f}}{\mathbf{f} \cdot \mathbf{V}^{-1} \cdot \mathbf{f}} \quad \text{and} \quad \mathbf{f} = \frac{V \cdot \mathbf{h}^q}{\mathbf{h}^q \cdot \mathbf{V} \cdot \mathbf{h}^q}. \tag{2.12}$$

The column in Table 2 that is labeled $\hat{q}$ contains the holdings in portfolio $\mathbf{h}^{\hat{q}}$, where $\mathbf{h}^{\hat{q}}$ is $\mathbf{h}^q$ divided by the sum of the holdings, e_q. Thus the holdings $\mathbf{h}^{\hat{q}}$ sum to one.

Portfolio $\mathbf{h}^q$ has the highest Sharpe ratio *conditional* on the information $\mathfrak{J}$ and for any portfolio $\mathbf{h}^P$,

$$SR\langle \tilde{r}_P | \mathfrak{J} \rangle = \frac{f_P}{\sigma_P} \leq SR\langle \tilde{r}_P | \mathfrak{J} \rangle = \frac{f_q}{\sigma_q} = \sqrt{\mathbf{f}' \cdot \mathbf{V}^{-1} \cdot \mathbf{f}}. \tag{2.13}$$

Portfolio $\mathbf{h}^{\hat{q}}$ has the same Sharpe ratio as portfolio $\mathbf{h}^q$. The opportunity set, conditional on $\mathfrak{J}$ is depicted in Figure 2.[8]

---

[8] We are considering the case in which q is a net long portfolio; i.e. $e_q > 0$. From 2.6 $e_q > 0$ if and only if $f_{mv} > 0$; i.e. the minimum variance portfolio has positive conditional expected excess return. In that case we can find a *fully invested* portfolio $\mathbf{h}^{\hat{q}} = \mathbf{h}^q / e_q$ with maximum Sharpe Ratio conditional on the information $\mathfrak{J}$; i.e. $SR\langle \tilde{r}_q | \mathfrak{J} \rangle = SR\langle \tilde{r}_{\hat{q}} | \mathfrak{J} \rangle$.

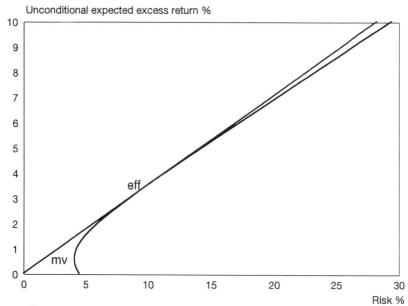

Figure 2: The Conditional Efficient Frontier

The curved line in Figure 2 represents the achievable results conditional expected excess return and risk combinations for fully invested portfolios (portfolios with $e_P = 1$). The straight line has a slope equal to $SR\langle \tilde{r}_q | \Im \rangle$. The straight line is tangent to the curved line at the point $(f_{\hat{q}}, \sigma_{\hat{q}}) = (f_q / e_q, \sigma_q / e_q)$. It may not be evident from the graph, but the portfolio $\mathbf{h}_{\text{eff}}$ that was presumed efficient with the unconditional returns is within the curved boundary in Figure 2.

The exceptional forecast: $\alpha = E\langle \tilde{\mathbf{r}} | \Im \rangle - E\langle \tilde{\mathbf{r}} \rangle = \mathbf{f} - \mu$.

Here[9] is the difference between the conditional and unconditional forecasts. The values in our example are in Table 3.

Let $\mathbf{h}^a$ designate the characteristic portfolio of $\alpha = \mathbf{f} - \mu$.

$$\mathbf{h}^a = \frac{\mathbf{V}^{-1} \cdot \alpha}{\alpha' \cdot \mathbf{V}^{-1} \cdot \alpha} \text{ and } \alpha = \frac{\mathbf{V} \cdot \mathbf{h}^a}{\mathbf{h}^a \cdot \mathbf{V} \mathbf{h}^a}. \tag{2.14}$$

In the column marked **a*0.01** in Table 2 there is a version of $\mathbf{h}^a$ scaled to have an α of 1.00%.

---

[9]We use the term α loosely. The residual return for any portfolio P is $\tilde{\theta}_P = \tilde{r}_P - \beta_P \cdot \tilde{r}_{\text{eff}}$. The proper use of α is as the expected residual return; i.e. $\alpha_P = E\langle \tilde{\theta}_P | \Im \rangle$. By definition $\tilde{\theta}_{\text{eff}} = 0$, so $\alpha_{\text{eff}} = 0$. However, in the paper we are using $\alpha_P = E\langle \tilde{r}_P | \Im \rangle - E\langle \tilde{r}_P \rangle$ which allows for the possibility that $\alpha_{\text{eff}} = E\langle \tilde{r}_{\text{eff}} | \Im \rangle - E\langle \tilde{r}_{\text{eff}} \rangle \neq 0$.

Country – Asset	μ	forecast	α	β
United States Stocks	4.98%	4.65%	−0.33%	1.42
Japanese Stocks	5.37%	3.61%	−1.76%	1.54
United Kingdom Stocks	5.67%	8.70%	3.03%	1.62
German Stocks	4.76%	6.96%	2.20%	1.36
United States Bonds	1.05%	0.39%	−0.66%	0.30
Japanese Bonds	0.93%	0.38%	−0.55%	0.27
United Kingdom Bonds	0.84%	1.66%	0.83%	0.24
German Bonds	0.60%	0.93%	0.33%	0.17

Table 3: Important asset characteristics

Since $\alpha = \mathbf{f} - \mu$ is the difference of $\mathbf{f}$ and μ, portfolio $\mathbf{h}^a$ will be a mixture of $\mathbf{h}^q$ and $\mathbf{h}^{\text{eff}}$. Starting with $\mathbf{V}^{-1} \cdot \mathbf{f} = \mathbf{V}^{-1} \cdot \mu + \mathbf{V}^{-1} \cdot \alpha$ and using (2.8), (2.12) and (2.14), yields

$$\left\{ \mathbf{f}' \cdot \mathbf{V}^{-1} \cdot \mathbf{f} \right\} \cdot \mathbf{h}^q = \left\{ \mu' \cdot \mathbf{V}^{-1} \cdot \mu' \right\} \cdot \left\{ \frac{\mathbf{h}^{\text{eff}}}{\mu_{\text{eff}}} \right\} + \left\{ \alpha' \cdot \mathbf{V}^{-1} \cdot \alpha \right\} \cdot \mathbf{h}^a. \quad (2.15)$$

Recall that $SR\langle \tilde{r}_q | \mathfrak{I} \rangle^2 = \mathbf{f}' \cdot \mathbf{V}^{-1} \cdot \mathbf{f}$ and $SR\langle r_{\widetilde{\text{eff}}} \rangle^2 = \mu \cdot \mathbf{V}^{-1} \cdot \mu$. There is a similar interpretation for $\{\alpha' \cdot \mathbf{V}^{-1} \cdot \alpha\}$.

Define the information ratio[10] of $\mathbf{h}^a$ as $IR\langle \tilde{r}_a | \mathfrak{I} \rangle^2$ which we will call simply IR^2

$$IR^2 = \left\{ \alpha' \cdot \mathbf{V}^{-1} \cdot \alpha \right\} = \frac{1}{\mathbf{h}^a \cdot \mathbf{V} \cdot \mathbf{h}^a} = \frac{1}{\omega^2} \quad (2.16)$$

Then (2.15) becomes

$$SR\langle \tilde{r}_q | \mathfrak{I} \rangle^2 \cdot \mathbf{h}^q = SR\langle r_{\text{eff}} \rangle^2 \cdot \left\{ \frac{\mathbf{h}^{\text{eff}}}{\mu_{\text{eff}}} \right\} + IR^2 \cdot \mathbf{h}^a. \quad (2.17)$$

Equation (2.17) will be useful after we have discussed the optimization that governs our portfolio choice.

3 The Optimization Problem

Our problem is in the classical tradition of mean variance portfolio selection.

$$\begin{array}{ll} \text{Maximize} \quad \mathbf{f}' \cdot \mathbf{h} - \frac{\lambda}{2} \cdot \mathbf{h}' \cdot \mathbf{V} \cdot \mathbf{h} & \\ \mathbf{e} \cdot \mathbf{h} = 1 & \text{fully invested} \\ \mathbf{b}^L \leq \mathbf{A} \cdot \mathbf{h} \leq \mathbf{b}^U & \text{additional constraints} \\ \mathbf{h}^L \leq \mathbf{h} \leq \mathbf{h}^U & \text{bounds on holdings} \end{array} \quad (3.1)$$

---

[10]We are sacrificing accuracy for simplicity with the term information ratio. A correct definition of the information ratio is $IR\langle \tilde{r}_P | \mathfrak{I} \rangle = SR\langle \tilde{r}_P - \beta_P \cdot \tilde{r}_{\text{eff}} | \mathfrak{I} \rangle$ where $\tilde{\theta}_P = \tilde{r}_P - \beta_P \cdot \tilde{r}_{\text{eff}}$ is the residual return.

Country – Asset	Lower	eff	Upper
United States Stocks	9.00%	24.00%	39.00%
Japanese Stocks	3.00%	18.00%	33.00%
United Kingdom Stocks	0.00%	12.00%	27.00%
German Stocks	0.00%	6.00 %	21.00%
United States Bonds	1.00%	16.00%	31.00%
Japanese Bonds	0.00%	12.00%	27.00%
United Kingdom Bonds	0.00%	8.00%	23.00%
German Bonds	0.00%	4.00%	19.00%

Table 4: Upper and Lower bounds on asset holdings

The example has two additional constraints:

- The total amount of stock must be at least 40% and at most 80%.

- The total investment in Europe, through UK stocks and bonds and German stocks and bonds, must be at least 10% and at most 50%.

Thus

$$\mathbf{A} = \begin{Bmatrix} 1 & 1 & 1 & 1 & 0 & 0 & 0 & 0 \\ 0 & 0 & 1 & 1 & 0 & 0 & 1 & 1 \end{Bmatrix} \tag{3.2}$$

$$\mathbf{b}^L = \begin{Bmatrix} 0.4 \\ 0.1 \end{Bmatrix} \qquad \mathbf{b}^U = \begin{Bmatrix} 0.8 \\ 0.5 \end{Bmatrix}$$

The upper and lower limits on the portfolio holdings are determined by two rules:

- no holding can differ from the holding in the presumed efficient portfolio eff by more than 15%, and

- no holding can be less than zero.

Applying these two principles gives the upper and lower bounds in Table 4.

The problem is completely specified by setting the risk aversion assumed to be $\lambda = 3.57$. This choice is discussed further in Section 5.

Using the forecast returns from Table 3 and solving problem (3.1) gives the optimal solution in Table 5.

The active position in Table 5 is the difference between the optimal portfolio and the presumed efficient portfolio. The optimal portfolio has 50% in European assets and is bumping up against the maximum. The optimal

Country – Asset	opt	eff	active
United States Stocks	18.81%	24.00%	−5.19%
Japanese Stocks	3.77%	18.00%	−14.23%
United Kingdom Stocks	27.00%	12.00%	15.00%
German Stocks	14.42%	6.00 %	8.42%
United States Bonds	9.35%	16.00%	−6.65%
Japanese Bonds	18.07%	12.00%	6.07%
United Kingdom Bonds	8.58%	8.00%	0.58%
German Bonds	0.00%	4.00%	−4.00%
Unconditional Expected Return	3.70%	3.50%	0.20%
Conditional Expected Return	4.61%	3.51%	1.11%
Alpha	0.92%	0.01%	0.91%
Forecast Standard Deviation	11.35%	9.90%	4.43%
Percent in Stock	64%	60%	4%
Percent in Europe	50%	30%	20%
Unconditional Sharpe Ratio	0.326	0.353	0.044
Conditional Sharpe Ratio	0.407	0.354	0.250

Table 5: The Optimal Solution and Active Position

portfolio only has 64% in stock so the stock holding constraint is slack. Two asset categories are against their bounds: the maximum of 27% in UK stock and zero in German bonds.

The optimal portfolio is the most important but not the sole output from the portfolio optimization. Other information available from the optimization will help interpret the positions taken in the optimal portfolio, modify the forecasting process and even change the constraints and bounds placed on portfolio choice.

4 The First Order Conditions: the Modified Forecast

The optimal solution of (3.1) implies a *modified forecast* $\mathbf{f}^*$. The modified forecast leads directly to the optimal portfolio in the following sense; if the objective was $\mathbf{f}^* \cdot \mathbf{h} - \frac{\lambda}{2} \cdot \mathbf{h}' \cdot \mathbf{V} \cdot \mathbf{h}$ with no constraints or bounds then $\mathbf{h}^{\mathrm{opt}}$ would be the optimal portfolio. The difference between $\mathbf{f}$ and $\mathbf{f}^*$ can be attributed to the constraints and bounds used to limit our portfolio choice.

This approach puts our research in providing a superior forecast, the difference $\alpha = \mathbf{f} - \mu$, on the same plane with the constraints and bounds that limit the portfolio, the difference $\eta = \mathbf{f}^* - \mathbf{f}$. The key to the analysis of $\mathbf{f}^* - \mathbf{f}$

is the first order conditions for the optimization problem (3.1).

The first order conditions associate multipliers, θ, $\mathbf{p}^U$, $\mathbf{p}^L$, $\mathbf{s}^U$, $\mathbf{s}^L$, with each constraint. The multipliers and companion constraints are

$$
\begin{aligned}
\mathbf{e}' \cdot \mathbf{h} &= 1; & \theta & & (4.1)\\
\mathbf{a}^j \cdot \mathbf{h} &\leq b_j^u; & p_j^U &\geq 0 \\
-\mathbf{a}^j \cdot \mathbf{h} &\leq -b_j^L; & p_j^L &\geq 0 \\
h_n &\leq h_n^U; & s_n^U &\geq 0 \\
-h_n &\leq -h_n^L; & s_N^L &\geq 0.
\end{aligned}
$$

The KKT conditions are

$$
\lambda \cdot \mathbf{V}\mathbf{h}^{\text{opt}} = \mathbf{f} - \left\{\theta \cdot \mathbf{e} + \mathbf{A}' \cdot \left\{\mathbf{p}^U - \mathbf{p}^L\right\} + \mathbf{q}^U - \mathbf{q}^L\right\} \tag{4.2}
$$

along with the usual complementarity conditions.[11]

A conditional expected excess return forecast of

$$
\mathbf{f}^* = \lambda \cdot \mathbf{V} \cdot \mathbf{h}^{\text{opt}} \tag{4.3}
$$

means the KKT conditions (4.2) would be satisfied by $\mathbf{h}^{\text{opt}}$, $\mathbf{f}^*$ and setting all the multipliers, θ, $\mathbf{p}^U$, $\mathbf{p}^L$, $\mathbf{s}^U$, $\mathbf{s}^L$, equal to zero. We can view the portfolio selection process in two equivalent ways. The traditional way is to solve problem 4.1 using forecast $\mathbf{f}$. An alternative would be to first modify the forecast from $\mathbf{f}$ to $\mathbf{f}^*$ and then solve the unconstrained problem of maximizing $\mathbf{f}^* \cdot \mathbf{h} - \frac{\lambda}{2} \cdot \mathbf{h}' \cdot \mathbf{V} \cdot \mathbf{h}$.

It is interesting to see how the forecasts are modified. The modified forecast for asset n is

$$
f_n^* = f_n + s_n^L - s_n^U - \left\{\theta + \sum_{j=1,J} a_n^j \cdot \left\{p_j^U - p_j^L\right\}\right\}. \tag{4.4}
$$

Table 6 details the changes for the eight assets in our example. The restriction on stock holding to between 40% and 80% of the portfolio has no influence since the optimal portfolio holds 64% stock. The two assets at their bounds are German Bonds, (lower bound) and UK Stocks (upper bound). A boost of 0.23% was needed to keep our German bond holding from falling below zero. For UK stocks a penalty of -0.15% is enough to keep the position at its upper bound.

---

[11]The complementarity conditions are

$$
\begin{aligned}
\{b_j^U - \mathbf{a}^j \cdot \mathbf{h}^{\text{opt}}\} \cdot p_j^U &= 0; \quad p_j^U \geq 0, \quad b_j^U - \mathbf{a}^j \cdot \mathbf{h}^{\text{opt}} \geq 0 \quad j = 1,2,\ldots J \\
\{\mathbf{a}^j \cdot \mathbf{h}^{\text{opt}} - b_j^L\} \cdot p_j^L &= 0; \quad p_j^L \geq 0, \quad \mathbf{a}^j \cdot \mathbf{h}^{\text{opt}} - b_j^L \geq 0 \quad j = 1,2,\ldots J \\
\{h_n^U - h_n^{\text{opt}}\} \cdot s_n^U &= 0; \quad s_n^U \geq 0, \quad h_n^U - h_n^{\text{opt}} \geq 0 \quad n = 1,2,\ldots,N \\
\{h_n^{\text{opt}} - h_n^L\} \cdot s_n^L &= 0; \quad s_n^L \geq 0, \quad h_n^{\text{opt}} - h_n^L \geq 0 \quad n = 1,2,\ldots,N
\end{aligned}
$$

Asset	Initial Forecast	Lower Bound	Upper Bound	Fully Invested	Stock Holding	Europe Holding	Modified Forecast
USA-S	4.65%	0.00%	0.00%	0.47%	0.00%	0.00%	5.12%
JPN-S	3.61%	0.00%	0.00%	0.47%	0.00%	0.00%	4.08%
UKD-S	8.70%	0.00%	-0.15%	0.47%	0.00%	−0.89%	8.13%
GER-S	6.96%	0.00%	0.00%	0.47%	0.00%	−0.89%	6.54%
USA-B	0.39%	0.00%	0.00%	0.47%	0.00%	0.00%	0.85%
JPN-B	0.38%	0.00%	0.00%	0.47%	0.00%	0.00%	0.85%
UKD-B	1.66%	0.00%	0.00%	0.47%	0.00%	−0.89%	1.24%
GER-B	0.93%	0.23%	0.00%	0.47%	0.00%	−0.89%	0.73%

Table 6: How constraints and bounds modify the effective forecast

The 0.47% on the fully invested constraint means that we need an across the board boost in expected return to stay fully invested. In fact, if we allow cash as an investment and replace the full investment (no cash!) constraint, $e' \cdot h = 1$, with $e' \cdot h \leq 1$, we only hold 71% of the portfolio in the eight assets and would make up the other 29% by holding cash. However, boosting all the expected excess returns by 0.47% will lead us to full investment even with the constraint $e' \cdot h \leq 1$.

The largest impact is from the Europe constraint. If that constraint were relaxed the optimal portfolio would hold 77% in European assets, with the big change being a shift from US and Japanese bonds toward UK stock (up to 23%) and German bonds (up to 7.4%). To avoid this the expected returns on all European assets are reduced by 0.89%. This reduction is just enough to lower the demand for European assets to its upper limit of 50%.

We can do the same analysis in terms of α. Table 7, below, shows the α produced by our research as well as the effective one, $\alpha_n^* = f_n^* - \mu_n$. The adjustments due to the constraints and bounds are nearly of the same order of magnitude as the α produced by our research. The 4.79% spread between the highest α (UK Stock) and the lowest (Japanese Stock) shrinks to a 3.75% spread after adjustment.

This type of analysis is useful in the context of a single example and it is a useful tool for analyzing the research process. We should ask if the constraints and bounds are there for a mandated reason, e.g. a lower holdings limit of zero, or if are they safeguards against unrealistic research. If they are the latter, then it may be better to deal with the cause of the problem rather than its effects. If we are consistently making large adjustments, then we might ask what is it about the alpha or conditional expected return generating process that is so much at odds with our notion of the proper shape and form of a portfolio? Either the research is unrealistic and should be reined in or the

Asset	research α	adjustment	effective α
USA-S	-0.33%	0.47%	0.13%
JPN-S	-1.76%	0.47%	-1.30%
UKD-S	3.03%	-0.58%	2.45%
GER-S	2.20%	-0.42%	1.78%
USA-B	-0.66%	0.47%	-0.20%
JPN-B	-0.55%	0.47%	-0.09%
UKD-B	0.83%	-0.42%	0.40%
GER-B	0.33%	-0.20%	0.14%

Table 7: Changes in α attributed to constraints and bounds

constraints are unnecessarily tight and should be loosened. Constraints and bounds are not a cure for unrealistic research; we should root out the cause instead of treating the symptoms.

5 The First Order Conditions: Attribution of Holdings

In Section 4 we used the KKT conditions to examine the interactions between our forecasts of exceptional return and the modifications to those forecasts in a constrained portfolio selection process. In this section we will look at the asset holdings and show how they are influenced in three ways: by our prior assumption of the efficiency of portfolio $\mathbf{h}^{\text{eff}}$, by our exceptional return forecast α, and by the bounds and constraints in the portfolio selection process.

To do this we return to the characteristic portfolios of $\mathbf{f}$, $\mathbf{e}$, μ and α developed in Section 2. We will need the characteristic portfolios of the additional constraints mentioned in the optimization problem 3.1. For any constraint $b_j^L \leq \mathbf{a}^j \cdot \mathbf{h} \leq b_j^U$, the characteristic portfolio is

$$\mathbf{h}^j = \frac{\mathbf{V}^{-1} \cdot \mathbf{a}^j}{\mathbf{a}^j \cdot \mathbf{V}^{-1} \cdot \mathbf{a}^j}; \ \sigma_j^2 = \frac{1}{\mathbf{a}^j \cdot \mathbf{V}^{-1} \cdot \mathbf{a}^j} = \mathbf{h}^j \cdot \mathbf{V} \cdot \mathbf{h}^j \text{ and } \sigma_j^2 \cdot \mathbf{a}^j = \mathbf{V} \cdot \mathbf{h}^j. \ (5.1)$$

Table 8 contains the characteristic portfolios for the two constraints in our example. The holdings, by and large, make sense. There is a stock versus bonds flavor to the Stocks portfolio and a distinct Europe versus non Europe in the Europe portfolio. Since bonds are less risky than stocks, there is a natural tendency to take larger positions with bonds in these portfolios. The exception is with UK equity. One would anticipate a larger, indeed a positive, position in the Stock portfolio. The subtle interactions of correlations are at

Country – Asset	Stocks	Europe
Unites States Stocks	42%	−3%
Japanese Stocks	28%	−3%
United Kingdom Stocks	−4%	−4%
German Stocks	34%	4%
United States Bonds	−51%	4%
Japanese Bonds	−45%	4%
United Kingdom Bonds	−8%	56%
German Bonds	−56%	45%

Table 8: The characteristic portfolios of the constraints

work here. The relatively high volatility of the UK stocks give us an incentive to reduce its holding in any characteristic portfolio. We can make an effective and lower risk substitute for UK equity with UK bonds and US equity and German equity. That is indeed happening, as one can see from the inflated positions in US and German equity and the relatively low position in UK bonds.

The goal is to transform the KKT conditions into a form, equation (5.8), that explains the holdings in the optimal portfolio. The first step is to add an initial complication by writing the multipliers in a slightly contrived form. This temporary inconvenience will lead to simplifications as we approach our goal, equation (5.8).

$$
\begin{aligned}
\mathbf{e}' \cdot \mathbf{h} &= 1 \; ; \quad \theta = \lambda \cdot \sigma_{\text{mv}}^2 \cdot \hat{\theta} \\
\mathbf{a}^j \cdot \mathbf{h} &\le b_j^U \; ; \quad p_j^U = \lambda \cdot \sigma_j^2 \cdot \hat{p}_j^U \\
-\mathbf{a}^j \cdot \mathbf{h} &\le -b_j^L \; ; \quad p_j^L = \lambda \cdot \sigma_j^2 \cdot \hat{p}_j^L \\
h_n &\le h_n^U \; ; \quad s_n^U = \lambda \cdot \hat{s}_n^U \\
-h_n &\le -h_n^L \; ; \quad s_n^L = \lambda \cdot \hat{s}_n^L
\end{aligned}
\tag{5.2}
$$

The KKT conditions become

$$
\mathbf{f} - \lambda \cdot \mathbf{V} \cdot \mathbf{h}^{\text{opt}} = \lambda \cdot \mathbf{e}\sigma_{\text{mv}}^2 \cdot \hat{\theta} + \lambda \cdot \sum_{j=1,J} \sigma_j^2 \cdot \mathbf{a}^j \cdot \left\{ \hat{p}_j^U - \hat{p}_j^L \right\} h + \lambda \cdot \left\{ \hat{\mathbf{s}}^U - \hat{\mathbf{s}}^L \right\}. \tag{5.3}
$$

We know: from (2.12) and (2.13) that $\mathbf{f} = SR\langle \tilde{r}_q | \mathfrak{J} \rangle^2 \cdot \mathbf{V} \cdot \mathbf{h}^q$, from (2.7) that $\sigma_{\text{mv}}^2 \cdot \mathbf{e} = \mathbf{V}\mathbf{h}^{\text{mv}}$, and from (5.1) that $\sigma_j^2 \cdot \mathbf{a}^j = \mathbf{V} \cdot \mathbf{h}^j$. After we make these substitutions, and divide by λ, we can rewrite (5.3) as

$$
\frac{SR\langle \tilde{r}_q | \mathfrak{J} \rangle^2}{\lambda} \cdot \mathbf{V} \cdot \mathbf{h}^q - \mathbf{V} \cdot \mathbf{h}^{\text{mv}} \cdot \hat{\theta} - \sum_{j=1,J} \mathbf{V} \cdot \mathbf{h}^j \cdot \left\{ \hat{p}_j^U - \hat{p}_j^L \right\} h - \left\{ \mathbf{s}^{\hat{U}} - \hat{\mathbf{s}}^L \right\} = \mathbf{V} \cdot \mathbf{h}^{\text{opt}}.
$$

$$
\tag{5.4}
$$

Multiply (5.4) by $\mathbf{V}^{-1}$, to yield

$$\mathbf{h}^{\mathrm{opt}} = \frac{SR\langle\tilde{r}_q|\mathfrak{I}\rangle^2}{\lambda} \cdot \mathbf{h}^q - \hat{\theta} \cdot \mathbf{h}^{\mathrm{mv}} - \sum_{j=1,J} \left\{\hat{p}_j^U - \hat{p}_j^L\right\} \cdot \mathbf{h}^j - \mathbf{V}^{-1} \cdot \left\{\hat{\mathbf{s}}^U - \hat{\mathbf{s}}^L\right\}. \quad (5.5)$$

The next step is to define[12]

$$\mathbf{h}^{ub} = \mathbf{V}^{-1} \cdot \hat{\mathbf{s}}^U; \quad \mathbf{h}^{lb} = \mathbf{V}^{-1} \cdot \hat{\mathbf{s}}^L, \quad (5.6)$$

so (5.5) becomes

$$\mathbf{h}^{\mathrm{opt}} = \frac{SR\langle\tilde{r}_q|\mathfrak{I}\rangle^2}{\lambda} \cdot \mathbf{h}^q - \hat{\theta} \cdot \mathbf{h}^{\mathrm{mv}} - \sum_{j=1,J} \left\{\hat{p}_j^U - \hat{p}_j^L\right\} \cdot \mathbf{h}^j - \mathbf{h}^{ub} + \mathbf{h}^{lb}. \quad (5.7)$$

The process in completed by using (2.17) to split the holdings $\mathbf{h}^q$ into its component parts.

$$\mathbf{h}^{\mathrm{opt}} = \frac{SR\langle\tilde{r}_{\mathrm{eff}}\rangle^2}{\lambda \cdot \mu_{\mathrm{eff}}} \cdot \mathbf{h}^{\mathrm{eff}} + \frac{IR^2}{\lambda} \cdot \mathbf{h}^a + \mathbf{h} \quad \text{where}$$

$$\mathbf{h} = \left\{\mathbf{h}^{lb} - \mathbf{h}^{ub} - \hat{\theta} \cdot \mathbf{h}^{\mathrm{mv}} - \sum_{j=1,J} \left\{\hat{p}_j^U - \hat{p}_j^L\right\} \cdot \mathbf{h}^j\right\}. \quad (5.8)$$

This is an instructive breakdown:

- holding $((SR\langle\tilde{r}_{\mathrm{eff}}\rangle^2/(\lambda \cdot \mu_{\mathrm{eff}})) \cdot \mathbf{h}^{\mathrm{eff}}$ due to our presumed efficient portfolio,

- holding $((IR^2)/(\lambda)) \cdot \mathbf{h}^a$ due to our research into exceptional returns,

- holding $\mathbf{h}$ due to the constraints and upper and lower bounds.

This echoes the decomposition of effective return forecast, f into $f^* = \mu + \alpha + \eta$, mentioned at the start of Section 4.

The coefficient $SR\langle\tilde{r}_{\mathrm{eff}}\rangle/(\lambda \cdot \mu_{\mathrm{eff}})$ indicates the weight one places on the presumed efficient portfolio. In the absence of any special information , i.e. $\alpha = 0$, we should have $\mathbf{h}^{\mathrm{opt}} = \mathbf{h}^{\mathrm{eff}}$. If $\mathbf{h}^{\mathrm{eff}}$ satisfies all constraints and bounds, then

$$\lambda = \frac{SR\langle\tilde{r}_{\mathrm{eff}}\rangle^2}{\mu_{\mathrm{eff}}} = \frac{\mu_{\mathrm{eff}}}{\sigma_{\mathrm{eff}}^2} = \frac{SR\langle\tilde{r}_{\mathrm{eff}}\rangle}{\sigma_{\mathrm{eff}}}. \quad (5.9)$$

Equation (5.9) is satisfied in the example: $\lambda = 3.57, \mu_{\mathrm{eff}} = 3.50\%, \sigma_{\mathrm{eff}} = 9.90\%$.

---

[12]We could push the point further and write $\mathbf{h}^{ub} = \sum_{n=1}^{N} \hat{s}_n^U \cdot \left\{\mathbf{V}^{-1} \cdot \mathbf{u}^n\right\}$ where $\mathbf{u}^n$ is the nth unit vector. The characteristic portfolio of the first unit vector is $\left\{1 \ -\hat{\mathbf{h}}^1\right\}$ where $\hat{\mathbf{h}}^1$ is the portfolio of assets 2–N that has minimum tracking error relative to asset 1.

The second term, $\{IR^2/\lambda\} \cdot \mathbf{h}^a$, can be written as follows

$$\frac{IR^2}{\lambda} \cdot \mathbf{h}^a = \frac{IR}{\lambda} \cdot \left\{\frac{\mathbf{h}^\alpha}{\omega}\right\}. \tag{5.10}$$

The holdings $\mathbf{h}^\alpha/\omega$ have a predicted standard deviation of 100%. Thus the expression IR/λ can be interpreted as the amount of risk, often called the *active* risk, that we intend to occur in pursuit of exceptional information. In our example, with $\lambda = 3.57$ and an information ratio of $IR = 0.2255$, this is 6.32%. The active risk level depends, as it should, directly on the perceived quality of that information as measured by IR and inversely with the degree of risk aversion λ. The ratio IR/λ is a forecast of the active risk. The *realized* active risk will differ for two reasons: of least importance, it is a forecast and no realization will exactly equal the forecast value, the second and more important reason is that the constraints will tend to limit the amount of active risk in the portfolio and thus IR/λ will tend to be an upper limit on the amount of active risk the portfolio experiences.

The third part, $\mathbf{h}$, due to the constraints and bounds, is a measure of the self imposed limitations put on the portfolio selection process. It depends upon the type and severity of the bounds and constraints,[13] while the first two terms are independent of any bounds or constraints.

The holdings attributed to these three sources for our example are shown in Table 9. The combined effect of the constraints and bounds is similar in scale to the effect of the αs. Our enthusiasm (24%) for UK bonds is negated (-23%), while our enthusiasm for UK stocks (12%) is reinforced (+3%).

Table 10 gives the breakdown of the constraint and bounds holdings by source. The stock constraint is slack so there is no contribution. The Europe constraint is tight, to get under the 50% upper limit we reduce the holdings of UK and German bonds by 30% and 37% respectively.

To meet the lower limit of 0% on German bonds we boost their holdings by 24% and hedge this increase with a short of German stock and UK bonds and a long position in UK stock.

This framework is particularly useful when the αs come from multiple sources. Suppose that the α is generated by K submodels

$$\alpha = \sum_{k=1,K} \alpha^k \tag{5.11}$$

---

[13]The structure can provide insight. Let $Q(1, \mathbf{b}^U, -\mathbf{b}^L, \mathbf{h}^U, -\mathbf{h}^L)$ be the optimal value of the optimization problem as a function of the constraint levels. Q is convex and the KKT multipliers $\theta, \mathbf{p}^U, \mathbf{p}^L, \mathbf{s}^U, \mathbf{s}^L$ are (sub) gradients of Q. Q is approximately quadratic, in fact locally quadratic, so the gradients $\theta, \mathbf{p}^U, \mathbf{p}^L, \mathbf{s}^U, \mathbf{s}^L$ are approximately (locally) linear in $\mathbf{b}^U, -\mathbf{b}^L, \mathbf{h}^U, -\mathbf{h}^L$. Thus weights on the characteristic portfolios will change in something like a piece-wise linear fashions if we move the constraint limits $\mathbf{b}^U, -\mathbf{b}^L, \mathbf{h}^U, -\mathbf{h}^L$.

Country – Asset	eff	alpha	con & bnd	total
United States Stocks	24%	−9%	4%	19%
Japanese Stocks	18%	−14%	0%	4%
United Kingdom Stocks	12%	12%	3%	27%
German Stocks	6%	18%	−9%	14%
United States Bonds	16%	−25%	18%	9%
Japanese Bonds	12%	−15%	21%	18%
United Kingdom Bonds	8%	24%	−23%	9%
German Bonds	4%	−10%	6%	0%
Unconditional Expected Return	3.50%	0.01%	0.19%	3.70%
Conditional Expected Return	3.51%	1.43%	−0.33%	4.61%
Alpha	0.01%	1.43%	−0.52%	0.92%
Forecast Standard Deviation	9.90%	6.32%	2.93%	11.35%
Percent in Stock	60%	6%	−2%	64%
Percent in Europe	30%	43%	−23%	50%
Unconditional Sharpe Ratio	0.353	0.001	0.065	0.326
Conditional Sharpe Ratio	0.354	0.227	−0.111	0.407

Table 9: Contribution to Portfolio Holdings

	Constraints			Bounds		
	Holding	Stocks	Europe	Lower	Upper	Total
USA-S	1%	0%	2%	1%	1%	4%
JPN-S	−2%	0%	2%	0%	0%	0%
UKD-S	−3%	0%	3%	8%	5%	3%
GER-S	0%	0%	−2%	−11%	4%	−9%
USA-B	21%	0%	−3%	0%	0%	18%
JAP-B	23%	0%	−3%	0%	0%	21%
UKD-B	21%	0%	−37%	−14%	7%	−23%
GER-B	18%	0%	−30%	24%	−6%	6%

Table 10: Contribution to the portfolio holdings by the constraints
and bounds

For each submodel $\alpha^1, \alpha^2, \ldots, \alpha^K$ there is a characteristic portfolio $\mathbf{h}^{\alpha^k}$; $k = 1, 2, \ldots, K$, an (ex-ante) information ratio $IR_k = \sqrt{\alpha^k \cdot \mathbf{V}^{-1} \cdot \alpha^k}$, and we know the risk $\omega_k = 1/IR_k = \sqrt{\mathbf{h}^{\alpha^k} \cdot \mathbf{V} \cdot \mathbf{h}^{\alpha^k}}$ of the characteristic portfolio. The overall (ex-ante) information ratio[14] is $IR^2 = \alpha' \cdot \mathbf{V}^{-1} \cdot \alpha$. The charac-

---

[14]We say that α^ℓ and α^k are uncorrelated if their characteristic portfolios are uncorrelated. This occurs if $\mathbf{h}^{\alpha^{\ell'}} \cdot \mathbf{V} \cdot \mathbf{h}^{\alpha^k} = 0$ or equivalently $\alpha^{\ell'} \cdot \mathbf{V}^{-1} \cdot \alpha^k = 0$. If the αs are all uncorrelated then $IR^2 = \sum_{k=1,K} IR_k^2$.

teristic portfolio of the αs is

$$\mathbf{h}^\alpha = \sum_{k=1,K} \mathbf{h}^{\alpha^k} \cdot c_k, \tag{5.12}$$

where $c_k = (IR_k^2)/(IR^2)$, and $\mathbf{h}^{\alpha^j} = (\mathbf{V}^{-1} \cdot \alpha^j)/(IR_j^2)$. This leads to

$$\frac{IR}{\lambda} \cdot \left\{ \frac{\mathbf{h}^\alpha}{\omega} \right\} = \sum_{k=1,K} \left\{ \frac{IR_k}{\lambda} \right\} \cdot \left\{ \frac{\mathbf{h}^{\alpha^k}}{\omega_k} \right\}, \tag{5.13}$$

where the risk of $\left\{ \mathbf{h}^{\alpha^k}/\omega_k \right\}$ is 100% and thus IR_k/λ is the active risk we are willing to take along the kth dimension and IR/λ is the overall active risk we are willing to endure.

6   Benchmark Linked Portfolio Selection

Institutional investment managers are frequently asked to manage relative to a benchmark. This type of mandate allows for decentralization of the investment management process: manager #1 can work relative to benchmark #1, manager #2 relative to benchmark #2, etc. with the owner of the funds orchestrating the process so the benchmarks mesh into a desired portfolio. This arrangement takes advantage of specialization; manager #1 may be an expert in international bonds and manager #2 and expert in international stocks, etc. In effect, the owner of the funds takes responsibility for the performance of the benchmark and each investment manager takes responsibility for the differential performance[15] of the portfolio and their benchmark.

We can devise a portfolio selection method that is in step with this decentralized procedure if we adopt the following premise: in the absence of any evidence of exceptional return, we will hold the benchmark portfolio. Thus our default is the benchmark and the evidence of exceptional return will cause us to move away from that default position.

Suppose the benchmark portfolio has holdings $\mathbf{h}^{\mathrm{bmk}}$; the investment manager would solve problem (6.1).

$$\text{Maximize } \alpha' \cdot \mathbf{h} - \frac{\lambda_{bl}}{2} \cdot \left\{ \mathbf{h} - \mathbf{h}^{\mathrm{bmk}} \right\}' \cdot \mathbf{V} \cdot \left\{ \mathbf{h} - \mathbf{h}^{\mathrm{bmk}} \right\}$$

$$\begin{array}{ll} \mathbf{e}' \cdot \mathbf{h} = 1 & \text{fully invested} \\ \mathbf{b}^L \leq \mathbf{A} \cdot \mathbf{h} \leq \mathbf{b}^U & \text{additional constraints} \\ \mathbf{h}^L \leq \mathbf{h} \leq \mathbf{h}^U & \text{bounds on holdings} \end{array} \tag{6.1}$$

The term $\left\{ \mathbf{h} - \mathbf{h}^{\mathrm{bmk}} \right\}' \cdot \mathbf{V} \cdot \left\{ \mathbf{h} - \mathbf{h}^{\mathrm{bmk}} \right\}$ is known as the *active variance* and its square root is called the *active risk*. The risk aversion λ_{bl} can be and

---

[15]Usually called the *active return*.

most likely will be different from the risk aversion used in the traditional case. In the event where all of our exceptional return forecasts are zero, $\alpha = 0$, the optimal solution to this problem (presuming that the benchmark is a feasible solution) will be to hold the benchmark.

If we mimic the analysis of Section 5, equations (5.8), (5.9) and (5.10), we get the following breakdown for the holdings in the optimal portfolio.

$$\mathbf{h}^{\text{opt}} = \mathbf{h}^{\text{bmk}} + \frac{IR}{\lambda_{bl}} \cdot \left\{ \frac{\mathbf{h}^a}{\omega} \right\} + \mathbf{h}, \text{ where}$$

$$\mathbf{h} = \mathbf{h}^{lb} - \mathbf{h}^{ub} - \hat{\theta} \cdot \mathbf{h}^{\text{mv}} - \sum_{j=1,J} \left\{ \hat{p}_j^U - \hat{p}_j^L \right\} \cdot \mathbf{h}^j \qquad (6.2)$$

We should contrast this with (5.8) in the case where the benchmark and presumed efficient portfolio coincide $\mathbf{h}^{\text{eff}} = \mathbf{h}^{\text{bmk}}$. In the conventional case the risk aversion plays two roles; to determine our exposure $SR\langle \tilde{r}_{\text{eff}} \rangle / \lambda \cdot \sigma_{\text{eff}}$ to the presumed efficient portfolio in (5.9) and to regulate the amount of risk IR/λ that we undertake in pursuit of our exceptional return forecasts, (5.10). Thus we have one parameter chasing two goals; always a tricky job. In the benchmark linked case we have, in effect, hard wired the benchmark, our first objective, and the risk aversion plays a single role in the control of the active risk.[16] In our example, with an information ratio of 0.2255, we may want the active risk to be at most 3%, so $\lambda_{bl} = 7.52$ will do the trick.

7 Attribution and Performance

The holdings attribution detailed in the previous two sections can be used to attribute portfolio performance over a sequence of investment periods.

Equation (6.2) attributes portfolio holdings to three sources:

- the benchmark portfolio $\mathbf{h}^{\text{bmk}}$,

- the exceptional forecast $\mathbf{h}^a$, and

- the bounds and constraints $\mathbf{h}$.

To evaluate performance over a sequence of periods requires a time dimension.

$$\mathbf{h}^{\text{opt}}(t) = \mathbf{h}^{\text{bmk}}(t) + \frac{IR}{\lambda_{bl}} \cdot \left\{ \frac{\mathbf{h}^a(t)}{\omega} \right\} + \mathbf{h}(t). \qquad (7.1)$$

---

[16]There are machinations that can bring the unconstrained and benchmark linked optimizations into line. These fall into the 'don't try this at home' category. Recall that $\mathbf{h}^\alpha/\omega$ has risk 100% and note that the ratio is independent of the scale of alpha. One could fix the risk aversion in (3.1) so that $\lambda = SR\langle \tilde{r}_{\text{eff}} \rangle / \sigma_{\text{eff}}$ and then scale the alphas $\alpha^* = \{\lambda \cdot IR/\lambda_{bl}\} \cdot \alpha$. If we use these re-scaled alphas in (3.1) and $\mathbf{h}^{\text{eff}} = \mathbf{h}^{\text{bmk}}$ then the optimal solutions of 3.1 and (6.1) will coincide.

The holdings, $\mathbf{h}^{\text{bmk}}(t), \mathbf{h}^a(t), \mathbf{h}(t)$ are at the *beginning* of period t. Let $\mathbf{r}(t)$ be the excess return on the assets *during* period t. With these definitions, we can calculate the return in period t on a component by component basis.

$$r^{\text{opt}}(t) = r^{\text{bmk}}(t) + \frac{IR}{\lambda_{bl}} \cdot \left\{ \frac{r^a(t)}{\omega} \right\} + r(t). \tag{7.2}$$

where $r^{\text{opt}}(t) = \mathbf{r}'(t) \cdot \mathbf{h}^{\text{opt}}(t)$, etc. We can refine the definition of $r(t)$, with

$$r(t) = \left\{ r^{lb}(t) - r^{ub}(t) - \hat{\theta}(t) \cdot r^{\text{mv}}(t) - \sum_{j=1,J} \left\{ \hat{p}_j^U(t) - \hat{p}_j^L(t) \right\} \cdot r^j(t) \right\}. \tag{7.3}$$

If alpha is the sum of various components, (5.11), (5.13), then we attribute to the components[17]

$$\frac{IR}{\lambda} \cdot \left\{ \frac{r^a(t)}{\omega} \right\} = \sum_{k=1,K} \left\{ \frac{IR_k}{\lambda_{bl}} \right\} \cdot \left\{ \frac{r^{\alpha^k}(t)}{\omega_k} \right\}. \tag{7.4}$$

To illustrate we ran the eight asset model over the years 1985 through 1995. To get forecasts of exceptional return we took the realized exceptional returns and added a random component.[18] The alphas were then scaled so that the anticipated information ratio was 0.9. Since this was an ex-post analysis we used the realized covariance matrix.[19] Table 11 gives summary statistics on the realized returns in main categories.

Returns in Table 11 are in excess of the cash rate. The first three columns summarize the sequence of returns on the three broad categories: (i) the benchmark portfolio, (ii) the alphas, and (iii) the constraints. The fourth column, labeled Port, has the summary statistics for the portfolio returns which are the sum of the first three columns. The last column, labeled Active, has the summary statistics for the difference between the portfolio and the benchmark (this is the same as the sum of the alpha and the constraint components).

The realized information ratio of the alphas is 0.96 which is close to the input estimate of $IR = 0.9$. In this benchmark linked application we used a

---

[17]For simplicity we have assumed that the ex-ante information ratios from each component are constant in time.

[18]The random component was scaled so the correlation of our forecast with the residual return would be 0.1. This correlation is know as the *IC* or information coefficient and can be related to a signal/noise ratio. If the noise has a standard deviation that is Y times greater than the standard deviation of the return then the *IC* will be $IC = 1/\sqrt{1 + Y^2}$

[19]The contrast between our forecast of covariance and the realized covariance over the eleven year period may change the result. Since we were interested in attribution rather than a simulating an investment strategy, we used the realized covariance and thus eliminated a possible source of confusion. Comparing this with another run using the ex-ante risk forecast will indicate the value of a good risk forecast.

Attribute	Bench	Alphas	Con	Port	Active
Annual Average	5.55%	5.22%	−2.56%	8.21%	2.66%
Annual Stdev	9.01%	5.42%	3.39%	8.53%	3.15%
Monthly Min.	−10.80%	−3.18%	−3.52%	−8.32%	−3.11%
Monthly Median	0.68%	0.29%	−0.18%	0.70%	0.17%
Monthly Max.	7.49%	9.57%	2.62%	6.96%	6.05%
Annual SR	0.62	0.96	−0.75	0.96	0.84

Table 11. Performance of the Portfolio by Component: January 1985 through December 1995

Attribute	Stock	Europe	Full Inv.	Bounds	Con.
Annual Average	−0.10%	−0.18%	−0.65%	−1.62%	−2.56%
Annual Stdev	0.24%	0.39%	1.51%	3.22%	3.39%
Monthly Min.	−0.69%	−0.65%	−1.28%	−2.88%	−3.52%
Monthly Median	0.00%	0.00%	−0.01%	−0.05%	−0.18%
Monthly Max.	0.00%	0.47%	1.53%	2.58%	2.62%
Annual SR	−0.41	−0.47	−0.43	−0.50	−0.75

Table 12: Attribution of return to the constraints and bounds.

risk aversion of $\lambda_{bl} = 18$, which is about five times greater that the absolute risk aversion cited above (3.57). The volatility of the alpha component is close to the prediction, $IR/\lambda_{bl} = 0.9/18 = 5\%$.

Over this eleven year period the benchmark portfolio did remarkably well producing a Sharpe ratio of 0.62. This boom followed the 1973-81 bust.

The potential for adding value, $IR = 0.9$, is large. However, some is value lost in implementation.[20] The combined effect of all restrictions drops the potential information ratio from 0.96 to 0.82, not a very serious shortfall. There is a drop in return of 2.56% per year and a corresponding decrease in active risk from 5.42% for the alphas to 3.15% for the constrained portfolios.

Table 12 presents a finer breakdown of the return that is attributed to constraints and bounds.

The average and standard deviation of the contribution is a good way of judging overall impact. The standard deviation for the contributions from the Stock and Europe constraints are very small, 0.24% and 0.39% compared with the like numbers for the full investment, 1.51% and bounds, 3.22%. Indeed, a re-run the analysis with only the full investment constraint and lower bounds

---

[20]For portfolios with a large number of assets you can anticipate losing about 40% of the information ratio due to the imposition of a "no short sales" restriction. With a small number of assets the lower bound of zero does not cause as much of an information loss in implementation.

of zero yields roughly the same result. In that rerun, the realized active risk is higher at 3.72%, and the active return is higher as well at 3.05%. The information ratio for the active return, at 0.82, is roughly the same.

8 Transactions Costs and Turnover

After expected return and risk, transactions costs are the third most important ingredient in building portfolios.

To model transactions we must introduce a new player; the *initial* portfolio $\mathbf{h}^I$. There are three ways that portfolio $\mathbf{h}^I$ can be used with an optimization:

- by deducting the transactions costs $tc\left\{\mathbf{h}^I,\mathbf{h}\right\}$ of moving from portfolio $\mathbf{h}^I$ to portfolio $\mathbf{h}$ from the objective,

- by placing limits on the amount of turnover $to\left\{\mathbf{h}^I,\mathbf{h}\right\}$ involved in moving from portfolio $\mathbf{h}^I$ to portfolio $\mathbf{h}$.

- by placing limits on the amount of trading in each particular asset; e.g. $-0.05 \le h_n - h_n^I \le 0.05$ if we can purchase or sell at most 5% of portfolio value in asset n. These transactions constraints can be folded into the overall holdings constraints.

The function $to\left\{\mathbf{h}^I,\mathbf{h}\right\}$ is convex in $\mathbf{h}$ and we *assume*[21] that $tc\left\{\mathbf{h}^I,\mathbf{h}\right\}$ is also convex in $\mathbf{h}$. We illustrate with a portfolio selection problem that contains both transactions costs and a turnover restriction.

$$\text{Maximize} \quad \mathbf{f}' \cdot \mathbf{h} - \tfrac{\lambda}{2} \cdot \mathbf{h}' \cdot \mathbf{V} \cdot \mathbf{h} - tc\left\{\mathbf{h}^I,\mathbf{h}\right\}$$

$$\begin{array}{ll}
\mathbf{e} \cdot \mathbf{h} = 1 & \text{fully invested} \\
\mathbf{b}^L \le \mathbf{A} \cdot \mathbf{h} \le \mathbf{b}^U & \text{additional constraints} \\
to\left\{\mathbf{h}^I,\mathbf{h}\right\} \le to^U & \text{limit on turnover} \\
\mathbf{h}^L \le \mathbf{h} \le \mathbf{h}^U & \text{bounds on holdings}
\end{array}$$

(8.1)

Let ϕ be the multiplier associated with the turnover constraint. The KKT conditions are

$$\mathbf{f} - \lambda \cdot \mathbf{V} \cdot \mathbf{h}^{\mathrm{opt}} - \mathbf{g} = \lambda \cdot \mathbf{e} \sigma_{\mathrm{mv}}^2 \cdot \hat{\theta} + \lambda \sum_{j=1,J} \cdot \mathbf{a}^j \cdot \left\{\hat{p}_j^U - \hat{p}_j^L\right\} + \lambda \cdot \left\{\hat{s}^U - \hat{s}^L\right\} + \phi \cdot \mathbf{d}, \quad (8.2)$$

---

[21] One usually assumes that $tc\left\{\mathbf{h}^I,\mathbf{h}\right\}$ is piecewise linear, $tc\left\{\mathbf{h}^I,\mathbf{h}\right\} = \sum_{n=1,N} tp_n \cdot \left\{h_n - h_n^I\right\}^+ + \sum_{n=1,N} ts_n \cdot \left\{h_n^I - h_n\right\}^+$, where tp_n and ts_n are the purchase and sales costs respectively and $\{x\}^+ = \max[0, x]$. The condition, $tp_n + ts_n \ge 0$, for all n assures convexity. The (one way) turnover is given by one half the total purchases plus total sales; $to\left\{\mathbf{h}^I,\mathbf{h}\right\} = 0.5 \cdot \left\{\sum_{n=1,N} \left\{h_n - h_n^I\right\}^+ + \sum_{n=1,N} \left\{h_n^I - h_n\right\}^+\right\}$.

where $\mathbf{g}$ is a (sub-)gradient of $tc\left\{\mathbf{h}^I, \mathbf{h}\right\}$ and $\mathbf{d}$ a (sub-)gradient[22] of $to\left\{\mathbf{h}^I, \mathbf{h}\right\}$. The first order conditions, expressed in the terms of Section 5 are[23]

$$\mathbf{h}^{\mathrm{opt}} = \mathbf{h}^{\mathrm{eff}} + \frac{IR}{\lambda} \cdot \left\{\frac{\mathbf{h}^a}{\omega}\right\} + \mathbf{h}^{tc} + \mathbf{h}, \tag{8.3}$$

where

$$\mathbf{h} = \mathbf{h}^{lb} - \mathbf{h}^{ub} - \hat{\theta} \cdot \mathbf{h}^{\mathrm{mv}} - \sum_{j=1,J}\left\{\hat{p}_j^U - \hat{p}_j^L\right\} \cdot \mathbf{h}^j, \tag{8.4}$$

and

$$\mathbf{h}^{tc} = -\left\{\mathbf{h}^{\mathrm{cost}} + \phi \cdot \mathbf{h}^{to}\right\}, \tag{8.5}$$

with[24]

$$\mathbf{h}^{\mathrm{cost}} = \frac{\mathbf{V}^{-1} \cdot \mathbf{g}}{\lambda}, \quad \text{and} \quad \mathbf{h}^{to} = \frac{\mathbf{V}^{-1} \cdot \mathbf{d}}{\lambda}. \tag{8.6}$$

The conditions are similar to (5.8) with the addition of the $\mathbf{h}^{tc}, \mathbf{h}^{to}$ and $\mathbf{h}^{\mathrm{cost}}$ terms. This additional effect will keep the optimal portfolio closer to $\mathbf{h}^I$ than it would be in the absence of transactions costs and turnover restrictions.

Table 13 shows the initial portfolio and the unit costs of buying and selling. There was no turnover constraint.

The optimal portfolio and attribution of holdings are in Table 14. If transactions costs were zero, there would be 47.5% one way turnover as we move from the initial portfolio in Table 13 to the optimal portfolio contained in Table 5. When we take the transactions costs into consideration the one way turnover drops to 31.7%.

Since we started with 100% in stock we end with a portfolio that contains 72% in stock rather than the 64% in stock that we would have selected (Table 5.). The European content restriction is still at its 50% upper limit. However, we are not pushing as hard against that restriction as we were before the transactions cost were introduced. Our initial portfolio starts with

---

[22]With the transactions cost and turnover functions as specified above we have

$$\left\{\begin{array}{ll} d_n = 0.5 & \text{if } h_n > h_n^I \\ d_n = -0.5 & \text{if } h_n < h_n^I \\ -0.5 < d_n < 0.5 & \text{if } h_n = h_n^I \end{array}\right\} \quad \text{and} \quad \left\{\begin{array}{ll} g_n = tp_n & \text{if } h_n > h_n^I \\ g_n = -ts_n & \text{if } h_n < h_n^I \\ -ts_n < g_n < tp_n & \text{if } h_n = h_n^I \end{array}\right\}.$$

For the assets that are purchased $h_n > h_n^I$ or sold $h_n < h_n^I$, there is no ambiguity in the choice of d_n or g_n. If there is no trade, then (7.2) can be used to unambiguously fix $g_n + \phi \cdot d_n$. We can arbitrarily set $-ts_n \leq g_n \leq tp_n$ and $-0.5 \leq d_n \leq 0.5$ to be consistent with the fixed sum $g_n + \phi \cdot d_n$.

[23]We have assumed that $\mu_{\mathrm{eff}}/\sigma_{\mathrm{eff}}^2 = \lambda$.

[24]This is equivalent to starting with a transactions costs

$$\left\{\begin{array}{l} tp_n + 0.5 \cdot \phi \text{ for purchases} \\ ts_n + 0.5 \cdot \phi \text{ for sales} \end{array}\right\}.$$

Asset	Cost	Initial
USA-S	0.50%	40%
JPN-S	0.50%	30%
UK-S	0.50%	20%
GER-S	0.50%	10%
USA-B	0.30%	0%
JPN-B	0.30%	0%
UK-B	0.30%	0%
GER-B	0.30%	0%

Table 13: Assumption on Transactions Costs and the Initial Portfolio

Country – Asset	eff	α	con& bnd	t-cost	total
United States Stocks	24%	−9%	2%	13%	30%
Japanese Stocks	18%	−14%	−2%	6%	8%
United Kingdom Stocks	12%	12%	2%	−2%	23%
German Stocks	6%	18%	−7%	−5%	11%
United States Bonds	16%	−25%	34%	−24%	1%
Japanese Bonds	12%	−15%	34%	−20%	11%
United Kingdom Bonds	8%	24%	−4%	−12%	16%
German Bonds	4%	−10%	13%	−8%	0%
Unconditional Expected Return	3.50%	0.01%	0.51%	0.00%	4.02%
Conditional Expected Return	3.51%	1.43%	0.03%	−0.18%	4.78%
α	0.01%	1.43%	−0.48%	−0.18%	0.76%
Forecast Standard Deviation	9.90%	6.32%	3.83%	3.00%	11.93%
Per cent in Stock	60%	6%	−4%	11%	72%
Per cent in Europe	30%	43%	4%	−27%	50%
Unconditional Sharpe Ratio	0.353	0.001	0.134	0.000	0.337
Conditional Sharpe Ratio	0.354	0.227	0.008	−0.062	0.401

Table 14: The Optimal Portfolio Attributed to α, Constraints and t-costs

30% in Europe, thus increasing the European is costly and thus less attractive. In the same vein we can see how the transactions costs are reducing our position in bonds and, by and large, adding to our position in stock. The exceptions are UK and German equity where we want to hold more than the initial allocations of 20% and 10%.

9 Conclusion

The most important information obtained from a portfolio optimization is the composition of the optimal portfolio. There is, however, a wealth of additional information that tells you why that particular portfolio was selected. In this paper we have taken that information, which is available in any mathematical program, and pointed out how it can lead to investment insights.

In our first approach, Section 4, we looked at the *expected asset returns* and broke them out into three parts: (i) a prior expectation μ, (ii) α a change in the expectation due to our research and (iii) a modification η of the expectations due to the constraints and bounds we place on the asset choice. The analysis reveals a great deal about the research process, the portfolio selection process and how the two work together. In particular, it serves as an alarm to warn that a portfolio selection process with multiple constraints that are usually very tight (high multiplier) is a strong indication that the research process is producing unrealistic forecasts. In that case a re-examination of the forecasting process may be needed.

In our second approach, Section 5, we used the notion of a characteristic portfolio along with the KKT conditions to parse out the *holdings* in the optimal portfolio and show how those holdings can be attributed to our prior assumptions of portfolio efficiency, our special information and the constraints and bounds in our portfolio selection methodology. We then, in Section 7, showed how this idea could be used in a sequence of periods to look at *performance* in terms of these components of the optimal portfolio.

We also showed how these notions could be employed in more complicated problems where we use a benchmark linked objective (Section 6) and where transactions costs are a concern (Section 8).

If understanding what you are doing is just as important as doing it ,then we have provided insight and a powerful tool to practitioners of the art of systematic investing.

References

V.K. Chopra and W.T. Ziemba (1993). 'The effects of Errors in Means, Variances, and Covariances on Optimal Portfolio Choice', *Journal of Portfolio Management* **19**, Winter, 6–11.

R. Grinold and R. Kahn (1995). *Active Portfolio Management: Quantitative Theory and Applications*, Probus Press.

P. Jorion (1992). 'Portfolio Optimization in Practice', *Financial Analysts Journal* **48**, Jan./Feb., 68–74.

R. Michaud (1989). 'The Markowitz Optimization Enigma: is Optimization Optimal?', *Financial Analysts Journal* **45**, Jan./Feb.

P. Muller (1993). 'Empirical Tests of Biases in Equity Portfolio Optimization'. In *Financial Optimization*, S.A. Zenios, ed. (Cambridge University Press), 80–98.

A. Rudd and B. Rosenberg (1979). 'Realistic Portfolio Optimization', *TIMS Study in the Management Sciences* **11** 21–46.

W.F. Sharpe (1994). 'The Sharpe Ratio', *Journal of Investment Management* **21** (3), Fall, 49–58.

National versus Global Influences on Equity Returns*

Stan Beckers, Gregory Connor, and Ross Curds

Abstract

Simple factor models of worldwide equity returns are used to explore the level and trend in international capital market integration. Global influences and national influences are of roughly equal importance in explaining the common movements in equity returns. Significant evidence indicates a trend toward increasing integration within the European Union, but not worldwide.

A large proportion of international portfolio managers and pension fund trustees allocate their funds in a top-down fashion, first making a decision across countries and/or geographical regions and then selecting securities within the various countries or regions. Similarly, most financial analysts evaluate the health of a company's balance sheet within a national context rather than in comparison with similar companies in other markets. These practices reflect a segregationist view of world capital markets. In a recent trend toward global analysis, however, securities are categorized and/or selected according to their underlying characteristics, not according to the nationality of their market listing. The conventional wisdom, at least as reflected in current practice, seems to be that the level of worldwide capital market integration is not high but is slowly increasing over time.

Capital market integration can be defined in at least three ways. One definition focuses on the barriers to international investing, such as regulatory, fiscal, or administrative impediments. To the extent that all investors have equal access to all world securities, markets are fully integrated by this definition. A second approach focuses on the consistency of asset pricing across markets. Under this definition, markets are integrated if any two assets with the same level of risk and the same expected cash flows always have the same price irrespective of the markets in which they trade. A third approach concentrates on the correlations of security returns across different

*Reprinted, with permission, from *Financial Analysts Journal*, March/April 1996. Copyright 1996, Association for Investment Management and Research Charlottesville, VA. All rights reserved.

markets. Under this definition of integration, the comovements in security returns are linked to a set of common factors. If markets are fully integrated, then the factors explaining the correlations of returns will be international ones, with no role for national factors. This third, correlation-based definition of integration is the focus of this article.

The correlation-based definition of integration permits accurate and reliable empirical testing. An accurate test of asset pricing integration may require 100 years or more of returns data, whereas correlation-based analysis yields meaningful test statistics based on as little as three years of monthly returns.[1]

Our analysis follows along the lines of earlier work by Beckers *et al.* (1992), Grinold *et al.* (1989) and Heston and Rouwenhorst (1994, 1995). Grinold *et al.* and Beckers *et al.* used a fundamental factor model with factors for size, success, volatility, and yield, as well as a CAPM-style local market factor to characterize each stock. The factor exposures, derived from combinations of accounting data and time series returns data, are used in monthly cross-sectional regressions to derive the factor returns.

Heston and Rouwenhorst relied on a much simpler type of factor model. They used simple dummy variables to identify the industry and country affiliation of each stock. When these dummy variables are regressed on the cross-section of security returns, the estimated coefficients on the dummy variables are the implicit returns of country and industry factors. Translated into standard factor modeling terminology, the factor betas in the Heston and Rouwenhorst model are all equal to zero or unity and the regression coefficients correspond to country and industry factor returns.

In this study, we used the same factor modeling approach as Heston and Rouwenhorst. We extended their approach in several ways, however: We

---

[1] First, consider asset pricing integration, which involves measuring and comparing risk premiums of assets across national markets. Suppose that an asset has an observed mean excess return of 6% per annum and a known standard deviation of 30% per annum. In order to reject the null hypothesis that the expected excess return is zero, we need the square root of the number of observations times the ratio of mean excess return to standard deviation to exceed approximately 2.0. This calculation requires exactly 100 years of data. The test statistic is unaffected by whether we measure returns on an annual, monthly, or daily basis. The measurement problem in our case is actually harder than this, because we need to compare risk premiums and show that they are higher or lower in one country than in another, which is more difficult than showing that one risk premium is nonzero.

Next, consider correlation-based analysis. Suppose that a sample correlation between two assets is 0.4. Using an asymptotic approximation, under the null hypothesis that the true correlation is zero, this sample correlation is normally distributed with mean zero and standard deviation equal to the reciprocal of the square root of the number of observations. Given more than 25 monthly observations, we can reject the null hypothesis that the true correlation is zero at the 95% confidence level. This simple comparison is illustrative of the much greater power of tests based on correlation compared with tests based on measured risk premiums.

estimated and compared a set of factor models with the same basic structure but varying degrees of national versus international focus; and we present some new procedures for measuring integration (both in its level and trend through time) and provide new empirical findings.

One of our key models contains a global market factor, country factors, and global industry factors. In this model, the global market factor explains 21% of the typical equity return variance, country factors explain an additional 14%, and global industry factors an additional 4%. The two global influences (the global market factor and global industry factors), therefore, explain 25% between them. In some of our other factor model specifications (e.g., with local industries), the national influences slightly outweigh the global influences. Our general conclusion is that global and national influences are of roughly equal importance.

We investigated the trend toward greater worldwide integration by testing for increased explanatory power coming from the world market factor and global industry factors and for decreasing power from the country factors. We found only weak (and not statistically significant) evidence for increasing integration worldwide, but the evidence for the European Union (EU) is strong and statistically significant. Additional evidence of increasing EU integration is found in the increase in the correlations of EU country factors through time.

Data

Our study is based on monthly excess returns for December 1982 through February 1995, 147 months in total. The data base covers 19 countries from the developed world.[2] The sample each month consists of stocks that were part of the Financial Times Goldman Sachs World Index that month. The number of stocks in the sample varies across months; the average number is 2,123. For some parts of our research, we also considered the European Union (EU) separately in order to investigate whether economic integration is more pronounced in the EU than in the rest of the world. Our EU subset consists of 9 countries and has an average of 723 stocks. Stocks are classified as belonging to 1 country from among the 19 countries and 1 industry from among a set of 36 industries. We also tried seven broad economic sectors as an alternative to the 36 industries. The sector and industry classifications are those used in the Financial Times Goldman Sachs Index.

---

[2]The countries are Australia, Austria, Belgium, Canada, Denmark, France, Germany, Hong Kong, Ireland, Italy, Japan, the Netherlands, Norway, New Zealand, Spain, Sweden, Switzerland, the United Kingdom, and the United States. See BARRA (1992), for a list of the industries and sectors.

The Explanatory Power of National versus International Factors

We used simple factor models of returns with zero/one exposures to the explanatory variables (country and industry factors). First, we describe the model with country factors and global industry factors (what we call the countries + global industries model). All the other models have essentially the same form, so describing each in detail is not necessary. In all the models, we used returns in excess of the local risk-free rate and expressed in local currency. From the viewpoint of an investment practitioner, our return calculations correspond to fully hedged excess returns. From the viewpoint of a financial researcher, the distorting effects of currency movements are not included in the comovements.

In the countries + global industries model, the local excess return to each equity in a given month is divided into a global market return, a country factor return, a global industry return, and an asset-specific return; that is,

$$r_i = f^G + \sum_{h=1}^{M} \delta_{ih}^I f_h^I + \sum_{j=1}^{L} \delta_{ij}^C f_j^C + \varepsilon_i, \tag{1}$$

where

r_i = excess return to security i, $\quad i = 1, \dots, N$

f^G = return to the global market factor

f_h^I = return to Industry factor h, $\quad h = 1, \dots, M$

f_j^C = return to Country factor j, $\quad j = 1, \dots, L$

ε_i = asset-specific return to security i

δ_{ih}^I = 1 if security i is in industry h, 0 otherwise

δ_{ij}^C = 1 if security i is in country j, 0 otherwise

We estimated the factor returns for each month by applying ordinary least squares to the cross-section of returns using (1) subject to two linear constraints. That is, we found $\hat{f}^G, \hat{f}^I, \hat{f}^C$ to minimize $\sum_{i=1}^{N} \hat{\varepsilon}^2$ in (1) subject to

$$\sum_{i=1}^{N} \sum_{h=1}^{M} \delta_{ih}^I \hat{f}h^I \tag{2}$$

and

$$\sum_{i=1}^{N} \sum_{j=1}^{L} \delta_{ij}^C \hat{f}j^C \tag{3}$$

The two linear constraints, (2) and (3), imply that each month, the average worldwide effect of the country factors is zero and the average worldwide effect of the industry factors is zero. Adding the two equality restrictions implies that the country factor returns are measured net of the global market return. If security returns worldwide are mostly positive in a given month and German securities are also up but by less than worldwide securities generally, then the German factor return will be negative. The same holds for the industry factors: If security returns are generally positive worldwide and steel stocks are also up but by less than in most other industries, then the steel industry factor return will be negative.

The country and industry factors are neutralized returns because they are estimated simultaneously. So, for example, the German country factor return is neutralized with respect to the differing industrial composition of the German market compared with other markets. If the German stock market is up in a given month but all of the positive return can be attributed to the heavier representation of German stocks in particular industries that did well worldwide that month, then the German country factor returns will be zero. As long as no two countries in the sample have exactly the same proportions of firms in all industries, there is no identification problem in simultaneously estimating industry-neutralized country factors and country-neutralized industry factors.

In addition to the countries + global industries model, we estimated a set of alternative factor models with varying degrees of national versus international focus. By comparing the fit of these models, we hoped to learn about the relative importance of national versus international influences. Drop the country dummies from the countries + global industries model to produce the global industries-only model, or drop the industry dummies to produce a countries-only model. Also, changing the industry dummies produces a different industry factor return for each country. We called this model the local industries-only model to differentiate it from the global industries-only model. All models include the global market factor (that is, a cross-sectional intercept). The model with nothing except the intercept is called the global market model. We also tried the seven sectors in place of the 36 industries. This approach does not produce any additional information about the level or trend in market integration, but it gives a different specification of industry factors. The full list of models and the linear constraints necessary to identify each model are shown in Table A1 in the appendix.

Table 1 shows the findings for the models, reporting two measures of model

Table 1. Adjusted R^2 and EP for each Model, with and without October
1987 (data in parentheses exclude October 1987).

Model	No Industry or Sector Factors	Global Sector Factors	Global Industry Factors	Local Industry Factors
Average adjusted R^2				
No country factors	0.0000	0.0208	0.0522	0.2571
	(0.0000)	(0.0206)	(0.0517)	(0.2558)
Country factors	0.1760	0.1947	0.2189	0.3039
	(0.1746)	(0.1928)	(0.2170)	(0.3022)
Adjusted EP				
No country factors	0.2107	0.2282	0.2537	0.4308
	(0.1800)	(0.1977)	(0.2238)	(0.4058)
Country factors	0.3620	0.3778	0.3970	0.4695
	(0.3347)	(0.3503)	(0.3700)	(0.4453)

fit: the average R^2s and the EP (explanatory power) statistics.[3] (Both are
adjusted for degrees of freedom.) The message from the average R^2 statistics
confirms the general conclusions reported in Grinold *et al.* (1989), Beckers
et al. (1992) and Heston and Rouwenhorst (1994, 1995): National influences
dominate global influences. The countries-only model has a three times higher
average R^2 than the global industries-only model. Adding countries to the
global industries-only model increases the average R^2 by 16.67%, whereas
adding the global industries to the countries-only model increases the average
R^2 by 4.29%. Alternatively, compare the global industries-only model to the
local industries-only model: Using local industries in place of global industries
increases the average R^2 from 5.22% to 25.71%. The simple message from the
average R^2 analysis is that national influences (country factors and country-
specific industry factors) are much more important than global influences
in explaining equity returns. Note that, by construction, the average R^2 is
exactly zero for the global market model.

The EP statistic offers quite a different interpretation of the same regres-

---

[3]See Connor and Korajczyk (1993) and Connor (1995) for a discussion of the EP statistic. The average R^2 is given by the formula,

$$\text{Average } R^2 = 1 - \sum_{t=1}^{T} \frac{1}{T} \frac{\frac{1}{N}\sum_{i=1}^{N} \hat{\varepsilon}_{it}^2}{\frac{1}{N}\sum_{i=1}^{N}\left(r_{it} - \frac{1}{N}\sum_{i=1}^{N} r_{it}\right)^2}$$

whereas the EP statistic is

$$\text{EP } R^2 = 1 - \frac{\frac{1}{N}\frac{1}{T}\sum_{i=1}^{N}\sum_{t=1}^{T} \hat{\varepsilon}_{it}^2}{\frac{1}{N}\sum_{i=1}^{N}\frac{1}{T}\sum_{t=1}^{T}\left(r_{it} - \frac{1}{T}\sum_{i=1}^{T} r_{it}\right)^2}.$$

sion results. The EP of the global market model is 21.07%. Considering the countries + global industries model, we can decompose its explanatory power by viewing it as a combination of three models. The global industries add 4.30% explanatory power to the 21.07% explanatory power of the global market factor alone, whereas the countries add 14.33% to the global industries-only model. The global industries alone are the weakest influence, but the two global influences together (global market factor plus global industries) contributed 25.37% to the explanatory power of this model, whereas the country factors only contributed 14.33%.

It is notable that in comparing the *differences* between the explanatory power of the factor models, the EP and average R^2 statistics shown in Table 1 give similar findings. The observed differences between EP and average R^2 stem largely from the substantial explanatory power that the EP assigns to the global market factor. The EP is discussed in more detail in the Appendix. Table 1 shows that global industries are more powerful explanatory variables than global sectors and local industries are more powerful than either. The countries + local industries model has the highest average R^2 and highest EP.

The Countries + Global Industries Problem

We estimated the countries + global industries model on the full cross-section of 19 countries and also separately on the 9 European Union (EU) countries. The EU-only sample does not include six current members of the European Union: three of them (Greece, Luxembourg, Portugal) because we have no reliable data, and three (Austria, Finland, Sweden) because these countries joined the EU very recently and so are not relevant for our historical analysis. In the EU-only model, the 'global' market factor captures the EU-wide market movement. Similarly, the 'global' industry factors for the EU-only model are the industry factors measured across the nine EU countries.

The importance of a factor derives from its contribution to explaining asset return variance. Because each factor exposure is either zero or one, the variance of the factor return captures its influence on asset variance. For each group of factors (countries or industries), we analyzed the average of the factor return variances across all of the group.

Table 2 shows the average factor return variance for each of the categories of factors for both the worldwide and EU models. The inclusion of October 1987 increases the variance of the global market factor substantially, especially in the worldwide model. In the EU model, the global market factor explains the largest percentage of volatility and the country factors are a close second. The global industry factors lag well behind the other two. The worldwide model has stronger country factors than the EU model, a weaker

Table 2. Average Variance with and without October 1987 (data in parentheses exclude October 1987).

Sample	Variance of Global Market Factor	Average Variance of Country Factors	Average Variance of Industry Factors
Worldwide	0.001764	0.0015306	0.0002769
	(0.001369)	(0.0015096)	(0.0002748)
EU-only	0.002209	0.0012089	0.0006074
	(0.001849)	(0.0012027)	(0.0006013)

global market factor, and smaller global industry influences. These comparative findings argue that the EU is more integrated than the world generally, possibly as a result of the harmonization of economic, monetary, and fiscal policies during the past 20 years.

Next, we examined whether there is a trend in the relative explanatory powers of the three groups of factors. Recall that in our simple factor model with zero/one factor exposures, the explanatory power of a factor can be measured by the factor return variance. If the world (or EU) is becoming more integrated over time, the variance of the global factors (that is, the global market factor and industry factors) should be increasing and/or the variance of the country factors should be declining over time.

The relative explanatory power of a factor can increase, even though its return variance does not increase, if the average return variance across stocks decreases through time. Average return variance declined significantly during the past 12 years for both the worldwide and EU-only sample. Figure 1 shows the month-by-month variability of total return relative to its long-run average. It is clear from Figure 1 that, both for the EU and the worldwide sample, average return variability has declined over the sample period. Our measure of relative variability for the factors is adjusted for this decline in total return variability (see the appendix for the technical definitions).

Figure 2 shows the trends in the relative variability of the global market factor, global industry factors, and country factors for the worldwide model. The country factors show some tendency to decrease in importance through time, with a concomitant increase in the relevance of the global industry factors, but neither of these effects is significant at the 10% level. The effect of the global market factor changes very little through time. Therefore, the evidence for an increase in the importance of global influences in the worldwide

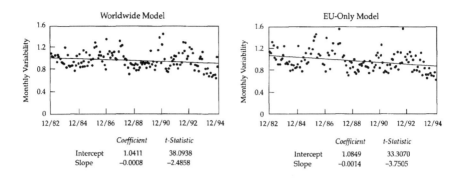

Figure 1. Trends in the Average Variance of Individual Equities (excludes October 1987)

model is weak at best.

Figure 3 repeats the same tests for the EU-only .sample. Here, the results are clearer and statistically significant: the EU-only global market factor increases in importance trough time, the country factors lose importance, and the global industry factors increase in importance. All of these effects point toward increasing market integration within the EU.

Another measure of the trend in European integration is based on the EU country factor returns from the worldwide model We estimated the correlations of the EU country factor returns over four subperiods of 37, 36, 36, and 36 months. Under the null hypothesis of no increased integration, the true correlation matrix should be the same in the subperiods. Under the alternative that the EU is becoming more closely integrated, the correlation of the country factors for the EU countries will be increasing. Table 3 shows the correlations of each country factor with the German country factor, which we use as the central point of this analysis. Most countries show a discernible trend toward increasing correlation over the four subperiods. Note that for all but two of the eight countries (Italy and Denmark), the correlation in the final subperiod is higher than the average of the four subperiod correlations. We take Table 3 as generally supportive of an increase in integration across these national markets over the period.

Figure 4 provides an analysis similar to Table 3 but based on 12 subperiods of 12 months each. For Figure 4, rather than using Germany as a central point, we examined the trend in the average of all the correlations between EU country factors. As in Table 3, the country factors are taken from the worldwide model estimation. The trend in the average correlation is positive and statistically significant, indicating increasing integration within the EU.

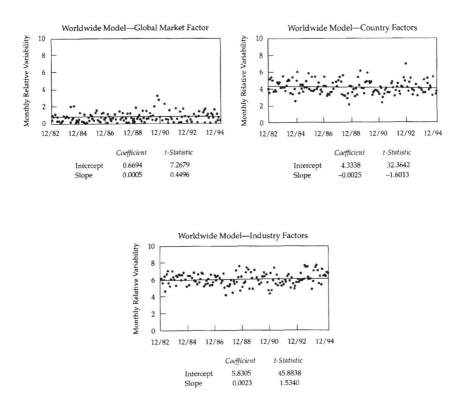

Figure 2. Trends in the Variability of the Global Market, Countries, and Global Industries Factors, Worldwide Sample (excludes October 1987)

Table 3. Correlations of European Union with Germany in Four Subperiods

Country	Dec. 1902–Nov. 1985	Dec. 1985–Dec. 1988 (excludes Oct. 1987)	Jan. 1989–Jan. 1992	Feb. 1992–Feb. 1995	Average of Four Subperiods
Belgium	0.2650	0.3666	0.4557	0.5269	0.4045
Denmark	0.1695	0.1366	0.3596	0.1269	0.1983
France	−0.2308	0.4862	0.4191	0.5349	0.3060
Ireland	0.0445	−0.1045	−0.0662	0.1037	−0.0060
Italy	0.0528	0.3273	0.4059	0.1533	0.2361
The Netherlands	0.0469	0.4313	0.4374	0.4524	0.3440
Spain	0.0162	−0.1964	0.0473	0.1918	0.0147
United Kingdom	−0.1426	−0.3540	0.1471	0.2013	−0.0363
Average	0.0277	0.1366	0.2757	0.2864	0.1826

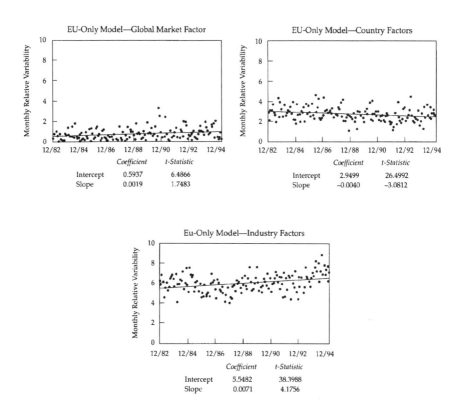

Figure 3. Trends in the Variability of the Global Market, Countries, and Global Industries Factors: EU-only Sample (excludes October 1987)

Summary

We used a simple approach to examine the influence of national and global factors on equity returns. We estimated factor models in which all securities have zero or one exposures to sector, industry, and/or country factors, with a unit exposure if the security belongs to the sector/industry/country and a zero otherwise. In addition to these zero/one exposures, all equities are assigned unit exposure to a global market factor. The returns to be explained are excess returns to individual equities in 19 countries, with all returns measured in local currency.

We compared the explanatory power of a variety of specifications. Industry factors outperformed sector factors (the sectors divide securities into broader categories than industries). Nation-specific industry factors have substantially more explanatory power than global industry factors. Country

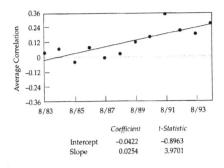

Figure 4. Trend in the Average Correlation of EU Country Factors
(excludes October 1987)

factors are strong, but the global market factor, capturing the tendency for all securities worldwide to move together, is even more powerful than the country factors. The best model (in terms of explanatory power, ignoring theoretical considerations) has a global market factor, country factors, and nation-specific industry factors.

We examined a particular specification in more detail to analyze the level and trend in capital market integration, both worldwide and within a sub-sample restricted to European Union countries. The specification has a global market factor, global industry factors, and country factors. The global market factor and global industry factors are more important in the subsample restricted to European Union countries than in the worldwide sample, and the country factors are less important. These comparative findings indicate that the European Union is more integrated than the world overall. We also examined the trend in the explanatory power of the three types of factors in this model. We found that in the European Union, the country factors are decreasing in importance while the global industry factors and global market factor are increasing in importance. This result indicates that the European Union is becoming more integrated through time. We found the same trends for the worldwide sample, but the trend coefficients are not statistically significant in this second case.[4]

Appendix

Table A1 shows the range of factor models we estimated and describes the equality constraints imposed to ensure statistical identification of the models.

---

[4]We would like to thank Lucie Chaumeton and Richard Grinold for helpful comments.

Table A1. List of Models

Model	No Industry or Sector Factors	Global Sector Factors	Global Industry Factors	Local Industry Factors
No country factors	Global Market Model No Constraints	Global Sectors-only Model Constraint B	Global Industries-only Model Constraint C	Local Industries-only Model Constraint D
Country factors	Countries-only Model Constraint A	Countries + Global Sectors Model Constraints A and B	Countries + Global Industries Model Constraints A and C	Countries + Local Industries Model Constraints A and D

Constraints: A = average country factor return equals zero. B = average global sector factor return equals zero. C = average global industry factor return equals zero. D = average local industry factor return equals zero in each country.

Two Measures of Explanatory Power

The factor model was estimated each month by cross-sectional regression, resulting in a time series of monthly regression statistics. We needed to combine these regression statistics into overall measures of explanatory power for the model. One measure of explanatory power is the time-series average of the R^2 statistics. The R^2 at each point in time gives the proportion of the cross-sectional variability of stock returns explained by the factors. Note that an intercept in a cross-sectional regression model has no ability to explain cross-sectional variance, because it merely removes the cross-sectional mean. In the context of our factor model, the intercept coefficient is the return of the global market factor. The R^2 at each point in time effectively assigns zero explanatory power to the global market factor. Taking the time-series average of the cross-sectional R^2 statistics, the explanatory power of the global market factor is lost. Because of this loss, the average R^2 can give a distorted picture of the influence of global versus local influences. For example, if the global market factor becomes more important while the explanatory power of the other factors (industries, countries) and the total asset variability remain the same, the average R^2 statistic *falls*.

An alternative to the average R^2 is the EP statistic. The EP statistic finds the average cross-sectional and time-series variability explained by the factors. Thus, the influence of the global market factor is included in the EP statistic. If the global market factor becomes more important and the explanatory power of the other factors (industries, countries) remains the same and total asset variability remains the same, the EP statistic increases. The EP statistic is therefore not subject to the distortion associated with the average R^2.

Measures of Relative Variability

For the global market factor, we can easily test for an increase in variance by dividing each month's realized squared global market factor return by its time-series variance measured over the whole sample period. If the true variance is increasing through time, then this ratio will tend to increase through time.

The same statistic can be applied to a vector of factors. Consider, for example, the vector of country factors, and suppose we want to test whether their generalized variance is declining through time. Delete one of the country factors. The reason for deleting one country factor is that the country factors are restricted to sum to zero each month. Deleting one of them eliminates this linear dependence between them. Because of the linear dependence, the choice of country factor deleted has no effect on this test statistic. Let $\hat{f}_i^C$ denote the realized returns on the remaining $L-1$ country factors at time t and $\hat{\Sigma}_C$ their covariance matrix estimated over the entire sample period. The vector analog of the realized squared return divided by variance is the quadratic product of the realized factor returns around the inverse of the covariance matrix $\hat{f}_i^{C\prime}\hat{\Sigma}_C^{-1}\hat{f}_i^C$. If the variance of the country factors is decreasing through time, then this ratio will tend to decrease through time. The same statistic can be applied to the industry factor returns.

We denoted these measures of time t variability for the global market factor, country factors, and industry factors by X_t^G, X_t^C and X_t^I:

$$X_t^G = \frac{1}{\hat{\sigma}_G^2}(f_t^G)^2$$
$$X_t^C = \hat{f}_t^C{}'\hat{\Sigma}_C^{-1}\hat{f}_t^C$$
$$X_t^I = \hat{f}_t^I{}'\hat{\Sigma}_I^{-1}\hat{f}_t^I$$

If the world capital market is becoming more integrated, X_t^G and X_t^I should increase through time and X_t^C should decline.

Let $\hat{\sigma}_r^2$ denote the cross-sectional average of the time-series variances of all the assets. Let X_t^r denote the cross-sectional average squared return in month t divided by $\hat{\sigma}_r^2$. If the average variability of total return is constant, then this ratio should be constant. Figure 1 shows the regression of the square root of this ratio against an intercept and time index. We used the square root because the distribution of this test statistic has fat tails; the square root is better behaved for regression analysis, although either formulation is statistically valid as an alternative hypothesis. Figure 1 indicates that total return variance is declining.

To adjust for the declining level of total return variance, we used the standardized measures X_t^G/X_t^r, X_t^C/X_t^r and X_t^I/X_t^r rather than X_t^G, X_t^C, and X_t^I to test for the trend in international integration. Figures 2 and 3 show the regressions of the square root of each of these standardized measures against an intercept and time index.

References

BARRA Inc., *The Global Equity Model Handbook*, 1992.

Beckers, Stan, Richard Grinold, Andrew Rudd, and Dan Stefek, 'The relative importance of common factors across the European equity markets,' *Journal of Banking in Finance* **1** (16) February, 1992, 75–95.

Connor, Gregory, 'The three types of factor models: a comparison of their explanatory power,' *Financial Analysts Journal* **51** (3) May–June 1995, 42–6.

Connor, Gregory and Robert A. Korajczyk, 'A test for the number of factors in an approximate factor model,' *Journal of Finance* **48** (4) September, 1993, 1263–91.

Grinold, Richard, Andrew Rudd, and Dan Stefek, 'Global factors: fact or fiction?' *Journal of Portfolio Management* **16** (1) Fall, 1989, 79–88

Heston, S.L. and K.G. Rouwenhorst, 'Does industrial structure explain the benefits of international diversification?' *Journal of Financial Economics* **36** (1) August, 1994, 3–27.

Heston S.L. and K.G. Rouwenhorst, 'Industry and country effects in international stock returns,' *Journal of Portfolio Management* **21** (3) Spring, 1995, 53–58.

A Global Stock and Bond Model*

Lucie Chaumeton, Gregory Connor, and Ross Curds

Abstract

Six fundamental risk factors (four for stocks and two for bonds) explain most of the common volatility of individual stocks and bonds worldwide. Some of the risk factors have a strong international component, and others are more purely national. The cross-national component of the risk factors tends to be stronger within the European Union than worldwide. The model proposed in this article can be used for integration of worldwide asset selection and asset allocation decisions.

Factor models are now widely used to support asset selection decisions. Global asset allocation, the allocation between stocks versus bonds and among nations, usually relies instead on correlation analysis of international equity and bond indexes. It would be preferable to have a single integrated framework for both asset selection and asset allocation. This framework would require a factor model applicable at an asset or country level, as well as at a global level, that covers both stocks and bonds.

We propose a simple and intuitive factor model of international stocks and bonds. It is built on six factors inspired by well-accepted research ideas: market, size, value, and duration factors for stocks, and yield curve shift and twist factors for bonds. Because this model is estimated using asset level data, it can be used to analyze each asset's sources of risk. Because it is estimated in a global framework, it can also be used for global asset allocation purposes. The estimates from our model allowed us to draw a number of conclusions about the sources of risk on worldwide stock and bond markets. We found that the global component of the factor model is stronger in bond markets than in stock markets and stronger in the European Union (EU) than worldwide. Two of the risk factors, the market factor for stocks and the shift factor for bonds, have strong global components, as well as nation-specific components. The other four risk factors are mostly nation specific. We also examined the relationships between the factor risks, both within national markets and on a global level.

---

*Reprinted, with permission, from *Financial Analysts Journal*, November/December 1996. Copyright 1996, Association for Investment Management and Research Charlottesville, VA. All rights reserved.

Data

Our data consist of monthly returns and fundamental descriptors for individual stocks and bonds from 13 countries: Australia, Belgium, Canada, Denmark, France, Germany, Italy, Japan, the Netherlands, Spain, Sweden, the United Kingdom, and the United States. The data cover 112 months, from February 1986 to May 1995. The number of stocks and bonds differs among months; in an average month, the worldwide sample consists of 487 bonds and 2,000 stocks.

The equity data consist of return series, market capitalizations, dividend yields, and book-to-price ratios each month for each stock. The bond data consist of return series and a standardized description of the cash flows of each bond. Our bond database provides, for each bond, an allocation of all future cash flows into fixed-date vertexes 0,1, 2, 3, 4, 5, 7,10, 20, and 30 years in the future. The cash flow allocation assumes that cash flows occur only on vertex dates. Cash flows occurring between vertexes are allocated proportionately to each vertex to preserve the bond's duration and present value.[1]

The returns for stocks and bonds are both expressed in local currencies and in excess of local one-month risk-free returns. In addition, we calculated for each country the end-of-month exchange rates relative to the US dollar. We used these rates to calculate the full set of cross-rates (e.g., the yen/pound exchange rate equals the ratio of the yen/dollar exchange rate to the pound/dollar exchange rate) over the same 112-month period (February 1986 to May 1995).

Six Fundamental Factors

Our factor model is designed to be simple, intuitive, statistically accurate, and reliable. None of the factors used in our model is entirely new; to select them, we relied on a whole range of findings from 20 years of factor modeling research in the academic and practitioner literature. For all factors, we used theoretically-defined measures of factor exposure, rather than estimating the exposures by time-series analysis.

The two bond market factors are standard (see Kahn 1995). We defined the shift exposure of a given bond as the percentage increase in its price, given a constant percentage decline in all yields. As is well known, this amount is equal to the modified duration of the bond. Let $B_{j,t}$ be the price of bond i at time t, $C_{i,t+s}$ be the cash flow of bond i at time $t + s$, and $y_{t,s}$ be the yield on an s-period pure discount bond at time t. Consider a constant percentage decrease in all yields in response to a random shock, dx so that $dy_{t,s}/dx = -(1 + y_{t,s})$ for all s. The shift exposure of a given bond is its

---

[1]For more details on cash flow allocation in this bond database, see BARRA (1991).

percentage price sensitivity to this parallel shift in yields; that is,

$$\text{SHIFT}_{i,t} = \left(\frac{dB_{i,t}}{dx}\right)\left(\frac{1}{B_{i,t}}\right).$$

Applying the modified version of the well-known results of Macaulay (1938) gives

$$\text{SHIFT}_{i,t} = \left(\frac{1}{B_{i,t}}\right)\sum_{s=1}^{T} s(1 + y_{t,s})^{-s}C_{i,t+s},$$

so that shift exposure is equal to the value-weighted time to maturity of the bond, or modified duration.[2]

Our second bond factor, twist, measures bond price responses to changes in the slope of the yield curve. We predefined a midpoint of the term structure, called *mid* (in our empirical work, we set *mid* equal to four years). We defined a shock to the term structure that causes yields longer than *mid* to increase and yields shorter than *mid* to decrease; the size of the change is proportional to the time difference from *mid*; that is,

$$\frac{dy_{t,s}}{dz} = (1 + y_{t,s})(s - mid).$$

We defined the twist exposure of a bond as the percentage price response to this rotational shock to yields:

$$\text{TWIST}_{i,t} = \left(\frac{dB_{i,t}}{dz}\right)\left(\frac{1}{B_{i,t}}\right).$$

Applying calculus to the present value formula for the price of a bond gives the following explicit formula for TWIST:

$$\text{TWIST}_{i,t} = \left(\frac{1}{B_{i,t}}\right)\sum_{s=1}^{T}(mid - s)s(1 + y_{t,s})^{-s}C_{i,t+s}.$$

The strongest factor in stock market returns is the tendency for all stocks in a given country to move together, the so-called market factor. Historically, exposure to this factor has been measured by market beta, usually estimated by time-series regression of each asset's return against a national equity index. We followed some recent research in replacing estimated market beta with a unit exposure to the market factor for every stock. This approach is the same as imposing the prior assumption that every stock has a market beta of 1.[3]

---

[2]For the original results, see Macaulay (1938). For a modern treatment, see Chapter 21, Elton and Gruber (1995).

[3]See Fama and French (1993), who showed that in a multifactor model such as ours, individual stock market betas differ very little from 1. Imposing the prior assumption that all market betas equal 1 eliminates a large amount of estimation error from the market betas (because we need not use time-series regression to estimate them) and sacrifices very little accuracy in the model.

A large number of empirical studies have established the importance of size and value as factors in stock returns (see Sharpe 1992; Capaul, Rowley and Sharpe 1993; and Fama and French 1993). We defined average capitalization as value-weighted average of the market capitalizations of all the firms in a country for a given month. We defined the size exposure of a stock in a given month as the logarithm of its market capitalization divided by average capitalization.

Dividing by the average capitalization of each country-month has two advantages. First, it adjusts for the tendency of average firm size to increase through time because of inflation and real growth. Second, it normalizes size exposure within each country, so that even a very small country has some 'large' firms. This adjustment has the disadvantage that size exposure is a country-relative quantity that needs to be interpreted carefully in our global model. In our model, a global investor may hold a 'large' stock in Belgium that is smaller than a 'small' stock in Germany.

We defined a stock's value exposure using its book-to-price ratio. We adjusted for differences in accounting standards across countries by dividing each company's book-to-price ratio by the capitalization-weighted average of book-to-price ratios for all stocks in the same country for all months in the sample period. We used each country's long-run average book-to-price ratio; we did not adjust for any time series variation in book-to-price ratios within each country.

Following earlier work, we defined stock duration as the reciprocal of the stock's dividend yield (see Bernstein and Tew 1991). Treating an equity as perpetuity with constant dividend growth, the reciprocal of dividend yield gives the value-weighted average maturity of the cash flows to the share which is the original definition of duration for bonds. Note that stock duration does not measure sensitivity to parallel shifts in interest rates. Stock duration is value-weighted average maturity, but it is not equal to interest rate sensitivity. We added 1% to each stock's stated dividend yield to adjust for non-dividend cash flows (such as corporate stock repurchases and cash-financed mergers).

Table 1 lists for each country the average number of stocks and bonds across the 112–month sample and the mean and standard deviation of each of the six factor exposures except market exposure (because market exposure is a 0/1 dummy, these sample statistics are not meaningful for it). Recall that the value and size exposures are standardized within each country but stock duration exposures are not. Standardizing stock duration within each country has some justification. Note in particular the long average duration of Japanese stocks. In our model, an investor choosing to hold Japanese stocks is also choosing to hold long-duration stocks. We believe that maintaining this feature of the model was better than eliminating it by standardizing stock duration within each country. The shift and twist exposures do not need to be standardized.

Relative Importance of National and Global Factors

We estimated two forms of the factor model: a global model, which has a single set of factor returns across all countries, and a nation-specific model, in which the factor returns are allowed to differ among countries. Table 2 shows the explanatory power of the global model, the nation-specific model, and individually for each of the nations within the nation-specific model. See the appendix for more details on the estimation of factor returns and on the calculation of explanatory power.

On the worldwide sample, our global factors explain 21% of stock volatility and 60% of bond volatility, an impressive result for such a simple and inhibitive model. Allowing factor returns to differ across nations explains an additional 13% of stock volatility and 35% of bond volatility. A large component of both stock and bond returns is therefore explained by global influences; in the case of bonds, the explanatory power of the global factors is almost two-thirds of that achieved with nation-specific factors.

The models are shown with two estimation universes. The first uses a worldwide sample; the second is restricted to the eight EU countries in our sample.[4] Global influences in this subsample are expected to be higher if the EU is more economically integrated than the world at large. The explanatory power of the global model is indeed higher within the EU than in the whole sample: 26% versus 21% for stocks and 70% versus 60% for bonds. Although global factors have more influence within the EU than worldwide, allowing nation-specific factors produces a smaller increase in explanatory power: 7% for stocks and 23% for bonds. Both the larger global component and the smaller increase from allowing nation-specific factors indicate that the EU is more integrated than the world generally.

The marginal explanatory power of each factor is measured in two ways. Table 3 presents the explanatory power of each factor used alone and the increase in explanatory power resulting from adding a factor to a model that already includes all the other factors.[5] Both measures of marginal explanatory power point to the market factor and shift factor as the main sources of risk for stocks and bonds, respectively. Used alone, the market factor explains

---

[4]Although Sweden is at present a member of the EU, we did not include it in our EU sample because it was not a member for most of the time period studied. The eight EU countries included in our sample are Belgium, Denmark, France, Germany, Italy, the Netherlands, Spain, and the United Kingdom. We use the term 'global model' regardless of the estimation universe. 'Regional model' would be a more accurate term in the case of the EU-only estimation universe.

[5]For a detailed discussion of this approach to measuring the power of individual factors, see Connor (1995).

Country	Number of Stocks	Number of Bonds	Size Mean	Size Standard Deviation	Value Mean	Value Standard Deviation	Stock Duration Mean	Stock Duration Standard Deviation	Shift Mean	Shift Standard Deviation	Twist Mean	Twist Standard Deviation
Australia	67.96	17.45	-1.700	1.155	1.327	1.269	29.433	25.192	3.730	1.577	-0.259	1.688
Belgium	46.21	18.29	-0.789	0.778	0.752	0.833	22.676	13.812	3.816	1.339	-0.190	1.433
Canada	95.26	42.51	-1.764	1.181	1.388	1.488	41.019	29.005	4.086	2.179	-0.608	2.304
Denmark	32.76	25.81	-1.045	0.979	1.396	1.155	47.859	23.692	2.460	1.751	0.965	1.691
France	105.49	26.67	-1.413	1.046	0.948	1.572	32.197	20.487	4.330	2.436	-0.716	2.621
Germany	59.91	73.93	-1.698	1.140	0.835	1.380	39.066	21.186	3.969	1.330	-0.314	1.381
Italy	70.28	36.54	-1.789	1.099	0.875	1.118	34.410	21.274	2.965	1.026	0.753	1.156
Japan	458.54	53.25	-1.954	1.063	1.169	0.616	60.919	14.186	5.244	2.022	-1.620	2.095
The Netherlands	31.57	38.96	-2.705	1.243	1.159	0.780	32.088	21.491	4.001	1.446	-0.414	1.410
Spain	36.44	12.15	-1.602	1.164	1.206	1.133	36.669	27.604	2.698	1.456	0.900	1.389
Sweden	26.03	7.00	-0.627	0.765	0.882	1.082	36.344	14.820	4.082	3.317	-0.215	2.451
United Kingdom	217.84	34.68	-2.087	1.181	1.240	1.593	24.206	12.414	5.164	2.392	-1.708	2.540
United States	511.22	159.10	-1.946	1.106	0.926	1.084	39.202	27.020	4.261	2.994	-0.790	.144

Note: The first two columns show the average number of stocks and bonds for each country across all months in the sample. The remaining columns show, across all months and all securities (either stocks or bonds), the means and standard deviations of the factor exposures (three for stocks and two for bonds) within each country. The market factor exposure is a 0/1 dummy, so the statistics are not shown (for each country, the market factor has a mean of 1 and a standard deviation of 0).

Table 1. Risk Exposures

Sample	Stocks		Bonds	
	Coefficient	Number of Months	Coefficient	Number of Months
Global				
Worldwide	0.214	112	0.603	112
European Union only	0.261	112	0.695	112
Nation Specific				
Worldwide	0.347	112	0.957	112
European Union only	0.332	112	0.923	112
Australia	0.384	112	0.971	85
Belgium	0.505	112	0.946	56
Canada	0.186	61	0.976	108
Denmark	0.375	112	0.954	99
France	0.419	112	0.960	92
Germany	0.490	112	0.942	88
Italy	0.542	112	0.916	39
Japan	0.448	112	0.942	98
The Netherlands	0.494	112	0.948	102
Spain	0.428	112	0.830	56
Sweden	0.520	112	NA	2
United Kingdom	0.406	112	0.898	91
United States	0.305	112	0.980	112

NA = not available.

Table 2. Explanatory Power of Global and Nation-Specific Models

18% of stock returns globally and 37% nationally, on average. Used alone, the shift factor explains 91% of bond returns nationally, on average, and 59% globally.

The stock duration and size factors have roughly equal importance for stock returns. Value is the weakest factor, explaining only 7% of stock returns globally and 19% nationally, on average, when used alone.

Relationships between the factors

The correlations between the factors provide an indication of the links between the types of pervasive risk. Table 4 shows these correlations for the global factors, and Table 5 shows the average correlations for nation-specific factors. In both tables, the correlations that are significantly different from

Model	Market	Size	Value	Stock Duration	Shift	Twist
			Factors Used Alone			
Global	0.184	0.143	0.073	0.153	0.586	0.308
Nation specific						
Australia	0.314	0.181	0.183	0.125	0.939	0.202
Belgium	0.452	0.260	0.141	0.388	0.924	0.272
Canada	0.116	0.119	0.093	0.104	0.943	0.379
Denmark	0.293	0.166	0.207	0.262	0.929	0.094
France	0.383	0.255	0.169	0.311	0.926	0.460
Germany	0.455	0.322	0.164	0.392	0.916	0.339
Italy	0.511	0.385	0.200	0.365	0.899	−0.035
Japan	0.402	0.351	0.325	0.381	0.931	0.668
The Netherlands	0.425	0.398	0.326	0.360	0.927	0.368
Spain	0.371	0.308	0.187	0.281	0.815	−0.093
Sweden	0.420	0.239	0.214	0.295	NA	NA
United Kingdom	0.370	0.321	0.160	0.280	0.864	0.561
United States	0.275	0.236	0.112	0.222	0.956	0.520
Average	0.368	0.272	0.191	0.290	0.914	0.311
			Factors Added to Model			
Global	0.028	0.006	0.005	0.015	0.294	0.016
Nation specific						
Australia	0.065	0.017	0.027	0.016	0.769	0.032
Belgium	0.077	0.017	0.001	0.030	0.674	0.022
Canada	0.022	0.018	0.023	0.009	0.597	0.032
Denmark	0.043	0.014	0.035	0.022	0.860	0.024
France	0.056	0.009	0.011	0.011	0.500	0.034
Germany	0.049	0.013	0.008	0.012	0.604	0.027
Italy	0.052	0.016	0.007	0.009	0.951	0.017
Japan	0.016	0.031	0.006	0.004	0.374	0.011
The Netherlands	0.036	0.020	0.024	0.015	0.580	0.021
Spain	0.065	0.020	0.022	0.009	0.923	0.015
Sweden	0.084	0.013	0.025	0.033	NA	NA
United Kingdom	0.038	0.016	0.009	0.005	0.337	0.031
United States	0.034	0.007	0.005	0.012	0.460	0.02
Average	0.019	0.016	0.016	0.014	0.627	0.02

NA = not available.

Table 3. Marginal Explanatory Power of each Factor used Alone and Added to Model using all other Factors

Factor	Market	Size	Value	Stock Duration	Shift	Twist
Market	1.000	0.097	−0.274*	−0.196	0.350	0.011
Size		1.000	−0.286	0.119	0.079	0.093
Value			1.000	0.223*	−0.259*	0.006
Stock duration				1.000	0.064	−0.134
Shift					1.000	−0.111
Twist						1.000

Significant at the 5% level.

Table 4. Correlations of the Global Model Factor Returns

zero with 95% confidence are marked with an asterisk.[6]

We observed a negative correlation between the market factor and value factor in stocks both globally and nationally. This negative correlation means that a value portfolio that is hedged against the other factor exposures will outperform when the market falls. The value factor and stock duration factor are positively correlated on a global level; the correlation is also positive on a national level but is not statistically significant. The market and stock duration factors are negatively correlated both nationally and globally.

The correlation between the two bond market factors is not economically meaningful because it is dependent on the 'midpoint' of the term structure defining the twist exposures. To fit the factor model best, the midpoint should be chosen so that this correlation is not too large in magnitude. A too-low value for the midpoint induces negative correlation between the shift and twist factors; a too-high value induces positive correlation. Because the global correlation is slightly negative and the average national correlation is slightly positive, the choice of four years as the midpoint seems appropriate.

Lastly, we considered the correlations between the stock market factors and the bond market factors. These correlations reflect interactions between two asset classes that are usually analyzed separately. As expected, we found that the market factor in stocks has a strong positive correlation with the shift factor in bonds. This comovement is observed homogeneously at both national and global levels. Bond and stock markets tend to move together,

---

[6]We used the standard large-sample approximation to the distribution of a time-series estimated correlation coefficient. That is, given that true correlation is zero, the estimated correlation coefficient is approximately normally distributed with mean equal to zero and standard error equal to $1/\sqrt{}$(the square root of the number of time-series observations). For example, with 112 months of data, to be significantly different from zero with 95% confidence, a correlation coefficient must have absolute value greater than 0.185. We used the same test for averages of correlation coefficients (as in Table 5), which is conservative because averaging tends to lower the true standard error.

Factor	Market	Size	Value	Stock Duration	Shift	Twist
Market	1.000	0.190*	−0.355*	−0.253*	0.408*	−0.043
Size		1.000	−0.079	0.027	0.058	−0.040
Value			1.000	0.179	0.038	0.038
Stock Duration				1.000	−0.020	0.024
Shift					1.000	0.015
Twist						1.000

Significant at the 5% level.

Table 5. Averages across Countries of the Correlations of the Factor Returns in the Nation-Specific Model.

and this interaction is captured by the correlation between their dominant risk factors.

It is revealing that the bond market shift factor and the stock market duration factor have a slight *negative* correlation, showing that theoretically defined stock duration has no useful relationship to interest rate sensitivity. For bonds, interest rate sensitivity and duration (value-weighted time to maturity) are empirically very close substitutes; for stocks, they are not.

Worldwide Integration of Security Market Risk

The results in Table 2 indicate a large global component to stock and bond market returns. They also reveal that the level of international integration is stronger across bond markets than across stock markets, and stronger within the EU than worldwide. An examination of the cross-country correlations of the factors from the nation-specific model allowed us to refine this analysis. If a particular factor is mostly 'global' in nature, then these cross-country correlations, shown in Tables 6 and 7, should be high.

Table 6 shows the average of the cross-country correlations for each factor and the percentage of these correlations that are positive. The market and shift factors have the highest average cross-national correlations at about 31% and 49%, respectively. All correlations between market factors are positive. For the shift factor, 92% of the correlations worldwide and 100% within the EU are positive.

As shown previously by Capaul, Rowley, and Sharpe (1993), the value factor is very nation specific. The average correlation is very close to zero, and only about half of the correlations are positive. The stock duration and size factors also appear to be mostly nation specific.

Factor	Worldwide Sample		European-Union-Only Sample	
	Average Correlation	Percentage Positive Correlations	Average Correlation	Percentage Positive Correlations
Market	0.306*	100.0	0.378*	100.0
Size	0.086	76.9	0.112	78.6
Value	0.007	55.1	−0.012	42.9
Stock duration	0.016	53.8	0.017	60.7
Shift	0.485*	92.3	0.599*	100.0
Twist	0.238*	78.2	0.309*	100.0

Significant at the 5% level.

Table 6. Average Correlation of each Factor across Countries in the Nation-Specific Model

For all factors but value, average correlation is higher and the percentage positive is higher within the EU than worldwide. This result confirms our earlier finding from Table 2 that EU capital markets are more integrated than those in the world at large.

We refined our analysis by examining the correlation of each country's factors with those of Germany, Japan, and the United States. The results are shown in Table 7. Germany and the United States were chosen to represent the core of two geopolitical blocs: the EU and North America. One would expect correlations between countries within these two geopolitical blocs to be higher than worldwide. Japan provides a counterpoint because it is a large economy not affiliated with either the EU or North American geopolitical bloc.

For the market and shift factors, all the correlations between Germany and other EU countries are positive and significant.

As expected from the results in Table 6, the correlations between value factors are low both within the EU and worldwide. Value is the only factor for which the correlation between Germany and France (the 'twin pillars' of the EU) is not significantly positive. The only case of a reliably positive value correlation might be between the United States and Canada. These two national markets are so closely integrated that even this (generally very local) stock market factor is correlated between them.

Correlations between market factors are generally high and significant. The US market factor has a significant positive correlation with the market factor from all countries except Sweden.

In contrast with the higher correlations generally observed between Germany and the EU and between the United States and Canada, the correlations

Factor	Australia	Belgium	Canada	Denmark	France	Germany	Italy	Japan	The Netherlands	Spain	Sweden	United Kingdom	United States
Market													
Germany	0.438*	0.378*	0.127	0.378*f	0.509	1.000	0.375*	0.148	0.374*	0.388*	0.177	0.345*	0.288*
Japan	0.061	0.215*	0.335*	0.108	0.309*	0.148	0.312*	1.000	0.136	0.375*	0.220*	0.275*	0.296
United States	0.332*	0.349*	0.382*	0.265*	0.417*	0.288*	0.285*	0.296*	0.412*	0.345*	0.174	0.635*	1.000
Size													
Germany	0.065	0.078	0.023	0.164	0.256*	1.000	0.056	0.038	0.191*	0.079	0.111	0.203*	0.019
Japan	-0.175	0.211*	0.153	-0.046	0.135	0.038	0.181	1.000	0.160	0.153	-0.040	0.199*	-0.139
United States	0.046	0.039	0.174	-0.154	0.218*	0.019	-0.054	-0.139	0.038	0.120	0.041	0.174	1.000
Value													
Germany	0.129	-0.008	-0.184	-0.022	-0.037	1.000	-0.169	0.043	0.071	-0.072	0.197*	-0.120	-0.112
Japan	-0.020	-0.131	0.087	-0.149	-0.023	0.043	-0.079	1.000	0.037	0.018	0.006	-0.057	0.022
United States	-0.051	0.003	0.281*	0.000	0.263*	-0.112	0.052	0.022	0.230*	0.090	0.024	0.147	1.000
Stock duration													
Germany	0.105	-0.080	-0.098	0.107	0.267*	1.000	0.083	0.131	-0.133	0.150	0.149	0.025	0.149
Japan	0.059	-0.048	0.090	-0.051	-0.0112	0.131	0.101	1.000	-0.036	0.000	-0.005	0.151	0.076
United States	0.159	0.028	0.062	-0.075	0.053	0.149	-0.078	0.076	-0.056	-0.021	-0.123	-0.001	1.000
Shift													
Germany	0.387*	0.774*	0.403*	0.583*	0.783*	1.000	0.561*	0.522*	0.960*	0.394*	—	0.625*	0.458*
Japan	0.165	0.527*	0.397*	0.248*	0.418*	0.522*	0.360*	1.000	0.511*	0.129	—	0.497*	0.470*
United States	0.367*	0.289*	0.688*	0.283	0.397*	0.458*	0.173	0.470*	0.450*	0.116	—	0.441*	1.000
Twist													
Germany	0.155	0.570*	-0.068	0.172	0.515*	1.000	0.186	0.103	0.566*	0.250	—	0.259*	0.048
Japan	0.142	0.062	0.080	0.155	0.317*	0.103	-0.149	1.000	0.133	0.014	—	0.161	0.032
United States	0.408*	-0.053	0.243*	0.082	-0.049	0.047	-0.092	0.032	0.094	-0.146	—	0.197	1.000

*Significant at the 5% level.

Table 7. Correlations of each Country's Factors with those from Germany, Japan, and the United States

with Japan are generally lower than average. For the market factor, 8 out of 12 correlations with Japan are below the worldwide average. For the shift and twist factors, 7 and 10 out of 11 correlations, respectively, are below the worldwide average.

The evidence from Tables 6 and 7 confirms our previous findings from Table 2 that the EU stock and bond markets form a geopolitical bloc within which integration is stronger than worldwide. The same holds for the United States and Canada. We found (again as in Table 2) that bond market factors are more international than stock market factors. Most of the international correlation of stocks comes from the positive correlations of the market factors. The value factor is notable for its particularly weak cross-national correlations.

Exchange Rate Returns and Stock and Bond Market Returns

In the analysis above, stock and bond returns are measured in local currencies and in excess of the local risk-free return. Working with local-currency excess returns is desirable for two reasons. First, it removes the distorting effects of currency changes and so allows clearer analysis of the worldwide comovements in stock and bond market returns. Second, it conforms to best practice, which views the currency decision as an independent 'overlay' on stock and bond decisions made in fully hedged terms (e.g., see Gastineau 1995). (Because of covered interest parity, excess local-currency returns are equivalent to fully hedged returns.)

We rounded out the analysis by relating the stock and bond market risk factors to exchange rate movements. For each country, we calculated the realized local-currency return to a basket of foreign currencies, with the basket weights proportional to each foreign country's national income. We call this amount the foreign exchange return.[7]

Table 8 shows the correlations between each country's foreign exchange return and its domestic stock and bond market factors. Only 18 of the 76 correlations are significant, and the correlations across countries or across factors have no consistent pattern. France, Japan, and the United States show the most significant coefficients, but the signs and magnitudes of the coefficients are too variegated to suggest any meaningful interpretation. As a general conclusion from Table 8, the relationship between the stock and

---

[7]For each month in a given year, we used the previous year's gross domestic product, measured in US dollars at the end-of-December exchange rate. Gastineau (1995) suggests adding the difference between the local and foreign risk-free returns to the currency returns. For correlation analysis, including the difference between the risk-free returns has no meaningful effect on the results. For simplicity, therefore, we show the results using pure currency returns and ignoring differences in risk-free returns.

Country	Market	Size	Value	Stock Duration	Shift	Twist
Australia	−0.291*	−0.072	0.002	−0.230*	−0.067	0.194
Belgium	0.215*	−0.225*	−0.067	0.000	0.219	−0.007
Canada	0.172	0.190	−0.241	−0.201	−0.173	−0.213
Denmark	0.269*	0.063	−0.172	−0.050	0.036	0.014
France	0.138	−0.060	0.187*	0.125	0.257*	0.296
Germany	0.118	−0.144	0.157	0.096	0.082	0.115
Italy	0.028	0.104	−0.041	−0.142	−0.026	0.059
Japan	−0.227	0.003	0.360*	0.189*	−0.338*	−0.044
The Netherlands	0.226*	−0.111	0.147	0.120	0.062	0.051
Spain	0.102	0.054	−0.127	0.051	0.130	0.128
Sweden	0.056	0.012	0.144	0.008	NA	NA
United Kingdom	0.095	−0.086	−0.044	0.029	0.049	0.326*
United States	−0.042	0.230*	−0.214*	−0.109	0.200*	0.152
Average	0.066	−0.003	0.007	−0.009	0.036	0.089

Significant at the 5% level.
NA = not available.

Table 8. Correlations between Foreign Exchange Returns and National Factor Returns

bond market factors and foreign exchange returns showed few, if any, reliable patterns. This finding justifies the approach taken throughout this study: the currency investment decision is best viewed separately from the stock and bond selection decision. Because local stock and bond market returns and currency movements have no reliable patterns of correlation, the global stock and bond decisions are best analyzed in local currency terms, and the currency decision should be studied separately.

Conclusion

The global portfolio management problem is multi-dimensional. It includes decisions on country allocation; stock versus bond allocation; and allocation across asset risk characteristics such as duration, size, and value, and individual asset selection.

The main attraction of a global stock and bond model is that it can help harmonize the investment decision process across all those dimensions. Such a model can provide a common framework for measuring risk and predicting return. Different types of assets in various countries can be analyzed within this single framework, and the work of country analysts, local portfolio managers and the asset allocation committee thereby can be made more coherent.

Using six simple and intuitive factors, we built a worldwide factor model of stock and bond returns. The global version of the model explains 21% of the return of the typical stock and 60% of the return of a typical bond. The two dominant factors, shift for bonds and the market factor for stocks, also happen to be the most global influences on stocks and bonds. Global asset allocation should weigh the global influence of these two factors against the diversifying potential of more local influences such as value, size, or twist.

Portfolio managers should take account of geopolitical blocs, within which a higher level of integration holds than worldwide. The two geopolitical blocs examined in this study are the EU and North America (represented by Canada and the United States). Within each of these blocs, cross-national correlations of the factors are higher than between blocs, so that a more 'regional' perspective is appropriate, particularly within bond markets.

Currency returns have no stable, reliable patterns of correlation with stock and bond market factor returns. The currency investment decision is best studied independently from the stock and bond investment decision because the influences on currency returns are quite distinct from those on stock and bond returns.

Appendix

We estimated the factor returns using a standard procedure. For each month, we regressed the cross-section of asset excess returns against the cross-section of factor exposures; the resulting regression coefficients are the factor returns for that month. We estimated the regressions separately for stocks and bonds. For the global model, the regression equations for stocks and bonds are

$$r_i = 1 \times f_{\text{mkt}} + \text{SIZE}_i \times f_{\text{size}} + \text{VALUE}_i \times f_{\text{value}} + \text{SDUR}_i \times f_{\text{sdur}} + \varepsilon_i$$
$$\text{for stocks;}$$
$$r_i = \text{SHIFT}_i \times f_{\text{shift}} + \text{TWIST}_i \times f_{\text{twist}} + \varepsilon_i \quad \text{for bonds.}$$

For the worldwide sample, these cross-sectional regressions include all stocks and bonds for which we have data in a given month, whereas for the EU-only sample, the regressions are limited to those securities in the eight EU member states.

For the nation-specific model, we allowed the factor returns to differ among each of the C countries in the sample (in particular, $C = 13$ for the worldwide sample and $C = 8$ for the EU sample):

$$r_i = \sum_{c=1}^{C} d_{i,c}(1 \times f_{\text{mkt},c} + \text{SIZE}_i \times f_{\text{size},c} + \text{VALUE}_i \times f_{\text{value},c}$$
$$+ \text{SDUR}_i \times f_{\text{sdur},c} + \varepsilon_i \quad \text{for stocks;}$$
$$r_i = \sum_{c=1}^{C} d_{i,c}(\text{SHIFT}_i \times f_{\text{shift},c} + \text{TWIST}_i \times f_{\text{twist},c} + \varepsilon_i \quad \text{for bonds.}$$

The dummy variable $d_{i,c}$ equals 1 if stock or bond i is in country c, and 0 otherwise.

The explanatory power is 1 minus the ratio of the average unexplained variance to average total variance; thus,

$$\text{EP} = 1 - (\text{DF})\frac{\sum_{i=1}^{N}\sum_{t=1}^{T}\hat{\varepsilon}_{i,t}^2}{\sum_{i=1}^{N}\sum_{t=1}^{T}(r_{i,t}-\bar{r}_i)^2}$$

where

$$\bar{r}_i = \frac{1}{T}\sum_{t=1}^{T}r_{i,t}.$$

The variable DF is a degrees of freedom correction that equals the total number of observations divided by the total number of observations minus the number of estimated factor returns. For asset return factor models, this measure of explanatory power is superior to the more conventional R^2 measure, as pointed out in earlier work (see Connor 1995 and Beckers $et\ al.$ 1996). We also calculated explanatory power for each individual country, using the estimates from the nation-specific model and restricting the cross-sectional sample to the stocks or bonds in a single country.[8]

References

BARRA. 1991. The Global Bond Risk Model, Technical Documentation. (January).

Beckers, Stan, Gregory Connor and Ross Curds. 1996. 'National versus Global Influences on Equity Returns,' *Financial Analysts Journal* **52** (2) (March/April), 31–39.

Bernstein, Richard and Bernard Tew. 1991. 'The Equity 'Yield Curve',' *Journal of Portfolio Management* **18** (1) (Fall), 35–39.

Capaul, Carlo, Ian Rowley and William F. Sharpe. 1993. 'International Value and Growth Stock Returns,' *Financial Analysts Journal* **49** (1) (January/February), 27–36.

Connor, Gregory. 1995. 'The Three Types of Factor Models: a Comparison of Their Explanatory Power,' *Financial Analysts Journal* **51** (3) (May/June), 42–46.

Elton, Edwin J. and Martin J. Gruber. 1995. *Modern Portfolio Theory and Investment Analysis*, 5th ed., Wiley.

Fama, Eugene F. and Kenneth R. French. 1993. 'Common Risk Factors in the Returns on Stocks and Bonds,' *Journal of Financial Economics* **33** (1) (February), 3–56.

---

[8]We would like to thank Stan Beckers and Ronald Kahn for helpful comments and Sam Wai-Cheng Lam for research assistance.

Gastineau, Gary L. 1995. 'The Currency Hedging Decision: a Search for Synthesis in Asset Allocation,' *Financial Analysts Journal* **51** (3) (May/June), 8–17.

Kahn, Ronald. 1995. 'Fixed Income Risk Modeling.' In *The Handbook of Fixed Income Securities*, 4th ed., edited by Frank J. Fabozzi and T. Dessa Fabozzi. Business One Irwin, 720–32.

Macaulay, F.R. 1938. *Some Theoretical Problems Suggested by the Movements of Interest Rates, Bond Yields, and Stock Prices in the United States since 1856*. Columbia University Press.

Sharpe, William F. 1992. 'Asset Allocation: Management Style and Performance Measurement,' *Journal of Portfolio Management* **18** (2) (Winter), 7–19.

Part IV
Dynamic Portfolio Models for Asset Allocation

On Timing The Market: the Empirical Probability Assessment Approach with an Inflation Adapter

Robert R. Grauer and Nils H. Hakansson

Abstract

Recent studies have documented varying degrees of predictability and mean reversion in stock returns but the question of how they might be exploited remains relatively open. This article applies dynamic portfolio theory to the construction and rebalancing of portfolios principally composed of stocks and cash or borrowing. Probability assessments were based on (all moments of) recent past returns, both in raw form and with an inflation adapter. The inflation adapter had little impact prior to the mid-sixties but markedly improved realized portfolio returns over the 1966–88 subperiod. Some excess returns exceeded one percent per quarter but the various performance tests applied reached contradictory conclusions concerning over which period the null hypothesis should be rejected.

1 Introduction

Recent studies have documented varying degrees of predictability of stock returns as well as mean-reversion of returns over intermediate to long holding periods (see e.g. Keim and Stambaugh (1986), French, Schwert, and Stambaugh (1987), Fama and French (1988a,b), and Poterba and Summers (1988))[1]. This raises the question of whether and how such patterns in returns might be exploited by a price-taking investor. The purpose of this article is to examine the extent to which (all moments of) recent past returns are useful (or not useful) in revising ones's portfolio while moving forward in time on the basis of the discrete-time dynamic investment model[2].

---

[1]Some schools of practitioners have long subscribed to the view that stock prices are at least partially predictable from past prices and related summary statistics via what is commonly called 'technical analysis'. While generally pooh-poohed by academics (see e.g. Malkiel (1990)), a recent study employing two simple and popular technical trading rules provides formal evidence in support of their value – see Brock, Lakonishok, and LeBaron (1992).

[2]For a review of alternative approaches, including stochastic programming, to the dynamic investment problem, see Mulvey and Ziemba (1995).

In earlier papers, Grauer and Hakansson (1982, 1984, 1986) applied the dynamic portfolio theory of Mossin (1968), Hakansson (1971, 1974), Leland (1972), Ross (1974), and Huberman and Ross (1983) in conjunction with the empirical probability assessment approach (EPAA) to construct and rebalance portfolios composed of US stocks, corporate bonds, government bonds, and a risk-free asset. Borrowing was ruled out in the first article, while margin purchases were permitted in the other two. The probability distributions used were naively estimated from *past* realized returns in the Ibbotson and Sinquefield and Ibbotson data bases, and both annual and quarterly holding periods were employed from the mid-thirties forward. The results of all three papers revealed that the gains from active diversification among the major asset categories were substantial, especially for the highly risk-averse strategies. (Two other papers, Grauer and Hakansson (1987) and Grauer and Hakansson (1995b), document even larger gains from international diversification and from the inclusion of real estate into the investment universe; see also Jorion (1989).) In addition, they found evidence of substantial use of, and gains from, margin purchases for the more risk-tolerant strategies from the mid-thirties to the mid-sixties. The third paper also showed that small stocks, while sometimes totally ignored, entered even the most risk-averse portfolios most of the time.

The empirical probability assessment approach may be modified by correcting for estimation error either in the means or in the variances (or both). In Grauer and Hakansson (1995a), a James–Stein, a Bayes–Stein, and a third estimator for adjusting the means were employed with mixed results compared to the no adjustment case. In Grauer and Hakansson (1995b), the desmoothing of the real estate input series resulted in a modest improvement.

In the present article, the raw joint empirical distribution approach to generating probability assessments is refined by the inclusion of an inflation adapter. The inflation adapter at any point in time is based on a simple regression of past returns on inflation. Specifically, the difference between the observed risk-free lending rate in the coming period and its average over the estimating period multiplied by the estimated inflation sensitivity coefficient is added to the raw probability distribution for each asset in the period. The effect is to change the projected mean returns, leaving all other moments unchanged.

Use of the inflation adapter substantially changed the portfolios selected, especially for the more risk-tolerant strategies during the highly inflationary 1966–82 sub-period. Superficially, the strategies give the appearance of intensified 'market timing' activities. In return space, the inflation adapter also led to uniformly higher geometric means and lower standard deviations of realized returns.

To minimize the noise factor in the performance measurement process, the model's portfolio selection possibilities is reduced to the two asset case (stocks and cash) with quarterly revision, for the most part. Stocks are represented by the Center for Research in Security Prices (CRSP) value-weighted index or market portfolio. Cash means either a long position in 90-day Treasury bills or borrowing at the call money rate +1%. A comparison with the performance of 130 mutual funds, and with the performance of active strategies generated from an expanded investment universe composed of long-term US government and corporate bonds, the S&P 500 index, and an index of small stocks, is also made for the 1968–82 period.

In all, seven tests of performance were employed: the Jensen test, the Henriksson–Merton (HM) test, the Treynor–Mazuy (TM) test, the paired *t*-test, and variants thereof. Overall, there is considerable evidence of significant positive abnormal returns or market timing ability with, and to some extent without, the inflation adapter. This is somewhat remarkable since only *past* realized returns were used as inputs. For example, based on the Jensen test, each of the risk tolerances that employed the inflation adapter yielded significant abnormal returns over the full 1934–88 period, as well as over the 1966–88 sub-period when the fourth quarter of 1987 was excluded. The market timing tests indicated that there was highly significant positive market timing ability in any sub-period beginning in 1966 that excluded the fourth quarter of 1987. Similarly, the paired *t*-tests showed that a number of the active strategies earned significantly higher returns than selected benchmarks over the 1934–88 period, as well as over the 1966–88 sub-period when the fourth quarter of 1987 was excluded from the sample.

2   Theory

Despite explosive development over three decades, and extensive application to the construction of equity portfolios, modern portfolio theory has found only modest use in the larger portfolio context, often referred to as the asset allocation problem – the choice of the proportions to be held in the major categories of common stocks and in different types of bonds, money market instruments, real estate, and foreign securities. There are several reasons for this. First, extant portfolio theory, being principally based on the mean-variance model, is single-period in nature, whereas the asset allocation problem accents the multiperiod, sequential nature of investment decisions[3]. On top of this, since the universe of interest extends well beyond common stocks, extant betas are too narrowly defined to be useful, and the appropriate betas are not easily estimated because of data problems concerning the market weights of bonds, for example. At the other extreme, continuous-time

---

[3]Tests showing the superiority of dynamic models appear in Carino, Kent, Myers, Stacy, Sylvanus, Turner, Watanabe, and Ziemba (1994).

portfolio theory is somewhat intractable in a world of non-trivial transactions costs, although discrete-time approximation applications have been made (see e.g. Brennan and Schwartz, this volume). Finally, many extant models rely heavily on narrow classes of theoretical (stationary) return distributions, with limited ability to capture the richness of joint, real-world stochastic processes.

There is, however, a middle category of investment models, usually classified under the heading of discrete-time dynamic portfolio theory, which has been largely ignored in portfolio selection applications. This is so despite the fact that these models have a strong foundation in theory and lend themselves naturally to the problem of rebalancing portfolios over many periods (up to 220 quarters in the present study). An additional virtue of these models is that they can handle general nonstationary return distributions.

To review, consider the simplest reinvestment problem, in which the market is perfect and returns are independent over time but otherwise arbitrary and not necessarily stationary. The investor has a preference function U_0 (with $U_0' > 0$, $U_0'' < 0$) defined on wealth w_0 at some terminal point (time 0). Let w_n denote the investor's wealth with n periods to go, r_{in} the return on asset i in period n, z_{in} the amount invested in asset i in period n (with $i = 1$ being the safe asset), and $U_n(W_n)$ the relevant (unknown) utility of wealth with n periods to go. At the end of period n (time $n - 1$), the investor's wealth is

$$w_{n-1}(z_n) = \sum_{i=2}^{M}(r_{in} - r_{1n})z_{in} + w_n(1 + r_{1n}),$$

where $z_n = (z_{1n}, \ldots, z_{Mn})$ and M is the number of securities.

Consider the portfolio problem with one period to go. The investor, with w_1 to invest, must solve

$$\max_{z_1|w_1} E\left[U_0\left(w_0\left(z_1\right)\right)\right] \equiv U_1\left(w_1\right).$$

Clearly, $U_1(w_1)$ represents the highest attainable expected utility level from capital level w_1 at time 1, and thus the 'derived' utility of w_1. Employing the induced utility function $U_1(w_1)$, the portfolio problem with two periods to go becomes

$$U_2(w_2) \equiv \max_{z_2|w_2} E\left[U_1\left(w_1(z_2)\right)\right].$$

Thus, with n periods to go, we obtain (the recursive equation)

$$U_n(w_n) = \max_{z_n|w_n} E\left[U_{n-1}\left(w_{n-1}(z_n)\right)\right], \quad n = 1, 2, \ldots .$$

Examining the above system, it is evident that the induced utility of current wealth, $U_n(w_n)$, generally depends on 'everything', namely the terminal utility function U_0, the joint distribution functions of future returns, and future interest rates. There is, however, a special case in which $U_n(w_n)$ depends only on U_0 . This occurs [Mossin (1968)] if and only if $U_0(w_0)$ is isoelastic, i.e., if and only if

$$U_0(w_0) = \frac{1}{\gamma}w^\gamma, \quad \gamma < 1.$$

(Note that for $\gamma = 0$, $U_0(w_0) = \ln w_0$.) $U_n(w_n)$ is now a positive linear transformation of $U_0(w_n)$, i.e., we can write

$$U_n(w_n) = \frac{1}{\gamma}w_n^\gamma. \tag{1}$$

For these preferences, the optimal investment policy $z_{n\gamma}^*(w_n)$ is proportional to wealth, i.e.,

$$z_{in\gamma}^*(w_n) = x_{in\gamma}^* w_n, \quad \text{all } i, \tag{2}$$

where the $x_{in\gamma}^*$ are constants. It is also completely *myopic* since it only depends on U_0 and the current period's return structure and *not* on returns beyond the current period. Both of these properties hold only for the family 1, which is also the only class of preferences exhibiting constant relative risk aversion[4]. Finally, 2 also implies that the utility of wealth relatives, $V_n(1+r_n)$, is of the same form only for this family, i.e.,

$$U_n(w_n) = \frac{1}{\gamma}w^\gamma \iff V_n(1 + r_n) = \frac{1}{\gamma}(1 + r_n)^\gamma.$$

While the above properties are interesting, they are clearly rather special. However, the isoelastic family's influence extends far beyond its numbers. As shown by Hakansson (1974) (see also Leland (1972), Ross (1974), and Huberman and Ross (1983)), there is a very broad class of terminal utility functions $U_0(w_0)$ for which the induced utility functions U_n converge to an isoelastic function, i.e., for which

$$U_n(w_n) \longrightarrow \frac{1}{\gamma}w^\gamma, \quad \text{for some } \gamma < 1. \tag{3}$$

Hakansson (1974) has also shown that 3 is usually accompanied by convergence in policy, i.e.,

$$z_n^* \longrightarrow x_{n\gamma}^* w_n.$$

Thus, the objectives given by 1 are quite robust and encompass a broad variety of different goal formulations for investors with intermediate to long-term investment horizons[5]. In particular, class 1 spans a continuum of risk attitudes all the way from risk neutrality ($\gamma = 1$) to infinite risk aversion ($\gamma = -\infty$).[6]

---

[4]This measure is defined as $-wU_n''(w)/U_n'(w)$ and equals $1 - \gamma$ for the class 1.

[5]The simple reinvestment formulation does ignore consumption of course.

[6]A plot of the functions $\frac{1}{\gamma}(1 + r)^\gamma$ for several values of γ was given in Grauer and Hakansson (1982, p. 42).

Having selected our model, we turn next to what we need to operate it. The major input to the model is an estimate of next period's *joint* return distribution for the risky asset categories[7]. In several previous studies, we based this estimate on the empirical probability assessment approach (EPAA). In this approach, the realized returns of the most recent n periods are recorded; each of the n *joint* realizations is then assumed to have probability $1/n$ of occurring in the coming period. Thus, estimates were obtained on a moving basis and used in raw form without adjustment of any kind. On the other hand, since the whole joint distribution was specified and used, there was no information loss; all moments and correlations were taken into account. The empirical distribution of the past n periods is optimal if the investor has no information about the form and parameters of the true distribution, but believes that this distribution went into effect n periods ago, see Bawa, Brown and Klein (1979, p. 160).

3   The Inflation Adjustment

It has been well documented that asset returns are sensitive to the rate of inflation, e.g. Coleman (1966), Fama (1975), Fama and Schwert (1977). In particular, there is strong evidence that US Treasury bills are a perfect hedge against anticipated inflation and that returns on long-term bonds and common stocks are negatively related to at least unanticipated inflation.

In this article, the empirical probability assessment approach is modified by the addition of an inflation adapter. This adapter is designed to adjust the raw distribution for anticipated inflation, as reflected in the three-month Treasury bill rate, at the time of investment, t.

Specifically, let $r_{i\tau}$ be the realized return on asset category i in period τ, $I_{\tau-}$ the realized inflation rate in the (three-month) period ending one month prior to the end of period (calendar quarter) τ (since the inflation rate is published with a lag), and r_{Lt} the risk-free lending rate in period t, for which the decision is to be made. The regression

$$r_{i\tau} = a_i^t + b_i^t I_{\tau-} + e_{i\tau}$$

is run for each i over the estimating period $t - n$ to $t - 1$ to obtain the 'rolling' estimated coefficients $\hat{a}_i^t$ and $\hat{b}_i^t$. We take as our base for estimating inflation's impact over the next period the difference between the observed risk-free lending rate, r_{Lt}, and its average, $\bar{r}_{Lt}$, over the estimating period, i.e., $r_{Lt} - \bar{r}_{Lt}$. The quantity

$$\hat{b}_i^t(r_{Lt} - \bar{r}_{Lt})$$

---

[7]For a comprehensive overview of the issues and problems associated with the estimation of return distributions see Bawa, Brown and Klein (1979).

is then added to the raw probability distribution for each asset i in period t. The effect is to change the anticipated mean returns, leaving all other moments unchanged.

4 Calculations

The model used can be summarized as follows. At the beginning of each period t, the investor chooses a portfolio, x_t, on the basis of some member, γ, of the family of utility functions for returns r given by

$$V(1+r) = \frac{1}{\gamma}(1+r)^\gamma. \tag{4}$$

This is equivalent to solving the following nonlinear programming problem in each period t:

$$\max_{x_t} \mathrm{E}\left[\frac{1}{\gamma}(1+r_t(x_t))^\gamma\right] = \max_{x_t} \sum_s \pi_{ts}\frac{1}{\gamma}(1+r_{ts}(x_t))^\gamma \tag{5}$$

subject to

$$x_{it} \geq 0, \quad x_{Lt} \geq 0, \quad x_{Bt} \leq 0, \qquad \text{all } i, \tag{6}$$

$$\sum_i x_{it} + x_{Lt} + x_{Bt} = 1, \tag{7}$$

$$\sum_i m_{it} x_{it} \leq 1, \tag{8}$$

$$\Pr(1 + r_t(x_t) \geq 0) = 1, \tag{9}$$

where

$r_{ts}(x_t) = \sum_i x_{it} r_{its} + x_{Lt} r_{Lt} + x_{Bt} r_{Bt}^d$, is the *(ex ante)* return on the portfolio in period t if state s occurs,

$\gamma \leq 1 =$ a parameter that remains fixed over time,

$x_{it} =$ the amount invested in risky asset category i in period t as a fraction of own capital,

$x_t = (x_{1t}, \ldots, x_{nt}, x_{Lt}, x_{Bt})$,

$r_{it} =$ the anticipated total return (dividend yield plus capital gains or losses) on asset category i in period t,

$r_{Lt} =$ the return on the risk-free asset in period t,

$r_{Bt}^d =$ the borrowing rate at the time of the decision at the beginning of period t,

m_{it} = the initial margin requirement for asset category i in period t expressed as a fraction, and

π_{ts} = the probability of state s at the end of period t, in which case the random return r_{it} will assume the value r_{its}.

Constraint 6 rules out short sales and 7 is the budget constraint. Constraint 8 serves to limit borrowing (when available) to the maximum permissible under the margin requirements that apply to the various asset categories. Finally, constraint 9 rules out any (*ex ante*) probability of bankruptcy. On the basis of the probability estimation method described earlier, the (sequential) solution of the portfolio problem may be described as follows. Suppose quarterly revision is used. Then, at the beginning of quarter t, the portfolio problem 5–9 for that quarter uses the following inputs – the (observable) risk-free return for quarter t, the (observable) call money rate $+1\%$ at the beginning of quarter t, and the (observable) realized returns for each of the risky assets for the previous n quarters, and the realized inflation rates for the previous n quarters. Each joint realization (whether inflation adjusted or not) in quarters $t - n$ through $t - 1$ is given probability $1/n$ of occurring in quarter t.

With these inputs, the portfolio weights for the various asset categories and the proportion of assets borrowed are calculated by solving the nonlinear programming system 5–9; the algorithm employed is described in Best (1975). At the end of quarter t, the realized returns on each of the risky assets are observed, along with the *realized* borrowing rate r_{Bt} (which may differ from the decision borrowing rate r_{Bt}^d)[8]. Then, using the weights selected at the beginning of the quarter, the realized return on the portfolio chosen for quarter t is recorded. The cycle is then repeated in all subsequent quarters[9].

All reported returns are gross of transaction costs and taxes and assume that the investor in question had no influence on prices. There are several reasons for this approach. First, we wish to follow precedent and keep the complications to a minimum. Second, the return series used as inputs and for comparisons also excludes transaction costs (for reinvestment of dividends) and taxes. Third, many investors are tax-exempt and various techniques are available for keeping transaction costs low. Finally, since the proper treatment of these items is nontrivial, they are better left to a later study.

---

[8]The realized borrowing rate r_{Bt} was calculated as the average of the monthly realized rates.

[9]If $n = 32$ under quarterly revision, then the first quarter for which a portfolio can be selected is $b + 32$, where b is the first quarter for which data is available.

5 Data

The risk-free asset was assumed to be 90-day US Treasury bills maturing at the end of the quarter; we used the *Survey of Current Business* and *The Wall Street Journal* as sources. The total returns on the value-weighted market portfolio were obtained from the monthly Center for Research in Securities Prices (CRSP) data file. Margin requirements for stocks were obtained from the *Federal Reserve Bulletin*[10].

As noted, the borrowing rate was assumed to be the call money rate +1%; for *decision* purposes (but not for rate of return calculations), the applicable beginning of period rate, r_{Bt}^d, was viewed as persisting throughout the period and thus as risk-free. For 1934–76, the call money rates were obtained from the *Survey of Current Business*; for later periods, *The Wall Street Journal* was the source.

As a benchmark against which to judge the performance of the active policies, we examined the performance of 130 open-ended mutual funds over a common time period using quarterly data from January 1968 to December 1982. The returns data include all dividends paid by the funds and are net of all management costs and fees. The data set, provided by David Modest and Bruce Lehmann, updates Henriksson's (1984) data base by two years and one month and includes an additional fourteen funds. (See Lehmann and Modest (1987) and Henriksson (1984) for more detailed descriptions of the data base.) For comparative purposes, we also examined the performance of the active strategies generated from an investment universe that included long-term US government bonds, long-term US corporate bonds, the S&P 500 index, and an index of small stocks. The source of this data set was *Stocks, Bonds, Bills, and Inflation 1989 Yearbook* published by Ibbotson Associates, Inc.

6 Results

Because of space limitations, only a portion of the results can be reported here. However, Tables 1 through 8 and Figures 1 through 4 provide a fairly representative sample of our findings.

---

[10]There was no practical way to take maintenance margins into account in our programs. In any case, it is evident from the results that they would come into play only for the more risk-tolerant strategies, even for them only occasionally, and that the net effect would be relatively neutral.

6.1   The Portfolio Returns

Table 1 shows the geometric means and standard deviations[11] of the realized annual returns for 16 strategies corresponding to γ's in 1 ranging from -75 (extremely risk averse) to 1 (risk neutral), with and without the inflation adapter, for the 55-year period 1934–88. The estimating period was 32 quarters. Recall that only two assets could be chosen in each period, the CRSP value-weighted index and a risk-free asset (Treasury bills). Panel A shows the results when no borrowing was permitted, while Panel B reports the returns when margin purchases were allowed. Finally, Panel C shows the return characteristics of various benchmarks: risk-free lending (RL), the CRSP value-weighted index (VW), inflation, and a set of fixed-weight (rebalancing) portfolios. Thus, V4 represents a portfolio which is always rebalanced to 40% in the index (VW) and 60% in risk-free lending RL at the beginning of each period. Similarly, V18 always invests 180% of its capital in the index by borrowing 80%, unless margin requirements put a lower cap on borrowed funds.

Figure 1 plots the geometric means and standard deviations of the realized annual returns for risk-free lending, inflation, and the value-weighted CRSP index (see squares), for the up and down-levered value-weighted CRSP index (see triangles), and, *for the borrowing case only*, for the 16 powers with the inflation adapter (see black dots), and for the 16 active strategies without the inflation adapter (see diamonds).

As Figure 1 shows (for the period 1934–88 when the compound inflation rate was 4.10 percent per annum), the benchmarks marginally outperformed the more risk-averse active strategies, while the less risk-averse active strategies clearly did better than the fixed-weight strategies. Moreover, Table 1 shows that in the inflation adapter case, with borrowing precluded, the nonnegative powers attained higher geometric mean returns than the market with less standard deviation. Furthermore, the returns with and without the inflation adapter were quite similar. However, with the exception of the -2 and -3 power strategies with borrowing permitted, the returns generated by the inflation adjusted strategies strictly 'dominated' the unadjusted strategies in the sense that each inflation adjusted strategy (with or without borrowing) earned a higher geometric mean return with equal to or less standard deviation than the corresponding unadjusted strategy.

Table 2 and Figure 2 show the results for the 1966–88 sub-period, which was characterized by a compound inflation rate of 5.96 percent per annum. In this period five observations stand out. First, the active strategies clearly dominated the benchmarks. Second, the differences between using and not

---

[11]The table reports the standard deviation of the log of unity plus the rate of return. This quantity is very similar to the standard deviation of the rate of return for return levels less than 25%.

Table 1. Geometric Means and Standard Deviations of Annual Returns for sixteen Power Policies with and without the Inflation Adapter, 1934–1988 Treasury Bills and CRSP Value-Weighted Index (Quarterly portfolio revision, 32–quarter estimating period)

With Inflation Adapter			Without Inflation Adapter		
Portfolio	Geom. Mean	Std. Dev.	Portfolio	Geom. Mean	Std. Dev.
Panel A: Borrowing Precluded					
Power −75	4.33	3.45	Power −75	4.32	3.45
Power −50	4.56	3.60	Power −50	4.54	3.62
Power −30	5.01	4.13	Power −30	4.97	4.19
Power −20	5.53	5.06	Power −20	5.48	5.17
Power −15	6.04	6.08	Power −15	6.00	6.18
Power −10	6.89	7.69	Power −10	6.82	7.97
Power −7	7.54	8.94	Power −7	7.48	9.12
Power −5	8.31	9.78	Power −5	8.22	9.86
Power −3	9.32	10.44	Power −3	9.19	10.66
Power −2	10.03	10.67	Power −2	9.95	11.07
Power −1	10.66	11.12	Power −1	10.54	11.66
Power 0	11.40	11.98	Power 0	11.05	12.58
Power .25	11.52	12.23	Power .25	11.13	12.90
Power .5	11.79	12.90	Power .5	11.27	13.72
Power .75	11.72	14.22	Power .75	10.92	14.95
Power 1	12.12	14.36	Power 1	11.22	15.10
Panel B: Borrowing Permitted					
Power −5	8.81	11.57	Power −5	8.61	11.93
Power −3	10.02	14.16	Power −3	10.09	14.29
Power −2	10.85	15.29	Power −2	11.04	15.31
Power −1	12.26	16.83	Power −1	12.04	17.12
Power 0	14.27	18.35	Power 0	13.78	18.78
Power .25	14.74	19.38	Power .25	14.19	19.78
Power .5	15.11	21.32	Power .5	14.57	22.01
Power .75	15.73	23.85	Power .75	15.21	24.79
Power 1	15.82	27.58	Power 1	15.57	28.05
Panel C: Benchmarks					
RL	3.86	3.44	V12	11.74	20.45
V2	5.54	4.18	V14	12.18	24.11
V4	7.12	6.88	V16	12.46	27.87
V6	8.58	10.08	V18	12.58	31.74
V8	9.93	13.44	V20	12.51	35.76
VW	11.16	16.89	Inflation	4.10	3.73

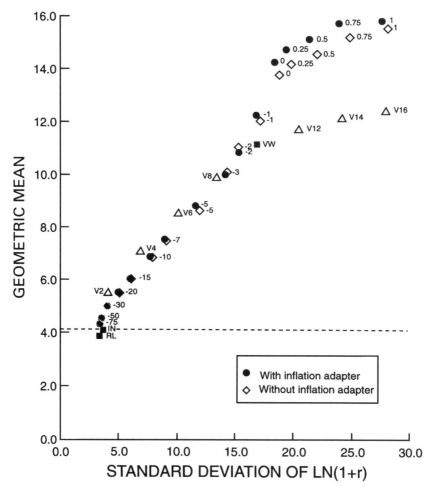

Figure 1. Geometric Means and Standard Deviations of Annual Returns for the Power Policies with and without the Inflation Adapter, Borrowing Permitted, 1934–1988 Treasury Bills and CRSP Value-Weighted Index (Quarterly portfolio revision, 32–quarter estimating period)

using the inflation adapter were somewhat larger. Third, with the exception of the −2 power with borrowing permitted, the inflation-adjusted strategies strictly 'dominated' the unadjusted strategies. Fourth, all the inflation adjusted active strategies less risk averse than the −7 power earned higher geometric mean rates of return, coupled with less variability, than the CRSP

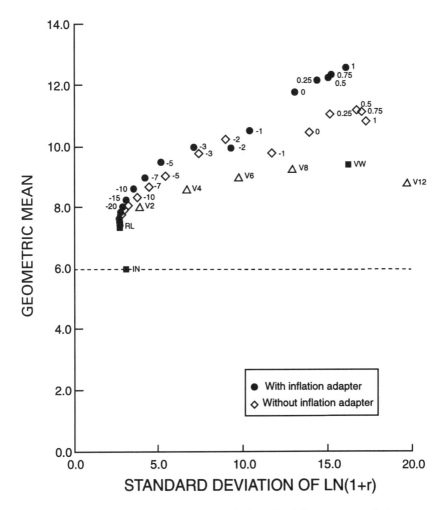

Figure 2. Geometric Means and Standard Deviations of Annual Returns for the Power Policies with and without the Inflation Adapter, Borrowing Permitted, 1966–1988 Treasury Bills and CRSP Value-Weighted Index (Quarterly portfolio revision, 32–quarter estimating period)

value-weighted index. Fifth, all the fixed-weight strategies that borrowed attained smaller geometric mean returns, with more variability, than the CRSP value-weighted index.

Table 2. Geometric Means and Standard Deviations of Annual Returns for sixteen Power Policies with and without the Inflation Adapter, 1966–1988 Treasury Bills and CRSP Value-Weighted Index (Quarterly portfolio revision, 32–quarter estimating period)

	With Inflation Adapter			Without Inflation Adapter		
Portfolio	Geom. Mean	Std. Dev.	Portfolio		Geom. Mean	Std. Dev.

Panel A: Borrowing Precluded

Power −75	7.59	2.61	Power	−75	7.54	2.63
Power −50	7.68	2.63	Power	−50	7.62	2.66
Power −30	7.85	2.71	Power	−30	7.76	2.77
Power −20	8.06	2.86	Power	−20	7.92	2.95
Power −15	8.26	3.04	Power	−15	8.08	3.17
Power −10	8.60	3.50	Power	−10	8.36	3.68
Power −7	8.98	4.17	Power	−7	8.68	4.40
Power −5	9.48	5.13	Power	−5	9.04	5.37
Power −3	10.33	6.69	Power	−3	9.73	7.43
Power −2	10.83	7.74	Power	−2	10.45	8.76
Power −1	11.00	8.10	Power	−1	10.93	9.60
Power 0	11.25	8.58	Power	0	10.67	10.17
Power .25	11.24	8.71	Power	.25	10.50	10.37
Power .5	11.22	8.79	Power	.5	10.08	10.68
Power .75	11.32	8.75	Power	.75	9.75	10.92
Power 1	11.30	8.74	Power	1	9.56	11.03

Panel B: Borrowing Permitted

Power −3	9.99	6.98	Power	−3	9.80	7.32
Power −2	9.96	9.27	Power	−2	10.20	8.94
Power −1	10.51	10.38	Power	−1	9.81	11.66
Power 0	11.78	13.07	Power	0	10.50	13.87
Power .25	12.17	14.39	Power	.25	11.07	15.12
Power .5	12.26	15.00	Power	.5	11.20	16.66
Power .75	12.36	15.15	Power	.75	11.14	16.91
Power 1	12.59	16.02	Power	1	10.86	17.22

Panel C: Benchmarks

RL	7.39	2.60	V12	8.80	19.62
V2	8.07	3.82	V14	8.03	23.18
V4	8.61	6.59	V16	7.09	26.86
V6	9.02	9.67	V18	5.98	30.72
V8	9.29	12.89	V20	4.68	34.77
VW	9.42	16.19	Inflation	5.96	3.04

6.2 The Investment Policies

Space does not permit a full analysis of the differences in investment policies. The following provides selected descriptive statistics for selected strategies with borrowing permitted, with and without the inflation adapter, over the 220 (92) quarters from 1934–88 (1966–88). Over the full period, the policies with (without) the inflation adapter were out of the market over 20 (between 13 and 16) percent of the time. But whenever funds were allocated to the market, the average commitment was greater with the inflation adapter. By way of contrast, during the 92 quarters from 1966–88, the policies with (without) the inflation adapter were out of the market approximately 49 (between 30 and 37) percent of the time. On the other hand, when funds were allocated to the market, the average commitment was again greater when the inflation adapter was in use than when it was not.

Table 3 provides more detail on the investment policies and rates of return of the logarithmic utility strategy, with and without the inflation adapter, over the 1966–88 period when borrowing is permitted. The inflation adapter caused this investor to behave much more conservatively from mid-1966 through 1970 for example – and at other times more aggressively.

7 Statistical Tests

There are a number of commonly accepted procedures for testing for abnormal investment performance:

(i) Jensen's (1968) test of selectivity, or microforecasting;

(ii) Merton's (1981), Henriksson and Merton's (1981) and Treynor and Mazuy's (1966) tests of market timing, or macroforecasting; and

(iii) a paired t-test of the difference in investment returns.

The first three of these tests embody both statistical and economic assumptions about the way assets are priced.

To compute the Jensen measure for portfolio j, we ran the regression

$$r_{jt} - r_{Lt} = \alpha_j + \beta_j(r_{mt} - r_{Lt}) + e_{jt}, \tag{10}$$

where r_{mt} is taken to be the CRSP value-weighted index and r_{Lt} is the return on three-month Treasury bills. The intercept α_j is the measure of investment performance[12]. Positive (negative) values indicate superior (inferior) perfor-

---

[12]While the Jensen measure is usually associated with the ability of a portfolio manager to select under-priced securities, a positive alpha may signify successful market timing if there is a quadratic relation between excess returns on the portfolio and the market as postulated in the Treynor–Mazuy test of market timing (see, for example, Sharpe and Alexander (1990) p. 755).

Table 3. Investment Policies and Returns of a Logarithmic Investor with and without the Inflation Adapter, Borrowing Permitted, 1966–1988 (Quarterly portfolio revision, 32–quarter estimating period)

	With Inflation Adapter					Without Inflation Adapter			
Date	Return	Lend	Borrow	VW	Date	Return	Lend	Borrow	VW
1966-1	−3.75		−0.43	1.43	1966-1	−3.75		−0.43	1.43
1966-2	−6.33	1.00	−0.43	1.43	1966-2	−6.33		−0.43	1.43
1966-3	1.21	1.00			1966–3	−13.89		−0.43	1.43
1966-4	1.35	1.00			1966-4	7.91		−0.22	1.22
1967-1	1.19	1.00			1967-1	14.40		−0.02	1.02
1967-2	2.23			1.00	1967-2	2.49		−0.43	1.43
1967-3	3.28	0.67		0.33	1967-3	10.24		−0.43	1.43
1967-4	0.98			1.00	1967-4	0.68		−0.43	1.43
1968-1	1.27	1.00			1968-1	−9.59		−0.43	1.43
1968-2	1.34	1.00			1968-2	17.54		−0.43	1.43
1968-3	1.35	1.00			1968-3	4.04		−0.25	1.25
1968-4	1.70	0.76		0.24	1968-4	3.16		−0.25	1.25
1969-1	1.54	1.00			1969-1	−3.97		−0.25	1.25
1969-2	1.54	1.00			1969-2	−3.95			1.00
1969-3	1.75	1.00			1969-3	−3.78			1.00
1969-4	1.76	1.00			1969-4	0.67	0.21		0.79
1970-1	1.98	1.00			1970-1	1.98	1.00		
1970-2	1.62	1.00			1970-2	−11.93	0.37		0.63
1970-3	1.62	1.00			1970-3	13.85	0.24		0.75
1970-4	1.48	1.00			1970-4	10.00			1.00
1971-1	16.30		−0.54	1.54	1971-1	11.20			1.00
1971-2	−0.81		−0.54	1.54	1971-2	−0.81		−0.54	1.54
1971-3	1.00	0.82		0.18	1971-3	−0.77		−008	1.08
1971-4	6.42		−0.54	1.54	1971-4	4.75			1.00
1972-1	10.64		−0.82	1.82	1972-1	7.96		−0.29	1.29
1972-2	−1.00		−0.82	1.82	1972-2	−0.29		−0.30	1.30
1972-3	4.03		−0.82	1.82	1972-3	2.92			1.00
1972-4	10.23		−0.55	1.55	1972-4	7.21			1.00
1973-1	−3.68	0.39		0.61	1973-1	−6.95			1.00
1973-2	1.57	1.00			1973-2	−3.44	0.43		0.57
1973-3	2.00	1.00			1973-3	2.00	1.00		
1973-4	1.79	1.00			1973-4	1.79	1.00		
1974-1	1.94	1.00			1974-1	1.94	1.00		
1974-2	2.06	1.00			1974-2	2.06	1.00		
1974-3	1.94	1.00			1974-3	1.94	1.00		
1974-4	1.81	1.00			1974-4	1.81	1.00		
1975-1	1.62	1.00			1975-1	1.62	1.00		
1975-2	1.42	1.00			1975-2	1.42	1.00		
1975-3	1.54	1.00			1975-3	1.54	1.00		
1975-4	1.52	1.00			1975-4	1.52	1.00		
1976-1	15.16	0.06		0.94	1976-1	1.24	1.00		
1976-2	2.68		−0.03	1.03	1976-2	1.90	0.53		0.47
1976-3	1.83	0.09		0.91	1976-3	1.37	0.90		0.10
1976-4	4.08			1.00	1976-4	1.60	0.87		0.13

Table 3 (cont.).

	With Inflation Adapter					Without Inflation Adapter			
Date	Return	Lend	Borrow	VW	Date	Return	Lend	Borrow	VW
1977-1	−12.22		−0.69	1.69	1977-1	−1.27	0.69		0.31
1977-2	5.35		−0.44	1.44	1977-2	1.58	0.86		0.14
1977-3	−2.91			1.00	1977-3	0.07	0.72		0.28
1977-4	1.47	1.00			1977-4	1.40	0.91		0.09
1978-1	1.52	1.00			1978-1	1.23	0.95		0.05
1978-2	1.61	1.00			1978-2	1.61	1.00		
1978-3	1.75	1.00			1978-3	6.89	0.29		0.71
1978-4	1.99	1.00			1978-4	0.49	0.81		0.19
1979-1	2.43	1.00			1979-1	2.43	1.00		
1979-2	2.37	1.00			1979-2	2.37	1.00		
1979-3	2.15	1.00			1979-3	2.15	1.00		
1979-4	2.52	1.00			1979-4	2.52	1.00		
1980-1	2.96	1.00			1980-1	2.96	1.00		
1980-2	3.68	1.00			1980-2	3.68	1.00		
1980-3	1.97	1.00			1980-3	1.97	1.00		
1980-4	2.85	1.00			1980-4	2.85	1.00		
1981-1	3.73	1.00			1981-1	3.73	1.00		
1981-2	3.24	1.00			1981-2	3.24	1.00		
1981-3	3.73	1.00			1981-3	3.73	1.00		
1981-4	3.77	1.00			1981-4	3.77	1.00		
1982-1	2.86	1.00			1982-1	2.86	1.00		
1982-2	3.46	1.00			1982-2	3.46	1.00		
1982-3	3.33	1.00			1982-3	3.33	1.00		
1982-4	22.71		−0.25	1.25	1982-4	22.29		−0.23	1.23
1983-1	12.91		−0.37	1.37	1983-1	10.55		−0.05	1.05
1983-2	18.04		−0.66	1.66	1983-2	17.63		−0.62	1.62
1983-3	−0.72		−0.25	1.25	1983-3	−1.07		−0.37	1.37
1983-4	−3.06		−1.00	2.00	1983-4	−3.06		−1.00	2.00
1984-1	−5.51		−0.44	1.44	1984-1	−5.32		−0.41	1.41
1984-2	−2.41			1.00	1984-2	−2.41			1.00
1984-3	9.63			1.00	1984-3	9.63			1.00
1984-4	1.97	0.11		0.89	1984-4	1.89			1.00
1985-1	13.09		−0.49	1.49	1985-1	10.89		−0.18	1.18
1985-2	12.30		−1.00	2.00	1985-2	12.30		−1.00	2.00
1985-3	−11.36		−1.00	2.00	1985-3	−11.36		−1.00	2.00
1985-4	30.94		−1.00	2.00	1985-4	30.94		−1.00	2.00
1986-1	25.42		−1.00	2.00	1986-1	25.42		−1.00	2.00
1986-2	8.53		−1.00	2.00	1986-2	8.53		−1.00	2.00
1986-3	−15.30		−1.00	2.00	1986-3	−15.30		−1.00	2.00
1986-4	6.44		−1.00	2.00	1986-4	6.44		−1.00	2.00
1987-1	36.80		−1.00	2.00	1987-1	36.80		−1.00	2.00
1987-2	5.77		−1.00	2.00	1987-2	5.77		−1.00	2.00
1987-3	10.35		−1.00	2.00	1987-3	10.35		−1.00	2.00
1987-4	−46.97		−1.00	2.00	1987-4	−46.97		−1.00	2.00
1988-1	11.90		−1.00	2.00	1988-1	10.94		−0.80	1.80
1988-2	10.68		−1.00	2.00	1988-2	10.68		−1.00	2.00
1988-3	−1.69		−1.00	2.00	1988-3	−1.62		−0.97	1.97
1988-4	2.91		−1.00	2.00	1988-4	2.83		−0.48	1.48

mance. The null hypothesis is that there is no superior investment performance and the alternative hypothesis is that there is. Thus, we report results of one-tailed tests for the Jensen test, as well as for the market timing and the paired t-test of differences in investment returns[13].

Next we considered the Henriksson–Merton test for market timing

$$r_{jt} - r_{Lt} = \alpha_j + \beta_{1j}(r_{mt} - r_{Lt}) + \beta_{2j}y_t + e_{jt}, \tag{11}$$

where $y_t = \max(0, r_{Lt} - r_{mt})$ may be interpreted as the payoff associated with a put option on the market portfolio with exercise price r_{Lt}. In this test it is assumed, in essence, that the investor places funds in equities when he expects an up-market and removes them when he expects a down-market. We may interpret α_j as a measure of microforecasting, β_{1j} as the up-market beta, and β_{2j} as the difference between the up and down-market betas. The null hypothesis of no timing ability is that $\beta_{2j} = 0$. We ran the regression as given as well as corrected for heteroscedasticity, using both White's (1980) correction and the correction suggested by Henriksson and Merton.

We also employed the Treynor–Mazuy test of market timing ability based on the regression

$$r_{jt} - r_{Lt} = \alpha_j + \beta_{1j}(r_{mt} - r_{Lt}) + \beta_{2j}(r_{mt} - r_{Lt})^2 + e_{jt}. \tag{12}$$

The null hypothesis of no timing ability is $\beta_{2j} = 0$. We again ran the regression both uncorrected as well as corrected for heteroscedasticity using White's (1980) correction. Finally, we turn to the paired t-test of differences in investment returns. Recall that terminal wealth w_0 in terms of beginning wealth w_n is given by

$$w_0 = w_n(1 + r_n)(1 + r_{n-1}) \ldots (1 + r_1) = w_n \exp\left[\sum_{t=1}^{n} \ln(1 + r_t)\right].$$

Since returns compound multiplicatively, we employ the paired t-test for dependent observations to the quarterly (and additive) variables $\ln(1 + r_t)$. Thus, to compare the return series $r_1^1, \ldots, r_n^1$ with the return series $r_1^2, \ldots, r_n^2$ for two different strategies, we calculate the statistic

$$t = \frac{\bar{d}}{\sigma(d)/\sqrt{n}}$$

where

$$\bar{d} = \sum_{t=1}^{n} \frac{\ln\left(1 + r_t^1\right) - \ln\left(1 + r_t^2\right)}{n}$$

---

[13]The Jensen measure is not without its critics – see, for example, Roll (1977, 1978), Dybvig and Ross (1985), Green (1986), and Grauer (1991). While a number of concerns have been expressed, perhaps the most important is a possible reversal in rankings when different proxies are used for the market portfolio.

and $\sigma(d)$ is the standard deviation of $\ln\left(1 + r_t^1\right) - \ln\left(1 + r_t^2\right)$ null hypothesis is

$$E\left[\ln\left(1 + r_t^1\right)\right] = E\left[\ln\left(1 + r_t^2\right)\right]$$

while the alternative hypothesis is that

$$E\left[\ln\left(1 + r_t^1\right)\right] > E\left[\ln\left(1 + r_t^2\right)\right].$$

8 Test Results

8.1 The Jensen Tests

Table 4 shows the results when Jensen's measure of performance is applied to the power strategies with the inflation adapter for the period 1934–88. The results when borrowing is precluded are reported in Panel A, and the results with borrowing permitted in Panel B. Figure 3 presents the corresponding average excess return-beta plot.

It should be noted at the outset that performance has rarely been measured over periods of this length (55 years in this case). The previous example that comes to mind is the Black, Jensen, and Scholes (1972) test of the capital asset pricing model. In that paper, the authors measured the performance of ten beta-ranked portfolios over the 1931–65 period and found that low (high) beta portfolios earned positive (negative) abnormal returns. These abnormal returns were small, however, and with three exceptions statistically insignificant.

In light of this and other findings, the results for the power strategies with the inflation adapter are surprising. First, as Table 4 shows, all 25 alphas are positive; all are statistically significant at the 5 percent level, and eleven are statistically significant at the 1 percent level. Note that many of the alphas run about 40 percent of total excess returns. Second, with some minor exceptions for the less risk-averse strategies, the larger the beta of a strategy, the larger its abnormal return. Note that beta is greater than one for only two strategies when borrowing is permitted. When borrowing is precluded, the largest beta among the risk-averse strategies is .667.

Next we turn to the 1966–88 sub-period, which, like the full period, spans the 'crash' of October 1987. Since outliers can have significant effects in small samples, we report the results of all statistical tests in this period both with the fourth quarter of 1987 included in, as well as excluded from, the sample. Table 5 shows the results for the power strategies with the inflation adapter, both with borrowing precluded (Panel A) and with borrowing permitted (Panel B). Several observations stand out. First, both the average excess returns and the betas are uniformly smaller than for the full 1934–88

Table 4. Results from Applying the Jensen Performance Test

$$r_{jt} - r_{Lt} = \alpha_j + \beta_j(r_{mt} - r_{Lt}) + e_{jt}$$

to Quarterly Portfolio Returns obtained with sixteen Power Policies using the Inflation Adapter, 1934–1988 Treasury Bills and CRSP Value-Weighted Index (Quarterly portfolio revision, 32–quarter estimating period)

Portfolio	Excess Return*	Alpha	Prob. Alpha = 0	Beta	R^2
Panel A: Borrowing Precluded					
Power −75	0.116	0.048	0.048	0.032	0.32
Power −50	0.172	0.072	0.048	0.048	0.32
Power −30	0.284	0.118	0.049	0.079	0.32
Power −20	0.419	0.173	0.049	0.117	0.31
Power −15	0.552	0.231	0.046	0.153	0.32
Power −10	0.775	0.328	0.035	0.213	0.34
Power −7	0.959	0.380	0.043	0.276	0.36
Power −5	1.163	0.486	0.024	0.323	0.39
Power −3	1.422	0.628	0.008	0.379	0.44
Power −2	1.601	0.736	0.003	0.413	0.47
Power −1	1.762	0.805	0.002	0.456	0.51
Power 0	1.962	0.841	0.001	0.535	0.59
Power .25	2.002	0.819	0.001	0.565	0.62
Power .5	2.085	0.802	0.002	0.612	0.65
Power .75	2.098	0.699	0.006	0.667	0.68
Power 1	2.210	0.700	0.005	0.721	0.72
Panel B: Borrowing Permitted					
Power −5	1.304	0.571	0.021	0.349	0.36
Power −3	1.681	0.695	0.027	0.470	0.39
Power −2	1.953	0.796	0.026	0.552	0.40
Power −1	2.332	1.009	0.010	0.631	0.44
Power 0	2.879	1.265	0.003	0.770	0.51
Power .25	3.044	1.276	0.003	0.843	0.54
Power .5	3.233	1.233	0.007	0.954	0.57
Power .75	3.531	1.206	0.012	1.109	0.62
Power 1	3.826	1.016	0.038	1.341	0.67

* (Average) excess return is measured in units of percent per quarter. The excess return on the CRSP value-weighted index was 2.10% per quarter.

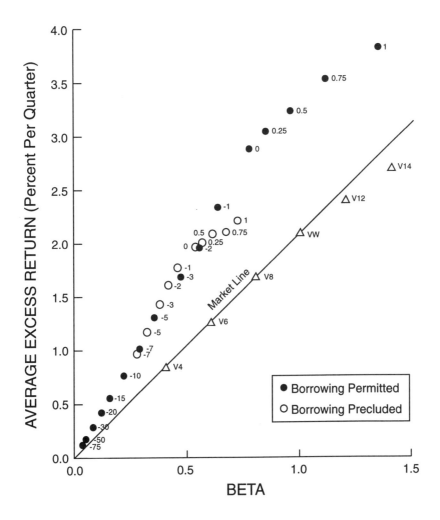

Figure 3. Average Quarterly Excess Returns and Betas for the
Power Policies with the Inflation Adapter, 1934–1988 Treasury
Bills and CRSP Value-Weighted Index (Quarterly portfolio revi-
sion, 32–quarter estimating period)

period. Second, when the fourth quarter of 1987 is excluded, the average
excess returns and the abnormal returns are uniformly larger, and the betas
are uniformly smaller, than when the fourth quarter is included. Third, all
the alphas are positive. When the fourth quarter of 1987 is included, none of
the alphas are statistically significant at the 5% level and their values are between 50 and 75 percent of the corresponding values when the fourth quarter

is excluded. However, when the fourth quarter of 1987 is excluded, all of the alphas are statistically significant at the 2.5% level.

To place the above results in perspective, we summarize the results for 130 mutual funds for the 1968–82 sub-period and contrast them with the results for the active strategies when the investment universe consists of either cash and stocks, or of cash, long-term US government bonds, long-term US corporate bonds, the S&P 500 index, and an index of small stocks. The average excess return-beta plot for the 130 mutual funds and the power policies based on the stocks and cash universe is shown in Figure 4. The average excess return on the market was .32 percent per quarter. The betas of the mutual funds averaged .95, ranging from .36 to 1.64. Fifty-seven (73) of the funds had positive (negative) abnormal returns. Furthermore, only 13 (22) of the positive (negative) alphas were statistically significant at the 5% level. On the other hand, the betas of the power strategies using the inflation adapter averaged only .12, ranging from .005 to .297. All 24 strategies exhibited positive abnormal returns. Moreover, the alphas averaged .58 percent per quarter and 16 of them were statistically significant at the 5% level.

In separate runs, with the investment universe consisting of cash, long-term US government bonds, long-term US corporate bonds, the S&P 500 index, and an index of small stocks, the active strategies produced even larger excess and abnormal returns (averaging .71 percent per quarter) and only marginally higher betas over the same (1968–82) period.

8.2   The Henriksson–Merton and Treynor–Mazuy Tests

While many have tried to time the market, the empirical evidence on the extent of success is mixed. For example, the evidence presented in Chang and Lewellen (1984), Henriksson (1984), Kon (1983), Treynor and Mazuy (1966), and confirmed in our sample of mutual funds, indicates that a greater number of the funds studied have exhibited negative timing ability than have displayed positive timing success. We were therefore not surprised that, over the full 1934–88 period, we found little evidence of market timing ability by the power strategies as measured by either the Henriksson–Merton or the Treynor–Mazuy test.

By way of contrast, Table 6 shows the results for the Henriksson–Merton test of market timing ability when the inflation adapter was employed during the 1966–88 sub-period. With all quarters included, the tests showed that 16 (8) strategies exhibited positive (negative) timing ability, with the negative sign concentrated among the more risk-averse powers. None of the strategies displayed statistically significant timing ability (although the less risk-averse powers without margin were not far off).

However, with the (outlier) fourth quarter of 1987 excluded, the results

Table 5. Results from Applying the Jensen Performance Test

$$r_{jt} - r_{Lt} = \alpha_j + \beta_j(r_{mt} - r_{Lt}) + e_{jt}$$

to Quarterly Portfolio Returns obtained with sixteen Power Policies using the Inflation Adapter, 1966–1988 Treasury Bills and CRSP Value-Weighted Index (Quarterly portfolio revision, 32–quarter estimating period)

| Portfolio | Results including 1987Q4 | | | | Results Excluding 1987Q4 | | | |
| | Excess | | Prob. | | Excess | | Prob. | |
	Return*	Alpha	Alpha=0	Beta	Return	Alpha	Alpha=0	Beta
Panel A: Borrowing Precluded								
Power −75	0.046	0.027	0.210	0.022	0.071	0.053	0.022	0.016
Power −50	0.068	0.040	0.210	0.033	0.106	0.079	0.022	0.023
Power −30	0.113	0.065	0.210	0.055	0.175	0.131	0.022	0.039
Power −20	0.166	0.096	0.211	0.081	0.258	0.193	0.022	0.057
Power −15	0.218	0.126	0.211	0.107	0.339	0.253	0.022	0.075
Power −10	0.317	0.183	0.212	0.155	0.493	0.369	0.022	0.109
Power −7	0.436	0.251	0.213	0.214	0.679	0.508	0.022	0.150
Power −5	0.575	0.355	0.163	0.254	0.846	0.635	0.017	0.185
Power −3	0.792	0.534	0.090	0.297	1.064	0.801	0.011	0.231
Power −2	0.923	0.641	0.066	0.325	1.197	0.899	0.010	0.261
Power −1	0.969	0.674	0.060	0.340	1.244	0.928	0.010	0.277
Power 0	1.039	0.724	0.053	0.363	1.315	0.970	0.010	0.302
Power .25	1.038	0.719	0.056	0.367	1.313	0.963	0.011	0.307
Power .5	1.035	0.714	0.058	0.370	1.311	0.957	0.012	0.310
Power .75	1.060	0.737	0.053	0.373	1.336	0.979	0.011	0.312
Power 1	1.055	0.732	0.055	0.373	1.331	0.974	0.011	0.313
Panel B: Borrowing Permitted								
Power −3	0.798	0.476	0.184	0.371	1.198	0.893	0.020	0.267
Power −2	0.938	0.525	0.221	0.476	1.472	1.089	0.023	0.336
Power −1	1.115	0.656	0.184	0.528	1.662	1.218	0.022	0.388
Power 0	1.483	0.950	0.116	0.614	2.034	1.485	0.017	0.481
Power .25	1.611	1.047	0.103	0.650	2.164	1.569	0.017	0.521
Power .5	1.652	1.070	0.103	0.670	2.205	1.586	0.020	0.542
Power .75	1.678	1.094	0.099	0.673	2.232	1.609	0.018	0.545
Power 1	1.761	1.149	0.095	0.704	2.315	1.653	0.020	0.579

* (Average) excess return is measured in units of percent per quarter. In the 1966–1988 period the excess return on the CRSP value-weighted index was 0.87% per quarter including the fourth quarter of 1987 and 1.14% per quarter excluding the fourth quarter of 1987.

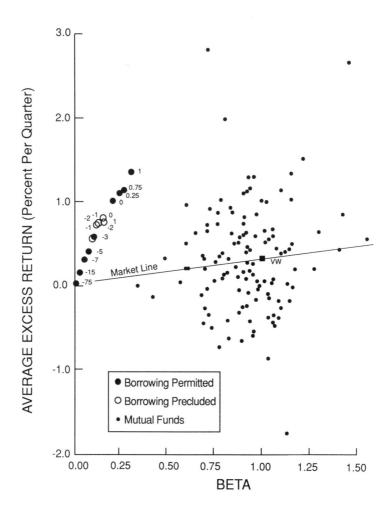

Figure 4. Average Quarterly Excess Returns and Betas for the
Power Policies with the Inflation Adapter and for 130 Mutual
Funds, 1968–1982 Treasury Bills and CRSP Value-Weighted Index
(Quarterly portfolio revision, 32–quarter estimating period)

differ substantially. First, there are substantial differences between the up and
down-market betas. The up-market betas are at least 1.7 times the the Jensen
betas estimated in Table 5; none of the down-market betas exceeds .095.
Second, some of the numbers now suggest timing ability of the statistically
significant variety. All 24 strategies exhibited positive timing at the 1% level,
both without correction as well as with the Henriksson–Merton correction for

Table 6. Summary of Results from Applying the Henriksson–Merton (HM) Market Timing Test

$$r_{jt} - r_{Lt} = \alpha_j + \beta_{1j}(r_{mt} - r_{Lt}) + \beta_{2j}y_t + e_{jt}$$

to Quarterly Portfolio Returns obtained with sixteen Power Policies using the Inflation Adapter, 1966–1988 Treasury Bills and CRSP Value-Weighted Index (Quarterly portfolio revision, 32–quarter estimating period)

| | Results including 1987Q4 | | | | | Results excluding 1987Q4 | | | | |
| | Up-Market Beta β_1 | Down-Market Beta $\beta_1 - \beta_2$ | No Corr. Prob. $\beta_2 = 0$ | White Corr. Prob. $\beta_2 = 0$ | HM Corr. Prob. $\beta_2 = 0$ | Up-Market Beta β_1 | Down-Market Beta $\beta_1 - \beta_2$ | No Corr. Prob. $\beta_2 = 0$ | White Corr. Prob. $\beta_2 = 0$ | HM Corr. Prob. $\beta_2 = 0$ |
Portfolio										
Panel A: Borrowing Precluded										
Power −75	0.022	0.023	0.47	0.48	0.47	0.028	0.003	0.01	0.04	0.00
Power −50	0.033	0.034	0.47	0.49	0.47	0.041	0.004	0.01	0.04	0.00
Power −30	0.054	0.056	0.47	0.49	0.47	0.068	0.006	0.01	0.04	0.00
Power −20	0.079	0.083	0.47	0.49	0.47	0.101	0.009	0.01	0.04	0.00
Power −15	0.104	0.109	0.47	0.49	0.47	0.132	0.012	0.01	0.04	0.00
Power −10	0.152	0.158	0.47	0.49	0.46	0.193	0.018	0.01	0.04	0.00
Power −7	0.209	0.218	0.47	0.49	0.46	0.265	0.024	0.01	0.04	0.00
Power −5	0.267	0.242	0.43	0.47	0.41	0.329	0.026	0.00	0.03	0.00
Power −3	0.348	0.247	0.24	0.37	0.24	0.410	0.034	0.00	0.02	0.00
Power −2	0.401	0.251	0.17	0.31	0.17	0.462	0.040	0.00	0.02	0.00
Power −1	0.422	0.260	0.15	0.30	0.16	0.483	0.051	0.00	0.02	0.00
Power 0	0.467	0.263	0.11	0.26	0.11	0.527	0.056	0.00	0.01	0.00
Power .25	0.473	0.265	0.10	0.26	0.11	0.533	0.060	0.00	0.01	0.00
Power .5	0.475	0.269	0.11	0.26	0.12	0.535	0.065	0.00	0.01	0.00
Power .75	0.477	0.272	0.11	0.26	0.12	0.536	0.068	0.00	0.01	0.00
Power 1	0.478	0.271	0.11	0.26	0.12	0.537	0.067	0.00	0.01	0.00
Panel B: Borrowing Permitted										
Power −3	0.381	0.361	0.46	0.48	0.44	0.473	0.042	0.00	0.04	0.00
Power −2	0.476	0.476	0.50	0.50	0.41	0.600	0.047	0.00	0.03	0.00
Power −1	0.566	0.491	0.39	0.45	0.37	0.692	0.057	0.00	0.03	0.00
Power 0	0.722	0.509	0.23	0.36	0.23	0.846	0.082	0.00	0.02	0.00
Power .25	0.792	0.513	0.18	0.32	0.19	0.916	0.089	0.00	0.01	0.00
Power .5	0.834	0.511	0.15	0.30	0.16	0.957	0.087	0.00	0.01	0.00
Power .75	0.836	0.516	0.15	0.30	0.16	0.959	0.093	0.00	0.01	0.00
Power 1	0.915	0.501	0.10	0.26	0.10	1.038	0.078	0.00	0.01	0.00

The probability that $\beta_2 = 0$ is reported three ways: with no correction for heteroskedasticity, with the White correction for heteroskedasticity, and with the Henriksson–Merton correction for heteroskedasticity.

heteroscedasticity. Based on the White correction for heteroscedasticity, 24 (1) strategies displayed positive timing ability at the 5% (1%) level.

Table 7 summarizes the results of the Henriksson–Merton and Treynor–Mazuy market timing tests for the 1966–88 sub-period with and without correcting for heteroscedasticity. When the fourth quarter of 1987 is excluded, all 24 alphas in the HM tests are negative (though not significantly so). This is consistent with the well-known observation (see e.g. Henriksson (1984) and

Jagannathan and Korajczyk (1986)) that the microforecasting and macrofore-casting measures are negatively correlated. This is troublesome in that the strategies under study here did not attempt to apply any microforecasting.

The Treynor–Mazuy results also lend support to market timing ability on the part of the 24 power strategies. When the fourth quarter of 1987 was included, 13 of the β_2-coefficients were positive. When the fourth quarter of 1987 was excluded, all 24 coefficients were statistically significant at the 1% percent level without correcting for heteroscedasticity, while 15 were statistically significant at the 5% level after White's correction for heteroscedasticity.

To add some perspective to the preceding analysis, we compare the results for 130 mutual funds over the 1968–82 period to the results for the active strategies when the investment universe consisted of either cash and stocks, or of cash, long-term US government bonds, long-term US corporate bonds, the S&P 500 index, and an index of small stocks. According to the Henriksson–Merton test, 60 (70) mutual funds displayed positive (negative) timing ability, but only 5 (11) of the associated return series were statistically significant at the 5% level. On the other hand, when the investment universe consisted of stocks and cash, all 24 power strategies exhibited positive timing ability at the 5% level. With the larger investment universe, 24 (21) of the power strategies displayed positive timing ability at the 5% (1%) level; remarkably, 19 of the 24 down-market betas were negative.

8.3   The Paired *t*-Tests

Recall that, with minor exceptions, the returns of the power strategies with the inflation adapter dominated the corresponding returns without the in-flation adapter, both for the full 1934–88 and for the 1966–88 sub-period. However, based on the paired *t*-test, none of these differences were statisti-cally significant.

The returns earned by the strategies using the inflation adapter were also compared with the returns of various other benchmarks. Results for selected pairs are shown in Table 8. Over the 1934–88 period, the very risk averse strategies −75, −50, and −30 generated higher returns than the risk-free asset at the 1% level of significance; the same is true for powers 0 and .25 when compared to fixed-weight strategy V6 but not when compared to policy V8. All other comparisons (except two) of active and fixed-weight strategies of similar risk yielded insignificant differences over the full 1934-88 period.

In the 1966–88 sub-period, statistical significance hinged to a high degree on the presence or absence of the (outlier) fourth quarter of 1987. When this quarter was *in*cluded, none of the comparisons between the active and the fixed-weight strategies of similar risk was found to be statistically significant, as Table 8 shows. However, when the fourth quarter of 1987 was *ex*cluded,

Table 7. Summary of Results from Applying the Henriksson–Merton (HM) Market Timing Test

$$r_{jt} - r_{Lt} = \alpha_j + \beta_{1j}(r_{mt} - r_{Lt}) + \beta_{2j}y_t + e_{jt}$$

and the Treynor–Mazuy (TM) Market Timing Test

$$r_{jt} - r_{Lt} = \alpha_j + \beta_{1j}(r_{mt} - r_{Lt}) + \beta_{2j}(r_{mt} - r_{Lt})^2 + e_{jt}$$

to Quarterly Portfolio Returns obtained with sixteen Power Policies using the Inflation Adapter, 1966–1988 Treasury Bills and CRSP Value-Weighted Index (Quarterly portfolio revision, 32–quarter estimating period)

	Results including 1987Q4				Results excluding 1987Q4			
	α		β_2		α		β_2	
	Pos	Neg	Pos	Neg	Pos	Neg	Pos	Neg
	HM Test with No Correction for Heteroskedasticity							
count	19	5	16	8	0	24	24	0
significant at 5%	0	0	0	0	0	0	24	0
significant at 1%	0	0	0	0	0	0	24	0
	HM Test with White Correction for Heteroskedasticity							
count	19	5	16	8	0	24	24	0
significant at 5%	0	0	0	0	0	0	24	0
significant at 1%	0	0	0	0	0	0	1	0
	HM Test with HM Correction for Heteroskedasticity							
count	21	3	16	8	0	24	24	0
significant at 5%	0	0	0	0	0	0	24	0
significant at 1%	0	0	0	0	0	0	24	0
	TM Test with No Correction for Heteroskedasticity							
count	24	0	13	11	24	0	24	0
significant at 5%	0	0	0	0	0	0	24	0
significant at 1%	0	0	0	0	0	0	24	0
	TM Test with White Correction for Heteroskedasticity							
count	24	0	13	11	24	0	24	0
significant at 5%	0	0	0	0	0	0	15	0
significant at 1%	0	0	0	0	0	0	0	0

Table 8. Results from Applying the Paired *t*-test of Differences in Investment Returns to Quarterly Portfolio Returns obtained with Selected Power Policies using the Inflation Adapter, 1934–88 and 1966–88 Treasury Bills and CRSP Value-Weighted Index (Quarterly portfolio revision, 32–quarter estimating period)

Power versus Benchmark	$\bar{d}$	Prob. $\bar{d}=0$	Power versus Benchmark	$\bar{d}$	Prob. $\bar{d}=0$	$\bar{d}$	Prob. $\bar{d}=0$
	1934–1988			1966–1988 Including 87Q4		Excluding 87Q4	
Panel A: Borrowing Precluded							
−75 vs. RL	0.0011	0.000	−75 vs. RL	0.0004	0.122	0.0007	0.009
−50 vs. RL	0.0017	0.001	−50 vs. RL	0.0007	0.124	0.0010	0.009
−30 vs. RL	0.0027	0.001	−30 vs. RL	0.0011	0.128	0.0017	0.009
			−20 vs. RL	0.0016	0.134	0.0025	0.009
−20 vs. V2	−0.0000	0.492					
−15 vs. V2	0.0012	0.187	−15 vs. V2	0.0004	0.403	0.0011	0.244
			−10 vs. V2	0.0012	0.296	0.0026	0.091
−10 vs. V4	−0.0005	0.396	−7 vs. V2	0.0021	0.255	0.0042	0.040
−7 vs. V4	0.0010	0.333	−5 vs. V2	0.0033	0.189	0.0057	0.020
−5 vs. V6	−0.0006	0.418	−3 vs. V4	0.0039	0.167	0.0058	0.054
−3 vs. V6	0.0017	0.276	−2 vs. V4	0.0051	0.117	0.0070	0.034
−2 vs. V6	0.0033	0.118	−1 vs. V4	0.0054	0.103	0.0074	0.029
1 vs. V6	0.0047	0.041	0 vs. V4	0.0060	0.086	0.0079	0.024
0 vs. V6	0.0064	0.008	.25 vs. V4	0.0060	0.089	0.0079	0.025
.25 vs. V6	0.0067	0.005	.5 vs. V4	0.0059	0.092	0.0079	0.027
			.75 vs. V4	0.0062	0.084	0.0081	0.024
0 vs. V8	0.0033	0.137	1 vs. V4	0.0061	0.086	0.0081	0.025
.25 vs. V8	0.0036	0.110					
.5 vs. V8	0.0042	0.068	.75 vs. V6	0.0052	0.146	0.0066	0.087
			1 vs. V6	0.0052	0.148	0.0065	0.088
.75 vs. VW	0.0024	0.253					
1 vs. VW	0.0034	0.161	1 vs. V8	0.0136	0.081	0.0053	0.192
Panel B: Borrowing Permitted							
−5 vs. V6	0.0005	0.432	−3 vs. V4	0.0032	0.284	0.0068	0.052
			−2 vs. V4	0.0031	0.344	0.0090	0.036
−3 vs. V8	0.0002	0.480					
−2 vs. V8	0.0021	0.317	−1 vs. V6	0.0034	0.337	0.0089	0.066
			0 vs. V6	0.0063	0.230	0.0118	0.035
−1 vs. VW	0.0025	0.303					
0 vs. VW	0.0069	0.069	0 vs. V8	0.0056	0.256	0.0106	0.070
.25 vs. VW	0.0079	0.046	.25 vs. V8	0.0065	0.230	0.0114	0.061
			.5 vs. V8	0.0067	0.226	0.0116	0.061
0 vs. V12	0.0056	0.137	.75 vs. V8	0.0069	0.218	0.0119	0.058
.25 vs. V12	0.0066	0.095	1 vs. V8	0.0074	0.206	0.0124	0.055
.5 vs. V12	0.0074	0.074					
			.75 vs. VW	0.0066	0.238	0.0109	0.097
.75 vs. V14	0.0078	0.080	1 vs. VW	0.0072	0.224	0.0114	0.090
1 vs. V16	0.0074	0.113					

many differences were statistically significant, especially in the no borrowing case. Over this sub-period, the very risk averse strategies again outdistanced the risk-free asset at the 1% level. In addition, many of the rather risk-tolerant policies outperformed various comparable fixed-weight portfolios at the 5% level of significance; this was not true for the most risk-tolerant strategies, however.

9 Concluding Remarks

Several conclusions emerge. First, the results suggest that adding a simple inflation adapter to the empirical probability assessment approach, which uses only the past to (naively) forecast the future, provides an improvement in applications of discrete-time dynamic investment theory. This improvement was especially notable in the sub-periods beginning in 1966, when inflation changes were most significant, especially when the fourth quarter of 1987 was excluded from the sample.

Second, the simple inflation adapter used appears to be biased toward conservatism (see Figures 1, 2 and 4). Recall that when probability estimates are unbiased, the logarithmic strategy (power 0) should asymptotically have the highest geometric mean.

Third, the empirical probability assessment approach employing the inflation adapter performed well when compared to the up and down-levered value-weighted market portfolio, with several statistically significant excess returns to its credit. The model had only two assets at its disposal with which to implement timing: cash and the value-weighted market.

Fourth, the various standard portfolio performance measures provided somewhat contradictory results. The Jensen test rated the model's performance very high over the full period, but rated it lower for the 1966–88 sub-period. The Henriksson–Merton and the Treynor–Mazuy tests, on the other hand, saw the full period as quite average, but gave a highly favorable report on the 1966–88 sub-period when the fourth quarter of 1987 was excluded. The paired t-test weighed in somewhere in between.

The reader should also be reminded of the limitations of the study. The model used focuses on sequential reinvestment only, without concern for intermediate consumption; even though its birth occurred in the mid-seventies, it was applied as far back as 1934. The latter statement also applies at least partially to the data base used. The joint probability estimates were based, on a moving basis, on the most recent eight years only[14]. All investors were assumed to be strict price-takers. Transactions costs and taxes were ignored (as in the underlying returns series); turnover, however, was

---

[14]Use of a ten-year estimating period produced similar results.

low (see e.g. Table 3). Finally, maintenance margins were ignored whenever leverage was used. Nevertheless, the simple inflation adapter used in this study has substantial power. One potential explanation is that the empirical assessment approach, aided by the inflation adapter, is able to exploit the kind of intermediate to long-term mean reversion documented by e.g. Fama and French (1988a,b) and Poterba and Summers (1988) – and to capture some of the fruits of technical analysis reported in Brock, Lakonishok, and LeBaron (1992).

Acknowledgements

Earlier versions of this article were presented at the Western Finance Association meetings, the European Finance Association meetings, and the American Finance Association meetings in New Orleans. The authors would like to thank the participants, especially Peter Bossaerts, John Herzog, and Robert Korajczyk for helpful comments.

Financial support from the Social Sciences and Humanities Research Council of Canada, and the most capable research assistance of Simon Ng, Jean-Marc Potier, Marie Racine, and Frederick Shen, is gratefully acknowledged.

References

V. Bawa, S. Brown, and R. Klein. (1979) *Estimation Risk and Optimal Portfolio Choice*, North-Holland.

M.J. Best. (1975) 'A Feasible Conjugate Direction Method to Solve Linearly Constrained Optimization Problems.' *Journal of Optimization Theory and Applications* **16** (July) 25–38.

F. Black, M.C. Jensen, and M. Scholes. (1972) 'The Capital Asset Pricing Model: Some Empirical Tests.' In *Studies in the Theory of Capital Markets*, M.C. Jensen (ed.), Praeger Publishers.

M.J. Brennan and E. Schwartz. 'The Use of Treasury Bill Futures in Strategic Asset Allocation Programs.' This volume, 205–228.

W. Brock, J. Lakonishok, and B. LeBaron. (1992) 'Simple Technical Trading Rules and the Stochastic Properties of Stock Returns.' *Journal of Finance* **47** (December) 1731–1764.

E.C. Chang and W.G. Lewellen. (1984) 'Market Timing and Mutual Fund Investment Performance.' *Journal of Business* **57** (January) 57–72.

D.R. Carino, T. Kent, D.H. Myers, C. Stacy, M. Sylvanus, A.L. Turner, K. Watanabe, and W.T. Ziemba. (1994) 'The Russell-Yasuda Kasai Model: An Asset/Liability Model for a Japanese Insurance Company Using Multistage Stochastic Programming.' *Interfaces* **24** (January–February 1994), 29–49. Reprinted in this volume, 609–633.

S. Coleman. (1966) 'Inflation and Stock Market in Past 50 Years Here and Abroad.' *The Commercial and Financial Chronicle*, November 3.

P.H. Dybvig and S. Ross. (1985) 'The Analytics of Performance Measurement Using a Security Market Line.' *Journal of Finance* **40** (June) 401–416.

E. Fama. (1975) 'Short-Term Interest Rates as Predictors of Inflation.' *American Economic Review* **65** (June) 269–282.

E. Fama and K.R. French. (1988) 'Permanent and Temporary Components of Stock Prices.' *Journal of Political Economy* **96** 246–273.

E. Fama and K. French. (1988) 'Dividend Yields and Expected Stock Returns.' *Journal of Financial Economics* **22** (October) 3–25.

E. Fama and W. Schwert. (1977) 'Asset Returns and Inflation.' *Journal of Financial Economics* **5** (November 1977) 115–146.

K.R. French, G.W. Schwert, and R.F. Stambaugh. (1987) 'Expected Stock Returns and Volatility.' *Journal of Financial Economics* **19** (September) 3–29.

R.R. Grauer. (1991) 'Further Ambiguity When Performance is Measured by the Security Market Line.' *Financial Review* **26** (November) 569–585.

R.R. Grauer and N.H. Hakansson. (1982) 'Higher Return, Lower Risk: Historical Returns on Long-Run, Actively Managed Portfolios of Stocks, Bonds and Bills, 1936–1978.' *Financial Analysts Journal* **38** (March–April) 39–53.

R.R. Grauer and N.H. Hakansson. (1985) 'Returns on Levered, Actively Managed Long-Run Portfolios of Stocks, Bonds, and Bills, 1934–1984.' *Financial Analysts Journal* **41** (September–October) 24–43.

R.R. Grauer and N.H. Hakansson. (1986) 'A Half Century of Returns on Levered and Unlevered Portfolios of Stocks, Bonds, and Bills, with and without Small Stocks.' *Journal of Business* **59** (April) 287–318.

R.R. Grauer and N.H. Hakansson. (1987) 'Gains from International Diversification: 1968–85 Returns on Portfolios of Stocks and Bonds.' *Journal of Finance* **42** (July) 721–739.

R.R. Grauer and N.H. Hakansson. (1995a) 'Stein and CAPM Estimators of the Means in Asset Allocation.' *International Review of Financial Analysis* **4** 35–66.

R.R. Grauer and N.H. Hakansson. (1995b) 'Gains from Diversifying into Real Estate: Three Decades of Portfolio Returns based on the Dynamic Investment Model.' *Real Estate Economics* **23** 117–159.

R. Green. (1986) 'Benchmark Portfolio Inefficiency and Deviations from the Security Market Line.' *Journal of Finance* **41** (June) 295–312.

N.H. Hakansson. (1971) 'On Optimal Myopic Portfolio Policies, With and Without Serial Correlation of Yields.' *Journal of Business* **44** (July) 324–34.

N.H. Hakansson. (1974) 'Convergence to Isoelastic Utility and Policy in Multi-period Portfolio Choice.' *Journal of Financial Economics* **1** (September) 201–24.

R.D. Henriksson. (1984) 'Market Timing and Mutual Fund Performance: An Empirical Investigation.' *Journal of Business* **57** (January) 73–96.

R.D. Henriksson and R.C. Merton. (1981) 'On Market Timing and Investment Performance II. Statistical Procedures for Evaluating Forecasting Skills.' *Journal of Business* **54** (October) 513–533.

G. Huberman and S. Ross. (1983) 'Portfolio Turnpike Theorems, Risk Aversion and Regularly Varying Utility Functions.' *Econometrica* **51** (September) 1104–19.

Ibbotson Associates. (1989) *Stocks, Bonds, Bills and Inflation: 1989 Yearbook*, Ibbotson Associates.

R. Jagannathan and R. Korajczyk. (1986) 'Assessing the Market Timing Performance of Managed Portfolios.' *Journal of Business* **59** (April) 217–235.

M.C. Jensen. (1968) 'The Performance of Mutual Funds in the Period 1945–1964.' *Journal of Finance* **23** (March) 389–416.

P. Jorion. (1989) 'Asset Allocation with Hedged and Unhedged Foreign Stocks and Bonds.' *Journal of Portfolio Management* **15** (Summer) 49–54.

D.B. Keim and R.F. Stambaugh. (1986) 'Predicting Returns in the Stock and Bond Markets.' *Journal of Financial Economics* **17** (December) 357–390.

S. Kon. (1983) 'The Market-timing Performance of Mutual Fund Managers.' *Journal of Business* **56** (July) 323–347.

B.N. Lehmann and D.M. Modest. (1987) 'Mutual Fund Performance Evaluation: A Comparison of Benchmarks and Benchmark Comparisons.' *Journal of Finance* **42** (June) 233–265.

H. Leland. (1972) 'On Turnpike Portfolios.' In *Mathematical Methods in Investment and Finance*, G.P. Szegö and K. Shell (eds.), North-Holland.

B.G. Malkiel. (1990) *A Random Walk Down Wall Street*, 5th edition, Norton.

R.C. Merton. (1981) 'On Market Timing and Investment Performance I: An Equilibrium Theory of Value for Market Forecasts.' *Journal of Business* (July) 363–406.

J. Mossin. (1968) 'Optimal Multiperiod Portfolio Policies.' *Journal of Business* **41** (April) 215–229.

J. Mulvey and W.T. Ziemba. (1995) 'Asset and Liability Allocation in a Global Environment.' In *Handbooks in Operations Research and Management Science: Finance*, Vol. 9, R. Jarrow, V. Maksimovic, and W.T Ziemba, (eds.), Elsevier.

J. Poterba and L. Summers. (1988) 'Mean Reversion in Stock Prices: Evidence and Implications.' *Journal of Financial Economics* **22** (October) 27–60.

R. Roll. (1977) 'A Critique of the Asset Pricing Theory's Test, Part I: On Past and Potential Testability of the Theory.' *Journal of Financial Economics* **4** (March) 129–176.

R. Roll. (1978) 'Ambiguity When Performance is Measured by the Securities Market Line.' *Journal of Finance* **33** (September) 1051–1070.

S. Ross. (1974) 'Portfolio Turnpike Theorems for Constant Policies.' *Journal of Financial Economics* **1** (July) 171–98.

W.F. Sharpe and G.J. Alexander. (1990) *Investments*, 4th edition, Prentice-Hall.

J.L. Treynor and K. Mazuy. (1966) 'Can Mutual Funds Outguess the Market?' *Harvard Business Review* **44** (July-August) 131–136.

H. White. (1980) 'A Heteroscedasticity-Consistent Covariance Matrix Estimator and a Direct Test for Heteroscedasticity.' *Econometrica* **48** (May) 817–838.

Multiperiod Asset Allocation with Derivative Assets

David R. Cariño and Andrew L. Turner

Abstract

Asset allocation is the process of dividing an investment fund among major asset classes such as equities, bonds, cash, etc. Typically, asset allocation decisions are made in a static fashion – by assuming constant rebalancing to a fixed mix and by assuming stationary return distributions. Stochastic programming provides a useful framework for developing dynamic asset allocation strategies. Dynamic strategies can create substantial skewness in the probability distribution of wealth. Derivative assets can also create skewed distributions, even within a single period. Skewness would confound the conventional asset allocation approach. In this paper, we describe an asset allocation model that treats multiple periods and asymmetric return distributions appropriately and demonstrate its use by examples.

Acknowledgments

The authors thank Yuan-An Fan, Steve Murray, Celine Stacy, and Mike Sylvanus for developing and implementing the models in this paper.

1 Introduction

Constructing portfolios of conventional asset classes such as equities, bonds, and cash is a common activity. The techniques of mean-variance analysis are so widely-used for this task that most investors associate the term "asset allocation" with "mean-variance portfolio selection." In this paradigm, an asset mix is chosen that minimizes portfolio variance at a given level of expected return (see Markowitz 1952, 1959). Nearly every introductory investments text discusses the basic ideas of mean-variance efficient frontiers (e.g., see Sharpe and Alexander 1990). Furthermore, software implementations of mean-variance analysis are commonplace.

Although there are great insights to be obtained from the approach, mean-variance analysis, like most paradigms, has limitations. Analysts often tolerate the limitations and compensate for them in ad-hoc ways in order to use familiar software tools.

Two of its limitations are (1) the single-period nature of the analysis, and (2) the symmetric definition of risk. These limitations need no longer be tolerated. New algorithms and software have recently been developed that directly deal with these aspects of portfolio construction. In this article we describe such a model, which we call the multiperiod asset allocation model.

1.1 Single-period nature of mean-variance analysis

One characteristic of the mean-variance approach is the static nature of the analysis. The inputs to the analysis consist of means and variances (or standard deviations) of asset returns, and correlations among asset returns. These parameters are often called 'input assumptions' and an analyst might create them from a combination of statistical modelling and subjective assessment.

In principle, these assumptions represent the analyst's belief about the probability distribution of future asset returns. Analysts are often vague, however, on the period of time over which the probability distribution is meant to apply. Do the assumptions represent our views over the next month, the next year, or the next five years? If we expect that asset returns in the short term (next month, say) will be lower than a more 'normal' expectation, which set of expectations should we use in the analysis?

The applicability of the input assumptions is related to the applicability of the asset mix determined from the analysis. Again, users of mean-variance analysis are often vague on the holding period for which the mix is being determined. The technology of mean-variance analysis explicitly allows for only a single period to be optimized. Do we intend this single period to be one month, one year, or five years?

Few investors would buy and hold a single portfolio for five years. It is more common to specify an asset mix and rebalance periodically back to that mix. Such a strategy is called a 'fixed-mixed' strategy (e.g., Mulvey and Thorlacius 1998, and Berger and Mulvey, 1998). If we expect to rebalance, should the single period of the analysis coincide with the expected time to the next rebalancing? Should the single set of asset input assumptions apply only up to the expected rebalancing time or should the input assumptions be long-term assumptions? How do transaction costs affect the analysis?

Users of mean-variance analysis are often in a quandary over questions such as these, because the analysis fundamentally does not distinguish multiple future periods. There is a timeless aspect to the approach that is difficult to adapt to dynamic circumstances.

The multiperiod asset allocation model described in this paper explicitly includes multiple future periods in the analysis. Questions such as the above have clear answers in a multiperiod approach.

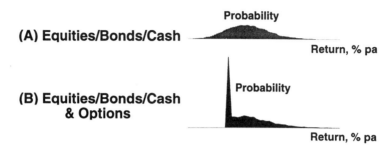

Figure 1: Two distributions with identical means and variances but different skewness.

1.2 Symmetric risk measure in mean-variance analysis

Another important shortcoming arises when we wish to include options in the analysis. With options, return distributions can be deliberately skewed. For example, the top panel of Figure 1 shows a symmetric probability distribution of the type that is typically assumed for conventional assets like equities. Buying put options on top of a position in equities can dramatically change the shape of the return distribution by limiting the possible magnitude of losses. The lower panel of Figure 1 shows a mixture of equities, cash, and options constructed so that the mean and variance of the distribution are the same as those of equities. Given that these two distributions have the same mean and variance, many investors would prefer the skewed shape over the symmetric shape since the possibility of large losses is limited (see Balzer 1994). Mean-variance analysis would ignore the skewness of the distribution and regard only the mean, variance, and correlations with other assets as the relevant asset characteristics.

Despite this limitation, the wide-spread adoption of mean-variance analysis combined with readily-available software implementing it leads some investors to inappropriately include options in the analysis. What is needed is a different paradigm and suitable software algorithms that can correctly handle derivatives.

1.3 Multiperiod analysis

More suitable approaches have existed for many years prior to the introduction of mean-variance analysis. For example, the expected-utility paradigm (see, e.g., Von Neumann and Morgenstern 1944, Samuelson 1969, and Merton 1969, 1971) has long been used by finance theorists for analysis. Recently, algorithms and computer programs have emerged that make such alternate approaches workable for practitioners. The algorithms come from the field of stochastic programming and, with the increasing speed of computers, a growing number of financial applications of these algorithms have been successfully implemented (see Cariño *et al.* 1994, Dempster and Ireland 1988, King and Warden 1994, Kusy and Ziemba 1986, Mulvey 1995, Zenios 1991, 1993, and other papers in this volume).

In this paper, we describe an asset allocation model that uses the stochastic programming approach to directly deal with multiple periods. We also show how options and other derivative return distributions can be properly handled in this framework. A key element of the approach is a measure of risk that, unlike variance, specifically distinguishes unexpected positive returns from unexpected negative returns. The risk measure, called *penalized shortfall*, can express risk in terms that are tangible and meaningful to decision makers. Cariño and Fan (1993) and Cariño, Fan, Ankrim, and Bouchey (1995) give an introduction to the penalized shortfall concept (see also Kusy and Ziemba 1986). This paper gives more details of the formulation and use of these techniques implemented in software we call the Multiperiod Asset Allocation Model, or *MAAM*.

2   Formulation

2.1   Periods and stages

Asset allocation models are often vague on the time period over which the recommended mix ought to apply. The stochastic programming approach requires that the analyst be specific about the time points at which the portfolio is expected to be revised or rebalanced. The dates at which we expect to make portfolio revisions are called decision *stages*, and the time intervals between stages are called *periods*.

A typical period structure is depicted in Figure 2. Here, the first period is one month in length, the second period is two months, and the third period is three months in length. Although a one-period model has two stages, a two-period model has three stages, etc., we label the initial date as stage $t = 0$ so that we may refer to 'stage t' and 'end of period t' interchangeably.

The choice of a period structure depends on several considerations. The initial stage decision is the decision that we are most interested in ('What should we hold today?'). The subsequent stages represent dates at which we expect to reexamine the allocation and possibly revise the portfolio. The period lengths may therefore be dictated by anticipated events such as the dates at which options expire, or a periodic (e.g., quarterly, annual, etc.) management review. The period lengths might also be suggested by our ability to forecast asset returns. If an analyst has models that forecast changes

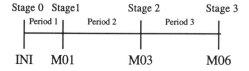

Figure 2: A typical period structure.

in expected returns, then the periods can be made consistent with the time periods over which expected returns are forecasted. For example, one might have different expectations over the next one month, the next three months, and the next six months, which would suggest using the period structure of Figure 2.

Periods beyond the first period can be of longer lengths than the first. The analyst might be primarily interested in the initial decision, but also want to take account of what might happen beyond the next revision date. However, the model will be re-run when that date actually arrives, so the analyst can be less concerned with great modelling detail in later periods than in earlier periods. The number of periods affects the solution time of the model; more periods means longer solution times. If we had unlimited computing resources, we would use many periods. Since we do not, we attempt to reduce the number of periods until an acceptable computation time is achieved.

2.2   Scenarios

Given a period structure, we model asset returns by means of discrete *scenarios*. A single scenario is a sequence of joint outcomes of asset returns. The collection of scenarios may be thought of as forming an event tree as depicted in Figure 3. The nodes (or events) are labeled so that we may associate a set of asset return outcomes with each node. To each node we also associate a probability of the node's occurrence; the conditional probabilities of all branches immediately emanating from a given node must sum to one.

Although scenario trees may have arbitrary structure, we usually use trees having some amount of symmetry. The number of branches emanating from a given node provides a key dimension of the tree; if that number is the same for all nodes at a given stage, then the tree is *balanced*. We summarize the structure of a balanced tree by specifying the number of branches per node at each stage. For example, the tree of Figure 3 has a structure of 3×2 branches per node, for a total of six scenarios. We commonly use tree structures of $50 \times 20 \times 10$, for a total of 10,000 scenarios.

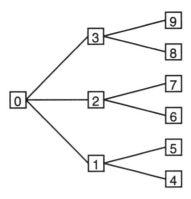

Figure 3: A scenario tree.

2.3   Asset Returns

At each node of the tree, we specify a joint outcome of asset returns. These joint outcomes are usually created by sampling from some specified probability distribution. The probability distribution could, in turn, be created from some sophisticated forecasting model of asset returns, or it could be a simple distribution specified by a few parameters.

Many analysts are accustomed to preparing tables of means, variances, and correlations of asset class returns for mean-variance analysis. Such tables, often referred to as "input assumptions," might come from some other modelling efforts by the investor. For example, Turner and Hensel (1993), Chopra, Hensel, and Turner (1993), and Hensel and Turner (1998) suggest massaging the inputs to control the sensitivity to forecast errors. Table 1 shows a typical set of input assumptions for six assets developed using these techniques. By making an appropriate distribution assumption, such as lognormality, we can create scenarios from inputs such as these.

Although it is common to make just one set of assumptions as above, the multiperiod approach allows the flexibility to specify a different set of assumptions for each period of the model. Hence, the analyst is required to be explicit about the time periods over which the assumptions are to apply.

The scenario outcomes generated by sampling can be enumerated in a table; for example Table 2 shows the node-by-node outcomes for a 3 × 2 tree.

Conventional asset classes (like equities, bonds, and cash) are adequately modelled by lognormal sampling. Derivative asset classes (containing options, for example), need not be created by sampling. As the terminology suggests, there is a clear relationship between the return of a derivative asset and that of its underlying security. In the scenario approach, we derive the returns of derivative assets from the known relationship.

Asset Classes	Expected Return	Standard Deviation	Correlations					
			US Equity Large Cap	US Equity Small Cap	Non-US Equity Unhedged	Emerging Markets Unhedged	US Bonds	US Cash
US Equity Large Cap	11.0	17.0	1.0					
US Equity Small Cap	11.0	25.0	0.8	1.0				
Non-US Equity Unhedged	11.0	21.0	0.5	0.3	1.0			
Emerging Mkts Unhedged	11.0	25.0	0.3	0.3	0.3	1.0		
US Bonds	7.0	7.0	0.4	0.3	0.2	0.0	1.0	
US Cash	5.7	1.0	0.0	0.0	0.0	0.0	0.3	1.0

Table 1: Typical asset class assumptions

Node	Conditional Probability	US Equity Large Cap	US Equity Small Cap	Non-US Equity	Emerging Markets	US Bonds	US Cash
1	.3333	.032955	.341701	.041221	.279216	.027300	-.014084
2	.3333	-.091184	.049939	.109955	.082171	-.128904	.024156
3	.3333	.090754	.534592	.120825	.082171	-.128904	.024156
4	.5000	.035930	.056592	-.000627	-.304342	.061070	.000830
5	.5000	.119713	-.130465	.193180	.519016	.069383	.028540
6	.5000	.461739	.392537	.116938	.360205	.089025	.050224
7	.5000	.245134	.122433	.568656	.180286	.110467	.092815
8	.5000	-.090453	-.292077	-.292757	.001132	.129944	.121655
9	.5000	.041096	.054468	.118764	-.048986	.065222	.088793

Table 2: Example scenario outcomes listed by node

For example, to model simple options, we could use the relation

$$\text{Option Return} = \frac{\text{Final Value of Option}}{\text{Option Price}} - 1, \qquad (2.1)$$

where the final option value is a function of the underlying asset return, and the option price is computed from an appropriate pricing formula (e.g., Black–Scholes).

The above equation represents the return to the funds invested only in the option itself. As such, the range of the returns can be extremely large — far beyond that of conventional asset classes. Since computational instabilities may result from such large returns, we instead model asset classes in which options are combined with some other asset class like cash. For example, we can consider an at-the-money equity call option combined with sufficient cash that is expected to be worth the strike price at maturity; this might be termed a fully-collateralized call option. For an option collateralized at $x\%$, the return is

$$\frac{(x\%) \cdot \text{Strike} + \text{Final Value of Option}}{\frac{(x\%) \cdot \text{Strike}}{(1+r)} + \text{Option Price}} - 1, \tag{2.2}$$

where r is the risk-free cash rate appropriate for the maturity of the option. The final value of the option depends on the return of the underlying asset class. Similar formulas can be written for other derivative strategies.

2.4   Asset accumulation and revision

An essential element of a multiperiod model is the description of how assets accumulate over time. This description is summarized in a set of equations, which are presented in Appendix A.

A key feature of the model equations is the explicit modelling of transactions at every decision point, that is, every node in the scenario tree. The model includes purchase and sale variables for each asset class at each node to which transaction costs apply. A budget constraint applies at each node, ensuring that the total cash flow in (from sales and contributions to the fund) equals the total cash flow out (from purchases and disbursements from the fund).

2.5   Objective function

The objective function trades off the desire for more wealth against the desire to reduce embarrassments. We express this as

$$\text{maximize} \quad E\left[\text{Final Wealth} - \frac{\text{Accumulated Penalized Shortfalls}}{\text{Risk Tolerance}}\right]. \tag{2.3}$$

The objective is to maximize expected wealth less a measure of risk. We define risk in terms of quantifiable outcomes that are undesirable from the investor's viewpoint. An example of a quantifiable outcome would be a shortfall in wealth below a given threshold. More generally, we call such outcomes *embarrassments* and there may be more than one type of embarrassment.

Embarrassments are penalized in the objective function through a convex cost function. Figure 4 shows an example of a cost function for shortfalls below a threshold return. In the figure, we begin to penalize returns that fall below 0%. Penalty costs increase at a faster rate below -5%, and at an even faster rate if return falls below -10%. The slopes in the figure are set to approximate a quadratic function; the resulting risk measure is sometimes known as 'target semivariance' (see Balzer 1994). In general, cost functions can be set to any convex shape.

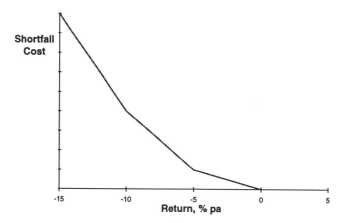

Figure 4: A shortfall cost function.

Thresholds like these often arise naturally. For example, many funds have liability reserving targets, below which the fund experiences increasing discomfort. There might also be statutory minimums, below which regulatory penalties apply. Penalized shortfall provides a useful framework for these situations (see Cariño and Fan 1993, and Cariño, Fan, Ankrim, and Bouchey 1995).

3   A Three-Period Example

To illustrate the multiperiod character of the model, consider a three-period example.

Consider an investment fund with a simple commitment to pay out all of its accumulated value at the end of five years. The fund manager may revise the portfolio annually. Suppose the manager would prefer more wealth over less wealth at the horizon, but would be particularly concerned if the wealth at the horizon fell far short of what could be earned by a 'safe' investment in cash over the five years.

We set a period structure with period lengths of 1-2-2 years, for an overall planning horizon of five years. The first period length of one year is set to coincide with the next possible revision date. The subsequent periods (of two years each) are set, as described earlier, to conserve computational effort. There are six asset classes and we use the input assumptions in Table 1. We assume that asset returns are independent between periods and that the same input assumptions apply to all three periods.

We specify a single embarrassment, wealth shortfall at the horizon. The

wealth threshold is set at 121.7 at year 5. From an initial wealth of 100, this
target represents a compound rate of return of 4% per annum. This target
is below the cash expected return and would represent a reasonable point
at which significant discomfort might be felt by the manager with a goal of
exceeding the cash return. Additional breakpoints of the cost function are
set at 95%, 90%, and 85% of the threshold.

Even with stationary asset return assumptions, this simple type of embarrassment based on an absolute wealth target creates a significant opportunity
for a dynamic strategy to excel. With more complicated objectives (more embarrassments, for example), optimal strategies other than those illustrated by
this example would apply.

We generate scenarios by sampling. Figure 5 shows some scenarios that
might be created for three assets, cash, bonds, and equities. Only 24 scenarios
are plotted, in a $4 \times 3 \times 2$ pattern.

A heavy line in the graphs depicts the 4% per annum wealth threshold.
An investment wholly in cash will ensure that outcomes above that level are
achieved, while 100% in either bonds or equities may result in shortfall at
the horizon. Clearly, the more volatile asset class, equities, has a greater
chance of shortfall. However, investment in equities also offers the possibility
of substantially higher wealth at the horizon than would be possible with
cash.

We compare two types of strategies. The first is a dynamic strategy indicated by the full optimization of the multiperiod model. The second type is
a fixed-mix strategy using mean-variance efficient portfolios. We take portfolios from the mean-variance efficient frontier, and assume that the allocations
are rebalanced back to that mix at each of the decision stages. The following results are generated by the model using 10,000 scenarios in a pattern of
$50 \times 20 \times 10$. The results are shown in Figure 6. By varying the weight in
the objective on the shortfall cost, we obtain the optimal tradeoff between
expected wealth and expected shortfall cost. The figure shows an 'efficient
frontier' using the penalized shortfall risk measure. The figure also shows
the expected wealth and shortfall costs for the range of fixed-mix strategies
obtained by holding mean-variance-efficient portfolios.

We can observe that for any given fixed-mix strategy, there is a dynamic
strategy that has either the same expected wealth and lower shortfall cost,
or the same shortfall cost and higher expected wealth. Table 3 gives the
expected wealth and expected shortfall costs for the three strategies circled
in Figure 6. Of course, the optimal strategy depends on the risk measure
specified.

How does the dynamic strategy achieve lower shortfall cost or higher expected wealth? If we look at the initial portfolio allocations, shown in Table 3
and Figure 7, we see that the portfolio allocation for the optimal strategy

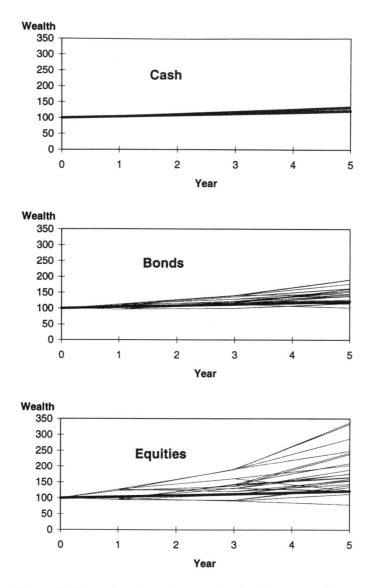

Figure 5: Examples of wealth scenarios for three asset classes.

starts out very similar to the fixed mix A having the same expected return. The split between equities and fixed-income assets is about 60%/40%. The fixed mix strategy B having the same risk has a much lower allocation to equities.

If we look at allocations at the end of year 1, we can get a feel for how the

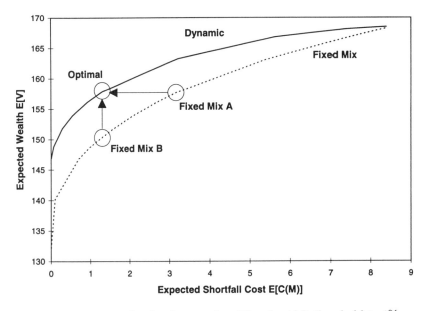

Figure 6: Return and risk of strategies. The shortfall threshold is 4% pa.

Strategy	Initial Equity/Bond Allocation (%)	Expected Wealth at Year 5	Expected Penalized Shortfall
Optimal	59/41	158 (9.6% pa)	1.3
Fixed Mix A	64/36	158 (9.6% pa)	3.2
Fixed Mix B	46/54	150 (8.5% pa)	1.3

Table 3: Expected wealth and shortfall costs of strategies.

dynamic strategy works. The allocations depend on the outcome of the first year's returns. In general, the optimal allocations can be related in a complex way to the outcomes. However, in this example, if we sort the outcomes by the total portfolio wealth, we find that the allocations basically shift to less volatile assets as the excess over the wealth target is reduced. This is shown in Figure 8, where the allocations have been averaged over ranges of first-year

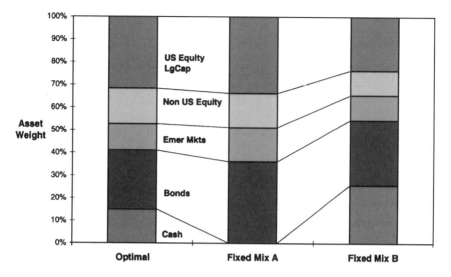

Figure 7: Initial portfolios of three strategies.

wealth outcomes.

The effect of dynamic allocation is to alter the overall probability distribution of final wealth, shown in Figure 9. In this example, after the initial allocation, there are just two reallocation opportunities modelled: at the end of year 1 and the end of year 3. Despite this very infrequent portfolio revision, there is a noticeable pattern of asymmetry in the final wealth outcomes. The optimal dynamic strategy, while it has the same expected wealth as the fixed mix A, has reduced the probability of falling below the 4% threshold. Furthermore, it has increased the probability of attaining very high final wealth.

Comparing the dynamic strategy to fixed mix B having equal expected shortfall cost, Figure 9 shows the effect of moving to a more aggressive portfolio when a cushion has been built up over the target 4% return. Adhering to a constant mix throughout the horizon leads to a lower expected wealth.

The results illustrated by this example should not be interpreted as 'selling after a market downturn.' It is important to note that the expected asset class return distributions year by year are identical. The model does not suggest that if an asset class has gone down, it is a good time to sell that asset. The reason for moving to a less aggressive portfolio is because there is a fixed wealth hurdle. The model suggests that by taking advantage of the opportunities to adapt the asset mix given the current wealth level, the

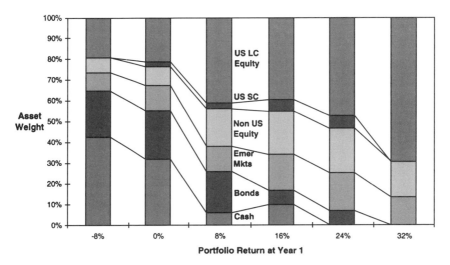

Figure 8: Contingent allocations at year one.

chances of exceeding the hurdle are increased. The strategy suggested by the model is best for this particular type of goal.

Other types of goals could lead to other dynamic strategies. The desire for interim cash outflows prior to the horizon, for instance, weighted against the desire for long-term wealth, could lead to strategies that exhibit dramatic dynamic behaviour as the dates of the desired flows are approached and passed. Since the goals in the example of this section need not apply to all investors, the multiperiod model gives the user great flexibility in expressing an investor's preferences. Only in a model which allows such flexibility can the implications of such preferences be explored.

4   A Call-Option Example

The skewed wealth distribution of Figure 9 is typical of dynamic strategies for achieving wealth targets. The particular example displays noticeable skewness using conventional assets with only three revision opportunities. Another example in which skewness is important is when option assets are explicitly included in the strategy. To illustrate the use of the model with derivative assets, consider a single-period example with three conventional assets (equi-

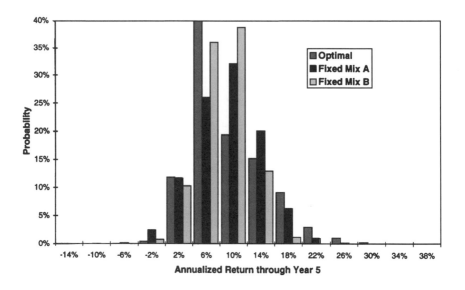

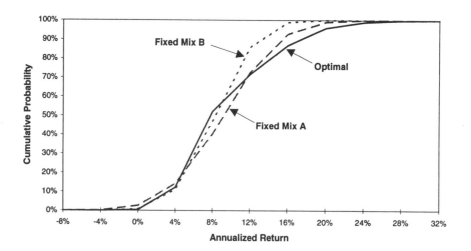

Figure 9: Probability distributions of final wealth at year 5.

	Expected Return	Standard Deviation	Correlations		
			Equities	Bonds	Cash
Equities	11%	17%	1		
Bonds	7%	7%	0.4	1.0	
Cash	5%	1%	0.0	0.3	1.0

Table 4: Assumptions for single-period example.

Strike Price	Expected Return	Standard Deviation
90	9.9%	15.7%
95	9.2%	14.7%
100	8.5%	13.3%
105	7.7%	11.7%
110	7.0%	10.0%

Table 5: Expected return and volatility of derived asset classes.

ties, bonds, and cash) plus five derived asset classes. The derived asset classes consist of call options on equities at a given strike price combined with sufficient cash to collateralize the strike price. This combination, calls plus cash, is equivalent (by put-call parity) to holding equities and buying put options. Table 4 gives the assumptions for the conventional asset classes. Using the formula given earlier (in Section 2.3) for a fully collateralized call option, we can compute the expected return and volatility of the derived asset classes. These are shown in Table 5.

Figure 10 shows the return distributions of the eight asset classes. We see a distinctive characteristic: the possible losses are limited below a certain point, depending on the strike price. The derived asset classes offer equity-like returns, with protection from losses. Of course, the cost of the protection is paid in terms of expected return; expected return decreases as the floor is raised.

The derivative assets offer a way to create skewed return distributions over a single period. If portfolios were formed using these assets in a mean-variance analysis, the skewness would be ignored. When we form portfolios having the least penalized shortfall at varying expected returns, we obtain the return distributions shown in Figure 11. We see the pattern of increasing skewness as more downside protection is applied. As noted above, the cost of greater

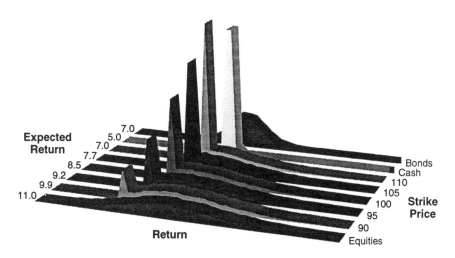

Figure 10: Option asset class return distributions.

protection is paid for in terms of lower expected return.

5   Conclusion

The conventional asset allocation approach is incompatible with a multiperiod
view of the world, whether that view arises from forecasts that vary with the
horizon, or from goals and other circumstances that vary over time. Such
goals can lead to asymmetric return distributions, which also arise when options are available for inclusion in the mix. The multiperiod asset allocation
model has the flexibility and expressiveness needed to properly handle these
situations.

Appendix A: Formulation Equations

Asset revision at $t = 0$

We denote by $\overline{X}_{n0}$ the market value of the portfolio initially held in asset
n, and let P_{n0} and S_{n0} be the market values of asset n purchased and sold,

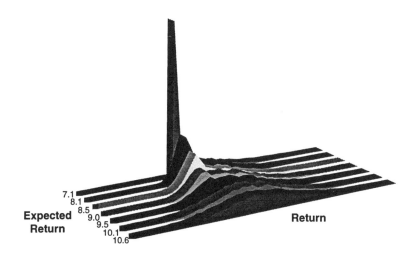

Figure 11: Optimal portfolio return distributions including option asset classes.

respectively, at the initial decision stage $t = 0$. For each of N asset classes, there is an equation relating the market value X_{n0} after revision to $\overline{X}_{n0}$:

$$P_{n0} - S_{n0} - X_{n0} = -\overline{X}_{n0}, \qquad n = 1, \dots, N.$$

Nonnegativity and asset bounds at $t = 0$

Values of purchases and sales necessarily are nonnegative and there may be bounds on the holdings after revision. These constraints are

$$P_{n0} \geq 0, \quad S_{n0} \geq 0, \qquad n = 1, \dots, N.$$

$$Xl_{n0} \leq X_{n0} \leq Xu_{n0}, \qquad n = 1, \dots, N.$$

Budget constraint at $t = 0$

Transaction costs are a_n and b_n per unit of market value purchased and sold, respectively. For a purchase amount P_{n0}, the amount of cash paid out is $(1 + a_n)P_{n0}$. Similarly, for a sale amount S_{n0}, the amount received is $(1 - b_n)S_{n0}$. The total proceeds from sales must equal the total paid out from purchases, namely

$$\sum_{n=1}^{N}(1 + a_n)P_{n0} - \sum_{n=1}^{N}(1 - b_n)S_{n0} = 0.$$

Asset accumulation and revision at $t = 1, \ldots, T - 1$

Similar relationships hold at subsequent stages. However, after $t = 0$, all variables are contingent on the scenario path, or node-dependent. Let $R_{nt}(s_t)$ be the return of asset n in period t in node s_t. The market value before revision at stage t is $(1 + R_{nt}(s_t))X_{nt-1}(a(s_t))$, where $a(s_t)$ denotes the parent (or ancestor) node of s_t. The market value after revision at stage t is

$$(1 + R_{nt}(s_t))X_{nt-1}(a(s_t)) + P_{nt}(s_t) - S_{nt}(s_t) - X_{nt}(s_t) = 0,$$

$$n = 1, \ldots, N, \quad s_t = 1, \ldots, S_t.$$

Wealth definition at $t = 1, \ldots, T - 1$.

The total market value is denoted V_t and is the sum of the holdings in each asset class:

$$\sum_{n=1}^{N} X_{nt}(s_t) - V_t(s_t) = 0, \qquad s_t = 1, \ldots, S_t.$$

Nonnegativity and asset bounds at $t = 1, \ldots, T - 1$

Purchases and sales must be nonnegative and holdings may be constrained within bounds:

$$P_{nt}(s_t) \geq 0, S_{nt}(s_t) \geq 0, \qquad n = 1, \ldots, N, s_t = 1, \ldots S_t$$

$$Xl_{nt} \leq X_{nt}(s_t) \leq Xu_{nt}, \qquad n = 1, \ldots, N, s_t = 1, \ldots S_t$$

Budget constraint at $t = 1, \ldots, T - 1$.

There could be net cash flows $C_t(s_t)$ into the porfolio, which may be scenario-dependent. These cash flows are incorporated into the budget at each node.

$$\sum_{n=1}^{N} (1 + a_n)P_{nt}(s_t) - \sum_{n=1}^{N} (1 - b_n)S_{nt}(s_t) = C_t(s_t), \qquad s_t = 1, \ldots, S_t$$

Asset accumulation and wealth definition at $t = T$

$$\sum_{n=1}^{N} (1 + R_{nT}(s_T))X_{nT-1}(a(s_T)) - V_T(s_T) = 0 \qquad s_T = 1, \ldots, S_T.$$

Allocation group restrictions at $t = 0, \ldots, T-1$

There can be restrictions on groups of assets. An example of this might be: domestic equities must be at least 30% of total global equities.

$$\kappa_l \sum_{n \in I} X_{nt}(s_t) \leq \sum_{i \in J} X_{it}(s_t) \leq \kappa_u \sum_{n \in I} X_{nt}(s_t), \qquad s_t = 1, \ldots, S_t,$$

where I and J are sets of asset classes.

Objective function

The objective function trades off the desire for more wealth against the desire to reduce embarrassments.

$$\text{maximise} \quad E\left[V_T - \frac{1}{\lambda} \sum_{t=1}^{T} (1+\gamma)^{N_T - N_t} v_t \sum_k u_k w_{kt} c_k(M_{kt}) \right] =$$

$$\sum_{s_T = 1}^{S_T} p_T(s_T) V_T(s_T) - \frac{1}{\lambda} \sum_{t=1}^{T} (1+\gamma)^{N_T - N_t} v_t \sum_{s_t = 1}^{S_t} p_t(s_t) \sum_{k=1}^{K} u_k w_{kt} c_k(M_{kt}(s_t))$$

The objective is to maximize expected wealth less a measure of risk. We define risk in terms of quantifiable outcomes that are undesirable from the investor's viewpoint. An example of a quantifiable outcome would be a shorfall in wealth below a given threshold. More generally, we call such outcomes embarrassments, denoted M_{kt}, and there may be more than one type of embarrassment.

Embarrassments are penalized in the objective function through a cost function $c_k(M_{kt})$. The various types of embarrassments are weighted relative to each other and relative to the ending wealth by a variety of parameters. The parameters are u_k, v_t, w_{kt}. There is a discount rate parameter γ and an overall risk tolerance parameter λ.

Embarrassments

Some specific types of embarrassments are useful.

1. Wealth Threshold

   The most straightforward type of embarrassment is based on specified wealth thresholds:

   $$V_t(s_t) + M_{1t}(s_t) \geq \alpha L_t, \qquad s_t = 1, \ldots, S_t,$$

   $$M_{1t}(s_t) \geq 0, \qquad s_t = 1, \ldots, S_t,$$

where $L_t, t = 1, \ldots, T$ and α are input parameters.

Cost function slopes are derived from input data for α. For example, given the input data

	Target Level	Higher Cost	Severe Cost
Target Wealth Ratio	110%	100%	95%

the segment slopes are

$$s_1 = 1.0,$$
$$s_2 = 2.5$$
$$s_3 = 3.5$$

2. Benchmark Portfolio Wealth Target

   Instead of absolute wealth, the thresholds can be specified relative to the cumulative wealth earned by a benchmark portfolio:

$$V_t(s_t) + M_{2t}(s_t) \geq \alpha \prod_{i=t}^{1} \left(1 + \left(\sum_{n=1}^{N} \bar{W}_n R_{ni}(\rho_i(s_t)) \right) + \beta_i \right) V_0, \quad s_t = 1, \ldots, S_t$$

$$M_{2t}(s_t) \geq 0, \quad s_t = 1, \ldots, S_t$$

   where

$$\rho_i(s_t) = a(\cdots a(a(s_t))), \quad \rho_t(s_t) = s_t$$

   and where $\alpha, \beta_i, i = 1, \ldots, T$ and $\bar{W}_n, n = 1, \ldots, N$ are input parameters.

   Benchmark weights must satisfy either $\sum_{n=1}^{N} \bar{W}_n = 1.0$ or 0.

3. Benchmark Portfolio Rate of Return Target

   Instead of cumulative wealth, this embarrassment assigns penalties to period by period shortfalls in the rate of return:

$$\sum_{n=1}^{N} R_{nt}(s_t) X_{nt-1}(a(s_t)) - \left(\sum_{n=1}^{N} \bar{W}_n R_{nt}(s_t) + \delta_t \right) V_{t-1}(a(s_t)) + M_{3t}(s_t) \geq 0,$$

$$M_{3t} \geq 0, s_t = 1, \ldots, S_t.$$

   where $\delta_t, t = 1, \ldots, T$ and $\bar{W}_n, n = 1, \ldots, N$ are input parameters. Benchmark weights must satisfy either $\sum_{n=1}^{N} \bar{W}_n = 1.0$ or 0.

References

L. Balzer. (1994) 'Measuring Investment Risk: A Review,' *Journal of Investing* **3** (3) (Fall), 47–58.

A.J. Berger and J.M. Mulvey. (1998) 'Asset and Liability Management for Individual Investors, this volume, 634–665.

D.R. Cariño and Y. Fan. (1993) 'Alternative Risk Measures for Asset Allocation,' *Gestion Collective internationale*, No. 2 (July/August), 47–51.

D.R. Cariño, Y. Fan, E. Ankrim, and P. Bouchey. (1995) 'Alternative Risk Measures for Asset Allocation,' *Russell Research Commentary*, (December), Frank Russell Company.

D.R. Cariño, T. Kent, D.H. Myers, C. Stacy, M. Sylvanus, A.L. Turner, K. Watanabe, and W.T. Ziemba. (1994) 'The Russell–Yasuda Kasai Model: An Asset/Liability Model for a Japanese Insurance Company Using Multistage Stochastic Programming,' *Interfaces* **24**, (1) (January–February), 29–49. Reprinted in this volume, 609–633.

V.K. Chopra, C.R. Hensel, and A.L. Turner. (1993) 'Massaging Mean-Variance Inputs: Returns from Alternative Global Investment Strategies in the 1980s,' *Management Science* **39** (7) (July), 845–855.

M.A.H. Dempster and A.M. Ireland. (1988) 'A financial expert decision support system.' In *Mathematical Models for Decision Support*, B. Mitra (ed.) NATO ASI Series, Vol. F48, Springer-Verlag, 415–440.

C.R. Hensel and A.L. Turner. (1998) 'Making Superior Asset Allocation Decisions in a Mean-Variance Framework,' this volume, 62–83.

A.J. King, and T. Warden. (1994) 'Stochastic Programming for Strategic Portfolio Management,' 15th International Symposium on Mathematical Programming, Ann Arbor, August 1994.

M.I. Kusy and W.T. Ziemba. (1986) 'A Bank Asset and Liability Management Model,' *Operations Research* **34** (3) 356–376.

H.M. Markowitz. (1952) 'Portfolio Selection,' *Journal of Finance* **7** (1) (March), 77–91.

H.M. Markowitz. (1959) *Portfolio Selection: Efficient Diversification of Investments*, Cowles Foundation Monograph 16, Yale University Press.

R.C. Merton. (1969) 'Lifetime Portfolio Selection under Uncertainty: The Continuous-Time Case,' *Review of Economic Statistics* **51**, 247–257.

R.C. Merton. (1971) 'Optimum Consumption and Portfolio Rules in a Continuous-Time Model,' *Journal of Economic Theory,* **3**, 373–413.

J.M. Mulvey. (1995) 'Financial Planning via Multi-Stage Stochastic Optimization,' Statistics and Operations Research Technical Report SOR–94–09, Princeton University. To appear in *ORSA Journal on Computing*.

J.M. Mulvey and E. Thorlacius. (1998) 'The Towers Perrin Global Capital Market Scenario Generation System: CAP:Link,' this volume, 286–312.

P.A. Samuelson. (1969) 'Lifetime Portfolio Selection by Dynamic Stochastic Programming,' *Review of Economics and Statistics* **51** (August), 239–46.

W.F. Sharpe and G.J. Alexander. (1990) *Investments*, 4th edition, Prentice-Hall.

A.L. Turner and C.R. Hensel. (1993) 'Were the Returns from Stocks and Bonds of Different Countries Really Different in the 1980s?' *Management Science* **39** (7) (July), 835–844.

J. von Neumann and O. Morgenstern. (1944) *Theory of Games and Economic Behavior*, Princeton University Press. 2d ed., 1947; 3d ed., 1953.

S.A. Zenios. (1991) 'Massively Parallel Computations for Financial Planning Under Uncertainty.' Chapter 18 in *Very Large Scale Computation in the 21st Century*, J. Mesirov (ed.), SIAM, 273–294.

S.A. Zenios. (1993) (ed.) *Financial Optimization*, Cambridge University Press.

The Use of Treasury Bill Futures in Strategic Asset Allocation Programs

Michael J. Brennan and Eduardo S. Schwartz

1 Introduction

Asset allocation models are designed to improve investment results by varying the allocation of a portfolio between broad asset classes over time[1], in response to changing assessments of expected returns (and risk) on the different asset classes. In *tactical* asset allocation the objective function is defined over the one-period (typically one quarter) return on the portfolio. Tactical asset allocation models have been quite successful[2], and gained greatly in popularity following the Crash of 1987.

However, as Brennan, Schwartz and Lagnado (1996) have argued, the analysis of Merton (1971) implies that the tactical asset allocation strategy, when employed over more than one period, is logically flawed[3]. This is because it fails to take account of the possibility of future changes in the investment opportunity set. Yet it is the fact that the opportunity set *does* change over time which provides the basis for the tactical asset allocation strategy.

The importance of taking account of changes in the opportunity set may be illustrated with a simple example. Consider an investor who is investing over a 10 year horizon and has a utility function defined over wealth at the horizon. Suppose that the investor has the choice between investing in short term bonds or a single 10 year pure discount bond. If interest rates vary stochastically, the return on the 10 year bond will appear risky from the viewpoint of an investor who is concerned only with wealth at the end of one year. However, from the viewpoint of the 10 year investor the 10 year bond is actually riskless. This is because when the price of the 10 year bond falls, future investment opportunities, in the shape of expected future interest rates, improve, and in fact, from the viewpoint of our 10 year investor who

---

[1] Portfolio risk for a large investor depends mainly on allocations to broad asset classes rather than the selection of individual securities. See Mulvey and Ziemba (1995).

[2] See, for example, Evnine and Henriksson (1987).

[3] Except in the case of a logarithmic utility function for which the analysis of Hakansson (1971), for example, shows that the myopic behaviour is optimal even in the face of stochastically evolving investment opportunities.

other words, the 10 year bond, while risky in terms of its effect on wealth at the end of one year, has the additional desirable characteristic of allowing the investor to *hedge* against changes in future investment opportunities. This second aspect of long term bonds is entirely missed by tactical asset allocation models, which assume that the investor is concerned solely with wealth at the end of one period and therefore does not value the hedging characteristics of different financial instruments.

Brennan, Schwartz and Lagnado (1996) have coined the term *strategic asset allocation* to refer to dynamic asset allocation strategies that take account, not only of the time-variation in expected returns on different asset classes, but also of the investor's horizon. They show that investment strategies are significantly affected by the investor's time horizon when the investment opportunity set is described by three state variables, the short term interest rate, the long term interest rate, and the dividend yield on the stock market portfolio. This paper extends that analysis by allowing the investor to take long and short positions in short term interest rate futures as well as in bonds, stock and cash. The significance of the short term interest rate futures contract is that it allows the investor to hedge against unanticipated changes in the short term rate of interest. A position in long term bonds allows the investor to hedge perfectly against changes in the long term rate and a position in stock allows the investor to hedge almost perfectly against changes in the dividend yield of the stock market portfolio because most of the change in the dividend yield is associated with the return in the stock market portfolio. Without this contract the investor is able to hedge against changes in only two of the three state variables. We demonstrate that this new investment opportunity, which allows the investor to hedge against changes in the future investment opportunity set associated with changes in the short term rate of interest, leads to a significant improvement in expected utility. In fact, under our assumptions and with historical parameter estimates, an investor with a 19 year horizon would be willing to pay as much as $3.04 per dollar of his initial wealth in order to have the right to trade in Treasury Bill futures as well as stock, bonds and cash.

2   Time-Variation in Expected Returns

While time-variation in the expected returns on short-term cash instruments is visible to the naked eye, and time variation in the expected returns on other fixed income instruments may reasonably be inferred by extension, the random walk model of stock price behavior has nevertheless continued to hold the ascendancy among academics until relatively recently[5]. This is despite

---

[5]Bossaerts and Hillion (1994), and Pesaran and Timmerman (1995), argue that the best model for predicting monthly stock returns changes over time, possibly as the result of learning.

the fact that it is more than 20 years since Lintner (1975) reported evidence that expected returns on stocks vary over time in a manner that is related to the short term interest rate; his results have been confirmed more recently by Keim and Stambaugh (1986), and others. [6]. Therefore the first variable that we employ as a predictor of asset returns is the short term riskless interest rate. Of course, this variable not only predicts stock returns, but it is also the (forecast of the) return on cash, which is an important asset class in its own right. The second variable that we use as a predictor of expected stock returns is the dividend yield on a stock index portfolio, defined as the ratio of the past 12 months' dividends divided by the current stock price. This variable is more controversial. Fama and French (1988), and Campbell and Shiller (1988), have provided evidence of a relation between dividend yields and subsequent stock returns. Their findings have been criticized by Nelson and Kim (1993) as being the result of small sample bias, and although Hodrick (1992) concludes after a lengthy study that *'the estimates and the Monte Carlo studies support the conclusion that changes in dividend yields forecast significant expected changes in expected stock returns'* (p383), Goetzmann and Jorion (1993) claim to show that his results do not take proper account of the fact that the regressor (the dividend yield) behaves like a lagged dependent variable because it depends on lagged returns. We nevertheless include it. The third variable that we include in our analysis is the yield on a long term bond, which is intended to proxy for the yield on a consol bond. In conjunction with the short term interest rate, this is equivalent to including a *'term premium* or slope of the yield curve as a predictor of stock returns. Campbell (1987), Fama and French (1989) and Keim and Stambaugh (1986) include such a variable, although Hodrick (1992) claims that it has no marginal explanatory power for stock returns in the presence of the dividend yield and the Treasury Bill rate. Our primary reason for including the long term interest rate is its role as a predictor of changes in the short term interest rate which has previously been documented by Brennan and Schwartz (1982), and Fama (1976). In summary, the three variables that we use to predict asset returns are the short rate, r, the long rate, l, and the dividend yield on the stock portfolio, δ. These state variables are assumed to follow a joint continuous time Markov process.

Thus, denoting the instantaneous rate of return on the stock portfolio by dS/S, the state variables and the stock return are assumed to follow a joint

---

[6]See also Campbell (1987), and Pesaran and Timmerman (1995). Hodrick (1987) reports that the confidence level of the test statistic for rejecting the null of 'no effect of Treasury Bill returns on stock returns' exceeds 0.999 for the period 1952 to 1987. Attempts to account for this empirical regularity include Fama (1981), Schwert (1981), Geske and Roll (1983), and Stulz (1986). Since the short term interest rate is closely related to the rate of inflation, stock returns are negatively related to the expected rate of inflation also. Kaul (1987) argues that this phenomenon is due not only to a negative relation between inflation and real activity but also to the counter-cyclical responses of the monetary authorities in the post war era.

stochastic process of the form:

$$\frac{dS}{S} = \mu_s dt + \sigma_s dz_s \qquad (2.1)$$

$$dr = \mu_r dt + \sigma_r dz_r \qquad (2.2)$$

$$dl = \mu_l dt + \sigma_l dz_l \qquad (2.3)$$

$$d\delta = \mu_\delta dt + \sigma_\delta dz_\delta. \qquad (2.4)$$

where the parameters $\mu_i, \sigma_i (i = r, l, \delta, S)$ are at most functions of the state variables r, l, δ, and dz_i are increments to Wiener processes. The correlation coefficients between the increments to the Wiener processes are denoted by $\rho_{r\delta}$ etc.

The empirical data that are used in the analysis that follows are taken from the period January 1976 to December 1994. This sample period was chosen because data on interest rate futures were available only from 1976. The specific data series are as follows:

Stock return: the monthly rate of return on the CRSP Value-Weighted Index.

- *Dividend Yield*: the sum of the past 12 months' dividends on the CRSP Value-Weighted Index divided by the value of the index at the end of the previous month.

- *Short (term interest) rate*: the yield on a one month Treasury Bill as of the beginning of each month, taken from the CRSP Government Bond Files[7].

- *Long (term interest) rate*: the (continuously compounded) yield to maturity on the longest maturity, taxable, non-callable government bond excluding flower bonds, as of the beginning of each month. This was taken from the CRSP Government Bond Files.

- *Return on the long term bond*: the monthly return including coupons on the bond used to compute the long rate.

In addition, data were obtained from Knight-Ridder on the International Monetary Market 3-month Treasury Bill futures contract. Monthly futures returns were computed using the month-end prices of the nearest to maturity contract that had more than one month to maturity when entered into[8].

---

[7]We discovered an anomaly in this file concerning the 1 month rate for April 30, 1987. The value for the 1 month rate on this date was 2.507%; the yields on the bills maturing 1 week earlier and 1 week later averaged 4.815%. This average value was substituted for the observed value for the empirical analysis.

[8]The quotes were converted into prices, and 'returns' were computed.

3   The Portfolio Problem

The investor is assumed to be able to invest in three asset classes as well as to take positions in interest rate futures. The asset classes assumed to be available to the investor are cash with sure rate of return, r; stock, whose rate of return is given by equation (2.1); and consol bonds. The price of a consol bond, $B(l)$, is inversely proportional to its yield, l. The total return on a consol bond is the sum of the yield and the price change; then a simple application of Ito's Lemma implies that the instantaneous total return on the consol bond is given by;

$$\frac{dB}{B} + ldt = \left(1 - \frac{\mu_l}{l} + \frac{\sigma_l^2}{l^2}\right)dt - \frac{\sigma_l}{l}dz_l. \tag{3.1}$$

It is assumed that the futures price of the Treasury Bill depends only on the short term interest rate so that the proportional change in the futures price may be written as:

$$\frac{dF}{F} = \mu_F dt + f(r)\sigma_r dz_r. \tag{3.2}$$

Define x as the proportion of the investment portfolio that is invested in stock, y as the proportion that is invested in the consol bond, and let z denote the notional value of the futures position expressed as a proportion of wealth[9]. Then the stochastic process for wealth, W, is:

$$
\begin{aligned}
\frac{dW}{W} &= \left[x\frac{dS}{S} + y\left(\frac{dB}{B} + ldt\right) + z\frac{dF}{F} + (1 - x - y)rdt\right] \\
&= \left[x(\mu_s - r) + y\left(1 - r - \frac{\mu_l}{l} + \frac{\sigma_l^2}{l^2}\right) + z\mu_F + r\right]dt \\
&\quad + \left[x^2\sigma_s^2 + y^2\frac{\sigma_l^2}{l^2} + z^2 f^2\sigma_r^2 - \frac{2xy\sigma_l\sigma_s\rho_{sl}}{l} + \right. \\
&\quad\quad \left. +2xz f\sigma_s\sigma_r\rho_{rS} - \frac{2yz f\sigma_r\sigma_l\rho_r l}{l}\right]^{1/2}dz_w \\
&\equiv \mu_W dt + \sigma_W dz_W
\end{aligned}
\tag{3.3}
$$

The Bellman equation for the investor's stochastic optimal control problem is

$$\max_{x,y,z} E[dV] = 0 \tag{3.4}$$

---

[9]We use the term '*notional value*' to remind the reader that a futures position requires no initial investment. If Q is the notional value of a futures position then the number of futures contracts held is Q/F where F is the futures price.

Or,

$$\max_{x,y,z} \Big[V_W \mu_W W + V_r \mu_r + V_l \mu_l + V_\delta \mu_\delta - V_\tau$$

$$+\frac{1}{2} V_{WW} \sigma^2{}_W W^2 + \frac{1}{2} V_{rr} \sigma^2{}_r + \frac{1}{2} V_{ll} \sigma^2{}_l + \frac{1}{2} V_{\delta\delta} \sigma^2{}_\delta$$

$$+V_{W\delta} W \sigma_\delta \sigma_W \rho_{\delta W} + V_{rl} \sigma_r \sigma_l \rho_{rl} + V_{r\delta} \sigma_\delta \sigma_r \rho_{r\delta}$$

$$V_{Wr} W \sigma_r \sigma_W \rho_{rW} + V_{Wl} W \sigma_l \sigma_W \rho_{lW} + V_{l\delta} \sigma_l \sigma_\delta \rho_{l\delta} \Big] = 0 \qquad (3.5)$$

The control problem (3.5) has four state variables including W. To reduce the number of state variables, we assume that utility is of the isoelastic form so that;

$$V(r, l, \delta, W, 0) = \frac{1}{\gamma} W^\gamma, \qquad \text{for } \gamma < 1 \qquad (3.6)$$

Then we can verify that $V(r, l, \delta, W, \tau)$ may be written as $\gamma^{-1} W^\gamma v(r, l, \delta, \tau)$, where:

$$v(r, l, \delta, 0) = 1 \qquad (3.7)$$

and

$$\max_{x,y,z} \Big[\mu_W v + \frac{1}{\gamma}\mu_r v_r + \frac{1}{\gamma}\mu_l v_l + \frac{1}{\gamma}\mu_\delta v_\delta - \frac{1}{\gamma} v_\tau$$

$$+\frac{1}{2}(\gamma-1)\sigma_W^2 v + \frac{1}{2\gamma}\sigma_r^2 v_{rr} + \frac{1}{2\gamma}\sigma_l^2 v_{ll} + \frac{1}{2\gamma}\sigma_\delta^2$$

$$+\sigma_W \sigma_r \rho_{rW} v_r + \sigma_W \sigma_l \rho_{lW} v_l + \sigma_W \sigma_\delta \rho_{\delta W} v_\delta$$

$$+\frac{1}{\gamma}\sigma_r \sigma_l \rho_{rl} v_{rl} + \frac{1}{\gamma}\sigma_r \sigma_\delta \rho_{r\delta} v_{r\delta} + \frac{1}{\gamma}\sigma_l \sigma_\delta \rho_{l\delta} v_{l\delta} \Big] = 0 \qquad (3.8)$$

where

$$\sigma_W \sigma_r \rho_{rW} \equiv \sigma_{Wr} = x\sigma_S \sigma_r \rho_{rS} - y\frac{\sigma_l}{l}\sigma_r \rho_{rl} + zf\sigma_r^2$$

$$\sigma_W \sigma_l \rho_{lW} \equiv \sigma_{Wl} = x\sigma_S \sigma_l \rho_{lS} - y\frac{\sigma_l^2}{l} + zf\sigma_r \sigma_l \rho_{rl}$$

$$\sigma_W \sigma_\delta \rho_{\delta W} \equiv \sigma_{W\delta} = x\sigma_S \sigma_\delta \rho_{S\delta} - y\frac{\sigma_l}{l}\sigma_\delta \rho_{\delta l} + zf\sigma_r \sigma_\delta \rho_{r\delta} \qquad (3.9)$$

Substituting for μ_W and collecting terms, we have finally:

$$\max_{x,y,z} \Big\{ v\Big[x(\mu_S - r) + y\Big(l - r - \frac{\mu_l}{l} + \frac{\sigma_l^2}{l^2}\Big) + z\mu_F + r$$

$$+\frac{1}{2}(\gamma-1)\Big(x^2\sigma_S^2 + y^2\frac{\sigma_\delta^2}{l^2} + z^2 f^2 \sigma_r^2 - \frac{2xy\sigma_S \sigma_l \rho_{Sl}}{l}$$

$$+2xzf\sigma_{Sr} - \frac{2yzf\sigma_{rl}}{l}\Big)\Big]$$

$$+v_r \Big[\frac{1}{\gamma}\mu_r + x\sigma_{Sr} - \frac{y}{l}\sigma_{rl} + zf\sigma_r^2\Big] + v_l \Big[\frac{1}{\gamma}\mu_l + x\sigma_{Sl} - \frac{y}{l}\sigma_l^2 + zf\sigma_{rl}\Big]$$

$$+v_\delta \left[\frac{1}{\gamma}\mu_\delta + x\sigma_{S\delta} - \frac{y}{l}\sigma_{l\delta} + zf\sigma_{r\delta}\right] - \frac{1}{\gamma}v_\tau$$

$$+\frac{1}{\gamma}\left[\frac{1}{2}v_{rr}\sigma_r^2 + \frac{1}{2}v_{ll}\sigma_l^2 + \frac{1}{2}v_{\delta\delta}\sigma_\delta^2 + v_{rl}\sigma_{rl} + v_{r\delta}\sigma_{r\delta} + v_{l\delta}\sigma_{l\delta}\right]\Bigg\} \quad (3.10)$$

The first order conditions for a maximum in (3.10) imply that the optimal controls, $x^* \equiv x^*(r,l,\delta,\tau)$, $y^* \equiv y^*(r,l,\delta,\tau)$, and $z^* \equiv z^*(r,l,\delta,\tau)$ are given by:

$$\begin{bmatrix} \sigma_S^2 & -\frac{\sigma_{Sl}}{l} & \sigma_{Sr} \\ -\frac{\sigma_{Sl}}{l} & \frac{\sigma_l^2}{l^2} & -\frac{\sigma_{rl}}{l} \\ \sigma_{Sr} & -\frac{\sigma_{rl}}{l} & \sigma_r^2 \end{bmatrix}\begin{bmatrix} x^* \\ y^* \\ z^*f \end{bmatrix} = \begin{bmatrix} \text{rhs}_1 \\ \text{rhs}_2 \\ \text{rhs}_3 \end{bmatrix} \quad (3.11)$$

where

$$\text{rhs}_1 = -\frac{1}{\gamma-1}\left[\mu_S - r + \frac{v_r}{v}\sigma_{Sr} + \frac{v_l}{v}\sigma_{Sl} + \frac{v_\delta}{v}\sigma_{S\delta}\right]$$

$$\text{rhs}_2 = -\frac{1}{\gamma-1}\left[l - r - \frac{\mu_l}{l} + \frac{\sigma_l^2}{l^2} - \frac{v_r}{v}\frac{\sigma_{rl}}{l} - \frac{v_l}{v}\frac{\sigma_l^2}{l} - \frac{v_\delta}{v}\frac{\sigma_{l\delta}}{l}\right]$$

$$\text{rhs}_3 = -\frac{1}{\gamma-1}\left[\frac{\mu_F}{f} + \frac{v_r}{v}\sigma_r^2 + \frac{v_l}{v}\sigma_{rl} + \frac{v_\delta}{v}\sigma_{r\delta}\right]$$

To determine z^*, the optimal position in the futures, it is necessary to define μ_F, the drift of the futures price process, and $f(r)$, the sensitivity of the change in the futures price to the change in the short term rate, r. One way to do this would be to use an equilibrium model of bond pricing such as the Cox–Ingersoll–Ross (1985) model or the Brennan–Schwartz (1982) model to determine the drift and sensitivity. However, both of these models require that a market risk aversion parameter be assessed, in addition to the parameters of the interest rate process. Therefore, for the sensitivity we adopted the simpler expedient of using the duration model[10]. Insofar as there is little maturity effect in interest rates out to the maturity of the underlying Treasury Bill, we expect that this will be quite accurate. Thus define i as the relevant interest rate. Then the price of a futures contract on a Bill maturing in T^* years, which is deliverable in T years, is given by $F(T,T^*) = \exp\{i(T - T^*)\}$. This implies that $dF/di = (T - T^*)F$. In our application $T - T^*$ is one quarter so that $f = -0.25$, and we identify i with the short rate, r.

For μ_F, the proportional drift in the futures price process, there are several possible modelling strategies. The first, which we refer to as *Model 1*, recognizes that μ_F is simply a risk premium associated with changes in the short rate, and assumes that the risk premium is constant over time; we take the risk premium as equal to zero, so that $\mu_F = 0$. *Model 2* assumes that $\mu_F = f\mu_r$; this is consistent with the assumption of the duration model that

[10]See Ingersoll (1979).

the term structure of interest rates is flat so that the yield underlying the futures price is the current short rate. However, unlike *Model 1*, it does not allow for the possibility that the futures price will reflect the expected change in the short rate (represented by the drift in equation (2.2)). A third possibility is to allow μ_F to be a linear function of r, l and δ, and to estimate the functional form as in equations (2.1)–(2.4) – we refer to this as *Model 3*[11].

In order to implement the model, it is necessary first to estimate the parameters of the stochastic process for the state variables. We consider that next.

4 Estimation of the Stochastic Process

Following Brennan, Schwartz and Lagnado (1996), we assume that the expected returns on stocks and bonds, and the drifts of the dividend yield and short rate, are linear functions of the three state variables, r, l, and δ, while the volatility of each state variable is assumed to be proportional to its current level, and the volatility of the stock rate of return is taken as constant. This implies from equation (3.1) that the drift of the long rate is a non-linear function of the state variables, being equal to the product of l and a linear function of the state variables. This specification implies that the joint stochastic process may be written as:

$$\frac{dS}{S} = (a_{S1} + a_{S2}\delta + a_{S3}r + a_{S4}l)dt + \sigma_S dz_S \tag{4.1}$$

$$dr = (a_{r1} + a_{r2}\delta + a_{r3}r + a_{r4}l)dt + r\sigma_r dz_r \tag{4.2}$$

$$dl = l(a_{l1} + a_{l2}\delta + a_{l3}r + a_{l4}l)dt + l\sigma_l dz_l \tag{4.3}$$

$$d\delta = (a_{\delta1} + a_{\delta2}\delta + a_{\delta3}r + a_{\delta4}l)dt + \delta\sigma_\delta dz_\delta \tag{4.4}$$

The dividend yield is defined as the sum of the past 12 months' dividends divided by the current level of the stock index, S. The specification (4.4) must therefore be regarded as an approximation since the stochastic process for lagged dividends is not modelled explicitly; to have done so would have introduced a fourth state variable into the analysis which would have considerably increased the difficulty of solving the control problem. However, we expect that the stochastic increment to the dividend yield will have a strong negative correlation with the return on the stock, since most of the stock return is accounted for by price changes.

The joint stochastic process was estimated on monthly data for the period January 1976 to December 1994, using a discrete approximation to the continuous process. The time-paths of r, l and δ during the sample period are shown in Figure 1. It is apparent that there is the most variability in

---

[11]We continue to assume that the proportional innovation in the futures price is perfectly correlated with innovations in the short rate. See Shiller (1979) for a similar finding.

STATE VARIABLES

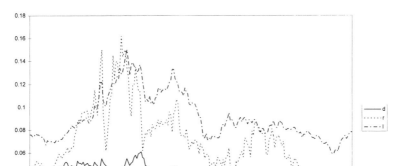

Figure 1. Monthly time series of stock dividend yield (d), short term interest rate (r), and long term interest rate (l)

the short rate, followed by the long rate, while the dividend yield fluctuates between 2% and 6%.

The system of equations (4.1)–(4.4) was estimated by non-linear seemingly unrelated regression using TSP. Table 1 reports the regression estimates. As previous investigators have found, the expected return on common stocks is negatively related to the current level of the short rate and positively related to the level of the dividend yield, but is not significantly related to the long rate (and therefore to the slope of the yield curve). As Brennan and Schwartz (1982) have found, the change in the short rate is negatively related to its current level and positively but not significantly related to the level of the long rate. The change in the long rate is negatively related to its current level and positively related to the short rate at conventional levels of significance[12]. The change in the dividend yield is negatively related to its current level, so that it shows mean reversion; in addition, it is positively related to the short rate[13]. Table 2 reports the correlations between the innovations in the state variables. As anticipated, the innovation, or unexpected change, in the dividend yield is very highly negatively correlated with the innovation in stock returns; it is also positively correlated with the innovation in the long

---

[12]This implies that the risk premium on long-term bonds, like that on stocks, is *negatively* related to the level of the short rate.

[13]This corresponds to the negative coefficient of the stock return on the short rate, since a negative stock return is associated with an increase in the dividend yield.

	Constant	δ	r	l	σ
dS/S	−0.037	1.904	−0.451	0.025	0.041
	(2.25)	(3.95)	(2.69)	(0.11)	
dr	−0.008	0.234	−0.124	0.074	0.114
	(2.48)	(2.54)	(3.36)	(1.60)	
dl	0.033	−0.284	0.328	−0.520	0.038
	(1.99)	(0.59)	(2.28)	(2.33)	
$d\delta$	0.002	−0.072	0.021	−0.004	0.044
	(2.56)	(3.18)	(2.72)	(0.42)	

(t-statistics in parentheses)
Log Likelihood: 1670

Table 1. The Estimated Stochastic Process for the State Variables and the Stock Return. January 1976–December 1994.

	Stock Return	r	l	δ
Stock return	1.0			
r	−0.022	1.0		
l	−0.358	0.362	1.0	
δ	−0.944	0.068	0.326	1.0

Table 2: Correlations of State Variables and Stock Return Innovations. January 1976–December 1994

rate, because the innovation in the long rate is negatively correlated with the innovation in stock returns. The innovations in the long and short rates have a correlation of 0.36, while the correlation between the stock return and the innovation in the short rate is only −0.02.

While the focus of this paper is on the optimal investment policy assuming that the parameters of the stochastic process are known, we shall also perform an out of sample experiment in which the policy used for the second half of the sample period is computed using parameter estimates derived from the first half. Therefore, Tables 3 and 4 report the parameter estimates obtained for the two halves of the sample period separately[14].

---

[14]To conserve space we do not report the correlations of the innovations for the subperiods.

	Constant	δ	r	l	σ
dS/S	−0.080	2.322	−0.574	0.328	0.039
	(2.60)	(3.21)	(2.77)	(1.27)	
dr	−0.0003	0.0466	−0.112	0.078	0.109
	(0.05)	(0.29)	(1.92)	(1.27)	
dl	0.076	−1.154	0.474	−0.588	0.039
	(2.31)	(1.52)	(2.24)	(2.21)	
$d\delta$	0.005	−0.106	0.032	−0.024	0.041
	(3.10)	(2.93)	(3.06)	(1.86)	

(t-statistics in parentheses)
Log Likelihood: 854

Table 3. The Estimated Stochastic Process for the State Variables and the Stock Return for the First Half of the Sample Period: January 1976–June 1985.

	Constant	δ	r	l	σ
dS/S	−0.071	6.602	−0.759	−1.329	0.040
	(1.89)	(4.47)	(2.32)	(2.13)	
dr	−0.008	0.451	−0.153	0.003	0.118
	(1.70)	(2.14)	(3.10)	(0.04)	
dl	0.081	−0.321	0.225	−1.049	0.035
	(1.99)	(0.59)	0.78	1.86	
$d\delta$	0.003	−0.232	0.028	0.047	0.044
	(1.92)	(4.16)	(2.16)	(2.03)	

(t-statistics in parentheses)
Log Likelihood: 835

Table 4: The Estimated Stochastic Process for the State Variables and the Stock Return for the Second Half of the Sample Period: July 1985–December 1994.

5 Empirical Analysis of the Role of Futures

The stochastic optimal control problem was solved[15] using the parameter estimates reported in Table 1 and a value of the risk aversion parameter, γ, of −5.0. This high value for the risk aversion parameter was chosen because we take the estimated parameters of the stochastic process as known, and with a low risk aversion parameter there is a tendency for the model to take unrea-

---

[15]For a discussion of the solution procedure see Brennan, Schwartz and Lagnado (1996).

sonably aggressive positions[16]. The holdings of bonds, stock and cash were restricted to between zero and 100%, and the absolute value of the nominal value of the futures position was restricted to be less than ten times wealth. The proportional drift of the futures price was first computed according to *Model 1*.

5.1   Asset Proportions

The stochastic optimal control problem yields the optimal portfolio position x^*, y^*, z^*, as functions of the state variables r, l, δ and time to maturity, τ. Then assuming that the horizon was December 31 1994, the strategy functions were combined with the values of the state variables for each month of the sample period to yield times series of the optimal portfolio positions in the different asset classes. Figure 2 shows the optimal positions in bonds, stock and futures[17] – the optimal cash position follows as a residual. There is considerable month to month variability in the portfolio positions – it is noticeable that the futures position tends to be negatively related to the bond position; this is what we should expect, since under *Model 1* the Treasury Bill future provides costless insurance against changes in the short rate and the short rate and the long rate are positively correlated[18]

Figure 3 and Table 5 relate the portfolio positions to the state variables. In Figure 3 the time series of the holdings of bonds and stock are plotted along with their expected returns computed from equations (3.1), (4.1) and (4.3) using the state variables and the parameter estimates in Table 1. The Figure shows that there is positive relation between the expected returns on both bonds and stock and the allocation to those asset classes. Table 5 reports the results of regressions of portfolio proportions[19] on the expected excess returns on bonds and stock. The stock and bond allocations are strongly positively related to the own expected returns and negatively related to the expected returns on the other asset class. The futures position is positively related to the expected returns on stock, and strongly negatively related to the expected returns on bonds. Most significantly, while the allocations to stock and futures are not significantly time dependent, the futures allocation is positively related to the time to the horizon. The nominal futures position

---

[16]For evidence on the importance of parameter errors in mean-variance analysis see Chopra and Ziemba (1993). We are currently working on incorporating estimation risk into the analysis.

[17]The proportional futures position is scaled by a factor of 5 for visual clarity.

[18]It has been suggested to us that perhaps the T-Bill futures contract improves welfare by relaxing the constraint of bond short sales. The fact that the futures position tends to be short when the bond position is long implies that this is not the main economic benefit of introducing a futures contract. In terms of volatility, an allocation to futures of 100% of wealth is roughly equivalent to a 100% investment in stock.

[19]We include in the regressions only those observations for which the portfolio allocation to the asset class was not at its upper or lower limit.

ASSET PROPORTIONS

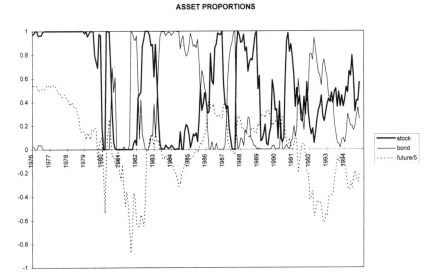

Figure 2. Monthly asset allocations to bonds and stocks, and futures position derived from *Model 1*, using in sample parameter values from Tables 1 and 2. Futures position divided by 5.

expressed as a proportion of wealth increases by about 6% for every year to maturity. Table 6 reports the results of regressing the futures position on the stock and bond allocations as well as time to maturity, using all the observations. The futures position is strongly negatively related to the bond allocation, and positively related to the stock allocation; the former is consistent with the correlation of -0.36 between bond returns and the innovation in the short rate. The positive correlation with the stock allocation is harder to explain; it maybe that this is an artifact of the constraint on the sum of the bond and stock allocations. The futures position increases with the time to the horizon.

For comparison, the optimal strategy without futures contracts was also computed. Figure 4 shows the time series allocations to bonds and stock with and without futures. The differences are often large: there is a tendency for the allocation to bonds to be larger when the opportunity set includes futures, but the effect of futures trading on the stock position is more ambiguous.

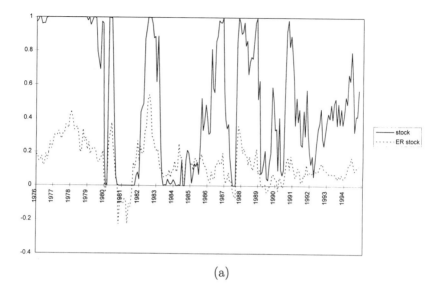

(a)

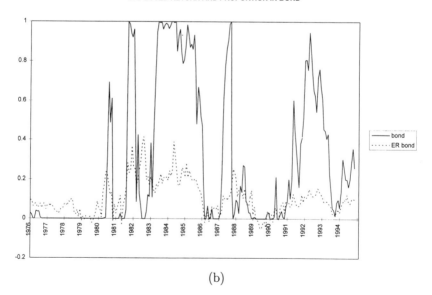

(b)

Figure 3. Monthly time series of allocation to stocks and bonds derived from *Model 1*, using in sample parameter values from Tables 1 and 2, and expected returns on stocks and bonds calculated using the coefficients in Table 1.

STOCK PROPORTIONS WITH AND WITHOUT FUTURES

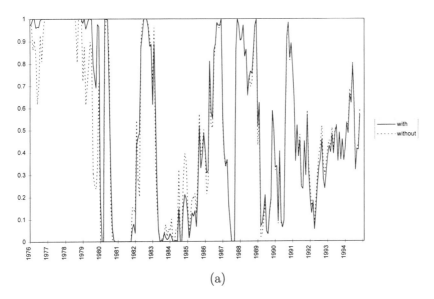

(a)

BOND PROPORTIONS WITH AND WITHOUT FUTURES

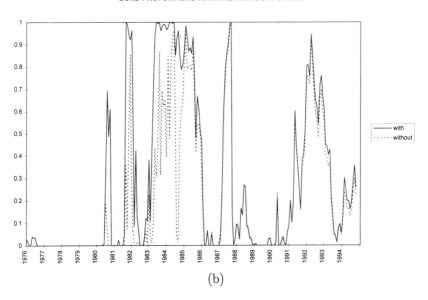

(b)

Figure 4. A comparison of the allocation to stocks and bonds when the opportunity set includes futures contracts and does not. The allocation is derived using *Model 1* when futures contracts are included. Based on in sample parameter estimates reported in Tables 1 and 2.

	Stock Allocation	Bond Allocation	Futures Position
R^2	0.83	0.78	0.44
NOBS	165	141	227
Intercept	0.373	0.235	−0.438
	(18.84)	(8.39)	(2.64)
Time to Maturity	0.000	−0.0004	0.0048
	(0.08)	(1.46)	(3.69)
Expected Excess Return on Stock	4.730	−3.497	6.081
	(26.97)	(16.63)	(8.47)
Expected Excess Return on Bond	−3.213	5.795	−9.985
	(20.27)	(20.61)	(9.78)

Table 5. Regressions of Asset Allocations on Time to Horizon, and Expected Excess Returns on Stocks and Bonds.

Intercept	Stock Allocation	Bond Allocation	Time to Horizon	R^2	Nobs
−0.685	1.007	−1.413	0.0061	0.41	227
(2.80)	(3.44)	(4.81)	(4.65)		

Table 6. Regression of Futures Position on Stock and Bond Allocations, and Time to Horizon

5.2 Investment Returns and the Value of Investment Opportunities

Figure 5 plots the time paths of wealth per dollar invested under the optimal strategies with and without futures, as well as an all-stock strategy. The figure also shows the certainty equivalent of wealth under the strategies. This is defined as the sure amount at the horizon that the investor would exchange for his current wealth *and the opportunity to invest up to the horizon*. Thus the difference between current wealth and the certainty equivalent represents the value of the remaining investment opportunities up to the horizon. The certainty equivalent, CE, is defined

$$\frac{1}{\gamma}(CE)^\gamma = V(W, r, l, \delta, \tau)$$

$$= \frac{1}{\gamma} W^\gamma v(r, l, \delta, \tau). \tag{5.1}$$

Hence

$$CE(W, r, l, \delta, \tau) = W[v(r, l, \delta, \tau)]^{1/\gamma}. \tag{5.2}$$

CERTAINTY EQUIVALENT AND WEALTH

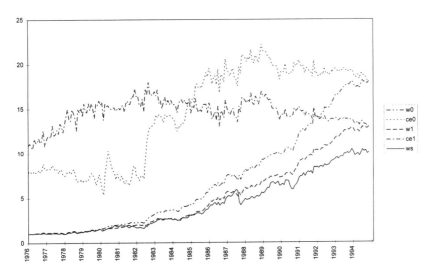

Figure 5. In sample certainty equivalents and wealth for asset allocation strategies with futures (ce1, w1) and without futures contracts (ce0, w0) and for an all stock strategy (ws). The allocation is derived using *model 1* when futures contracts are included. Based on out of sample parameter estimates reported in Table 3.

It is readily verified that since $E[dV] = 0$, $E[d(CE^\gamma] = 0$. Moreover, $CE(W, r, l, \delta, 0) = W$. Thus, under the optimal strategy the certainty equivalent raised to the power γ follows a random walk, and its terminal value is equal to the terminal wealth realized under the optimal strategy. Therefore, the accumulated volatility of the certainty equivalent over the life of the strategy is an appropriate measure of the risk of the strategy[20].

Not surprisingly, both asset allocation strategies substantially outperform the all-stock strategy. The introduction of futures leads to a substantial decrease in risk; as can be seen from the figure, the volatility of the certainty equivalent is much less when futures contracts are introduced[21]. The initial certainty equivalent for the with futures strategy is 10.90 versus only 7.86 for the without futures strategy; thus, the *ex ante* value of the futures opportu-

---

[20]Consider the volatilities of wealth and the certainty equivalent of wealth under a strategy of investing in a pure discount bond with maturity equal to the horizon. Under this riskless strategy the volatility of wealth will be positive while the volatility of the certainty equivalent will be zero.

[21]The volatility of the monthly proportional change in the certainty equivalent is 0.064 without futures and 0.052 with futures

nity is \$3.04 per dollar of initial wealth when the horizon is 19 years[22]. The fact that the final certainty equivalent (and therefore final wealth) is higher under the without futures strategy must be attributed to sampling variability.

The above results are obtained in sample and therefore overstate the advantages of the asset allocation strategies. Therefore, for comparison we show in Figure 6 the results of computing the optimal strategy from *Model 1* using the parameter values reported in Table 3 for the first half of the sample period, and then implementing this strategy for the second half of the sample period without further updating of the parameter values. We do this both with and without trading in futures contracts. The first thing to note is the downward trend in the certainty equivalents. This reflects the fact that the realized investment opportunities are not as good as the *ex ante* assessment of them which treats the parameter values as known. In addition, the certainty equivalent of the with futures strategy is above that of the without futures strategy, not only at inception[23], but throughout. While the final wealth under the pure stock strategy exceeds that under the asset allocation strategies, it should be noted that, not only are the parameters of the asset allocation strategies not updated, but also the pure stock strategy is much more risky[24].

To this point we have relied on the assumption of *Model 1* that the futures price process has zero drift. We consider next the other two models of the futures price process.

5.3   Alternative Models of Futures Price Behavior

Model 2 assumes that the proportional drift in the futures price is $\mu_F = f\mu_r$ where f is -0.25 for the T-Bill futures contract and μ_r is the drift in the short rate. *Model 3* requires that the conditional drift of the futures price be estimated by regressing the proportional change in the futures price on r, l, and δ. Table 7 reports the regression estimates for the whole sample period and two subperiods. It is evident that there is some degree of predictability in the futures price based mainly on the long rate.

Figure 7 shows the wealth and certainty equivalents yielded by the asset allocation strategies based on the three models of futures prices for the whole sample period, with parameters estimated in sample. The certainty equivalent for *Model 1* which assumes no predictability in the futures price is, not surprisingly, initially below the certainty equivalents for the other two mod-

---

[22]When the portfolio proportions are unconstrained the value rises to \$5.20. The fact that the value of the futures contract is higher when borrowing and short sales are allowed demonstrates that the primary benefit of the contract is not to circumvent these restrictions, but to improve hedging opportunities.

[23]The certainty equivalent of the with futures strategy must be at least as great of the without futures strategy when the wealth is the same for the two strategies.

[24]The standard deviation of the monthly return on the stock strategy is 4.4%, as compared with 1.5% and 1.6% for the two asset allocation strategies.

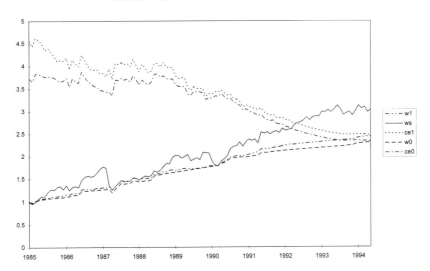

Figure 6. Out of sample certainty equivalents and wealth for as-set allocation strategies with futures (ce1, w1) and without futures contracts (ce0, w0) and for an all stock strategy (ws). The allocation is derived using *Model 1* when futures contracts are included. Based on out of sample parameter estimates reported in Table 3.

	Intercept	δ	r	l	R^2	Nobs
January 1976–	−0.002	0.040	−0.023	0.027	0.05	227
December	(2.95)	(1.67)	(0.027)	(2.50)		
1994						
January 1976–	−0.006	0.116	−0.031	0.034	0.07	114
June 1985	(2.93)	(2.46)	(2.35)	(2.08)		
July 1985–	−0.000	−0.068	0.002	0.032	0.06	113
December	(0.23)	(2.23)	(0.22)	(2.37)		
1994						

Table 7.   Regression of Proportional Change in Futures Price $(\Delta F/F)$ on State Variables δ, r and l.

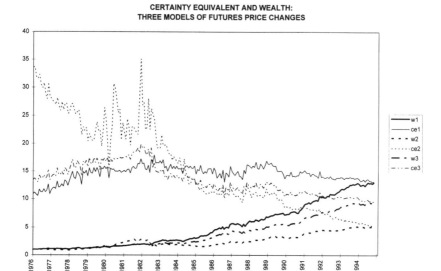

Figure 7. In sample certainty equivalents and wealth for asset allocation strategies with futures based on the three models of futures price changes. *Model 1* (ce1, w1), *Model 2* (ce2, w2), *Model 3* (ce3, w3). Based on out of sample parameter estimates reported in Tables 1 and 2.

els which do assume predictability. *Model 2*, which assumes that the futures price change is as predictable as the changes in the short rate itself, has the highest initial value of the certainty equivalent, but this declines rapidly, as the realized predictability turns out to be less than that assumed. Moreover, the spurious predictability of this model causes it to take large bets on futures, with the result that the return and the certainty equivalent are much more volatile than for the other two models, and for these models there is no downward trend in the certainty equivalent. *Model 1*, which assumes no predictability in the futures price yields the highest final wealth, and the difference between it and *Model 3* is statistically significant at the 5% level.

Figure 8 depicts the out-of-sample performance of the asset allocation strategies based on the three models of futures price behavior. The parameters are estimated over the first half of the sample period and the strategies are implemented over the second half. For all three models the certainty equiv-

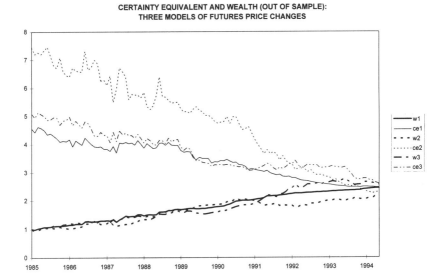

Figure 8. Out of sample certainty equivalents and wealth for asset allocation strategies with futures based on the three models of futures price changes. *Model 1* (ce1, w1), *Model 2* (ce2, w2), *Model 3* (ce3, w3). Based on out of sample parameter estimates reported in Table 3.

alents show a declining trend, reflecting the failure of the predictive models to perform as well out of sample; however, the rate of decline is greatest for *Model 2* with its unrealistic assumption that the futures price will move one for one with the short term rate. The performance of *Models 1 and 3* is almost indistinguishable.

6   Conclusion

In this article we have shown that the opportunity to trade in short-term interest rate futures can be valuable for an investor who has a long-term horizon. There appear to be three reasons for this. First, the futures contract allows the investor to hedge part of the risk of the long term bond, because innovations in the short rate are correlated with innovations in the long rate. Secondly, the short rate is an important determinant of the investor's opportunity set, both because it is the return on an asset class, and because it is an important predictor of stock returns. Therefore the futures contract allows the investor to hedge against adverse shifts in the investment opportunity set. Finally, insofar as the changes in the futures price are predictable, trading in this security allows the investor to place favorable bets on these price changes. We have compared three models of futures price changes. These all assume that the proportional innovation in the price is given by the (negative of the) duration of the underlying bill times the innovation in the short rate, but differ in their assumptions about the proportional drift. *Model 1* assumes that the drift is zero; *Model 2* assumes that it is equal to the (negative of the) duration of the underlying bill times the drift of the short rate. *Model 3* assumes that the drift can be written as a linear function of the three state variables, r, l and δ. We find that asset allocation strategies based on *Models 1 and 3* outperform those based on *Model 2* which assumes too much predictability in futures price changes.

Most of the analysis was conducted using in-sample parameter estimates of the stochastic process governing asset returns since our concern has been with comparing the theoretical value of the futures contract with a known stochastic process for returns. The tendency for the certainty equivalent of wealth to decline over time when the strategies are implemented out of sample points to the need to take account of estimation risk in constructing asset allocation strategies. This is the subject of ongoing research.

Acknowledgments

We are grateful to participants in seminars at the Newton Institute, Cambridge, the University of Chicago, the University of British Columbia, and particularly to George Constantinides and William Ziemba for comments on previous drafts of this article.

References

P. Bossaerts and P. Hillion, (1994), 'Implementing Statistical Criteria to Select Return Forecasting Models,' Working Paper, Caltech.

M.J. Brennan, E.S. Schwartz, and R. Lagnado, (1996), 'Strategic Asset Allocation,' *Journal of Economic Dynamics and Control* **21** 1377–1403.

M.J. Brennan, and E.S. Schwartz, (1982), 'An Equilibrium Model of Bond Pricing and a Test of Market Efficiency,' *Journal of Financial and Quantitative Analysis* **41** 301–329.

J. Campbell, (1987), 'Stock Returns and the Term Structure,' *Journal of Financial Economics* **18** 373–400.

J.Y. Campbell, and R.J. Shiller, (1988), 'The Dividend-Price Ratio and Expectations of Future Dividends and Discount Factors,' *Review of Financial Studies* **1** 195–228.

V.K. Chopra, and W.T. Ziemba, (1993), 'The Effect of Errors in Means, Variances, and Covariances on Optimal Portfolio Choice,' *Journal of Portfolio Management* **19** 6–11.

J.C. Cox, J.E. Ingersoll, and S.A. Ross, (1979), 'Duration and the Measurement of Basis Risk,' *Journal of Business* **52** 51–62.

J.C. Cox, J.E. Ingersoll, and S.A. Ross, (1985), 'A Theory of the Term Structure of Interest Rates,' *Econometrica* **53** 385–407.

J. Evnine, and R. Henriksson, (1987), 'Asset Allocation and Options,' *Journal of Portfolio Management* **14** 56–61.

E.F. Fama, (1976), 'Forward Rates as Predictors of Future Spot Rates,' *Journal of Financial Economics* **3** 361–377.

E.F. Fama, (1981), 'Stock Returns, Real Activity, Inflation and Money,' *American Economic Review* **71** 545–565.

E.F. Fama, and K.R. French, (1988), 'Dividend Yields and Expected Stock Returns,' *Journal of Financial Economics* **22** 3–26.

E.F. Fama, and K.R. French, (1989), 'Business Conditions and Expected Returns on Bonds and Stocks,' *Journal of Financial Economics* **25** 23–50.

R. Geske and R. Roll, (1983), 'The Fiscal and Monetary Linkage between Monetary Policy and Inflation,' *Journal of Finance* **39** 443–455.

W.N. Goetzmann, and P. Jorion, 1993, 'Testing the Predictive Power of Dividend Yields,' *Journal of Finance* **48** 663–680.

N.H. Hakansson, (1971), 'On Optimal Myopic Policies with and without Serial Correlation in Yields,' *Journal of Business* **44** 324–334.

R.J. Hodrick (1992), 'Dividend Yields and Expected Stock Returns,' *The Review of Financial Studies* **5** 357–386.

G. Kaul, (1987), 'Stock Returns and Inflation: the Role of the Monetary Sector,' *Journal of Financial Economics* **18** 253–276.

D. Keim and R. Stambaugh, (1986), 'Predicting Returns in the Stock and Bond Markets', *Journal of Financial Economics* **17** 357–390.

J. Lintner, (1975), 'Inflation and Security Returns,' *Journal of Finance* **30** 259–280.

R.C. Merton, (1971), 'Optimum Consumption and Portfolio Rules in a Continuous Time Model,' *Journal of Economic Theory* **3** 373–413.

J. Mulvey and W.T. Ziemba, (1995), 'Asset and Liability Management in a Global Environment.' In *Finance*, R.A. Jarrow, V. Maksimovic, and W.T. Ziemba (eds.), North Holland.

C. Nelson and M. Kim, (1993), 'Predictable Stock Returns: the Role of Small Sample Bias,' *Journal of Finance* **48** 641–661.

M.H. Pesaran and A. Timmerman, (1995), 'Predictability of Stock Returns: Robustness and Economic Significance,' *Journal of Finance* **50** 1201–1228.

G.W. Schwert, (1981), 'The Adjustment of Stock Prices to Information about Inflation,' *Journal of Finance* **36** 15–29.

R.J. Shiller, (1979), 'The Volatility of Long Term Interest Rates and Expectations Models of the Term Structure,' *Journal of Political Economy* **87** 1190–1219.

R. Stulz, (1986), 'Asset Pricing and Expected Inflation,' *Journal of Finance* **41** 209–224.

Part V
Scenario Generation Procedures

Barycentric Approximation of Stochastic Interest Rate Processes

Karl Frauendorfer and Michael Schürle

Abstract

The incorporation of single-factor interest rate models within the stochastic programming methodology is investigated and applied to multiperiod investment. Barycentric approximation is used for discretizing the stochastic factors and for generating scenario trees which take the various term structure movements into account. It is shown that employing the Vasicek model for the instantaneous rate process preserves convexity of the stochastic multistage program and, hence, guarantees information on the accuracy of the approximate investment strategies. To the contrary, the convexity of the program cannot be assessed if the square root process due to Cox-Ingersoll-Ross is used for the instantaneous rate. In this case, the approximate investment policies and their associated interest surplus may be accepted as estimates. Numerical results for 8-period and 6-period investment problems are discussed.

1 Introduction

The economic and financial environment in which many institutions operate have experienced heightened risk and increased opportunities of returns. Pricing theory, portfolio selection theory, and risk analysis provide sophisticated tools to analyze the effects of the interaction between time and uncertainty. The complexity of this interaction makes finance models one of the most important applications of probability and optimization theories.

Many investment instruments have emerged to meet the critical needs for more sophisticated and flexible tools of asset and liability management and to help shape and maintain the specified risk-return profile of portfolios. In the fixed-income market *interest rate risk, shape risk, volatility risk, liquidity risk*, and *credit risk* are the major types of risk an asset and liability manager is exposed to. *Interest rate risk* is the risk associated with a uniform increase in all default free rates. *Shape risk* is caused by changes of the term structure of interest rates. *Volatility risk* is present in interest rate derivatives and is measured by the standard deviation of the log price relatives. *Liquidity risk* is affected by changes in the bid-ask spread of securities. Empirical tests show that the bid-ask spread of bonds depends strongly on the maturity and trading volume. *Credit risk* occurs if payments are not received in the agreed time. Risks also come from prepayment options embedded in certain positions. Incorporating credit or prepayment risk in pricing models often causes nonconvexity of value functions of interest rate derivative securities and makes it harder to identify the sensitivities of returns. As an example consider mortgage backed securities where prepayment options and credit risk entail a nonconvex value function (see Gilkeson and Smith, 1992).

If the cash flow from assets and the cash flow from liabilities within a balance sheet coincide in maturity the institution is hedged against interest rate changes. In practice, such a matching can be hardly guaranteed because the cash flows are stochastic. Usually, large gaps among the different maturities arise. Therefore, institutions concerned with funding of and investing into interest rate derivative securities are exposed to substantial risk. A classical hedging strategy of fixed-income managers is the duration matching strategy, which states that immunization against interest rate risk is achieved when the duration of assets and liabilities coincide. However, only parallel shifts of the interest rate curve are taken into account. Duration-based concepts do not capture shape risk which is implied from an inverse or steepening term structure. Furthermore, hedging strategies work only if the marginal returns with respect to changes in the risk factors are constant and perfectly correlated over the planning period. Unfortunately, perfect correlation of marginal returns are unrealistic as well as that marginal returns do not change over time. This necessitates a periodic revision of funding and investment strategies with increasing transactions costs.

In the finance literature, intertemporal decision making is usually modeled as *dynamic stochastic control problems* over discrete or continuous time. The aim lies in finding optimal policies, implementable for each period as a function of the current state. The specified dynamics are supposed to depend on the state and control action. The optimal policy is derived by the dynamic programming technique, which amounts to the solution of Bellman's equation. The stochastic control problem is solved once for the entire planning horizon, yielding the optimal policy for each period with respect to the observed state and the underlying dynamics. The solvability of control type problems relies on finding closed form expressions for the solution of Bellman's equation. In case the functions involved are of a special analytical and differentiable form, a substitution of the state equation into the value function preserves differentiability, and the Bellman's equation often yields an optimal policy in closed form through differentiation.

Intertemporal decision making may also be modeled via *stochastic programming*, which was initiated by G.B. Dantzig and E.M.L. Beale in the mid-fifties (see Dantzig, 1991). The intention was to provide a methodology for planning under uncertainty with respect to given constraints taking into account correlations and the time dependency of marginal returns. Although stochastic programming started almost at the same time when Markowitz developed his mean-variance approach, it did not find enough importance with finance experts of that time. One of the reasons could have been that it has been too closely tied to large-scale programming. In recent years stochastic programming has received increasing attention in finance due to the achieved progress. At this place it is referred to the successful and valuable contributions of G.B. Dantzig, W.T. Ziemba, M.A.H. Dempster, J. Dupačová, J. Mulvey, S.A. Zenios, (see the papers in the bibliography and the references in them).

Stochastic multiperiod programming models applied to financial problem statements consist of a set of securities, a multiperiod finite planning horizon, the relevant factors and their stochastic evolvement which determine the price and cash flow dynamics, investors' objective or utility function, and the constraints which may comprise institutional requirements or regulatory guidelines for funding and investment activities. The model aims at today's optimal policy with respect to the underlying stochastic dynamics of prices and cash-flows. Today's optimal policy is determined with respect to anticipating optimality of the future decisions subject to the then observed information on prices and cash-flows. Contrary to the stochastic control problem, the stochastic multiperiod program is solved once in each period, taking into account the periodically updated forecasts of the dynamics. Employing stochastic multiperiod programming, a closed form expression of the optimal policy cannot be expected. The value functions lose their differentiability because of the constraints. At best, convexity of the value functions may be expected, at least if the constraints and the objective function are convex in their variables, and if additional assumptions hold for the functional relationship of risk factor processes and interest rate processes.

A key element in stochastic programming is the way stochastic processes may be analyzed and discretized with respect to the underlying decision structure. This amounts to the generation of scenario trees which approximate the uncertain evolvements adequately. The solution of the associated surrogate optimization problems provide the decision maker with an implementable policy and, in the convex case, with information on its accuracy.

We focus on the incorporation of one-factor interest rate models within the stochastic programming methodology applied to multiperiod investment. Barycentric approximation is used for discretizing the stochastic factors and for generating scenario trees which take the various term structure movements into account. In section 2 a numerical example illustrates how to model multiperiod investment activities as a linear stochastic multistage program. In section 3 the structural properties of the underlying stochastic program are outlined for the case that the reinvestment opportunities of stochastic interest payments are considered implicitely in the stochastic cash flow evolvement. This is followed by a rough outline of the barycentric approximation technique. Section 4 summarizes well-known one-factor interest rate models and recalls the functional relationship between the instantaneous rate (the single factor) and the term structure. In section 5 barycentric approximation provides the discretization of the stochastic instantaneous rate and corresponding scenario trees. Preliminary numerical results are discussed which correspond to 8-period and 6-period investment problems. In section 6 a summary and an outlook of further research activities is given.

2   An Illustrative Example for Multistage Investment

Consider an investor who invests \$1000 in deposits with maturities of 1, 2, 3, or 6 months. The investor follows a "buy and hold" strategy and intends to maximize the interest surplus within a planning horizon of 6 months. For the ease of exposition only nominal payments are considered without discounting cash flows. A long term investment is appropriate if the term structure is expected to fall or if the term structure moves from normal to inverse. A short term investment strategy is appropriate if the interest rates are expected to rise. Clearly, time and term structure movements play an important role for investment decisions. The impact of the dynamics has to be analyzed carefully for identifying the optimal strategy and assessing the risk one is exposed to.

Let the investment decisions be denoted $v_t^{\tau,+}$ with $(\tau \in \mathcal{D}^S = \{1,2,3,6\})$, representing deposits with maturity of 1, 2, 3, or 6 months at time $t = 0,\ldots,T$. The corresponding interest rates are denoted r_t^τ. c_t is the amount of cash invested in t, negative c_t represents a need for liquidity. Interests are paid monthly and the total interest surplus received at time s by investments in maturity τ is

$$\sum_{\tau \in \mathcal{D}^S} \sum_{t=s-\tau}^{s-1} r_t^\tau v_t^{\tau,+}.$$

Only interest rate payments within the period $[0,T]$ are considered. The optimization problem is

$$
\begin{aligned}
\max \quad & \sum_{s=1}^{T} \sum_{\tau \in \mathcal{D}^S} \sum_{t=s-\tau}^{s-1} r_t^\tau v_t^{\tau,+} \\
\text{s.t.} \quad & \sum_{\tau \in \mathcal{D}^S} \sum_{s=\hat{t}-\tau+1}^{\hat{t}} v_t^{\tau,+} \leq \sum_{s=1}^{\hat{t}} \sum_{\tau \in \mathcal{D}^S} \sum_{t=s-\tau}^{s-1} r_t^\tau v_t^{\tau,+} + \sum_{t=1}^{\hat{t}} c_t \qquad \forall \hat{t} \qquad (2.1) \\
& v_t^{\tau,+} \geq 0 \qquad \forall t, \tau \in \mathcal{D}^S.
\end{aligned}
$$

The constraint ensures that in each period $\hat{t}$ the total amount of deposits invested in all maturities does not exceed the sum of cumulated interest surplus up to period $\hat{t}$ and the volume invested or withdrawn up to $\hat{t}$. The interest rate gains are reinvested. In case an investor relies on estimates of each of the T future term structures $(r_t^\tau, \tau \in \mathcal{D}^S = \{1, 2, 3, 6\})$, and if he is aware of the future cash flows c_t, the investor will implement that strategy which represents the optimal solution of the above linear program.

c_t	t	maturity τ 1	2	3	6
1000	0	4.7	4.7	4.6	4.5
0	1	4.6	4.5	4.4	4.2
100	2	4.4	4.3	4.2	4.1
-200	3	4.3	4.2	4.1	4.1
0	4	4.2	4.1	4.0	4.2
0	5	4.0	4.0	4.1	4.2

$\Longrightarrow$

t	maturity τ 1	2	3	6
0		88.32		911.68
1	3.77			
2	195.86			
3				
4				3.42
5				3.43

a) Scenario 1

c_t	t	maturity τ 1	2	3	6
1000	0	4.7	4.7	4.6	4.5
0	1	4.7	4.6	4.5	4.4
100	2	4.6	4.5	4.5	4.4
-200	3	4.5	4.4	4.4	4.5
0	4	4.5	4.4	4.5	4.6
0	5	4.4	4.3	4.5	4.7

$\Longrightarrow$

t	maturity τ 1	2	3	6
0	1000.00			
1	1003.92			
2	1107.85			
3	912.10			
4				915.52
5				3.51

b) Scenario 2

c_t	t	maturity τ 1	2	3	6
1000	0	4.7	4.7	4.6	4.5
0	1	4.8	4.8	4.8	4.7
100	2	4.8	4.9	4.9	4.9
-200	3	4.9	4.9	4.8	4.8
0	4	4.8	4.7	4.7	4.7
0	5	4.7	4.7	4.6	4.5

$\Longrightarrow$

t	maturity τ 1	2	3	6
0	1000.00			
1	1003.92			
2	195.49			912.44
3				
4	3.73			
5	7.47			

c) Scenario 3

Table 1: Future interest rates and optimal investment policy

Estimates of future term structures, or future cash flows, respectively, are called *scenarios*. Table 1a lists the result of a numerical example for one particular scenario of interest rates and cash flow evolvement. The rates at $t = 0$ are inverse and represent observed current market data; the investment volume is $1000. The future interest rates corresponding to maturity of 1 and 2 months are falling steadily; 3- and 6-month rates fall sharply with 30 basis points at time $t = 1$, amounting to steepen the inverse shape. In the subsequent period a decrease of 10 BP is followed by an increase of 10 BP at time $t = 4$. Apparently, the shape of the term structure has changed from inverse to normal. The cash flow is positive at stage $t = 2$, negative at stage $t = 3$, and 0 in the remaining stages. Anticipating that interest rates and cash flows will evolve according to scenario 1, the optimal investment plan is to place 911.68 in deposits with maturity of 6 months. The residual amount to $1000 is placed in 2-month deposits which mature at $t = 3$. At this stage an investment in one month allows the investor to satisfy the required liquidity at $t = 4$. The interest payments received at stages 4 and 5 are invested in 6-month deposits. The cumulated interest surplus is 21.97.

Table 1b lists the second scenario of interest rates and cash flows. As before, the rates at $t = 0$ are inverse and represent observed current market data; the investment volume is $1000. The level of future interest with respect to short term maturities fall slightly over the entire planning horizon; long term rates fall 10 BP after the first period and then increase slightly. The fall of the term structure is less severe than in scenario 1. The cash flow evolves as in scenario 1. Investing in 1-month deposits at stages $0, 1, 2$ and 3 and afterwards in 6-month deposits turns out to be optimal. The total interest surplus is 22.55. Apparently, an investor may benefit from the inverse shape of the term structure as long as the rates do not fall significantly.

Table 1c lists a third scenario. The cash flow stream coincides with that in the previous scenarios. The term structure is slightly rising up to stage 2 and then falls again to the current level. Investing in 1-month deposits at stages $0, 1$ and at stage 2 in 6-month deposits is optimal. A sufficient amount has to be placed in one month so that the need of liquidity can be supplied. The total interest surplus is 23.66.

Clearly, investors will not focus on a single scenario for identifying their strategies. Instead, they will rather derive a set of scenarios with associated probabilities on which they base their investments. For extending the above example in this respect let the probabilities of the scenarios $1, 2$, and 3 be given by $p_1 = 0.3$, $p_2 = 0.5$ and $p_3 = 0.2$. The scenarios are indexed by $k = 1, 2, 3$. This amounts to denoting the term structures $r_{t,k}^{\tau}$ and the investment decisions $v_{t,k}^{\tau,+}$, both depending on the scenario. The cash flow evolves equally in all three scenarios. Observe further that they all start from the same term structure which is currently $(t = 0)$ observed in the market; then the three

t	Scenario 1 maturity τ			Scenario 2 maturity τ			Scenario 3 maturity τ		
	1	2	6	1	2	6	1	2	6
0		1000			1000			1000	
1	3.92			3.92			3.92		
2	1107.85			1107.85			195.49		912.36
3	911.91			912.10					
4	915.18					915.52	3.73		
5			918.38			3.51	7.47		
o_k		21.60			22.55			23.58	
p_k		0.3			0.5			0.2	

Table 2: Optimal nonanticipative strategy and objective values o_k.

scenarios branch without sharing any common history in the following stages. Due to nonanticipativity this requires that the investment decision at $t = 0$ has to be unique, i.e. $v_{0,1}^{\tau,+} = v_{0,2}^{\tau,+} = v_{0,3}^{\tau,+}$, $\tau \in \mathcal{D}^S$. For any nonanticipative policy the expected interest surplus is

$$\sum_{k=1}^{3} p_k \sum_{s=1}^{T} \sum_{\tau \in \mathcal{D}^S} \sum_{t=s-\tau}^{s-1} r_{t,k}^{\tau} v_{t,k}^{\tau,+}.$$

To maximize the expected interest surplus with respect to feasible, nonanticipative policies amounts to the solution of the following multistage program:

$$\text{maximize} \qquad \sum_{k=1}^{3} p_k \sum_{s=1}^{T} \sum_{\tau \in \mathcal{D}^S} \sum_{t=s-\tau}^{s-1} r_{t,k}^{\tau} v_{t,k}^{\tau,+}$$

$$\text{s.t.} \qquad \sum_{\tau \in \mathcal{D}^S} \sum_{s=\hat{t}-\tau+1}^{\hat{t}} v_{t,k}^{\tau,+} \leq \sum_{s=1}^{\hat{t}} \sum_{\tau \in \mathcal{D}^S} \sum_{t=s-\tau}^{s-1} r_{t,k}^{\tau} v_{t,k}^{\tau,+} + \sum_{t=1}^{\hat{t}} c_t, \quad \forall \hat{t}, k$$

$$v_{0,1}^{\tau,+} = v_{0,2}^{\tau,+} = v_{0,3}^{\tau,+} \quad \tau \in \mathcal{D}^S$$

$$v_{t,k}^{\tau,+} \geq 0 \quad \forall t, k, \tau \in \mathcal{D}^S.$$

$$(2.2)$$

Table 2 shows the optimal nonanticipative solution with an expected profit of $0.3 \cdot 21.60 + 0.5 \cdot 22.55 + 0.2 \cdot 23.58 = 22.47$.

3 Valuation and Optimization of Intertemporal Decisions

The example illustrates the structural properties inherent in a dynamic investment process. Similar argumentation holds also for optimal funding, e.g., for refinancing mortgages in banking, or for financing risky projects in industry. The key issue in optimal decision making under uncertainty is how to incorporate the evolvement of the stochastic data over time adequately.

The necessity to pose additional restrictions on the decisions to make them implementable motivates the usage of *stochastic multistage programming*. Such a model allows periodic decision making over finitely many stages $t = 0, 1, \ldots, T$ within the planning horizon $[0, T]$. Herein, the set of securities available for investment comprise bonds with different maturities. For simplicity, the maturities τ are given in months. Taking into account the liquidity within set of maturities $\mathcal{D} = \{1, 2, \ldots, 12, \ldots, 36, \ldots, 60, \ldots, 120, \ldots, D\}$, only a subset of standard maturities $\mathcal{D}^S$ (e.g., $\mathcal{D}^S = \{1, 2, 3, 6, 9, 12, 24, 36, 60, 84, 120\}$) are considered for investments. Furthermore, it is assumed that the bonds are held until maturity, so that changes in the price of a bond during the holding period may be relaxed. This is a common assumption of bank treasuries due to the decision making process which requires investing the monthly income taking into account the new information revealed during the periods, and which allows for compensating unfavourable past decisions. Clearly, the profit of the investment decisions is determined by the interest payments received. The risk of price changes is taken into account implicitly within the first periods of the planning horizon. In the final periods only the interest surplus contribute to the profit, as favourable or less favourable investment decisions after T periods are not considered.

3.1   Stochastic multistage investment

Let the amount of investment in bonds with maturity τ at time t be denoted by $v_t^{\tau,+}$; v_t^τ is the investment volume with maturity τ at time t. Clearly, v_t^τ satisfies

$$v_t^\tau = v_{t-1}^{\tau+1} + v_t^{\tau,+} \qquad t = 0, 1, \ldots, T; \ \tau \in \mathcal{D}^S,$$

and

$$v_t^\tau = v_{t-1}^{\tau+1} \qquad t = 0, 1, \ldots, T; \ \tau \in \mathcal{D} - \mathcal{D}^S.$$

The total investment volume at t is given by

$$x_t = \sum_{\tau \in \mathcal{D}} v_t^\tau \qquad t = 0, 1, \ldots, T.$$

Its stochastic evolvement over time is determined by the volume change $\xi_t \in \mathbb{R}$,

$$x_t = x_{t-1} + \xi_t \qquad t = 1, 2, \ldots, T.$$

It is stressed that per period t there are D deterministic constraints with totally $[3|\mathcal{D}^S| + 2(|\mathcal{D}| - |\mathcal{D}^S|) + D + 1]$ nonzero coefficients and one stochastic constraint with two nonzeros. This demonstrates the sparsity of the $(D +$

$2) \times 2(D+1)$ matrix within period t and over the entire planning horizon. Observe, that the nonzeros in the matrix consist only of $+1$ and -1, which indicate numerical stability even for large problems (see Frauendorfer *et al.* 1996).

The cumulated interest surplus achieved by the investment decision $v_t^{\tau,+}$ at t is given by the interest rate associated with maturity τ. Let the term structure of interest rates with respect to maturities in $\mathcal{D}^S$ be denoted $\hat{\rho} \in \mathbb{R}^{|\mathcal{D}^S|}$. Clearly, the term structure depends on the stochastic value of the underlying K-dimensional risk factor $\eta_t \in \mathbb{R}^K$. At $t = 0$ the risk factor η_0 and, hence, the current term structure $\hat{\rho}(\eta_0)$ is assumed to be known. Let the cumulated interest surplus during $[t, T]$ of one unit investment in maturity τ be denoted by $\rho_t(\eta_t, \tau)$. The functional relations $\rho_t : \mathbb{R}^K \times \mathcal{D}^S \to \mathbb{R}$ are given through proper K-factor interest rate models which evaluate the term structure at t. The underlying risk factor dynamics are given probabilistically through stochastic processes and will help identify the joint probability measure P of (η, ξ) on a Borel space $(\mathbb{R}^{T(K+1)}, \mathcal{B})$, incorporating the stochastic dependencies of term structure $\hat{\rho}(\eta_t)$ and investment volume ξ_t. The expected interest payments are then

$$\langle \rho_0(\eta_0), v_0^+ \rangle + \int \sum_{t=1}^{T} \langle \rho_t(\eta_t), v_t^+ \rangle dP(\eta, \xi)$$

with $\eta := (\eta_1, \eta_2, \ldots, \eta_T), \xi := (\xi_1, \xi_2, \ldots, \xi_T), v_t := (v_t^\tau; \tau \in \mathcal{D}), v_t^+ := (v_t^{\tau,+}; \tau \in \mathcal{D}^S)$ and with

$$\langle \rho_t(\eta_t), v_t^+ \rangle = \sum_{\tau \in \mathcal{D}^S} \rho_t(\eta_t, \tau) \cdot v_t^{\tau,+}$$

denoting the inner product on $\mathbb{R}^{|\mathcal{D}^S|}$. The sequences of observations $\eta^t := (\eta_1, \eta_2, \ldots, \eta_t), \xi^t := (\xi^1, \xi^2, \ldots, \xi^t)$, of state variables $v^t := (v_0, v_1, \ldots, v_t)$, $x^t := (x_0, x_1, \ldots, x_t)$, and of control variables $v_t^+ := (v_0^+, v_1^+, \ldots, v_t^+)$ determine the investor's performance. State and control variables have to be taken measurably (nonanticipatively) with respect to the observed evolvement of the stochastic factors. Nonanticipativity is required in order to ensure that the sets of feasible decisions depend only on past decisions and observed realisations of stochastic data.

The corresponding stochastic multistage program which maximizes the expected interest payments is

$$\max \quad \langle \rho_0(\eta_0), v_0^+ \rangle + \int \sum_{t=1}^{T} \langle \rho_t(\eta_t), v_t^+ \rangle dP(\eta, \xi)$$

$$
\begin{array}{llr}
v_t^\tau - v_{t-1}^{\tau+1} - v_t^{\tau,+} & = 0 & t = 0, 1, \ldots, T; \forall \tau \in \mathcal{D}^S \\
v_t^\tau - v_{t-1}^{\tau+1} & = 0 & t = 0, 1, \ldots, T; \forall \tau \notin \mathcal{D}^S \quad (3.1) \\
x_t - \sum_{\tau \in \mathcal{D}} v_t^\tau & = 0 & t = 0, 1, \ldots, T \\
x_t - x_{t-1} & = \xi_t & t = 1, 2, \ldots, T \\
v_t^+ \geq 0, v_t, x_t \text{ nonanticipative} & & t = 0, 1, \ldots, T.
\end{array}
$$

Negative subscripts of variables indicate decisions of the past; nonpositive subscripts of the stochastic data indicate data of the past, currently observed data, respectively. The left side of the constraints is deterministic. Contrary to the deterministic example in the previous section the interest rate payments are not considered explicitly in the constraint multifunction. These stochastic cash flows may be taken into account implicitly through the stochastic volume change. Therefore, the interest sensitive risk factors appear only in the objective. Relaxing the fact that the distribution function of the conditional probabilities $P(\cdot|\eta^{t-1},\xi^{t-1})$ are dependent of states and control actions preserves the convexity of the stochastic multistage program (3.1).

3.2 Saddle property of value functions

In financial modeling, intertemporal portfolio selection is usually modeled via control type approaches. The solvability of the associated Bellman's equation depends heavily on the differentiability of the value function. Contrary to control type problems stochastic multistage programs benefit from convexity. Furthermore, the latter focus on the implementation of the optimal decision associated with $t = 0$. The decisions in future periods $t = 1, \ldots, T$ will unlikely be implemented; rather they provide the valuation of optimal decisions at t conditioned on the evolvement of risk factors and volume up to $t - 1$. Herewith, the valuation of future decisions may be seen as interest rate risk measurement of today's investment decisions. If the correlation of term structure and volume is taken into account adequately, one may even argue that a kind of prepayment risk is incorporated. Formally, this amounts to the dynamic representation of the above program (3.1):

$$\phi_t(v_t, \eta^t, \xi^t) :=$$
$$\max\langle \rho_t(\eta_t), v_t^+ \rangle + \int \phi_{t+1}(v_{t+1}, \eta^{t+1}, \xi^{t+1}) dP(\eta_{t+1}, \xi_{t+1}|\eta^t, \xi^t)$$

s.t. $$f_t(x_t, x_{t-1}, v_{t-1}, v_t, v_t^+, \xi_t) = 0 \qquad (3.2)$$
$$\vdots$$
$$f_T(x_T, x_{T-1}, v_{T-1}, v_T, v_T^+, \xi_T) = 0,$$

where $\phi_{T+1}(\cdot) := 0$. For the ease of exposition, the convex constraint multifunction corresponding to period t in (3.2) is written as

$$f_t(x_t, x_{t-1}, v_{t-1}, v_t, v_t^+, \xi_t) = 0.$$

Referring to Frauendorfer (1994a), one may state the following:

Proposition: If the distribution function of $P(\cdot|\eta^t, \xi^t)$ depends linearly on the past (η^t, ξ^t), and if, additionally, $\rho_t(\cdot), t = 1, \ldots, T$ are convex functions in the risk factors η_t, then the value functions $\phi_t(v_t, \eta^t, \xi^t)$ are saddle functions, convex in η_t and concave in (v_t, ξ_t).

Proof: Problem (3.2) is a stochastic linear multistage program and is covered by the assumptions of Proposition 2.1 in Frauendorfer (1994a).

The *entire convex case* obtains when the assumptions of the proposition are fulfilled. The *convex case* is the situation where the distributions are independent of the policy; this implies convexity of the value functions in the policy only. Using the above conventions, a *scenario tree* with associated scenario probabilities is defined as

$$
\begin{aligned}
\mathcal{A} & := \{(\eta^T, \xi^T) \in \mathbb{R}^{K+1} | \ (\eta_t, \xi_t) \in \mathcal{A}_t(\eta^{t-1}, \xi^{t-1}) \ \forall t \geq 1\} \\
q(\eta^T, \xi^T) & := \prod_{t=1}^{T} q_t(\eta_t, \xi_t | \eta^{t-1}, \xi^{t-1})
\end{aligned}
\tag{3.3}
$$

and represents an approximation of the discrete-time process $(\eta_t, \xi_t; \ t = 1, \ldots, T)$. Here, $\mathcal{A}_t(\eta^{t-1}, \xi^{t-1})$ denotes the set of finitely many outcomes for (η_t, ξ_t) conditioned on η^{t-1}, ξ^{t-1}; $Q_t(\cdot | \eta^{t-1}, \xi^{t-1})$ is the corresponding conditional discrete probability measure. The associated discrete probability space $(\mathbb{R}^{K+1}, \mathcal{B}, Q_t)$ represents an approximation of $(\mathbb{R}^{K+1}, \mathcal{B}, P_t)$.

The discretization of the stochastic evolvements allows for treating the associated stochastic program as a deterministic equivalent to (3.1), which in turn represents a sparse large-scale mathematical program. Mathematical programming algorithms may be applied, which consider the sparsity, the continuity of the state and control variables, and the linearity of the constraints and objective function. Benefitting from increasing efficiency of those mathematical programming algorithms (see Spedicato, 1994), stochastic programming has received increasing attention in finance in recent years.

3.3   Barycentric Approximation

The key element is the way the discretization of $Q_t(\cdot | \eta^{t-1}, \xi^{t-1})$ is performed. The underlying saddle structure of the value function motivates the application of *barycentric approximation*, which optimizes the discretization of the stochastic evolvement of investment volume and interest rates in the sense of Frauendorfer (1994b). In the entire convex case, barycentric approximation provides lower and upper bounds for the expectation of a saddle function. For those familiar with the Jensen-Inequality (see Jensen, 1906) and the Edmundson-Madansky-Inequality (see Madansky, 1960), which both allow for bounding the expectation of convex functions, it is stressed that barycentric approximation is a generalization of these inequalities which is applicable to saddle functions of correlated stochastic data. Applied to our situation, this scheme covers the outcomes η_t of the K-dimensional risk factor by a regular simplex Θ_t and the outcomes ξ_t by an interval (i.e., a one-dimensional simplex) Ξ_t. Both Θ_t and Ξ_t may depend on the observations (η^{t-1}, ξ^{t-1}). When $P_t(\cdot | \eta^{t-1}, \xi^{t-1})$ has unbounded support one has to ensure that $P_t(\Theta_t \times \Xi_t | \eta^{t-1}, \xi^{t-1}) \geq 1 - \epsilon$ for sufficiently small $\epsilon > 0$ and then substitutes $P_t(\cdot|\eta^{t-1}, \xi^{t-1})$ by its normalized truncation. Let the vertices of $\Theta_t \subset \mathbb{R}^K$ and $\Xi_t \subset \mathbb{R}$ be denoted a_{ν_t}, $\nu_t = 0, \ldots, K$, and b_{μ_t}, $\mu_t = 0, 1$. The probability measure P_t induces mass distributions $\mathcal{M}_{\nu_t}$ with associated *generalized barycenters* ξ_{ν_t} on the 1-dimensional simplices $\{a_{\nu_t}\} \times \Xi_t$. As for $\nu_t = 0, \ldots, K$ the mass distributions $\mathcal{M}_{\nu_t}$ add up to a probability distribution, one gets a discrete probability measure Q_t^u on $\Theta_t \times \Xi_t$ when assigned probability $\mathcal{M}_{\nu_t}(\{a_{\nu_t}\} \times \Xi_t)$ to the points (a_{ν_t}, ξ_{ν_t}). Due to orthogonality, the probability measure P_t induces mass distributions $\mathcal{M}_{\mu_t}$ with associated *generalized barycenters* a_{μ_t} on the (K-dimensional) simplices $\Theta_t \times \{b_{\mu_t}\}$ for $\mu_t = 0, 1$. Again, the two mass distributions $\mathcal{M}_0$ and $\mathcal{M}_1$ add up to a probability distribution, yielding a discrete probability measure Q_t^l on $\Theta_t \times \Xi_t$, when assigned probabilities $\mathcal{M}_{\mu_t}(\Theta_t \times \{b_{\mu_t}\})$ to the points $(\eta_{\mu_t}, b_{\mu_t})$.

This way, two discrete probability measures Q_t^l and Q_t^u are derived which represent the solutions of two corresponding generalized moment problems. Such discrete probability measures are called *extremal*. The advantageous feature from a computational viewpoint is that generalized barycenters η_μ, $\mu = 0, 1$, and ξ_ν, $\nu = 0, \ldots, K$, and their probabilities $q(\eta_\mu)$ and $q(\xi_\nu)$ are completely determined by the first moments of η and ξ, and by the bilinear cross moments $\mathrm{E}(\eta_k \cdot \xi)$, $k = 1, \ldots, K$. To see this, the formulae for the generalized barycenters η_μ, $\mu = 0, 1$, and ξ_ν, $\nu = 0, \ldots, K$, and the associated positive probabilities $q(\xi_\nu)$ and $q(\eta_\mu)$ are recalled with respect to a probability space $(\Theta \times \Xi, \mathcal{B}, P)$.

$$
\begin{aligned}
\eta_\mu &:= \quad \sum_\nu a_\nu \frac{\int \gamma_\mu(\xi) \cdot \lambda_\nu(\eta) dP}{\int \gamma_\mu(\xi) dP} & \mu &= 0, 1 \\[2mm]
q(\eta_\mu) &:= \quad \int \gamma_\mu(\xi) dP(\eta, \xi) & \mu &= 0, 1 \\[2mm]
\xi_\nu &:= \quad \sum_\mu b_\mu \frac{\int \gamma_\mu(\xi) \cdot \lambda_\nu(\eta) dP}{\int \lambda_\nu(\eta) dP} & \nu &= 0, \ldots, K \\[2mm]
q(\xi_\nu) &:= \quad \int \lambda_\nu(\eta) dP & \nu &= 0, \ldots, K.
\end{aligned}
\tag{3.4}
$$

$\gamma_\mu(\xi)$ and $\lambda_\nu(\eta)$ denote the barycentric weights of η and ξ with respect to the simplices Θ and Ξ. The barycentric weights are linear in their components so that the integrand $\gamma_\mu(\xi) \cdot \lambda_\nu(\eta)$ is a bilinear function in (η, ξ). Hence, the generalized barycenters are completely determined by the first moments of η and ξ, and by the K cross moments

$$
\int \xi \cdot \eta_k \, dP \qquad k = 1, \ldots, K.
$$

Recalling that the covariance of two random outcomes is given by the first moments and the corresponding cross moment, it is observed that the extremal probability measure incorporates implicitly the covariances between the volume and each of the K risk factors. Cross moments and, hence, covariances among the risk factors are not taken into account. In case one-factor interest rate models are employed, the evaluation of the barycenters simplifies considerably. Clearly, this amounts to the evaluation of $E\eta\xi$, $E\eta_k$, and $E\xi$.

Arguing from a dual viewpoint (see Frauendorfer, 1992), the multidimensional integral of the value functions ϕ_t is approximated by a sum of at most two-dimensional integrals, whose integrands are bilinear. The approximation can be improved by partitioning the simplices. In case the subsimplices become arbitrarily small weak convergence of the extremal measures to P, and, hence, the covergence of lower and upper bounds to the expectation of the value function is guaranteed.

3.4 Barycentric scenario trees

According to Frauendorfer (1994b), barycentric approximation can be applied dynamically over time, building *barycentric scenario trees* $\mathcal{A}^l$ and $\mathcal{A}^u$ with their path probabilities of type (3.3). The associated discretization Q^l and Q^u of the stochastic risk factor dynamics amounts to the evaluation of the corresponding barycentric term structures. Substituting in (3.1) P by the probability measures Q^l and Q^u, stochastic multistage problems are obtained, whose optimal values represent lower and upper bounds for (3.1) in the entire convex case, i.e., if the assumptions of the above proposition hold. The optimal investment decisions of the surrogate stochastic multistage programs may be accepted as approximates for today's optimal policy. In the entire convex case the accuracy is quantifiable by the difference of lower and upper bounds. This way, the decision maker can improve the approximation through simplicial refinements in case the accuracy is not satisfying. In the nonconvex case barycentric approximation is still applicable; however, information on the accuracy can be derived only in a probabilistic sense.

4 Stochastic term structure models

Since the seminal papers of Vasicek (1977) and Cox, Ingersoll, and Ross (1985), numerous models have been developed for evaluating the term structure of interest rates. Commonly, the stochastic evolvement of the underlying interest rate factors are modelled as continuous time processes which help derive the current term structure as well as prices of various interest rate derivative securities. These models have become very popular due to their tractability. Herein, in the context of stochastic programming, they will be employed for approximating the dynamics of the term structures by means of a distinguished set of scenarios.

The dynamics of the term structure are driven by stochastic factors. According to the number of these factors, the literature differs one-, two-, or multifactor models. Commonly, one-factor models take the instantaneous

rate as the only source of uncertainty for deriving the term structure. The
drawback of working with a single factor is that only a limited variety of
shapes for the term structure is taken into account. There is also empirical evidence that volatility and stochastic evolvement of the various interest
rates can be better explained by more than one source of uncertainty (see,
e.g., Canabarro, 1993, or Litterman and Scheinkman, 1991). This motivated
the elaboration of multifactor models. In many of these, the instantaneous
rate remains incorporated, additionally to others which are chosen from a
broad variety. In the two-factor model proposed by Brennan and Schwartz
(1979), the second factor is the continuously compounded yield on a consol
bond. In recent approaches, particularly in the two-factor models by Longstaff
and Schwartz (1992) and Fong and Vasicek (1991), the second factor is the
volatility of the instantaneous rate. Such models are intuitively appealing to
practitioners which are interested in pricing interest rate derivatives. Herein,
starting the analysis of interest rate processes and their discretization, the
investigations are restricted to one-factor models.

   Apart from the number of factors employed for characterizing the dynamics, interest rate models are distinguished between *equilibrium-based* and
arbitrage-free approaches. The equilibrium approach starts from assumptions
on the underlying economy, the preferences of a representative investor, and
the stochastic dynamics of state variables. The instantaneous rate as well
as the prices of all contingent claims are then derived from general equilibrium conditions. In the arbitrage-free approach, the stochastic process of the
factors are given exogenously. The term structure is evaluated under the condition that no arbitrage opportunities exist. However, both approaches are
essentially equivalent and no-arbitrage models can also be embedded in an
equilibrium approach (see, e.g., Bühler, 1995).

4.1   One-Factor Dynamics

Consider a one-factor model with the instantaneous rate η_t as the single source
of uncertainty. Let $R(t,T)$ denote the interest rate associated with maturity
T. Recalling that a discount bond is a bond with a single cash flow of unity at
maturity and with no intermediate interest payments, $R(t,T)$ is the internal
rate of return at time t on a discount bond that matures at date T. Then,

$$\eta_t = \lim_{T \to t} R(t,T).$$

The relation between the price $D(t,T)$ of a discount bond maturing at T and
the corresponding yield $R(t,T)$ is

$$D(t,T) = e^{-R(t,T)\tau} \qquad \text{or} \qquad R(t,T) = -\frac{1}{\tau}\ln D(t,T),$$

where $\tau = T - t$. Hence, the term structure at time t is representable by the prices of discount bonds as well as by the interest rate curve.

The instantaneous rate η_t is supposed to follow a diffusion process with a deterministic drift a and the instantaneous variance b^2, where both may depend on the interest rate level and on time. The underlying stochastic differential equation is given by

$$d\eta_t = a(\eta_t, t)dt + b(\eta_t, t)dz_t, \tag{4.1}$$

with dz_t denoting the increment of a Wiener process during a small time intervall dt. As η_t is the only source of uncertainty for all $t < T$, the term structure is completely determined by its evolvement between time t and T. Therefore, the discount bond price or the associated internal rate of return are written as a function of η_t:

$$D(t, T, \eta_t) = e^{-R(t,T,\eta_t)\tau}. \tag{4.2}$$

Applying Ito's Lemma to (4.1) yields a parabolic differential equation for the discount bond price function:

$$\frac{dD(t, T, \eta_t)}{D(t, T, \eta_t)} = \mu(t, T, \eta_t)dt + \sigma(t, T, \eta_t)dz_t, \tag{4.3}$$

where

$$\mu(t, T, \eta_t) = \frac{1}{D}\left[\frac{\partial D}{\partial t} + a\frac{\partial D}{\partial \eta_t} + \frac{1}{2}b^2\frac{\partial^2 D}{\partial \eta_t^2}\right], \tag{4.4}$$

$$\sigma(t, T, \eta_t) = \frac{1}{D}b\frac{\partial D}{\partial \eta_t}. \tag{4.5}$$

Now, a hedge portfolio may be constructed consisting of two independent discount bonds with maturities T_1, T_2 and volatilities $\sigma(t, T_1, \eta_t) \neq \sigma(t, T_2, \eta_t)$ $\forall t, T_1, T_2, \eta_t; T_1 \neq T_2$. The bonds are weighted so that the portfolio is riskless, i.e., $w_1\sigma_1 + w_2\sigma_2 = 0$. By the no-arbitrage condition, its rate of return must coincide with the instantaneous rate η_t. Thus, the following equations must hold:

$$\begin{aligned} w_1 \;+\; w_2 &= 1 \\ w_1\sigma_1 + w_2\sigma_2 &= 0 \\ w_1\mu_1 + w_2\mu_2 &= \eta_t. \end{aligned} \tag{4.6}$$

This amounts to

$$\frac{\mu(t, T_1, \eta_t) - \eta_t}{\sigma(t, T_1, \eta_t)} = \frac{\mu(t, T_2, \eta_t) - \eta_t}{\sigma(t, T_2, \eta_t)}, \tag{4.7}$$

and hence to the conclusion that the ratio

$$\frac{\mu(t, T, \eta_t) - \eta_t}{\sigma(t, T, \eta_t)} =: \Upsilon(t, \eta_t), \qquad T \geq t \tag{4.8}$$

Reference	Process Specification
(1) Dothan (1978)	$d\eta_t = \sigma \eta_t dz_t$
(2) Vasicek (1977)	$d\eta_t = \alpha(\gamma - \eta_t)dt + \sigma dz_t$
(3) Cox-Ingersoll-Ross (1985)	$d\eta_t = \alpha(\gamma - \eta_t)dt + \sigma\sqrt{\eta_t}dz_t$
(4) Courtadon (1982)	$d\eta_t = \alpha(\gamma - \eta_t)dt + \sigma\eta_t dz_t$

Table 3: Specifications of the instantaneous rate process.

is independent of T. It is called *market price of risk* since it specifies the change in the expected instantaneous rate of return on a bond for an additional unit of risk, similar to the risk premium for β in the Capital Asset Pricing Model. Transforming equation (4.8) to

$$\mu(t, T, \eta_t) - \eta_t = \Upsilon(t, \eta_t)\sigma(t, T, \eta_t), \qquad (4.9)$$

substituting for $\mu(\cdot)$, $\sigma(\cdot)$ from (4.4)–(4.5) and rearranging the terms properly, a partial differential equation is obtained that must be satisfied by the prices of all discount bonds:

$$\frac{\partial D}{\partial t} + (a - b\Upsilon)\frac{\partial D}{\partial \eta_t} + \frac{1}{2}b^2\frac{\partial^2 D}{\partial \eta_t^2} - \eta_t D = 0, \qquad T \geq t. \qquad (4.10)$$

The boundary condition $D(T, T, \eta_T) = 1$ is due to the fact that discount bonds have a single payoff at maturity and allows for solving the partial differential equation. Therefore, (4.10) provides a complete specification of the interest rate curve

$$R(t, T, \eta_t) = -\frac{\ln D(t, T, \eta_t)}{T - t}. \qquad (4.11)$$

The instantaneous rate η_t follows a process of the form

$$d\eta_t = \alpha'(\gamma' - \eta_t)dt + \sigma\eta_t^\delta dZ_t, \qquad (4.12)$$

where α', γ', $\sigma > 0$ and $\delta \geq 0$ are constants with $\alpha'\gamma' > 0$. Assuming that there exists a unique equivalent martingale probability measure the process takes the form

$$d\eta_t = \alpha(\gamma - \eta_t)dt + \sigma\eta_t^\delta dz_t, \qquad (4.13)$$

where $\alpha(\gamma - \eta_t)$ is the *risk-adjusted* drift term and z_t represents a Brownian motion under the martingale measure. The process (4.13) describes the evolvement of the instantaneous rate η_t in a risk-neutral world. The modelling of η_t under the risk-neutral measure has the advantage that the market price of risk Υ does not appear in any of the subsequent formulae.

Depending on the properties of the diffusion process (4.13), equation (4.10) may be solved either analytically or numerically. Table 3 shows several process specifications that have been proposed in literature. All of whose, except for the Dothan model, differ only in the instantaneous standard deviations (here σ, $\sigma\sqrt{\eta_t}$, and $\sigma\eta_t$). This has an impact on the density functions of the instantaneous rate at t, given the initial starting value η_0. For instance, in the Vasicek approach, the density function $f(\eta_s|\eta_t)$, $t < s$, is normally distributed, and in the CIR-model the distribution will approach a gamma distribution as t becomes large. Stationary density functions exist in all cases (see, e.g., Fischer, 1984). In the specifications (2)–(4) from Table 3, there is a deterministic drift term $\alpha(\gamma - \eta_t)$ that forces the process towards its long-run average γ at a speed determined by α. This property is known as *mean reversion*. Hence, the models described above take into account that interest rates fluctuate within a certain range. Economically, this can be explained by the control of monetary policy.

Closed-form solutions of the fundamental differential equation (4.10) exist only for the Vasicek and CIR-models. A drawback of the former approach can be seen in the fact that the specification of the instantaneous short rate process, also known as Ornstein-Uhlenbeck process, ignores the empirical evidence of a positive correlation between interest rates and the level of volatility (see, e.g., Chan et al., 1992). Further, it allows interest rates to become negative, whereas in the latter approach, negative rates are precluded and the origin is even inaccessible if $2\alpha\gamma \geq \sigma^2$. In addition, the instantaneous standard deviation $\sigma\sqrt{\eta_t}$ depends on the square root of the current short rate (for details, see CIR, 1985).

4.2 Analytical solvability

Under the (risk-adjusted) process specification (4.13), the fundamental pricing equation is

$$\frac{\partial D}{\partial t} + (\alpha(\gamma - \eta_t))\frac{\partial D}{\partial \eta_t} + \frac{1}{2}(\sigma\eta_t^\delta)^2\frac{\partial^2 D}{\partial \eta_t^2} - \eta_t D = 0. \qquad (4.14)$$

The solution of (4.14) takes the form

$$D(t, T, \eta_t) = A(\tau)e^{-\eta_t B(\tau)}, \qquad \tau = T - t. \qquad (4.15)$$

After inserting the partial derivatives of (4.15) into (4.14) and further rearrangemants, one obtains

$$A(\tau) = \exp\left[\frac{(B(\tau) - \tau)(\alpha^2\gamma - \sigma^2/2)}{\alpha^2} - \frac{\sigma^2 B^2(\tau)}{4\alpha}\right] \qquad (4.16)$$

$$B(\tau) = (1 - e^{-\alpha\tau})/\alpha \qquad (4.17)$$

for the Vasicek model ($\delta = 0$) subject to the boundery conditions $A(0) = 1$ and $B(0) = 0$. Using these results, the term structure of interest rates (4.11) may be rewritten as:

$$R(t, T, \eta_t) = \frac{1}{\tau}\left[\eta_t B(\tau) - \ln A(\tau)\right], \qquad \tau = T - t. \qquad (4.18)$$

Note that $R(t, T, \eta_t)$ is a linear function of η_t. Since η_t is normally distributed due to the properties of the Ornstein-Uhlenbeck process, it follows that $R(t, T, \eta_t)$ is also normally distributed. The yield of a bond with infinite maturity is independent of the current instantaneous rate

$$R(\infty) = R(t, \infty, \eta_t) = \gamma - \frac{1}{2}\frac{\sigma^2}{\alpha^2}, \qquad (4.19)$$

which implies that the long end of the interest rate curve is fix. For values of η_t smaller or equal to $R(\infty) - 0.25\,\sigma^2/\alpha^2$, the interest rate curve is normal. When η_t exceeds $R(\infty) + 0.5\,\sigma^2/\alpha^2$, it is inverse. For intermediate values, the interest rate curve is humped (see Vasicek, 1977).

In case of the CIR-model (1985) with $\delta = 0.5$, the solution of (4.10) is given by

$$A(\tau) = \left[\frac{2\psi e^{(\alpha+\psi)\tau/2}}{(\psi+\alpha)(e^{\psi\tau}-1)+2\psi}\right]^{2\alpha\gamma/\sigma^2} \qquad (4.20)$$

$$B(\tau) = \frac{2(e^{\psi\tau}-1)}{(\psi+\alpha)(e^{\psi\tau}-1)+2\psi} \qquad (4.21)$$

$$\psi = \sqrt{\alpha^2 + 2\sigma^2}. \qquad (4.22)$$

Again, the long end of the interest rate curve is constant:

$$R(\infty) = R(t, \infty, \eta_t) = \frac{2\alpha\gamma}{\psi+\alpha}. \qquad (4.23)$$

As in the Vasicek model, there are three shapes taken into account for the interest rate curve: monotonically increasing, monotonically decreasing, and humped. For the general case $\delta \notin \{0, 0.5\}$ in (4.13), no analytical solutions of the fundamental pricing equations (4.14) are known (see, e.g., Strickland, 1994).

5   Barycentric approximation of the instanteneous rate

The instantaneous rate process η_t considered in the Vasicek- and the CIR-model will be approximated next over $t = 1, \ldots, T$. The barycentric approximation technique described in the previous section is used. Two types of scenario trees for the instantaneous rate process will be derived which amount to approximations of the term structure dynamics at $t = 1, \ldots, T$.

The first step is the determination of a regular simplex Θ_t that covers the outcomes of η_t. Both may depend on observations η^{t-1}. Let this dependency be expressed by $\Theta_t|\eta^{t-1}$ and $\eta_t|\eta^{t-1}$. Since $\eta_t|\eta^{t-1}$ is one-dimensional, the simplex $\Theta_t|\eta^{t-1}$ is an interval with vertices denoted $a_t|\eta^{t-1}$ and $b_t|\eta^{t-1}$. The unbounded support of the normal distribution requires a truncation, which ensures that the mass distribution induced by the corresponding probability measure is covered by an interval of the form

$$\Theta_t|\cdot = [E(\eta_t|\cdot) - k\sigma(\eta^{t-1}), E(\eta_t|\cdot) + k\sigma(\eta^{t-1})] \tag{5.1}$$

suffiently accurate. The Tchebycheff Inequality ensures that for any small $\epsilon > 0$, $P_t(\Theta_t|\eta^{t-1}) \geq 1 - \epsilon$ for k is sufficiently large. Using the Minkowski operations on sets, one may write

$$\Theta_t|\eta^{t-1} = E(\eta_t|\eta^{t-1}) + \sigma(\eta^{t-1}) \cdot \Theta, \tag{5.2}$$

with $\Theta = [a, b] = [-k, k]$ for an adequate value of k, and $\sigma(\eta^{t-1})$ denoting the standard deviation of $\eta_t|\eta^{t-1}$.

Let Θ be partitioned into N subintervals $\Theta^{(i)}$, $i = 1, \ldots, N$, satisfying

$$\bigcup_{i=1}^{N} \Theta^{(i)} = \Theta \tag{5.3}$$

with $\Theta^{(i)} \cap \Theta^{(j)} = \emptyset$ for $i \neq j$. Let $f(\eta)$ denote the density function of η on Θ. For $i = 1, \ldots, N$, the barycenter $\eta^{(i)}$ of the subinterval $\Theta^{(i)}$ is the conditional expectation

$$\eta^{(i)} = E[\eta|\eta \in \Theta^{(i)}] = \int_{\Theta^{(i)}} \eta f(\eta) d\eta \ / \int_{\Theta^{(i)}} f(\eta) d\eta, \tag{5.4}$$

with associated probability

$$p^{(i)} = p(\eta \in \Theta^{(i)}) = \int_{\Theta^{(i)}} f(\eta) d\eta \ / \int_{\Theta} f(\eta) d\eta. \tag{5.5}$$

Analogously, the barycenters of $\eta_t|\eta^{t-1}$ with respect to $\Theta_t|\eta^{t-1}$ represent the conditional expectations and are obtained through

$$\eta_t^{(i)}|\eta^{t-1} = E(\eta_t \mid \eta_t \in \Theta_t^{(i)}|\eta^{t-1}). \tag{5.6}$$

Setting $\mathcal{A}_t^l(\eta^{t-1}) := \{\eta_t^{(i)}|\eta^{t-1}\}$ and $q_t^{(i)}(\eta_t^{(i)}|\eta^{t-1}) := p^{(i)}$, one may construct the *barycentric scenario tree* $\mathcal{A}^l$ with its path probabilities due to (3.3). This represents a first-type discretization of the instantaneous rate process.

The vector of barycentric weights $\lambda(\eta) \in \mathbb{R}^2$ with respect to some $\eta \in \Theta^{(i)} = [a^{(i)}, b^{(i)}]$ is obtained through

$$\begin{pmatrix} 1 & 1 \\ a^{(i)} & b^{(i)} \end{pmatrix} \cdot \lambda(\eta) = \begin{pmatrix} 1 \\ \eta \end{pmatrix}. \tag{5.7}$$

Clearly, for $\eta^{(i)} = E(\eta | \eta \in \Theta^{(i)})$ it follows that

$$
\begin{aligned}
\lambda_0(\eta^{(i)}) &= \frac{b^{(i)} - \eta^{(i)}}{b^{(i)} - a^{(i)}} \geq 0, \\
\lambda_1(\eta^{(i)}) &= \frac{\eta^{(i)} - a^{(i)}}{b^{(i)} - a^{(i)}} \geq 0.
\end{aligned} \tag{5.8}
$$

Setting

$$
\begin{aligned}
\mathcal{A}_t^u(\eta^{t-1}) &:= \{a_t^{(i)} | \eta^{t-1}, b_t^{(i)} | \eta^{t-1}\}, \\
q_t^{(i)}(a_t^{(i)} | \eta^{t-1}) &:= \lambda_0(\eta^{(i)}), \\
q_t^{(i)}(b_t^{(i)} | \eta^{t-1}) &:= \lambda_1(\eta^{(i)}),
\end{aligned}
$$

one may construct the *barycentric scenario tree* $\mathcal{A}^u$ with its path probabilities due to (3.3). This represents the second-type discretization of the instantaneous rate process.

5.1 Investment strategies based on the Vasicek model

Employing the scenario trees $\mathcal{A}^l$ and $\mathcal{A}^u$ for the approximation of the instantaneous rate process due to Vasicek at $t = 1, 2, \ldots, T$ allows for evaluating the associated dynamic evolvement of the interest rate curve, and, hence, for defining the objective function in program (3.1). For the ease of exposition, it is assumed that changes in investment volume ξ_t, $t = 1, 2, \ldots, T$, are deterministic.

One way to proceed is to approximate the Ornstein-Uhlenbeck process in discrete time:

$$
\eta_{t+1} = \eta_t + \alpha(\gamma - \eta_t) + \sigma U_t, \qquad U_t \sim N(0, 1). \tag{5.9}
$$

Since η_{t+1} depends linearly on the instantaneous rate η_t, the distribution function of $P(\eta_{t+1} | \eta^t)$ depends linearly on the past η^t. Furthermore, the coefficients $\rho_t(\eta_t, \tau)$ representing the cumulated interest surplus of one unit investment in maturity $\tau \in \mathcal{D}^S$ are also linear and therefore convex in η_t. These can be obtained directly from the spot rates $R(t, \tau, \eta_t)$. Hence, the corresponding value function $\phi_t(v_t, \eta^t)$ of the stochastic program is convex in η^t.

Another way is to use that the distribution of $\eta_s | \eta_t$ $(t < s)$ is analytically known for the Vasicek model. More precisely, the density function of the instantaneous rate at time s, conditional on its value at time t, is

$$
f(\eta_s | \eta_t) \propto pdf \left(\frac{\eta_s - \gamma - (\eta_t - \gamma) e^{-\alpha(s-t)}}{e^{-\alpha(s-t)} y(s, t)} \right) =: g(\eta_s | \eta_t), \tag{5.10}
$$

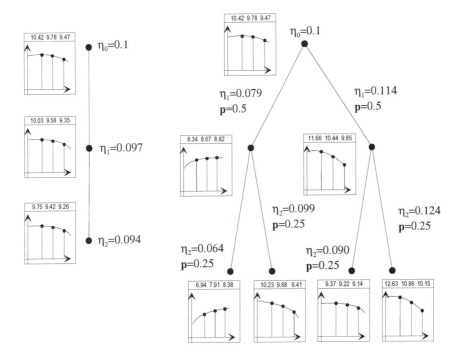

Figure 1: Scenario trees for the lower (left) and upper (right) approximations. The interest rates correspond to maturities of 1, 6, and 12 months.

where $pdf(\cdot)$ is the standard normal density function and

$$y^2(s,t) = \frac{\sigma^2}{2\alpha}\left[e^{2\alpha(s-t)} - 1\right].$$

Moreover, the expected value and the variance of η_s are given by

$$E(\eta_s|\eta_t) = \eta_t e^{-\alpha(s-t)} + \gamma(1 - e^{-\alpha(s-t)}) \tag{5.11}$$

$$Var(\eta_s|\eta_t) = \frac{\sigma^2}{2\alpha}(1 - e^{-2\alpha(s-t)}). \tag{5.12}$$

The fact that in the Vasicek model $\sigma(\eta^{t-1})$ depends only on the process parameters and not on the current value of η_t simplifies the evaluation of the barycentric scenario trees $\mathcal{A}^l$ and $\mathcal{A}^u$ considerably. Furthermore, the conditions of the proposition in section 3 are satisfied, which entails convexity of the value functions $\phi_t(v_t, \eta^t)$ in (v_t, η^t) for $t = 0, 1, \ldots, T$. This implies that approximate investment strategies are obtained with associated lower and upper bounds on the optimal profits, in case that the barycentric scenario trees are incorporated in the stochastic program.

Remark: In our particular one-dimensional situation the barycentric scenario trees are also obtainable in case that the Jensen-Inequality and the Edmundson-Madansky-Inequality are applied to the value function in each period. For details on the contributions of these inequalities within stochastic programming see, e.g., Madansky (1960), Avriel and Williams (1970), Ben-Tal and Hochman (1972), Huang, Ziemba and Ben-Tal (1977), or Kall and Stoyan (1982), Gassmann and Ziemba (1986), Birge and Wets (1987), Frauendorfer (1988), Edirisinghe and Ziemba (1994a,b), and Kall and Wallace (1994).

Example 5.1 *Let $\alpha = 0.3$, $\gamma = 0.087$, $\sigma = 0.01$, and the current value of the state variable $\eta_0 = 0.1$. Choose $\epsilon = 0.05$ for the simplical coverage. Furthermore, in equations (5.4) and (5.5) the density $f(\cdot)$ is replaced by the function $g(\cdot)$ from (5.10) for evaluating the barycenters and probabilities. Figure 1 shows the scenario trees for the upper and lower approximation and the resulting term structure for the first two periods with unpartitioned simplices.*

The objective values for $T = 8$ are 508.5834 for the lower approximation and 523.0054 for the upper approximation, yielding an accuracy of 2.836%. Next, the intervals of the first stage are partitioned into four subintervals, the simplices at the second stage into two subintervals. The corresponding scenario trees are shown in figure 2. The objective values for the lower and upper approximation are 508.9286 and 522.0943, respectively. The accuracy improved to 2.587%. This is a characteristic feature due to the monotonicity of the bounds: The approximation obtained from partitioning Θ into $N + 1$ subintervals is as least as good as the approximation obtained from a partition of Θ into N subintervals, provided that the value function $\phi_t(v_t, \eta^t)$ is convex in (v_t, η^t) and the sequence of partitions represents refinements.

5.2   Investment strategies based on the CIR model

The Cox-Ingersoll-Ross approach is considered next. The behavior of the underlying square root process (4.13) for $\delta = 0.5$ has some important properties, as pointed out in the original paper of CIR: In continuous time, negative interest rates are precluded. If the instantaneous rate approaches zero, the standard deviation $\sigma\sqrt{\eta_t}$ approaches zero, too. The mean-reverting term pulls the process towards its long term mean $\gamma > 0$. In addition, the absolute variance of the interest rate is proportional to the level of the rate, and there is a steady state distribution for the interest rate. The probability density of the instantaneous rate $\eta_s|\eta_t$ at time s, conditional on its value at the current time t, is given by

$$f(\eta_s|\eta_t) = ce^{-u-v}\left(\frac{v}{u}\right)^{q/2} I_q\left(\sqrt{2(uv)}\right), \qquad (5.13)$$

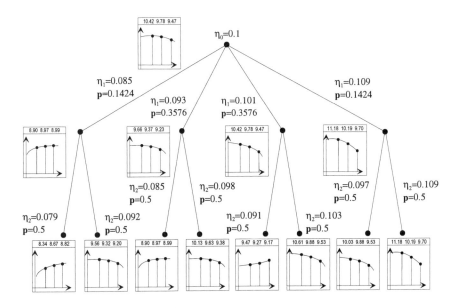

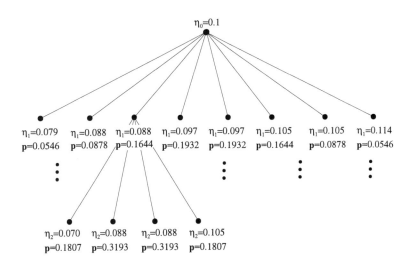

Figure 2: Refined scenario trees for the lower (top) and upper (bottom) approximation.

where

$$c = \frac{2\alpha}{\sigma^2(1 - e^{-\alpha(s-t)})}$$

$$u = c\eta_t e^{-\alpha(s-t)}$$

$$v = c\eta_s$$

$$q = \frac{2\alpha\gamma}{\sigma^2} - 1.$$

$I_q(\cdot)$ denotes the modified Bessel function of the first kind of order q. The distribution function of the random variable $\eta_s|\eta_t$ is the noncentral chi-square, $\chi^2[2c\eta_s; 2q + 2, 2u]$.

The expected value and the variance of $\eta_s|\eta_t$ are

$$E(\eta_s|\eta_t) = \eta_t e^{-\alpha(s-t)} + \gamma(1 - e^{-\alpha(s-t)}) \tag{5.14}$$

$$Var(\eta_s|\eta_t) = \eta_t \left(\frac{\sigma^2}{\alpha}\right)(e^{-\alpha(s-t)} - e^{-2\alpha(s-t)})$$

$$+ \gamma\left(\frac{\sigma^2}{2\alpha}\right)(1 - e^{-\alpha(s-t)})^2. \tag{5.15}$$

The CIR model with its discrete-time process specification of the factor η_t,

$$\eta_{t+1} = \eta_t + \alpha(\gamma - \eta_t) + \sigma\sqrt{\eta_t}U_t, \qquad U_t \sim N(0, 1), \tag{5.16}$$

reveals that due to the square-root term in (5.16) observations η_{t+1} at time $t+1$ and, hence, the distribution function of $P(\cdot|\eta^t)$ do not depend linearly on the up-to-date history η^t. In case scenarios are generated based on the time-discrete process (5.16), then there is a positive probability for $\eta_{t+1} < 0$ which is inconsistent with the model specification in continuous time. Furthermore, the density function (5.13) is not real-valued for $\eta_s|\eta_t < 0$. Thus, applying equations (5.4)–(5.5) for evaluating the barycentric scenario trees necessitates that the subintervals $\Theta^{(i)}|\eta^{t-1}$ are subsets of $\mathbb{R}_0^+$. The standard deviation $\sigma(\cdot)$ depends now on the current level of the η^{t-1}. Unlike the Vasicek model, where the simplices have been linearly transformed, the length of the interval $\Theta_t|\cdot$ varies from stage to stage as well as between the different nodes of the scenario tree for a fixed stage. As a consequence, the conditions of the proposition are not fulfilled. Therefore, the convexity of the value functions $\phi_t(v_t, \eta^t)$ in (v_t, η^t) for $t = 0, 1, \ldots, T$ is not ensured. The values of the approximate investment strategies may be accepted as estimates for the optimal profits.

Example 5.2 *Again, the investment problem (3.1) is considered. Interest rate scenarios are generated based on the CIR model with parameters $\alpha = 0.1779$, $\gamma = 0.0866$, $\sigma = 0.02$, and $\eta_0 = 0.1$. ϵ was set to 0.05 for the simplicial coverage. This problem was solved for a planning horizon $T = 6$*

and three different partitions: (i) unpartitioned for all stages, (ii) partitioned twice for the first two stages into identical subintervals, and (iii) each subinterval additionally partitioned into two subintervals for the first stage. The simplices and barycenters for the upper and lower scenario tree, $\mathcal{A}^l$ and $\mathcal{A}^u$, were obtained according to (5.2)–(5.5).

Partitions	Lower Bound	Upper Bound
(i)	277.7647	278.0060
(ii)	277.7663	277.9257
(iii)	277.7665	277.9210

Now, the values in the table above represent estimates for the objective values since the convexity of the value function is not guaranteed. However, in this particular situation, the value function $\phi_t(\cdot, \cdot)$ is almost linear and therefore, the convexity is not violated significantly. For this reason, the bounds converge with increasing numbers of partitions.

Employing the Vasicek model preserves convexity of the value functions in the instantaneous rate and in the investment policies. The values of the surrogate optimization problems represent lower and upper bounds on the optimal profit and help quantify the accuracy of the associated investment policy. The bounds converge monotonously to the optimal value in case that the support of the stochastic factors is refined successively and the corresponding subcells become arbitrarily small.

The Cox-Ingersoll-Ross model uses a square-root process which entails nonconvexity of the value-functions in the instantaneous rate. Convexity in the investment policies is preserved due to the fact that the conditional probability distributions remain independent of the policies. Hence, the surrogate optimization problems are still convex programs and allow for applying sophisticated convex programming algorithms. Unfortunately, the square-root term in the CIR process specification prohibits that information on the accuracy is received. The values may be accepted only as estimates which converge to the optimal profit in case of successive refinements. A careful analysis of the nonconvex value functions with respect to the subcells of the partition might provide information on the accuracy at least in a probabilistical sense. Investigations in these directions remain to be done.

6 Conclusions

Single-factor interest-rate models and stochastic programming have been applied to multiperiod investment. Obviously, a similar argumentation holds also for optimal funding, for refinancing mortgages, or for financing risky projects in industry. The key issue addressed is the incorporation of stochastic factor evolvements in the optimization process. Barcyentric approximation has been employed for discretizing the stochastic processes. Distinguished scenario trees covering various cash-flow and term structure movements provide the portfolio manager with approximate policies and corresponding estimates for the profits. In the entire convex case these estimates represent lower and upper bounds for the maximum interest surplus achievable within the planning horizon and hence, allow for quantifying the accuracy. In case the scenario trees are refined adequately, convergence of the estimates is ensured. Due to the curse of dimensionality, the computational effort might explode if a high number of partitions is required. This is particular the case when two or more factors are incorporated.

In the literature few investigations have been undertaken to specify stochastic evolvements for cash-flow and term structures jointly. Due to our experience with refinancing mortgages, it has to be stressed that the optimal refinancing policies strongly depend on the correlation between interest rates and mortgage volume. Hence, additional effort which result in a better understanding of the interaction between interest rate factors and cash-flow factors will yield a positive pay-off.

We have dealt with one-factor models for generating scenario trees in the context of stochastic programming, in particular with the Vasicek and the Cox-Ingersoll-Ross approach, which allow closed-form solutions of the fundamental differential equation and an analytical specification of the distribution. Furthermore, the Vasicek approach preserves the convexity of the value functions in the stochastic factors and policies jointly, in contrast to the CIR approach whose square-root term entail nonconvexity of the value function in the stochastic factors and convexity in the policies. Hence, the achieved accuracy may only be assessed with Vasicek. In the CIR setting a deeper analysis of the nonconvexity will likely lead to probabilistic accuracy statements. Fortunately, in both cases convergence of the bounds, estimates respectively, is ensured.

The integration of multiple factor models seems desirable since there is empirical evidence that the dynamics of interest rates can be better explained by more than one state variable. E. g., in the Brennan and Schwartz approach (1979, 1982), the yield on a consol bond with infinite maturity is chosen as the second factor, allowing the model to reflect that the long end of the term structure is variable. Schaefer and Schwartz (1984) use the same information about interest rates as Brennan and Schwartz, but replace the short rate by the spread between the long and the short rate. This change of variables allows to obtain an analytical solution of the fundamental term structure equation. Beside this, the spread is assumed to follow an Ornstein-Uhlenbeck process whereas the long rate process is of the square-root type, both being uncorrelated. Thus, there is only one risk factor violating the convexity of the value function. Unfortunately, in many of the most common multiple factor models the distributions do not depend linearly on the past history according to their process specifications. The entire convex case is left.

Litterman and Scheinkman (1991) reported that there are three major components driving the yield curve which are identified as the level, steepness, and curvature factors, and which explain most of the term structure movements. In an A&L management model different from that described in the previous sections, we implemented a three-dimensional correlated Wiener process for the evolvement of these risk factors. The interest rates of the various maturities are obtained from a function that is linear in the latter. In addition, stochastic right hand sides of constraints are considered by modelling the change in portfolio volume by a Wiener process. Therefore, the saddle property of the value function is given. In our experience, there is a broad variety of different strategies that is suggested by this model depending on the current shape and level of the term structure. In contrast to this, it has been observed that the policies obtained from the one-factor models described above tend to fluctuate between extreme values, either short or long maturities. Intuitively, this might be explained by the fact that the level of long rates remains relatively stable.

The three-factor model has been successfully implemented for the problem of funding variable rate mortgages in cooperation with a major Swiss bank. Significant savings have been reported with the stochastic optimization methodology compared to various constant mix policies. A case study using historical interest rate curves over a period of 60 months has shown that the volatility of the dynamic refinancing policy obtained by stochastic programming is less than that of various constant mix policies. However, the Wiener process model fails with respect to an accurate description of interest rate dynamics because it does not exhibit typical properties like mean-reversion. It is expected that incorporating multi-factor term structure models in a stochastic programming model and applying the barycentric approximation scheme will yield further improvements.

Research for this paper was supported by Grundlagenforschungsfonds *of the University of St. Gallen for Business Administration, Economics, Law and Social Sciences (HSG), Grant No. 121'417, and by* Schweizerischer Nationalfonds, *Grant No. 21-39'575.93.*

Bibliography

M. Avriel and A.C. Williams (1970). 'The Value of Information and Stochastic Programming.' *Operations Research* **18** 947–954.

A. Ben-Tal and E. Hochman (1972). 'More bounds on the expectation of a random variable.' *Journal of Applied Probability* **9** 803–812.

J. Birge and R.J.-B. Wets (1987). 'Computing Bounds for Stochastic Programming Problems by Means of a Generalized Moment Problem'. *Math. of OR* **12** 149–162.

J. Birge and R.J.-B. Wets (1991). 'Stochastic Programming, Part I, II.' In *Proceedings of the 5th International Conference in Stochastic Programming, Ann Arbor, Michigan, August 13–18. Annals of Operations Research*, 30–31.

M.J. Brennan and E.S. Schwartz (1979). 'A Continuous Time Approach to the Pricing of Bonds.' *Journal of Banking and Finance* **3** 133–155.

M.J. Brennan and E.S. Schwartz (1982). 'An Equilibrium Model of Bond Pricing and a Test of Market Efficiency.' *Journal of Financial and Quantitative Analysis* **17** 301–329.

A. Bühler (1995). *Einfaktormodelle der Fristenstruktur der Zinssätze – Theoretische und empirische Betrachtungen.* Verlag Paul Haupt.

E. Canabarro (1993). 'Comparing the Dynamic Accuracy of Yield-Curve-Based Interest Rate Contingent Claim Pricing Models'. *Journal of Financial Engineering* **2** 365–401.

D.R. Cariño, T. Kent, D.H. Myers, C. Stacy, M. Sylvanus, A.L. Turner, K. Watanabe, and W.T. Ziemba (1994). 'The Russell-Yasuada Kasai Financial Planning Model: An Asset/Liability Model for a Japanese Insurance Company Using Multistage Stochastic Programming'. *Interfaces* **24** 29–49. Reprinted in this volume.

K.C. Chan, G.A. Karolyi, F.A. Longstaff, and A.B. Sanders (1992). 'An Empirical Comparison of Alternative Models of the Short-Term Interest Rate.' *Journal of Finance* **47(3)** 1209–1227.

G. Courtadon (1982). 'The pricing of options on default-free bonds.' *Journal of Financial and Quantitative Analysis* **17** 75–100.

J.C. Cox, J.E. Ingersoll, and S.A. Ross (1985). 'A Theory of the Term Structure of Interest Rates.' *Econometrica* **53** 385–407.

H. Dahl, A. Meeraus, and S.A. Zenios (1992). 'Some Financial Optimization Models'. In *Financial Optimization*, S.A. Zenios (ed.), Cambridge University Press, 3–36.

G.B. Dantzig (1991). 'Linear Programming'. In *History of Mathematical Programming*, J.K. Lenstra, A.H.G. Rinnooy Kan, and A. Schrijver (eds.), CWI North-Holland, 19–31.

G.B. Dantzig and P.W. Glynn (1990). 'Parallel Processors for Planning under Uncertainty.' *Annals of Operations Research* **22** 1–21.

G.B. Dantzig and G. Infanger (1993). 'Multi-Stage Stochastic Linear Programs for Portfolio Optimization.' In *Financial Engineering*, H. Konno, D.G. Luenberger, and J.M. Mulvey (eds.) *Annals of Operations Research* **45** 59–76.

M.A.H. Dempster (1994). 'An EVPI-Based Sampling Algorithm for Large Scale Dynamic Stochastic Programming.' Presented at the 15th Intern. Symp. on Math. Programming, Ann Arbor, August 15-19, 1994.

L.U. Dothan (1978). 'On the Term Structure of Interest Rates.' *Journal of Financial Economics* **6** 59–69.

J. Dupačová (1992). 'Applications of Stochastic Programming in Finance.' In *Atti del XVI Convegno A.M.A.S.E.S.*, Treviso 1992, 13–30.

R.L. D'Ecclesia and S.A. Zenios (eds.) (1994). *Operations Research in Quantitative Finance.* Physica-Verlag.

N.C.P. Edirisinghe and W.T. Ziemba (1994a). 'Bounds for Two-Stage Stochastic Programs with Fixed Recourse.' *Math. of OR* **19** 292–313.

N.C.P. Edirisinghe and W.T. Ziemba (1994b). 'Bounding the Expectation of a Saddle Function with Application to Stochastic Programming.' *Math. of OR* **19** 314–340.

Y. Ermoliev and R.J.-B. Wets (1988). *Numerical Techniques for Stochastic Optimization.* Springer-Verlag.

E.O. Fischer and J. Zechner (1984). 'Diffusion Process Specification for Interest Rates – An Empirical Investigation for the FRG'. In *Risk and Capital*, K. Spremann and G. Bamberg, (eds.), Lecture Notes in Economics and Mathematical Systems **227**, Springer-Verlag, 64–73.

H.G. Fong and O.A. Vasicek (1991). 'Fixed-Income Volatility Management.' *Journal of Portfolio Management*, (Summer) 41–46.

K. Frauendorfer (1988). 'Solving SLP Recourse Problems with Arbitrary Multivariate Distributions – The Dependent Case.' *Math. of OR* **13** 377–394.

K. Frauendorfer (1992). *Stochastic Two-Stage Programming*, Lecture Notes in Economics and Mathematical Systems **392**, Springer-Verlag.

K. Frauendorfer (1994a). 'Multistage Stochastic Programming: Error Analysis for the Convex Case.' *ZOR – Mathematical Methods of Operations Research* **39** 93–122.

K. Frauendorfer (1994b). 'Barycentric Scenario Trees in Convex Multistage Stochastic Programming.' Working Paper, University of St. Gallen (accepted for publication in *Mathematical Programming*).

K. Frauendorfer, F. Härtel, M.F. Reiff, and M. Schürle (1996). 'SG-Portfolio Test Problems for Stochastic Multistage Linear Programming.' In *Operations Research 1995*, P. Kleinschmidt, A. Bachem, U. Derigs, D. Fischer, U. Leopold-Wildburger, and R. Mühring (eds.), Springer-Verlag, 102–107.

H.I. Gassmann and W.T. Ziemba (1986). 'A Tight Upper Bound for the Expectations of a Convex Function of a Multivariate Random Variable.' *Math. Prog. Study* **27** 39–53.

J.H. Gilkeson and S.D. Smith (1992). 'The Convexity Trap: Pitfalls in Financing Mortgage Portfolios and Related Securities.' *Economic Review*, 14–27.

B. Golub, M. Holmer, R. McKendall, L. Pohlman and S.A. Zenios (1995). 'Stochastic programming models for money management.' *European Journal of Operations Research* **85** 282–296.

R.S. Hiller and J. Eckstein (1994). 'Stochastic dedication: Designing fixed income portfolios using massively parallel Benders decomposition.' *Management Science* **39** 1422–1438.

M.R. Holmer (1994). 'The asset/liability management system at Fannie Mae.' *Interfaces* **24** 3–21.

C. Huang, W.T. Ziemba, and A. Ben-Tal (1977). 'Bounds on the Expectation of a Convex Function of a Random Variable: with Applications to Stochastic Programming.' *Operations Research* **25** 315–325.

J. Hull (1993). *Options, Futures, and Other Derivative Securities*, 2nd ed., Prentice-Hall.

J.E. Ingersoll, Jr. (1987). *Theory of Financial Decision Making* Rowman & Littlefield.

J. Jensen (1906). 'Sur les fonctions convexes et les inégalités entre les valeurs moyennes.' *Acta Math.* **30** 175–193.

P. Kall and D. Stoyan (1982). 'Solving Programming Problems with Recourse Including Error Bound.' *Mathematische Operationsforschung und Statistik, Ser. Opt. 13* 431–447.

P. Kall and S.W. Wallace (1994). *Stochastic Programming*, Wiley.

J.G. Kallberg, R.W. White, and W.T. Ziemba (1982). 'Short term financial planning under uncertainty.' *Management Science* **28** 670–682.

A.J. King (1993). 'Asymmetric Risk Measures and Tracking Models for Portfolio Optimization under Uncertainty.' *Annals of OR* **45** 163–177.

M.I. Kusy and W.T. Ziemba (1986). 'A Bank Asset and Liability Management Model.' *Operations Research* **34** 356–376.

J.K. Lenstra, A.H.G. Rinnooy Kan, A. Schrijver (1991). *History of Mathematical Programming*, CWI North-Holland.

R. Litterman and J. Scheinkman (1991). 'Common Factors Affecting Bond Returns.' *Journal of Fixed Income* **1** 54–62.

F.A. Longstaff and E.S. Schwartz (1992). 'Interest Rate Volatility and the Term Structure: A Two-Factor General Equilibrium Model.' *Journal of Finance* **47** 1259–1282.

A. Madansky (1960). 'Inequalities for Stochastic Linear Programming Problems.' *Management Science* **6** 197–204.

H. Markowitz (1987). *Mean-Variance Analysis in Portfolio Choice and Capital Markets*, Basil Blackwell.

R. McKendall, S.A. Zenios, and M. Holmer (1994). 'Stochastic Programming Models for Portfolio Optimization with Mortgage Backed Securities: Comprehensive Research Guide.' In *Operations Research Models in Quantitative Finance*, R.L. D'Ecclesia and S.A. Zenios (eds.), Physica-Verlag, 134–171.

R.C. Merton (1990). *Continuous-Time Finance* Basil Blackwell.

H. Müller (1988). 'Modern Portfolio Theory: Some Main Results.' *Astin Bulletin* **18** 127–145.

J.M. Mulvey (1994a). 'Integrating Assets and Liabilities for Large Financial Organizations.' In *Computational Economics*, Kluwer Academic Publishers.

J.M. Mulvey (1994b). 'Multi-Stage Financial Planning Systems.' In *Operations Research Models in Quantitative Finance*, R.L. D'Ecclesia and S.A. Zenios (eds.), Physica-Verlag, 18–35.

J.M. Mulvey and H. Vladimirou (1991). 'Applying the Progressive Hedging Algorithm to Stochastic Generalized Networks.' In *Stochastic Programming, Part II*. J. Birge and R.J.-B. Wets (eds.), *Annals of Operations Research* **31** 399–424.

J.M. Mulvey and H. Vladimirou (1992). 'Stochastic Network Programming for Financial Planning Problems.' *Management Science* **38** 1642–1664.

S.M. Robinson (1991). 'Extended Scenario Analysis.' In *Stochastic Programming, Part II*. J. Birge and R.J.-B. Wets (eds.), *Annals of OR* **31** 385–39.

R.T. Rockafellar (1989). 'Perturbation of Generalized Kuhn-Tucker Points in Finite-Dimensional Optimization.' Technical Report, University of Wisconsin.

R.T. Rockafellar (1990). 'Lagrangian optimization.' Technical Report, Dept. of Applied Mathematics, University of Washington.

R.T. Rockafellar and R.J.-B. Wets (1976). 'Nonanticipativity and L^1-Martingales in Stochastic Optimization Problems.' *Math. Prog. Study* **6** 170–187.

R.T. Rockafellar and R.J.-B. Wets (1986). 'A Lagrangian Finite Generation Technique for Solving Linear-Quadratic Problems in Stochastic Programming.' *Math. Prog. Study* **28** 63–93.

R.T. Rockafellar and R.J.-B. Wets (1990). 'Generalized Linear-Quadratic Problems of Deterministic and Stochastic Optimal Control in Discrete Time.' *SIAM J. of Control and Optimization* **28** 810–822.

R.T. Rockafellar and R.J.-B. Wets (1991). 'Scenarios and Policy Aggregation in Optimization under Uncertainty.' *Math. of OR* **16** 119–147.

A. Ruszczynski (1993). 'Parallel Decomposition of Multistage Stochastic Programming problems.' *Mathematical Programming* **58** 201–228.

S.M. Schaefer and E.S. Schwartz (1984). 'A Two-Factor Model of the Term Structure: Approximate Analytical Solution.' *Journal of Financial and Quantitative Analysis* **19** 413–424.

E. Spedicato (ed.) (1994). *Algorithms for Continuous Optimization – The State of the Art*, NATO ASI Series C, Vol. 434, 383–413, Kluwer.

C. Strickland (1994). 'A Comparision of Models of the Term Structure.' Working paper 93/46, University of Warwick, Financial Options Research Centre.

O. Vasicek (1977). 'An Equilibrium Characterization of the Term Structure.' *Journal of Financial Economics* **5** 177–188.

R.J.-B. Wets (1994). 'Challenges in Stochastic Programming.' Working Paper WP-94-32, International Institute for Applied Systems Analysis (IIASA).

S.A. Zenios (1991). 'Massively Parallel Computations for Financial Modeling under Uncertainty.' In *Very Large Scale Computing in the 21st Century'*, J. Mesirov (ed.), SIAM, Philadelphia, 273–294.

S.A. Zenios (1992). 'Asset/Liability Management under Uncertainty: The Case of Mortgage-Backed Securities.' Technical Report 92–08–05, Department of Decision Sciences, The Wharton School, University of Pennsylvania, Philadelphia, PA 19104–6366.

S.A. Zenios (ed.) (1992). *Financial Optimization*. Cambridge University Press.

S.A. Zenios and W.T. Ziemba (eds.) (1992). Focused Issue on Financial Modeling *Management Science* **38**, No. 11.

Postoptimality for Scenario Based Financial Planning Models with an Application to Bond Portfolio Management

Jitka Dupačová, Marida Bertocchi and Vittorio Moriggia

Abstract

The contamination technique is presented as a numerically tractable technique for postoptimality analysis and analysis of the robustness of the optimal value of various scenario based stochastic programs with respect to inclusion of additional 'out-of-sample' scenarios. Using results based on the initial selection of scenarios and those based on the alternative out-of-sample scenarios it provides bounds for the optimal value based on the pooled sample of scenarios of these groups. The application of the method to models supporting financial decision making is detailed for bond portfolio management and tracking models. Numerical experience is presented for a bond portfolio management model using data from the Italian bond market.

1 Scenario Based Models

The outcome of financial decisions depends on realization of numerous input values which are unknown to the decision maker at the time when the decision has to be taken. Examples include future prices or returns, interest rates, exchange rates, external cashflows including liabilities, prepayment rates, lapse behavior and future inflation. Given a set of forecasted values of these parameters from a scenario, one accepts a decision which is plausible under the assumed circumstances but which may be unacceptable for a different scenario. Another approach is to interpret the input parameters as random and to base the decisions on a stochastic programming model; we refer to the recent monographs by Kall and Wallace (1994), and Prékopa (1995) for a general information about stochastic programming and to collections Konno *et al.*, eds. (1993), Zenios, ed. (1993), Zenios and Ziemba, eds. (1992), surveys Dupačová (1991), Mulvey (1994) or to numerous papers on applications of stochastic programming in finance, e.g., Bradley and Crane (1972), Cariño *et al.* (1994), Dembo (1991), Dempster and Ireland (1988), Dert (1995), Dupačová and Bertocchi (1996), Kusy and Ziemba (1986), McKendall *et al.* (1994), Mulvey and Vladimirou (1992), Shapiro (1988), and Zenios (1991).

The numerical techniques designed for solving stochastic programming problems are mostly based on approximation of the distribution of the random parameters by a discrete scenario, obtained by sampling in the course of numerical solution or given in advance; cf. Ermoliev and Wets, eds. (1988). We shall consider here the latter approach; hence, we assume that there is a given *discrete* distribution P concentrated in a finite number of points, say, $\omega_1, \ldots, \omega_S$ with positive probabilities $p_s > 0 \quad \forall s$, $\sum_{s=1}^{S} p_s = 1$, that enter the coefficients and the function values in a known way. The atoms $\omega_1, \ldots, \omega_S$ are called *scenarios*.

The origin of scenarios can be very diverse; they may be from a truly discrete known distribution, be obtained in the course of a discretization/approximation scheme or by a limited sample information, or come from attempts to model uncertainty by means of scenarios obtained by a preliminary analysis of the problem and with probabilities of their occurence that may reflect an ad hoc belief or a subjective opinion of an expert.

One is interested in both the robustness of the obtained optimal solution and the optimal value of the objective function. The procedure should be robust in the sense that small perturbances of the input, i.e., of the chosen scenarios and of their probabilities, should alter the outcome only slightly so that the results obtained remain close to the unperturbed ones, and that somewhat larger perturbations do not cause a catastrophe. The importance of robust procedures increases with the complexity of the model and with its dimensionality.

We shall elaborate here the *contamination technique* which is, *inter alia*, suitable for analysis of influence of additional scenarios and for constructing the corresponding error bounds. We refer to Dupačová (1986, 1990) for the theoretical results, to Dupačová (1995) for an application in the field of scenario based multistage stochastic linear programs with fixed complete recourse, and to Dupačová (1996b, 1998) for an extension to problems in which the objective function is nonlinear in distribution P to cover, e.g., the case of mean-variance criterion or the robust optimization models by Mulvey *et al.* (1995).

The models considered in this paper can be put into the form:

$$\text{Minimize} \quad f(\mathbf{x}, P) \quad \text{on the set} \quad \mathcal{X} \subset R^n \tag{1.1}$$

where

f convex in $\mathbf{x}$ and linear in P;

P is the probability distribution of the random parameters $\omega \in \Omega$ that enter the problem; in the case of scenario based stochastic programs that we deal with in our applications, P is a discrete probability distribution and for a given set Ω of possible scenarios, this distribution is *fully determined* by the vector $\mathbf{p}$ of their probabilities. Accordingly, the objective function is linear in $\mathbf{p}$.

$\mathcal{X}$ a closed, nonempty set that *does not depend on* P; and

$\mathbf{x} \in \mathcal{X}$ the main, scenario independent decision variable, typically, the first stage decision.

Problems with $f(\mathbf{x}, \bullet)$ *linear* in P that are considered in this paper correspond to minimization of the expected value of a random outcome of the modeled decision process.

Example 1 Scenario based two-stage stochastic linear programs (SLP) with *random relatively complete recourse* appear in financial models that take into account random prices in connection with portfolio rebalancing or with conservation of cashflows, cf. Golub *et al.* (1993), McKendall *et al.* (1994), Zenios (1991).

They can be written as

$$\text{minimize} \quad \mathbf{c}^{\top}\mathbf{x} + \sum_{s=1}^{S} p_s \mathbf{q}_s^{\top} \mathbf{y}_s \tag{1.2}$$

subject to

$$
\begin{array}{llll}
\mathbf{Ax} & & = \mathbf{b} \\
\mathbf{T}_1\mathbf{x} & + \; \mathbf{W}_1\mathbf{y}_1 & & = \mathbf{h}_1 \\
\mathbf{T}_2\mathbf{x} & + & \mathbf{W}_2\mathbf{y}_2 & = \mathbf{h}_2 \\
\quad \vdots & & \ddots & \quad \vdots \\
\mathbf{T}_S\mathbf{x} & + & \cdots & \mathbf{W}_S\mathbf{y}_S & = \mathbf{h}_S
\end{array}
$$

$$\mathbf{x} \geq 0, \mathbf{y}_s \geq 0, s = 1, \ldots, S \tag{1.3}$$

where $\omega_s = [\mathbf{q}_s, \mathbf{T}_s, \mathbf{W}_s, \mathbf{h}_s], s = 1, \ldots, S$ are scenarios or atoms at which the probability distribution P is concentrated and $p_s \geq 0, s = 1, \ldots, S$ are their probabilities, $\sum_s p_s = 1$.

Example 2 Scenario based expected utility models use principle of the maximal expected utility, namely

$$\text{maximize} \quad \sum_{s=1}^{S} p_s u(g(\mathbf{x}, \omega_s)) \tag{1.4}$$

subject to $\mathbf{x} \in \mathcal{X}$. The function g is often defined as the optimal value of an auxilliary optimization problem that is related with a given scenario ω_s and a given initial decision $\mathbf{x}$. This optimal value can be the final wealth achieved by optimal management of a bond portfolio at the end of the pay-off period (see Sections 3 and 4) or the difference between the return of the portfolio and the index, see Worzel *et al.* (1994), etc. The choice of the utility function is restricted to concave nondecreasing functions and there are various types of utility functions which are popular in finance, such as isoelastic utility functions $u(W) = W^\gamma/\gamma$. The book by Ziemba and Vickson (1975) discusses the pros and cons of typical utility functions.

Also the two-stage stochastic linear program from Example 1 can be modified to an expected utility model:

$$\text{minimize} \quad -\sum_{s=1}^{S} p_s u(\mathbf{c}^\top \mathbf{x} + q(\mathbf{x}, \omega_s)) \tag{1.5}$$

on the set $\mathcal{X}$ and with

$$q(\mathbf{x}, \omega_s) = \min_{\mathbf{y}_s} \left\{ \mathbf{q}_s^\top \mathbf{y}_s \,|\, \mathbf{W}_s \mathbf{y}_s = \mathbf{h}_s - \mathbf{T}_s \mathbf{x}, \quad \mathbf{y}_s \geq 0 \right\} \tag{1.6}$$

Example 3 The tracking model related to (1.2), see Dembo (1991), can be formulated as follows: Let $v_s, s = 1, \ldots, S$ be the optimal values of the *individual* scenario problems

$$\text{minimize} \quad \mathbf{c}^\top \mathbf{x} + \mathbf{q}_s^\top \mathbf{y}_s \tag{1.7}$$

subject to

$$\begin{aligned}
\mathbf{A}\mathbf{x} & & &= \mathbf{b} \\
\mathbf{T}_s \mathbf{x} &+ \mathbf{W}_s \mathbf{y}_s &&= \mathbf{h}_s \\
& \mathbf{x} \geq 0, \ \mathbf{y}_s \geq 0. &&
\end{aligned} \tag{1.8}$$

Then the basic compromising or tracking model is

$$\text{minimize} \quad \sum_{s=1}^{S} p_s \left(\| \mathbf{c}^\top \mathbf{x} + \mathbf{q}_s^\top \mathbf{y}_s - v_s \| + \| \mathbf{T}_s \mathbf{x} + \mathbf{W}_s \mathbf{y}_s - \mathbf{h}_s \| \right) \tag{1.9}$$

subject to $\mathbf{x}$ and $\mathbf{y}_s \forall s$ that fulfil the 'hard' constraints

$$\mathbf{A}\mathbf{x} = \mathbf{b} \tag{1.10}$$

$$\mathbf{x} \geq 0, \ \mathbf{y}_s \geq 0, \qquad s = 1, \ldots, S. \tag{1.11}$$

The first and second stage solutions obtained by solving this problem track the optimal solutions of the individual scenario problems (1.7)–(1.8) as closely as possible. The norm in (1.9) can be in principle chosen in an arbitrary way; its choice influences the solution procedure.

Further examples that can be used to illustrate the general form of the considered problem (1.1) and to provide a motivation for our studies are scenario based multistage stochastic programs, see Dupačová (1995).

In these examples, we are interested in resistance of the obtained optimal decisions and of the optimal value with respect to the used input: for the given set of scenarios $\Omega = \{\omega_1, \ldots, \omega_S\}$ we want to study the influence of this choice of scenarios ω_s and of their probabilities as well as the influence of inclusion of additional scenarios on the optimal value of the objective function (1.2), (1.4) or (1.9). We exploit classical results of parametric linear and nonlinear programming together with the contamination technique of robust statistics. This is a tractable approach in situations when a straightforward application of standard postoptimality methods of linear programming is in general hardly manageable: even for a fixed sample size S, inclusion of an additional scenario means an extension of the system of equations, for instance those in problem (1.2)–(1.3), for a new block of second-stage constraints and for additional second-stage variables, etc.

In the next section, we shall briefly describe the contamination technique and provide the main result – bounds on the optimal value of the perturbed problem. This approach will be applied to the bond portfolio management problem which is an application of the expected utility model from Example 2 and to the tracking model of Example 3. Numerical results presented in the last section are based on an application of the bond portfolio management model to the Italian bond market.

2 Contamination Technique - The Basic Ideas

We present a brief summary of the contamination technique (cf. Dupačová (1986, 1991)) for the general form of stochastic programs (1.1) under assumptions that $\mathcal{X}$ is a given nonempty convex closed set of feasible solutions that does not depend on the probability distribution P and that the objective function f is convex in $\mathbf{x}$ and linear in P. Let $\varphi(P)$ denotes the minimal value of the objective function in (1.1) and let $\mathcal{X}(P)$ be the set of optimal solutions. We shall embed the problem (1.1) into a family of optimization problems parametrized by a *scalar* parameter λ. This family comes from contamination of the original probability distribution P by another *fixed* probability distribution Q, i.e., from using distributions P_λ of the form

$$P_\lambda = (1 - \lambda)P + \lambda Q \quad \text{with} \quad \lambda \in (0, 1) \tag{2.1}$$

in the objective function of (1.1) at the place of P. For fixed distributions P, Q the contaminated distribution P_λ depends only on λ and

$$f(\mathbf{x}, P_\lambda) := f_Q(\mathbf{x}, \lambda) \tag{2.2}$$

is the corresponding objective function which is a convex - concave function on $R^n \times [0,1]$. Let

$$\varphi(P_\lambda) = \varphi_Q(\lambda) = \inf_{\mathbf{x} \in \mathcal{X}} f_Q(\mathbf{x}, \lambda) \quad \text{and} \quad \mathcal{X}(P_\lambda) = \mathcal{X}_Q(\lambda) = \arg\min_{\mathbf{x} \in \mathcal{X}} f_Q(\mathbf{x}, \lambda)$$
(2.3)

be the optimal value function and the set of optimal solutions of the perturbed stochastic program

$$\text{minimize} \quad f(\mathbf{x}, P_\lambda) := f_Q(\mathbf{x}, \lambda) \quad \text{on the set} \quad \mathcal{X}.$$
(2.4)

There are various statements about persistence, stability and sensitivity for parametric programs of the above type:

- Under the additional assumption that the set $\mathcal{X}(P) := \mathcal{X}_Q(0)$ of optimal solutions of the original problem (1.1) is nonempty and bounded and that $\mathcal{X}(Q) = \mathcal{X}_Q(1) \neq \emptyset$, the function φ_Q is a finite concave function on $[0,1]$, continuous at $\lambda = 0$ (cf. Gol'shtein (1972), Theorem 15) and its value at $\lambda = 0$ equals the optimal value of (1.1):

$$\varphi_Q(0) = \min_{\mathbf{x} \in \mathcal{X}} f(\mathbf{x}, P) = \varphi(P)$$
(2.5)

- If the objective function f_Q is jointly continuous with respect to $\mathbf{x}$ and λ, its derivative exists with respect to λ at $\lambda = 0^+$ for all $\mathbf{x}$ from a neighborhood, say, $\mathcal{X}^*$ of $\mathcal{X}(P)$ and if the convergence of the difference quotients $\frac{1}{\lambda}[f_Q(\mathbf{x}, \lambda) - f_Q(\mathbf{x}, 0)]$ for $\lambda \to 0^+$ is uniform in $\mathbf{x}$ on $\mathcal{X}^*$, we can use a slight modification of Theorem 17 of Gol'shtein (1972) to get the derivative of the optimal value of the perturbed program (2.4) at $\lambda = 0^+$:

$$\varphi'_Q(0^+) = \frac{d}{d\lambda}\varphi_Q(0^+) = \min_{\mathbf{x} \in \mathcal{X}(P)} \frac{d}{d\lambda} f_Q(\mathbf{x}, 0^+).$$
(2.6)

When $f(\mathbf{x}, P)$ is linear in P,

$$f_Q(\mathbf{x}, \lambda) = (1 - \lambda) f(\mathbf{x}, P) + \lambda f(\mathbf{x}, Q)$$
(2.7)

is a linear function in λ and for an arbitrary fixed $\mathbf{x}$, the sequence of difference quotients is a stationary one. Then, (2.6) reduces to

$$\varphi'_Q(0^+) = \min_{\mathbf{x} \in \mathcal{X}(P)} [f(\mathbf{x}, Q) - f(\mathbf{x}, P)] = \min_{\mathbf{x} \in \mathcal{X}(P)} f(\mathbf{x}, Q) - \varphi(P).$$
(2.8)

In this special but important case the derivative equals the difference between the minimal expected cost of an optimal decision based on the initial distribution P if $Q \neq P$ applies and the minimal expected costs under P.

Using (2.8) and concavity of φ_Q on $[0,1]$ we can bound the considered perturbed optimal value function $\varphi_Q(\lambda)$:

$$(1-\lambda)\varphi_Q(0) + \lambda\varphi_Q(1) \leq \varphi_Q(\lambda) \leq \varphi_Q(0) + \lambda\varphi_Q'(0^+) \quad \forall \lambda \in [0,1] \quad (2.9)$$

and get bounds on the relative change of the perturbed optimal value due to contamination:

$$\varphi_Q(1) - \varphi_Q(0) \leq \frac{1}{\lambda}[\varphi_Q(\lambda) - \varphi_Q(0)] \leq \varphi_Q'(0^+) \quad \forall \lambda \in [0,1]. \quad (2.10)$$

These bounds can be written in terms of the probability distributions P, Q:

$$(1-\lambda)\varphi(P) + \lambda\varphi(Q) \leq \varphi(P_\lambda) \leq \varphi(P) + \lambda\varphi_Q'(0^+) \quad \forall \lambda \in [0,1], \quad (2.11)$$

$$\varphi(Q) - \varphi(P) \leq \frac{1}{\lambda}[\varphi(P_\lambda) - \varphi(P)] \leq \varphi_Q'(0^+) \quad \forall \lambda \in [0,1]. \quad (2.12)$$

The bounds (2.11) and (2.12) are based on the assumed properties of the objective function $f(\mathbf{x}, P)$ as a function of the probability distribution P *without any convexity assumptions concerning random coefficients* that enter the initial formulation of the analyzed stochastic program, such as (1.2)-(1.3), (1.4), (1.5)-(1.6) or (1.7)-(1.11).

When (1.1) has a unique optimal solution, say $\mathbf{x}(P)$ for the initial distribution P, the derivative (2.8) and the bounds (2.11) have the form

$$\varphi_Q'(0^+) = f(\mathbf{x}(P), Q) - \varphi(P) \quad (2.13)$$

$$(1-\lambda)\varphi(P) + \lambda\varphi(Q) \leq \varphi(P_\lambda) \leq (1-\lambda)\varphi(P) + \lambda f(\mathbf{x}(P), Q) \quad \forall \lambda \in [0,1] \quad (2.14)$$

so the additional numerical effort consists in solution of the stochastic program based on the alternative distribution Q and in evaluation of the function value of this program at the already known point $\mathbf{x}(P)$. If there are multiple optimal solutions the bounds (2.14) computed at an *arbitrary optimal solution* of the initial problem are valid bounds, but not necessarily the most tight ones.

Similarly, one can approximate the optimal value $\varphi(P_\lambda)$ using the solution $\mathbf{x}(Q)$ and the optimal value $\varphi(Q)$ of $\min_{\mathbf{x} \in \mathcal{X}} f(\mathbf{x}, Q)$ (provided that the set of optimal solutions $\mathcal{X}(Q) = \mathcal{X}_Q(1)$ is nonempty and bounded):

$$(1-\lambda)\varphi(P) + \lambda\varphi(Q) \leq \varphi(P_\lambda) \leq \lambda\varphi(Q) + (1-\lambda)f(\mathbf{x}(Q), P) \quad \forall \lambda \in [0,1] \quad (2.15)$$

so that

$$(1-\lambda)\varphi(P) + \lambda\varphi(Q)$$
$$\leq \varphi(P_\lambda) \leq \min\{(1-\lambda)\varphi(P) + \lambda f(\mathbf{x}(P), Q), \lambda\varphi(Q) + (1-\lambda)f(\mathbf{x}(Q), P)\},$$
$$\forall \lambda \in [0,1]. \quad (2.16)$$

The contamination technique is very flexible and it is a suitable tool for postoptimality analysis in various disparate situations. The choice of a degenerated distribution $Q = \delta(\omega_*) := Q_*$ concentrated at $\omega_* \notin \Omega$ corresponds to an additional scenario and (2.11), (2.12) or (2.14) provide an information about *the influence of including the additional scenario ω_** on the optimal outcome. Similarly, a degenerated distribution $Q_* = \delta(\omega_*)$ with $\omega_* \in \Omega$ models the case of *increasing probability of scenario ω_** and so on. The derivatives of the optimal value of the program perturbed by a degenerated contaminating distributions are related to the *influence curve* and they can be used to construct further characteristics of robustness acknowledged in robust statistics, cf. Hampel (1974).

Contamination by a distribution Q on Ω that gives the same expectation $E_Q \omega = E_P \omega$ is helpful in studying resistance with respect to changes of the sample in situations where the corresponding input information – the known fixed expectation of the random parameters ω – is to be preserved; see Dupačová (1996 b).

3   Application To Financial Decision Models

3.1   The bond portfolio management problem

The objective of the portfolio management model is to maximize the expected utility of the wealth at the end of a given planning period subject to securing the prescribed or uncertain future payments. An *active trading strategy*, which allows for rebalancing the portfolio, is permitted under constraints on conservation of holdings for each asset at each time period and on conservation of cashflows. The main factor which determines the prices, cashflows and other coefficients of the model is the evolution of the short term future interest rates. The possible sequences of interest rates can be determined ad hoc or using a probabilistic model of the term structure, e.g., Black *et al.* (1990). We assume that these scenarios of interest rates have been already selected, indexed by superscripts $s, s = 1, \ldots, S$, and their probabilities fixed as $p_s > 0, s = 1, \ldots, S, \quad \sum_s p_s = 1$.

We follow the notation introduced in Golub *et al.* (1993), see also Dupačová and Bertocchi (1996):

$j = 1, \ldots, J$ are indices of the considered bonds and T_j the dates of their maturities; the considered horizon for evaluation of prices is $T \geq \max_j T_j$;

$t = 0, \ldots, T_0$ is the considered discretization of the planning horizon;

b_j denote the initial holdings (in face value) of bond j;

b_0 is the initial holding in the riskless asset;

r_t^s is the short term interest rate valid in the time interval $(t, t+1]$ under scenario s;

f_{jt}^s is the cashflow generated from bond j at time t under scenario s expressed as a fraction of the face value;

ξ_{jt}^s and ζ_{jt}^s are the selling and purchasing prices of bond j at time t for scenario s obtained from the corresponding fair prices

$$P_{jt}^s = P_{jt}(\mathbf{r}^s) = \sum_{\tau=t+1}^{T} f_{j\tau}^s \prod_{h=t}^{\tau-1} (1 + r_h^s)^{-1} \qquad (3.1)$$

by subtracting or adding fixed transaction costs and spread; the initial prices ξ_{j0} and ζ_{j0} are known, i.e., scenario independent;

L_t is liability due at time t;

x_j/y_j are face values of bond j purchased / sold at the beginning of the planning period, i.e., at $t = 0$, nonnegative *first-stage decision variables*;

z_{j0} is the face value of bond j held in portfolio after the initial decisions x_j, y_j have been made and the auxiliary nonnegative variable y_0^+ denotes the initial surplus.

The second-stage decision variables on rebalancing, borrowing and reinvestment, $x_{jt}^s, y_{jt}^s, z_{jt}^s, y_t^{-s}, y_t^{+s}$ as well as the wealth $W_{T_0}^s$ at the end of the planning horizon depend on scenarios of interest rates.

The model is

$$\text{maximize} \quad \sum_s p_s u(W_{T_0}^s) \qquad (3.2)$$

subject to the first-stage constraints on conservation of holdings

$$y_j + z_{j0} = b_j + x_j \quad \forall j \qquad (3.3)$$

and on cashflow

$$y_0^+ + \sum_j \zeta_{j0} x_j = b_0 + \sum_j \xi_{j0} y_j \qquad (3.4)$$

subject to the second-stage constraints on conservation and holdings for individual interest rate scenarios

$$z_{jt}^s + y_{jt}^s = z_{j,t-1}^s + x_{jt}^s \quad \forall j, s, 1 \le t \le T_0 \qquad (3.5)$$

and on cashflow (including rebalancing the portfolio) at each time period $1 \le t \le T_0$

$$\sum_j \xi_{jt}^s y_{jt}^s + \sum_j f_{jt}^s z_{j,t-1}^s + (1 - \delta_1 + r_{t-1}^s) y_{t-1}^{+s} + y_t^{-s} =$$
$$L_t + \sum_j \zeta_{jt}^s x_{jt}^s + (1 + \delta_2 + r_{t-1}^s) y_{t-1}^{-s} + y_t^{+s} \quad \forall s, t \qquad (3.6)$$

under nonnegativity of all variables, with $y_0^{-s} = 0 \, \forall s, y_0^{+s} = y_0^+ \, \forall s$ and with

$$W_{T_0}^s = \sum_j \xi_{jT_0}^s z_{jT_0}^s + y_{T_0}^{+s} - \alpha y_{T_0}^{-s} \quad \forall s. \tag{3.7}$$

The multiplier α in (3.7) should be fixed. For instance, a pension plan assumes repeated application of the model with rolling horizon and values $\alpha > 1$ take into account the debt service in the future.

Thanks to the assumed possibility of reinvestments and of unlimited borrowing, the problem has always a feasible solution. The existence of optimal solutions is guaranteed for a large class of utility functions that are *increasing and concave* which will be assumed henceforth. From the point of view of stochastic programming, it is a *scenario based multiperiod two-stage model with random relatively complete recourse* and with additional nonlinearities due to the choice of the utility function.

The main output of the model is the optimal value of the objective function (the maximal expected utility of the final wealth) and the optimal values of the first-stage variables x_j, y_j, y_0^+ (and z_{j0}) for all j. They depend on the initial portfolio of bonds, on the model parameters ($\alpha, \delta_1, \delta_2$, transaction costs), on the chosen utility function, on the scheduled stream of liabilities, on the applied model of interest rates and the market data used to fit the model, and on the way how a modest number of scenarios has been selected out of a whole population. If this input is known and an initial trading strategy determined by scenario independent first-stage decision variables x_j, y_j, y_0^+ (and z_{j0}) for all j has been accepted, then the subsequent scenario dependent decisions have to be made in an optimal way regarding the goal of the model. It means that given the values of the first-stage variables y_0^+ and $\mathbf{x}, \mathbf{y}, \mathbf{z}$ with components $x_j, y_j, z_{j0} \forall j$, the maximal contribution of the portfolio management under the sth scenario to the value of the objective function is obtained as the value of the utility function computed for the maximal value of the wealth $W_{T_0}^s$ attainable for the sth scenario under the constraints of the model, i.e., the utility of the optimal value $W_{T_0}^{s*}$ of the linear program

maximize $W_{T_0}^s$ subject to

$$z_{jt}^s + y_{jt}^s = z_{j,t-1}^s + x_{jt}^s \quad \forall j, 1 \leq t \leq T_0$$

$$\sum_j \xi_{jt}^s y_{jt}^s + \sum_j f_{jt}^s z_{j,t-1}^s + (1 - \delta_1 + r_{t-1}^s) y_{t-1}^{+s} + y_t^{-s} =$$
$$L_t + \sum_j \zeta_{jt}^s x_{jt}^s + (1 + \delta_2 + r_{t-1}^s) y_{t-1}^{-s} + y_t^{+s}, \quad t = 1, \ldots, T_0, \tag{3.8}$$

under nonnegativity of all variables, with $y_0^{-s} = 0, y_0^{+s} = y_0^+, z_{j0}^s = z_{j0} \forall j$ and with

$$W_{T_0}^s = \sum_j \xi_{jT_0}^s z_{jT_0}^s + y_{T_0}^{+s} - \alpha y_{T_0}^{-s}. \tag{3.9}$$

Denote the corresponding maximal value $u(W_{T_0}^{s*})$ of the utility function by $U^s(\mathbf{x}, \mathbf{y}, \mathbf{z}, y_0^+)$. Using this notation we can rewrite the program (3.2)–(3.7) as

$$\text{maximize} \quad \sum_{s=1}^{S} p_s U^s(\mathbf{x}, \mathbf{y}, \mathbf{z}, y_0^+) \tag{3.10}$$

subject to nonnegativity constraints and subject to (3.3)–(3.4). Except for maximization at the place of minimization, this is already the form which fits the general framework (1.1). The objective function (3.10) is concave in the first-stage decision variables and linear in the initial probability measure P carried by S fixed scenarios indexed as $s = 1, \ldots, S$. Denote by $\varphi(P)$ the optimal value of (3.10) and by $\mathbf{x}(P), \mathbf{y}(P), \mathbf{z}(P), y_0^+(P)$ the optimal first-stage decision. For simplicity, assume that this optimal first-stage solution is unique.

Inclusion of other out-of-sample scenarios means to consider a convex mixture of two probability distributions: P that is carried by the initial scenarios indexed by $s = 1, \ldots, S$ with probabilities $p_s > 0, \sum_s p_s = 1$ and Q carried by the out-of-sample scenarios indexed by $\sigma = 1, \ldots, S'$ with probabilities $\pi_\sigma > 0, \sum_\sigma \pi_\sigma = 1$. Let λ denote the parameter that gives the contaminated distribution

$$P_\lambda = (1 - \lambda)P + \lambda Q, \quad 0 \le \lambda \le 1 \tag{3.11}$$

carried by the pooled sample of $S + S'$ scenarios that occur with probabilities $(1 - \lambda)p_1, \ldots, (1 - \lambda)p_S, \lambda\pi_1, \ldots, \lambda\pi_{S'}$. For the fixed initial distribution P and a fixed contaminating distribution Q for which the maximal value $\varphi(Q)$ of the objective function $\sum_\sigma \pi_\sigma U^\sigma(\mathbf{x}, \mathbf{y}, \mathbf{z}, y_0^+)$ is finite, the optimal value $\varphi(P_\lambda) := \varphi(\lambda)$ is a finite convex function on $[0,1]$ and its derivative at $\lambda = 0^+$ equals

$$\varphi'(0^+) = \sum_\sigma \pi_\sigma U^\sigma(\mathbf{x}(P), \mathbf{y}(P), \mathbf{z}(P), y_0^+(P)) - \varphi(P) \tag{3.12}$$

cf. (2.13); this should be substituted into the formula (2.11) multiplied by -1 to obtain the bounds for the optimal value $\varphi(P_\lambda)$ for the problem based on the pooled set of $S + S'$ scenarios:

$$(1 - \lambda)\varphi(P) \;+\; \lambda \sum_\sigma \pi_\sigma U^\sigma(\mathbf{x}(P), \mathbf{y}(P), \mathbf{z}(P), y_0^+(P))$$
$$\le \varphi(P_\lambda) \le (1 - \lambda)\varphi(P) + \lambda\varphi(Q) \tag{3.13}$$

for all $0 \le \lambda \le 1$. The lower and upper bound coincide if the optimal first-stage solution $\mathbf{x}(P), \mathbf{y}(P), \mathbf{z}(P), y_0^+(P)$ of the initial program (3.2)–(3.7) is optimal also for the corresponding program based on distribution Q carried by the additional S' scenarios indexed by $\sigma = 1, \ldots, S'$.

If, for instance, P is carried by S equally probable scenarios (sampled form a given population) and Q is carried by other S' equally probable scenarios sampled from the same population, it is natural to fix λ so that the pooled sample consists of $S + S'$ equally probable scenarios, again

$$\lambda = S'(S + S')^{-1}. \tag{3.14}$$

Hence, the bounds for the optimal value based on the pooled sample of size $S + S'$:

$$S(S + S')^{-1}\varphi(P) + S'(S + S')^{-1} \sum_\sigma \pi_\sigma U^\sigma(\mathbf{x}(P), \mathbf{y}(P), \mathbf{z}(P), y_0^+(P))$$
$$\leq \varphi(P_\lambda) \leq S(S + S')^{-1}\varphi(P) + S'(S + S')^{-1}\varphi(Q). \tag{3.15}$$

The additional numerical effort consists in solving the stochastic program

$$\text{maximize} \quad \sum_\sigma \pi_\sigma U^\sigma(\mathbf{x}, \mathbf{y}, \mathbf{z}, y_0^+) \tag{3.16}$$

subject to (3.3)–(3.4) and to nonnegativity constraints for the distribution Q carried by S' out-of-sample scenarios to obtain $\varphi(Q)$ and in evaluation and averaging the S' function values $U^\sigma(\mathbf{x}(P), \mathbf{y}(P), \mathbf{z}(P), y_0^+(P))$ for the new scenarios at the already obtained optimal first-stage solution; these are in fact the main numerical indicators which appear in various simulation studies of the portfolio performance under out-of-sample scenarios, cf. McKendall *et al.* (1994). For relatively large values of λ (or S'), it pays to use the more complicated lower bound according to (2.16); see Figure 1. The important special case of small λ is $S' = 1$, i.e., the inclusion of one additional scenario.

The bounds can be also used to derive simple rules on the influence of additional scenarios on the optimal value. For instance:

- If the derivative

$$\varphi'(0^+) = \sum_\sigma \pi_\sigma U^\sigma(\mathbf{x}(P), \mathbf{y}(P), \mathbf{z}(P), y_0^+(P)) - \varphi(P) > 0, \tag{3.17}$$

  then the optimal value $\varphi(P_\lambda)$ increases for all $0 < \lambda < 1$.

- If $\varphi(Q) < \varphi(P)$, the optimal value $\varphi(P_\lambda)$ decreases at $\lambda = 0^+$.

The postoptimality technique described here is independent of the method which was used to generate or to select the scenarios – the atoms of the distributions P and Q. It can be used without any problems for scenario dependent liabilities and cashflows what allows, for instance, to include the case of callable and puttable bonds and mortgage backed securities (cf. Golub *et al.* (1993) for the corresponding model formulation).

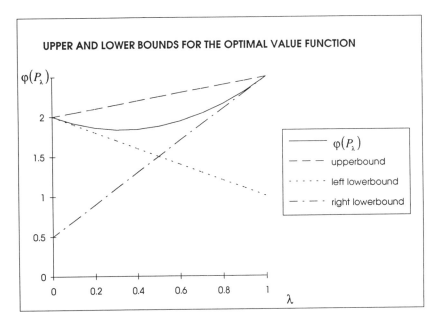

Figure 1

3.2 The tracking model

Assume the individual scenario problems (1.7) have been solved for all considered scenarios, i.e., for all sets of coefficients values $\omega_s = [\mathbf{q}_s, \mathbf{W}_s, \mathbf{T}_s, \mathbf{h}_s]$, $s = 1, \ldots, S$ so that the values $v_s \forall s$ needed for the tracking objective function (1.9) are known. The tracking model (1.9) is

$$\text{minimize} \quad \sum_{s=1}^{S} p_s g(\mathbf{x}, \mathbf{y}_s; \omega_s) \tag{3.18}$$

subject to

$$\mathbf{x} \in \mathcal{X} := \{\mathbf{x} | \mathbf{A}\mathbf{x} = \mathbf{b}, \mathbf{x} \geq \mathbf{0}\} \tag{3.19}$$

and $\mathbf{y}_s \geq 0 \quad \forall s$, with

$$g(\mathbf{x}, \mathbf{y}_s; \omega_s) = \|\mathbf{c}^\top \mathbf{x} + \mathbf{q}_s^\top \mathbf{y}_s - v_s\| + \|\mathbf{T}_s \mathbf{x} + \mathbf{W}_s \mathbf{y}_s - \mathbf{h}_s\|. \tag{3.20}$$

Let the alternative distribution Q be carried by scenarios $\omega_{S+1}, \ldots, \omega_{S+S'}$ with probabilities $\pi_{S+s} > 0, s = 1, \ldots, S', \sum_s \pi_{S+s} = 1$. Inclusion of additional scenarios brings along new variables $\mathbf{y}_{S+s}$. As we need to work with a fixed set of feasible solutions, one more reformulation is needed before application of the contamination technique discussed in Section 2.

Consider the pooled sample of scenarios $\omega_1, \ldots, \omega_S, \omega_{S+1}, \ldots, \omega_{S+S'}$; the initial distribution P assigns them probabilities $p_s, s = 1, \ldots, S$ and 0 for the remaining ones, the distribution Q assigns zero probabilities to the initial scenarios and probabilities $\pi_{S+s}, s = 1, \ldots, S'$, to the new ones. The perturbed problem carried by the pooled sample is

$$\text{minimize} \quad (1 - \lambda) \sum_{s=1}^{S} p_s g(\mathbf{x}, \mathbf{y}_s; \omega_s) + \lambda \sum_{s=1}^{S'} \pi_{S+s} g(\mathbf{x}, \mathbf{y}_{S+s}; \omega_{S+s}) \quad (3.21)$$

subject to nonnegativity of all variables $\mathbf{y}_s, s = 1, \ldots, S + S'$ and subject to (3.19). The derivative of its optimal value at $\lambda = 0^+$ can be computed according to (2.8) as

$$\varphi'_Q(0^+) = \min_{\mathbf{x}^*, \mathbf{y}_s^* \forall s} \sum_{s=1}^{S'} \pi_{S+s} g(\mathbf{x}^*, \mathbf{y}_{S+s}^*; \omega_{S+s}) - \varphi(P). \quad (3.22)$$

The minimization in (3.22) is carried over all optimal solutions of the augmented initial program (3.18)

$$\text{minimize} \quad \sum_{s=1}^{S} p_s g(\mathbf{x}, \mathbf{y}_s; \omega_s) + \sum_{s=1}^{S'} 0 * g(\mathbf{x}, \mathbf{y}_{S+s}; \omega_{S+s}) \quad (3.23)$$

subject to nonnegativity constraints on all variables and subject to (3.19). Due to the special form of (3.22), the minimization concerns the optimal $\mathbf{x}$-part of the solution and arbitrary nonnegative variables $\mathbf{y}_{S+s}, s = 1, \ldots, S'$.

Assume that the optimal $\mathbf{x}$-part of the solution of the initial problem (3.18)–(3.19) is unique, say, $\mathbf{x}(P)$. Then the derivative is obtained by solving S' minimization problems with objective functions

$$g(\mathbf{x}, \mathbf{y}_{S+s}; \omega_{S+s}) = \|\mathbf{c}^\top \mathbf{x}(P) + \mathbf{q}_{S+s}^\top \mathbf{y}_{S+s} - v_{S+s}\|$$
$$+ \|\mathbf{T}_{S+s}\mathbf{x}(P) + \mathbf{W}_{S+s}\mathbf{y}_{S+s} - \mathbf{h}_{S+s}\| \quad (3.24)$$

subject to $\mathbf{y}_{S+s} \geq 0$ and by taking an average of the obtained optimal values with weights π_{S+s}.

In the context of the postoptimality analysis for the tracking models it would be important to get some results concerning the optimal first-stage solutions $\mathbf{x}$. Indeed there exist theoretical results connected with contamination technique (see Dupačová (1986)) and also pathfollowing methods which can be implemented for parametric programs depending on a scalar parameter, as is λ in our case. However, up to now we are not ready to report any numerical experience in this direction.

4   Numerical Results

This section provides numerical results based on the contamination technique described by formulas (3.13) and (3.15) for the bond portfolio management

Bonds	Qt	coupon	redemp.	payment	dates	exercise	maturity
BTP36658	10	3.9375	100.1875	01/04	01/10		01/10/96
BTP36631	20	5.03125	99.5313	01/03	01/09		01/03/98
BTP12687	15	5.25	99.2312	01/01	01/07		01/01/2002
BTP36693	10	3.71875	99.3875	01/08	01/02		01/10/2004
BTP36665	5	3.9375	99.2188	01/05	01/11		01/11/2023
CTO13212	20	5.25	100.0000	20/01	20/07	20/01/95	20/01/98
CTO36608	20	5.25	99.9500	19/05	19/11	19/05/95	19/05/98

Table 1.

problem. We analyze the change in the optimal final wealth due to the following typical cases:

- parallel shifts in interest rates scenarios;

- doubling number of scenarios in the sampling strategy;

- doubling number of scenarios when scenarios are randomly generated.

To simulate the behavior of an investment portfolio of fixed income securities on the Italian bond market we use the model described by (3.2)–(3.7) with the linear utility function and within the time horizon of one year ($T_0 = 12$).

The initial portfolio and the term structure are related to September 1, 1994. We consider the same portfolio that was used in Bertocchi *et al.* (1996) for the sensitivity analysis of portfolio with respect to sampling strategies. It includes typical governmental bonds, paying semi-annual coupons and covering two year forward till 29 years maturities (the so-called BTPs) as well as puttable bonds (CTOs), paying semi-annual coupons with the maturity of 8 years and a possible exercise of the option in the 4th year or with the maturity of 6 years and an exercise at the 3rd year; see Table 1: note that the coupon yields and the redemption prices are after tax.

To estimate the term structure of interest rates we used the regression model of Bradley and Crane (1972) applied to the yields obtained by the market quotation of the BTPs on the relevant day; see Figure 2 for the term structure of interest rates. We refer to Dupačová *et al.* (1996) for a detailed discussion.

In this application, liabilities are not considered, liquidity can be obtained from the interbank market at a rate greater than that one at which surplus can be always reinvested; hence in (3.7) $\alpha = 1$. The additive transaction costs are fixed at ± 0.01, $\delta_2 = 0.025$ and δ_1 is 0 or .001.

The scenarios are based on data from Italian bond market generated according to the Black–Derman–Toy model and selected according to the simplified version of the nonrandom sampling strategy by Zenios and Shtilman

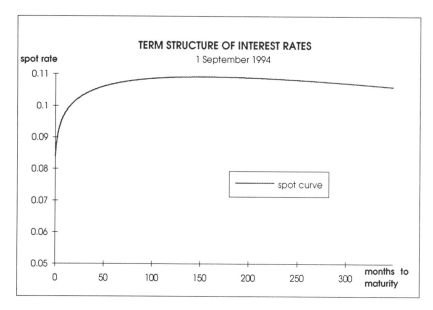

Figure 2

case	$l = 1, ..., L$	$L+1$	$l = L+2, ..., T_0$	$T_0 + 1$
B1	$s = 1, ..., 2^3$	$\omega^s_{L+1} = 0$	$\omega^s_l = 0$	$\omega^s_{T_0+1} = 1$
B2	$s = 1, ..., 2^3$	$\omega^s_{L+1} = 0$	$\omega^s_l = 0$	$\omega^s_{T_0+1} = 0$
B3	$s = 1, ..., 2^3$	$\omega^s_{L+1} = 0$	$\omega^s_l = 1$	$\omega^s_{T_0+1} = 1$
B4	$s = 1, ..., 2^3$	$\omega^s_{L+1} = 1$	$\omega^s_l = 1$	$\omega^s_{T_0+1} = 1$
C4	$s = 1, ..., 2^4$	$\omega^s_{L+1} = 1$	$\omega^s_l = 1$	$\omega^s_{T_0+1} = 1$

Table 2

(1993) as described in Dupačová and Bertocchi (1996) and in Bertocchi *et al.* (1996).

The binomial model produces scenarios that can be coded by 2^T ($T = 350$ for our data) binary fractions uniformly distributed in $[0, 1]$ and the sampling strategy chooses the number of periods L for which all possibilities (choices of zeros and ones on the first L positions) are fully covered. The remaining digits necessary to complete the full length paths were selected according to Table 2 and for $l > T_0 + 1$, the components ω^s_l alternate up and down (1 or 0) starting with the indicated value of $\omega^s_{T_0+1}$.

distribution P	distribution Q	Figure
B2st	B2	3
B2	B2st	3
B3	B4	4
B4	B3	4
D1	D2	5
D2	D1	5

Table 3

the rates based on the case B2 perturbed by the additive shift of -0.000355 (which corresponds to the shift of 5% of the current B2 rates).

The numerical results reported in the Figures are organized according to the scheme in Table 3. They consist of the optimal values and optimal initial strategies for the two alternative cases based on distributions P and Q, and they contain the average values $\sum_\sigma \pi_\sigma U^\sigma(\mathbf{x}(P), \mathbf{y}(P), \mathbf{z}(P), y_0^+(P))$ or $\sum_s p_s U^s(\mathbf{x}(Q), \mathbf{y}(Q), \mathbf{z}(Q), y_0^+(Q))$ under headings 'means of contam. solutions', the values of the directional derivatives and of the lower and upper bounds computed according to (3.13) for distinct values of λ and a graphical representations of these bounds.

The results in Figure 3 show that the choice of $\delta_1 \neq 0$ for which the return on cash investment is less than that in bonds influences the optimal initial strategy for scenario bed B2, making more valuable the investment in CTO36608 than in cash. For scenarios in B2st the optimal initial strategy does not change. The bounds for $\delta_1 = 0$ are similar to those for $\delta_1 \neq 0$.

Choice of the couple of scenario beds B2 and B2st allows comparison of the situations when a certain bed of scenarios, for example B2st, is changed so that all rates are increased of a fixed quantity, like in B2. The left lower bound and the upper bound in Figure 3 for contamination of B2st by B2 are very precise and show that the optimal final wealth is untouched by a parallel shift in interest rates. Similar tests of resistance of the optimal value with respect to a shift of interest rates or prices have appeared in several applications, see e.g. McKendall *et al.* (1994), but without giving the global bounds (2.15) or (2.16) for the optimal value of the perturbed problem.

The results on contamination between B3 and B4, see Figure 4, illustrate another possible application of the bounds, namely, for supporting decisions concerning the required number of scenarios (i.e., concerning the value of L in Table 2). For $\lambda = 0.5$, the example in Figure 4 gives the interval [11350.93, 11351.54] for the optimal value of C4 based on 2^4 scenarios – a union of scenarios from the beds B3 and B4. (Indeed, the true optimal value for C4 is 11350.97.) It means that using the double number of scenarios does not increase essentially the precision of the obtained approximate of the

1 September 1994

Bradley & Crane (NON-LINEAR)

utility function:		**linear**	
bid/ask spread:		.01	
$\delta 1$:		0.001	

from		*to*		means			upper	lower	
P	$\varphi(P)$	Q	$\varphi(Q)$	of contam. solutions		λ	bound $\varphi(P_\lambda)$	bound $\varphi(P_\lambda)$	$\varphi'(0^+)$
B2st	11795.52					0	11795.52	11795.52	-416.409
						0.1	11754.11	11753.88	
						0.3	11671.28	11670.6	
						0.5	11588.46	11587.32	
						0.7	11505.64	11504.03	
						0.9	11422.81	11420.75	
		B2	11381.4	11379.11166		1	11381.4	11379.11	
B2	11381.4					0	11381.4	11381.4	132.0186
						0.1	11422.81	11394.6	
						0.3	11505.64	11421.01	
						0.5	11588.46	11447.41	
						0.7	11671.28	11473.81	
						0.9	11754.11	11500.22	
		B2st	11795.52	11513.41498		1	11795.52	11513.42	

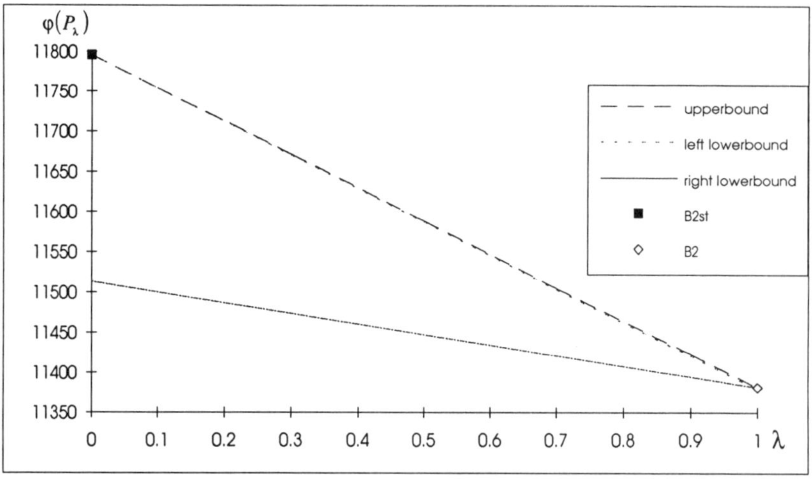

INITIAL STRATEGY: $\delta_1 = 0$

	B2st	B2
BTP36658	0	10
BTP36631	0	0
BTP12687	0	15
BTP36693	0	10
BTP36665	134.41	5
CTO13212	0	20
CTO36608	0	20
yPLUS	0	2565.53
yMINUS	0	0

INITIAL STRATEGY: $\delta_1 = 0.001$

	B2st	B2
BTP36658	0	10
BTP36631	0	0
BTP12687	0	15
BTP36693	0	10
BTP36665	134.41	5
CTO13212	0	20
CTO36608	0	44.85
yPLUS	0	0
yMINUS	0	0

Figure 3

1 September 1994

Bradley & Crane (NON-LINEAR)					utility function:		linear	
					bid/ask spread:		.01	
					$\delta 1$:		0.001	
from		*to*		*means*		*upper*	*lower*	
P	$\varphi(P)$	*Q*	$\varphi(Q)$	*of contam.*	λ	*bound*	*bound*	$\varphi'(0^{\cdot})$
				solutions		$\varphi(P_\lambda)$	$\varphi(P_\lambda)$	
B3	11353.94				0	11353.94	11353.94	-8.84001
					0.1	11353.46	11353.06	
					0.3	11352.5	11351.29	
					0.5	11351.54	11349.52	
					0.7	11350.58	11347.75	
					0.9	11349.62	11345.98	
		B4	11349.14	11345.10029	1	11349.14	11345.1	
B4	11349.14				0	11349.14	11349.14	3.576724
					0.1	11349.62	11349.5	
					0.3	11350.58	11350.21	
					0.5	11351.54	11350.93	
					0.7	11352.5	11351.64	
					0.9	11353.46	11352.36	
		B3	11353.94	11352.71972	1	11353.94	11352.72	

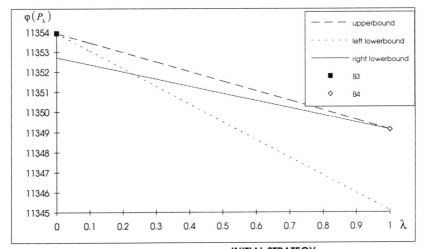

	INITIAL STRATEGY:			
	C4		B3	B4
C4 optimal value:	83.93	BTP36658	10	104.41
11350.97	0	BTP36631	0	0
	0	BTP12687	15	0
	0	BTP36693	0	0
	0	BTP36665	0	0
	20	CTO13212	20	0
	0	CTO36608	56.7	0
	0	yPLUS	0	0
	0	yMINUS	0	0

Figure 4

1 September 1994

Bradley & Crane (NON-LINEAR)

utility function:	linear
bid/ask spread:	.01
$\delta 1$:	0.001

from		*to*		means			upper	lower	
P	$\varphi(P)$	Q	$\varphi(Q)$	of contam. solutions		λ	bound $\varphi(P_\lambda)$	bound $\varphi(P_\lambda)$	$\varphi'(0^+)$
D1	11335.84					0	11335.84	11335.84	-8.04881
						0.1	11335.62	11335.04	
						0.3	11335.18	11333.43	
						0.5	11334.74	11331.82	
						0.7	11334.3	11330.21	
						0.9	11333.86	11328.6	
		D2	11333.64	11327.78939		1	11333.64	11327.79	
D2	11333.64					0	11333.64	11333.64	0.726076
						0.1	11333.86	11333.71	
						0.3	11334.3	11333.86	
						0.5	11334.74	11334	
						0.7	11335.18	11334.15	
						0.9	11335.62	11334.29	
		D1	11335.84	11334.36218		1	11335.84	11334.37	

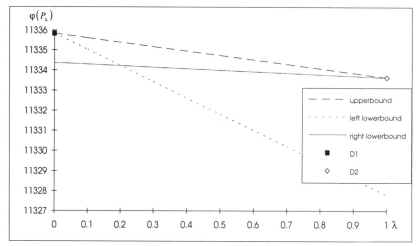

INITIAL STRATEGY:

	D1	D2
BTP36658	10	104.09
BTP36631	0	0
BTP12687	15	0
BTP36693	10	0
BTP36665	5	0
CTO13212	20	0
CTO36608	44.87	0
yPLUS	0	0
yMINUS	0	0

Figure 5

union of scenarios from the beds B3 and B4. (Indeed, the true optimal value for C4 is 11350.97.) It means that using the double number of scenarios does not increase essentially the precision of the obtained approximate of the optimal value for the full hypothetical problem which would be based on all 2^{350} possible scenarios of interest rates generated according to the Black–Derman–Toy model. However, the optimal initial investment strategies are quite different.

Finally, Figure 5 illustrates an application of the bounds to the case of the pooled sample based on randomly chosen scenarios, such as in experiments D1 and D2. Again, the bounds for the optimal value based on the pooled sample carried by 16 scenarios, as obtained with $\lambda = 0.5$ from the results of the two small problems for beds D1 and D2 of 8 randomly chosen scenarios, are very tight: [11334, 11334.74]. Exploitation of the more complicated bound of the type of (2.16) helped to increase the lower bound 11331.82 obtained according to (3.13) or (3.15); a similar observation holds true also for Figure 4.

Acknowledgements

This research was partly supported by the Grant Agency of the Czech Republic under grant No. 402/96/0420 and by CNR grants No. 94.00538.ct11 and 96.01313.ct10. Joint research was supported by the University of Bergamo and by the INCO-DC project No. 951139 of the European Community.

References

Bertocchi, M., Dupačová, J. and Moriggia, V. (1996) Sensitivity analysis of a bond portfolio model for the Italian market. Technical report **18**, University of Bergamo.

Black, F., Derman, E. and Toy, W. (1990) A one-factor model of interest rates and its application to treasury bond options. *Financial Analysts J.*, Jan./Feb., 33–39.

Bradley, S.P. and Crane, B.D. (1972) A dynamic model for bond portfolio management. *Manag. Sci.* **19**, 139–151.

Cariño, D.R. *et al.* (1994) The Russell–Yasuda Kasai model: An asset/liability model for a Japanese insurance company using multistage stochastic programming. *Interfaces* **24**, 29–49.

Dembo, R.S. (1991) Scenario optimization. *Annals of Oper. Res.* **30**, 63–80.

Dempster M.A.H. and Ireland, A.M. (1988) A financial expert decision system. In *Mathematical Models for Decision Support*, NATO ASI Ser. Vol. F48 (G. Mitra, ed.), Springer, 415–440.

Dert, C.L. (1995) *Asset Liability Management for Pension Funds. A Multistage Chance-Constrained Programming Approach.* PhD Dissertation, Erasmus University

Dupačová, J. (1986) Stability in stochastic programming with recourse. Contaminated distributions. *Math. Progr. Study* **27**, 133–144.

Dupačová, J. (1990) Stability and sensitivity analysis for stochastic programming. *Annals of Oper. Res.* **27**, 115–142.

Dupačová, J. (1991) Stochastic programming models in banking. Tutorial paper, IIASA Laxenburg. See also *Ekonomicko–Matematický Obzor* **27**, 201–234.

Dupačová, J. (1995) Postoptimality for multistage stochastic linear programs. *Annals of Oper. Res.* **56**, 65–78.

Dupačová, J. (1996a) Uncertainty about input data in portfolio management. In *Modelling Techniques for Financial Markets and Bank Management, Proc. of the 16–17th EWGFM Meeting*, Bergamo 1995 (M. Bertocchi, E. Cavalli and S. Komlosi, eds.), Physica Verlag, 17–33.

Dupačová, J. (1996b) Scenario based stochastic programs: Resistance with respect to sample. *Annals of Oper. Res.* **64**, 21–38.

Dupačová, J. (1998) Reflections on robust optimization. To appear in *Proc. of the 3rd GAMM/IFIP Workshop 'Stochastic Optimization'*, Neubiberg, June 17–20, Lecture Notes in Economics and Math. Systems, **458**.Springer.

Dupačová, J. and Bertocchi, M. (1996) Management of bond portfolios via stochastic programming – postoptimality and sensitivity analysis. In *System Modelling and Optimization, Proc. of the 17th IFIP TC7 Conference*, Prague 1995 (J. Doležal and J. Fidler, eds.), Chapman & Hall, 574–581.

Dupačová, J., Bertocchi, M. and Abaffy, J.(1996) Input analysis for a bond portfolio management problem. Technical Report, **24**, University of Bergamo.

Ermoliev, Yu. and Wets, R.J.-B., eds. (1988) *Numerical Techniques for Stochastic Optimization Problems.* Springer.

Gol'shtein, E.G. (1972) *Theory of Convex Programming.* Translations of Mathematical Monographs **36**, American Mathematical Society.

Golub, B. *et al.* (1993) A stochastic programming model for money management. Research report, **93-01-02**, Hermes Laboratory, The Wharton School, University of Pennsylvania.

Hampel, F.R. (1974) The influence curve and its role in robust estimation. *J. Am. Statist. Assoc.* **69**, 383–397.

Kall, P. and Wallace, S.W. (1994) *Stochastic Programming.* Wiley.

Konno, H., Luenberger, D.G. and Mulvey, J.M., eds. (1993) Financial Engineering. *Annals of Oper. Res.* **45**.

Kusy, M.I. and Ziemba, W.T. (1986) A bank asset and liability management model. *Oper. Res.* **34** 356–376.

McKendall, R., Zenios, S.A. and Holmer, M. (1994) Stochastic programming models for portfolio optimization with mortgage backed securities. Comprehensive research guide. In *Operations Research Models in Quantitative Finance* (R. L. D'Ecclesia and S. A. Zenios, eds.), Physica Verlag, 134–171.

Mulvey, J.M. (1994) Financial planning via multi-stage stochastic programs. In *Mathematical Programming – State of the Art 1994* (J.R. Birge and K.G. Murty, eds.), Univ. of Michigan, 151–171.

Mulvey, J.M., Vanderbei, R.J. and Zenios, S.A. (1995) Robust optimization of large scale systems. *Oper. Res.* **43**, 264–281.

Shapiro, J.F. (1988) Stochastic programming models for dedicated portfolio selection. In *Mathematical Models for Decision Support*, NATO ASI Series, Vol. F48 (B. Mitra, ed.), Springer, 587–611.

Prékopa, A. (1995) *Stochastic Programming.* Kluwer.

Worzel, K.J., Vassiadou–Zeniou, C. and Zenios, S.A. (1994) Integrated simulation and optimization models for tracking indices of fixed-income securities. *Oper. Res.* **42**, 223–233.

Zenios, S.A. (1991) Massively parallel computations for financial planning under uncertainty. In *Very Large Scale Computing in the 21-st Century* (J. Mesirov, ed.), SIAM, 273–294.

Zenios, S.A., ed. (1993) *Financial Optimization*, Cambridge University Press.

Zenios, S.A. and Shtilman, M.S. (1993) Constructing optimal samples from a binomial lattice. *Journal of Information & Optimization Sciences* **14**, 125–147.

Zenios, S.A. and Ziemba, W.T., eds. (1992) Focused Issue on Financial Modeling. *Manag. Sci.* **38**, No. 11.

Ziemba, W T. and Vickson, R.G., eds. (1975) *Stochastic Optimization Models in Finance*, Academic Press.

The Towers Perrin Global Capital Market Scenario Generation System

John M. Mulvey and A. Eric Thorlacius*

Abstract

Financial management requires a systematic approach for generating scenarios of future capital markets. Today's global environment demands that the scenarios link the economies of individual countries within a common framework. We describe a global scenario system, developed by Towers Perrin, based on a cascading set of stochastic differential equations. The system applies to financial systems for pension plans and insurance companies throughout the world. A case study illustrates the process.

1 Introduction

Towers Perrin, one of the world's largest actuarial consulting companies, employs a capital market scenario generation system, called CAP:Link, for helping its clients in understanding the risks and opportunities relating to capital market investments. The system produces a representative set of individual simulations – typically 500–1000. Each scenario contains key economic variables such as price and wage inflation, interest rates at different maturities (real and nominal), stock dividend yields and growth rates, and exchange rates through each year for a period of up to 20 years. We model returns on asset classes and liability projections consistent with the underlying economic factors, especially interest rates. The economic variables are simultaneously determined for multiple economies within a common global framework. Long-term asset and liability management is the primary application.

A variety of stochastic optimization models exist for multi-period financial analysis. Prominent examples include: Berger & Mulvey (1998), Boender, van Aalst & Heemskerk (1998), Cariño et al. (1994), Cariño & Turner (1998), Dempster (1998) and Wilkie (1995). These systems are designed around a set of simplifying assumptions that depend upon the target application. Stochastic financial simulations fall within three broad categories:

---

*This paper was written while the second author was at Towers Perrin

Prediction These systems forecast short-term market movements. It is not necessary to achieve perfect accuracy, only enough to realize an expected profit. Hedge funds and other leveraged traders often employ computerized prediction systems. The desired output is a single scenario for next period's economic variables. Prediction becomes more challenging as the length of the planning period increases since markets adapt to changing conditions. Most trading systems consider short-term horizons – hours or days. Hence, validity can be ascertained by historical back-testing the recommended investment strategies.

Pricing Derivative securities often possess complex price relationships with capital market variables, such as the pattern of interest rates movements. By capturing this information via models where the end-of-period outcome of the various alternatives can be compared through a set of scenarios, we can check for consistency in pricing. It often applies in the creation of custom deals wherein the parties exchange specific market risks. The arbitrage-free condition is important since the existence of arbitrage opportunities suggests inconsistency with traded securities.

Risk Analysis This activity evaluates the potential rewards and risks of various investment and liability management strategies. By managing investment decisions and considering liability issues as part of an integrated financial picture, significant risks can be avoided and opportunities for enhanced return created by accepting risks which may be more severe to other investors – due to a different liability structure. Viewing investment choices from the perspective of their ability to meet specified liability objectives changes the relative riskiness of investment alternatives. As an example, long term bonds which may be relatively risky to some short term investors are attractive to a pension plan with a specific long-term and fixed horizon. In contrast, when the liability structure depends upon inflation, index linked bonds becomes a conservative asset category. Alternatives like cash may be risky in this context because the inflation adjusted value at a long term horizon is uncertain.

Several obstacles stand in the way of successful applications of asset and liability management. First, industries, such as insurance and banks, have been controlled by regulations and legal restrictions, causing difficulty in modeling the dynamic aspects of the key economic variables with reference to the market values of assets and liabilities. As regulations and rules change across the globe, however, economic issues gain in importance, and as a consequence, the ability to model the stochastic elements improves. A second receding obstacle is the computational resources needed to solve the multi- period financial planning model. Today, powerful workstations, PCs, and efficient algorithms are available for solving financial optimization problems; for example, see Mulvey, Armstrong & Rothberg (1995).

CAP:Link, developed primarily for asset liability management (ALM), entails an ongoing process of information gathering, evaluation and action in order to maximize the organization's wealth over time. Asset and liability mixtures complicate the situation. It is common to either over emphasize the liability matching investment alternatives (via immunization or similar approaches) or to focus on expected return and volatility of asset return. The investor must balance expected asset return and confidence in meeting liability obligations. A case study (section 5) offers more ideas on asset liability management and CAP:Link's role in financial planning. In brief, CAP:Link portrays the relationships between modeled variables, their interactions through time, and the ranges and distributions of outcomes in a consistent manner. Investment strategies and assumption setting can be evaluated by means of representative scenarios.

The rest of the paper is organized as follows. The next section describes the overall structure of global CAP:Link. We emphasize the relationship of the single country modules within a global setting. Much experience has been gained by implementing CAP:Link in over 17 countries throughout North America, Europe and Asia. The global design links the single country modules in a consistent fashion. In addition, we model currencies between all pairs of countries, as described in section 2.4. Section 3 takes up some implementation issues, including the critical assumption setting and calibration elements. We present a case study in section 5 in which the scenarios form the basis for an analysis of a large pension plan. We show the process for evaluating the pension plan's health from several perspectives. Last, in section 6, we mention ideas for future work.

2   Model Structure

The global CAP:Link system forms a linked network of single country modules. Figure 1 illustrates the overall structure for four countries. The three major economic powers – the United States, Germany, and Japan – occupy a central role, with the remaining countries designated as home or other countries. We assume that the other countries are affected by, but do not impact, the economies of the three major countries. The basic stochastic differential equations are identical in each country, although the parameters reflect unique characteristics of each particular economy. Notice the direction of Figure 1 arcs. Additional countries can be readily included in the framework by increasing the number of other countries.

Within each country, the basic economic structure is illustrated in Figure 2. Variables at the top of the structure influence those below, but not vice-versa.

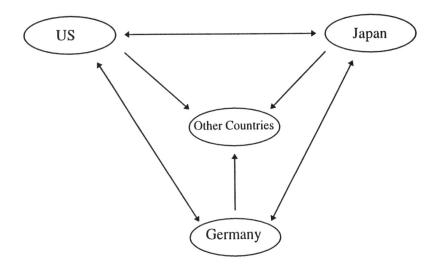

Figure 1: Triad of cornerstone countries and other countries. Direction of arcs shows the flow of information in the model.

This approach eases the task of calibrating parameters. The ordering does not reflect causality between economic variables, but rather captures significant co-movements. Linkages across countries occur at various levels of the model – for example, interest rates and stock returns. These connections will be discussed later. Roughly, the economic conditions in a single country are more or less affected by those of its neighboring countries and by its trading partners. The degree of interaction depends upon the country under review.

The structure is based on a cascade format – each sub-module within the system is possibly affected by modules above and equal to that module. Briefly, the first level consists of short and long interest rates, and price inflation. Interest rates are a key attribute in modeling asset returns and especially in coordinating the linkages between asset returns and liability investments. To calculate a pension plan's surplus, we must be able to discount the projected liability cash flows at a discount rate which is consistent with bond returns. Also, since dynamic relationships are essential in risk analysis, the interest rate model forms a critical element.

The second level entails real yields, currency exchange rates and wage inflation. At the third level, we focus on the components of equity returns: dividend yields and dividend growth. Returns for the remaining asset classes form the next level, with fixed income assets reflecting the term structure of interest rates and other mechanisms as discussed below. Each economic

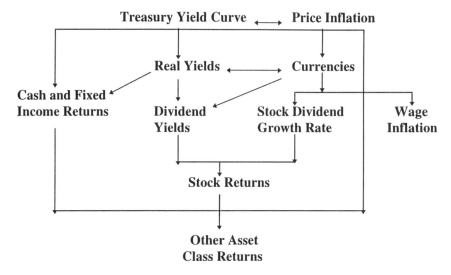

Figure 2: The cascade CAP:Link structure within a single country.
Each country in Global CAP:Link depicts a common heritage.

variable is projected by means of a stochastic differential equation – relating
the variable through time and with the stochastic elements of the equation
and, of course, to other variables and factors at the same or higher levels in
the cascade. See Mulvey (1996) for a discussion of the single country design.

2.1   Interest Rate Generation

The path of spot interest rates for non-callable government obligations sits
at the top of the cascade. Yield curve values are determined in a sequential
fashion. First, the short and long spot rates are computed by a variant of
the two–factor Brennan–Schwartz approach (1982). At its simplest within a
country, we assume that long and short interest rates link together through a
correlated white noise term and by means of a stabilizing term which keeps the
spread between the short and long rates under control. In addition, we link
the white noise terms across selected countries. The requisite equations are
listed below. All elements are indexed as vectors across the target countries.

Short rate: $dr_t = f_1(r_u - r_t)dt + f_2(r_u, r_t, l_u, l_t, p_u, p_t)dt + f_3(r_t)dZ_1$ (2.1)

Long rate: $dl_t = f_4(l_u - l_t)dt + f_5(r_u, r_t, l_u, l_t, p_u, p_t)dt + f_6(r_t)dZ_2$ (2.2)

where

r_u is the normative level of short interest rates,

r_t is the level of interest rates at time t,

l_u is the normative level of long interest rates,

l_t is the level of rates at time t,

p_u is the normative level of inflation and p_t is the level at time t, and

$f_1, \ldots, f_6$ are vector functions that depend upon various economic factors up to period t.

The random coefficient vectors – dZ_1 and dZ_2 – depict correlated Wiener terms.

These two vector diffusion equations provide the building blocks for the remaining spot interest rates, and the full yield curve. At any point t, the mid- rate is a function of the short and long rates. Other points on the spot rate curve are computed by smoothing a double exponential equation for the spot rate of interest.

2.2 Inflation Generation

The price inflation routine lies aside the interest rate model in the cascade structure. Price inflation at period t depends upon price inflation in previous time periods and on the current yield curve. Again, the economic time series is modeled as a diffusion process. Controls are placed on internal parameters and interest rates – the yield curve. The key diffusion equations are shown below. As before, these equations index over the individual countries (as vectors):

Price inflation: $\quad dp_t = f_7 dr_t + f_8(p_u, p_t, r_u, r_t, l_u, l_t)dt + f_9(v_t)dZ_3 \quad$ (2.3)

Stochastic volatility: $\quad dv_t = f_{10}(v_u - v_t)dt + f_{11}(v_t)dZ_4 \quad$ (2.4)

where

v_u is the normative level of inflation volatility and v_t is the level at time t, and

$f_7, \ldots, f_{11}$ are vector functions that depend upon various economic factors up to period t.

The random coefficient vectors – dZ_3 and dZ_4 – depict white noise terms consisting of parts reflecting local and global effects. In the US and other economies, price inflation is more volatile than interest rates. Part of the explanation is the lack of a traded security representing inflation.[1] Also, the *volatility* of inflation persists. Once inflation volatility increases, for example, through a economic shock such as the oil crisis, it remains high for some time before settling down to a more normal level. Modeling stochastic volatility requires the second diffusion equation (2.4).

Next in the cascade is wage inflation. This parameter connects to price inflation in a lagged and smoothed fashion. Wages are relatively slow to react to changes in price inflation, but they inevitably follow over time. Pension plan liabilities are directly tied to wage inflation.

2.3   Real Yields

Real interest rates are defined in two distinct ways. First, we equate real interest rates with the spread of nominal interest rates to current inflation. Some economies provide explicit real return bonds as traded securities issued by government bodies, for instance, index linked bonds in the US, UK and Canada. These securities provide a specified return over inflation for a fixed period. Real returns in a net of inflation framework are equivalent to nominal yields in a total return context. Regulatory bodies in some countries set valuation by discounting with real yield rates. Real yields relate to the movement of interest rates, current inflation, as well as expectations for future inflation. The equation for long term real yields is

$$dk_t = f_{12}(k_u, k_t, l_u, l_t, p_u, p_t)dl + f_{13}(p_u, p_t, k_u, k_t, l_u, l_t)dt + f_{14}(k_t)dZ_5 \quad (2.5)$$

where

k_u is the normative level of real yields and k_t is the level at time t, and

f_{12} and f_{14} are vector functions that depend upon various economic factors up to time t.

The random coefficient vector dZ_5 depicts correlated Wiener terms. To derive the complete real yield curve, we calculate a short-term real yield based on short nominal rates and current inflation – providing two points on the curve. Remaining points are determined by interpolating the structure of the nominal interest rate curve.

---

[1]The US government began selling index-linked bonds in January 1997. Thus, estimates of inflationary expectations are now available as traded securities.

2.4   Currencies

Several complication issues arise when modeling currency exchange rates. First, currencies must enforce the arbitrage free condition among spot exchange rates and among forward rates with differential interest rates. The second issue involves symmetry and numeraire independence; we must create a structure in which the distribution of currency returns from country A to B has the same distribution as returns from B to A. Both issues limit the form of the currency exchange models, especially when integrating three or more currencies. To avoid these problems, we focus on the strength of each country's currency. Exchange rate follows as the ratio between the strengths of any two counties. See Mahieu & Schotman (1994) for a similar approach to currencies. The absolute strength of any currency is a notional concept; the relative levels reflect the difference in the exchange rates.

The exchange rate equation between country i and j is the ratio: s_i/s_j, where s_i and s_j reflects the strength of currency i and j, respectively.

The equation that describes the development of the individual currency strengths is

$$ds_t = f_{15}(r_u, r_t, p_u, p_t, pp_t)dt + f_{16}dZ_6 \qquad (2.6)$$

where

pp_t is the average cost of goods in foreign countries relative to domestic cost and

f_{15} and f_{16} are vector functions that depend upon various economic factors up to period t.

The random coefficient vector dZ_6 depicts correlated Wiener terms. Addressing relative currency strengths involves not only current economic conditions but also historical trends. For example, an important long-term relationship is purchase power parity (PPP). The pp_t term depicts the pattern of inflation and currency movements in the projection up to time t. Purchase power parity states that exchange rates should keep pace with price inflation. Currency is an intermediary for the ultimate exchange of goods and services. The inflation differential between two economies approximates the change in the relative cost of goods in each country; currency being conduit for the ultimate goods and services having equal value. (Of course, transportation costs and other issues complicate the issue.) Currency movements should offset relative changes in inflation based on trading arguments. Empirical tests have validated the purchase power conditions over long time frames. Depending on the countries, reversion periods range from 2 years to over 12 years. Two issues are important for risk analysis – the strength of the relationship within individual scenarios and the strength of the relationship when considered across multiple simulations. CAP:Link includes both of these issues.

The system reflects a number of other currency issues relating to interest rates, real interest rates as well as other economic factors. For example, global investors are interested in maximizing their risk adjusted rewards. Over the past 25 years, a desirable currency strategy has been to place a portion of one's assets in countries that possess relatively high real interest rates. Currencies with high real interest rates rose as compared with those currencies with lower real interest rates. Under this theory, investments move to countries with high real interest rates – causing the currency to strengthen. These relationships can be readily modeled within the CAP:Link structure.

2.5   Stock Returns

We divide stock returns into its two elements: dividends and capital appreciation. Dividend yields equals the ratio of the annual rate of dividend payment divided by price. Given the initial dividend yield, an arbitrary price level can be set (say to 100) which specifies the current dividend payment level. Dividend growth through the period determines the income as well as the end of period dividend rate. By separating the base components, we can accurately depict cash income. Also, we found that for modeling purposes, due to their temporal stability, the decomposed structure provides more accurate linkages to the key economic factors – interest rates and inflation levels; see Poterba & Summers (1988).

The dividend growth equation is

$$dg_t = f_{17}(p_u, p_t, g_u, g_t)dt + f_{18}(g_t)dZ_7 \qquad (2.7)$$

where

> g_u is the normative level of dividend growth and g_t is the level at time t, and

> f_{17} and f_{18} are vector functions that depend upon various economic factors up to period t.

The random coefficient vector dZ_7 depicts correlated Wiener terms. We take advantage of the relatively strong relationship between dividend growth rates and inflation.

The dividend yield equation is

$$dy_t = f_{19}dr_t + f_{20}dl_t + f_{21}ds/s_t + f_{22}(y_u, y_t, r_u, r_t, k_u, k_t)dt + f_{23}(g_t)dZ_8 \quad (2.8)$$

where

> y_u is the normative level of dividend yield and y_t is the level at time t, and

$f_{19}, \ldots, f_{23}$ are vector functions that depend upon various economic factors up to period t.

The random coefficient vector dZ_8 depicts correlated Wiener terms. Dividend yield depends upon the movement of interest rate levels and currency exchange rates, among other factors. Also, there are longer-term relationships between the absolute level of real yields and a mean reverting tendency for dividend yields. These equations are first developed for the large capitalization markets: the US S&P 500, Germany's DAX index, and Japan's TOPIX indices. Next, we calculate stock returns for the remaining countries.

2.6   Asset Class Construction

The cascade structure provides a framework for building generic asset classes. The fundamental asset classes are already defined: cash, government bonds and large capitalized equities. Other fixed income assets are modeled via the appropriate spreads over government rates. CAP:Link addresses spreads and the underlying interest relationships of a given fixed income security. This approach extents to real return (or index linked) bonds based on the real yield curves. Given this methodology, we can calculate foreign asset returns by combining the foreign country local return with the currency exchange rate to calculate a total return in the currency of choice.

In addition, we model a number of asset classes outside the fundamental group. Prominent examples include: real estate; venture capital; alternative equity market segments such as small capitalization stocks; emerging market investments; catastrophe related securities; and derivatives. CAP:Link addresses these asset classes by providing tools to describe levels of volatility, serial correlation, relationships to interest rates and inflation, correlations to other asset classes, and relative return expectations. The tools represent our view of the dominant characteristics important to ALM – that is, interest rate and inflation relationships, diversification potential over different time horizons, and range of potential tradeoffs of risk and expected rewards.

2.7   Alternative Approaches

The mean-covariance model is the classic alternative to stochastic differential equations underlying CAP:Link. Typically, this model assumes either normal or lognormal time independent distributions. The user specifies expected values and variances of each asset category as well as covariances for all pairs of assets. The approach often appears in finance textbooks and software packages but suffers in realism for ALM since it avoids features such as mean reversion in interest rates. A second alternative is vector autoregressive (VAR). This approach forms a rolling regression analysis in which the independent variables – interest rates, inflation, asset returns, and liability returns in previous periods – determine the next period values for the economic variables as well as the asset returns and liabilities. Because of the unstructured nature of the process, VAR has considerable adaptability to changing economic conditions. It has been employed in Cariño *et al.* (1994) and others. The approach has proven effective for prediction and pricing applications. At times, however, VAR may produce divergent results for long-term risk analysis. In order to overcome this problem, VAR can be coupled with equilibrium conditions as discussed by Boender *et al.* (this volume). A third alternative has been developed by Wilkie (1995). The approach possesses a cascade structure similar to CAP:Link. However, there are substantial differences in the form of the equations, in the calibration process, and in the ordering of the variables in the cascade.

3   Implementation Details

Setting the parameters for the stochastic differential equations (2.1) through (2.8) presents a formidable task with conflicting objectives. We separate the process into two steps. In the first, called *assumption setting*, we determine the expected level of returns for asset classes and economic variables. The second, called *calibration*, calculates the parameter coefficients for the relationship and distribution characteristics, such as correlations, standard deviations, ranges, rates of convergence, spreads, and other statistics.

3.1   Assumption Setting

For our discussion, assumption setting refers to coefficients that affect risk premiums and expected rewards. Towers Perrin assesses a return premium appropriate to each element of risk. The first block focuses on inflation. The second determines the expected return on cash net of inflation. The third encompasses the spread of long term interest rates over short term interest rates. Next, corporate bonds require a spread for quality risk. Large cap stocks assess a spread of expected equity returns over a bond universe fund. Capital market line relationships (comparing expected return to volatility) are considered for consistency. This approach depends upon the assumption that each element of expected return should be appropriate to compensate for a risky characteristic. Determining the relative premiums is based on dissecting the historical returns into expectation elements (such as the initial level of interest rates) and valuation change elements (such as capital gains/losses due to changes in interest rates). An additional component relates initial condition values to normative assumptions.

Assumption setting is an iterative process: examining the reasonableness of the inputs and the results, analyzing the sensitivity of the recommendations, and modifying the parameters until the patterns are deemed acceptable by the economic staff.

4 Calibration Procedure

Calibration describes the elements relating to risks characteristics, such as distributional spreads. This step presents a more challenging technical task than assumption setting. Target values are determined for identified important relationship and distributional characteristics of the model based on an analysis of historical values and expert judgment. Calibration parameters are adjusted until the simulated statistics become sufficiently close to the target values. Performing this task uncovers new characteristics. Non-convex optimization tools can automate the calibration procedure (Mulvey, Rosenbaum & Shetty 1996), but in the end judgment is required to accept a satisfactory outcome or demand further investigation of a perplexing element.

4.1 Sampling Procedures

The stochastic differential equations (2.1) through (2.8) form the basis for generating scenarios – the main CAP:Link output. We employ a sampling procedure based on variance reduction methods. For example, we have found antithetic variates to be effective for many applications. Each scenario depicts a single realization of the stochastic equations, by sampling from the white noise terms, over the planning horizon. The number of scenarios depends upon the targeted application and the risk aversion of the investor. For example, a long-term planning horizon coupled with dynamic investment strategies, such as portfolio insurance, necessitates a relatively large set of scenarios to provide accurate estimates of risks. On the other hand, stable investment strategies, such as dynamically balanced, involve small market impact costs under volatile conditions and therefore require fewer scenarios to produce acceptable risk estimates. See Mulvey *et al.* (1998).

4.2 Precision Tests

A critical issue involves reliability of the model's recommendations, especially relating to risks and rewards. Accordingly, we conduct stress tests of the proposed strategies by means of out-of-sample analyses. First, we generate a new set of scenarios based on further sampling as described in the previous section, or by modifying the parameters of the stochastic equations and regenerating the results. The precision of the risk measures and valid confidence

Figure 3: The three stages for pension planning. The process entails a combination of feedback and revision in order to become comfortable with the recommendations.

limits are estimated by simulating the investment/liability strategies with the new data. The second risk estimates are compared with the original. If the two values are close, we deem the precision tests to be a success. Otherwise, we can employ a larger set for the optimization stage of ALM system.

By employing precision tests, we have found that 500 scenarios is adequate for many financial planning purposes. The number of scenarios will need to be substantially increased, however, when the investor displays extreme risk averse since the solution depends to a greater degree on rare events.

5   Illustration of Pension Planning

Consider the case of a hypothetical defined benefit pension plan in the US – the NEWCO company – which is in the process of developing an investment strategy for its portfolio mangers. We will discuss the planning process and demonstrate sample forms of analysis. The approach applies to other long term financial investment decisions (insurance companies, endowment funds, personal financial planning) with differing emphases (such as funding decisions).

The process consists of three stages. First, we determine the objectives for the pension plan and carry out the assumption setting tasks. In the second, we analyzes risks and rewards for alternative investment strategies. The final stage involves decision making and implementing the asset and liability allocation strategy. In practice, the process often entails a cycle – observations made from the analysis lead to modifications in the objectives and assumptions. The goal is to understand the financial dynamics, not to run the data through a standardized process and produce a single 'answer'.

5.1   Stage 1 : Objective and Assumption Setting

Setting goals for a defined benefit pension plan is often difficult due to the presence of multiple parties possessing differing interests: beneficiaries, with benefit security or even benefit enhancement; and plan sponsors, with cost levels and volatility. Taken from the sponsors' view, there are alternative bases on which to measure cost: contributions, expense levels for accounting, long term economic cost, impact on corporate income statement, etc. Differing actuarial bases are possible. Issues of time frame (1 year, 3 years, or longer) and risk measurements (standard deviation, volatility, or chances of failing to meet some minimum acceptable target) add complexity. To approach the problem, we focus on the major objectives and stakeholders. Having made a list of these items, we render projections in order to identify areas of concern. For our illustration, we examine projections of contributions and timings, funded ratio levels, and pension expense under current funding and investment policy.

Generating estimates of the future viability and surplus of the pension plan requires four elements: the scenario generation program; a package for projecting the liability cash-flows based on the plan's actuarial rules;[2] a multi-period investment and contribution management system (possibly with an optimization component);[3] and a system for calculating financial and accounting statistics. The four components must fit together with regard to structure and dependence on key assumptions.

We apply the Towers Perrin's system to assist NEWCO. The resulting estimates for the range of contributions over the next ten years appear in Figure 4. This projection shows that the likelihood of making contributions in any one of the next ten years is less than 50%. Since contributions are linked temporally, we must examine the distribution of cumulative contributions through the decade. Because contributions are strongly affected by funding policy, however, these numbers can be difficult to evaluate.[4] In some cases, inadequate cash flows increase the importance of the contributions.

The next issue to consider is the company's expenses. In the US, pension expenses depend upon accounting and other regulatory rules such as FAS 87, which requires plans to discount their liabilities at a close to market rate. Figure 5 depicts the range of possible expenses for NEWCO over the next ten years. We see that they are roughly 1–2% of the company's payroll. However, there is considerable uncertainty, given the range of investment returns and the company's contribution policy during the ensuing period. In fact, there

---

[2]Towers Perrin's system is called VALCAST.

[3]Towers Perrin's system, OPT:Link, employs Leon Lasdon's GRG package (Smith & Lasdon 1992).

[4]Contribution in the US is based on a complex set of regulatory rules and employee preferences.

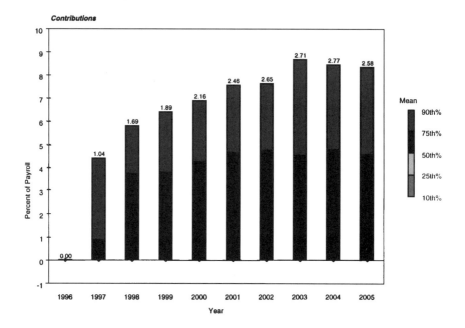

Figure 4: Distribution of contributions. The probability of a contribution in any year is less than 50%.

is a reasonable chance that expenses will be negative – indicating income to the firm.

As a consequence of the investment asset mix and the contribution policy, we can observe the health of the pension plan surplus over the 10 year planning horizon. One view of the surplus is the Accumulated Benefit Obligation (ABO) – roughly the amount of money needed to pay off the current promised liabilities at present value. Figure 6 shows the range of possible ABO numbers. Here, the 100% value indicates a fully funded pension plan on a ABO basis, or there is just enough asset value to compensate for liabilities at present value (discounting at current government non-callable rates). Amounts above 100% are in surplus, whereas amounts below 100% indicate a deficit. A company's required contribution depends upon the value of the ABO numbers.

Projecting the impact of today's decisions on a company's future wealth make sense for long term investors. A multiyear simulation tool based on discrete scenarios is essential. To make these projections, we must address

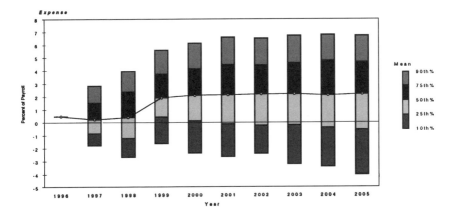

Figure 5: Distribution of expenses as a percentage of payroll.
Expenses can become negative in future years indicating income
to the firm.

the interaction of economic variables such as interest rates and inflation levels with capital market returns. Also, the individual patterns impact the
financial results, such as pension expenses. In addition to the economic and
capital market simulations, as provided by CAP:Link, we must convert this
information into the relevant financial statistics. The calculation ought to be
straightforward, but in practice many regulatory, timing, and other technical issues must be addressed. All financial projections presented have been
produced by Towers Perrin's FIN:Link system.

Contribution, expense, and other projections depend upon calibration assumptions. Similar to the objective setting process it is useful to examine the
projections to access their reasonableness and also to provide more insight
into the drivers behind the financial projections.

For NEWCO, we reflect the economic variables at the beginning of 1996
– adjusting the assumptions to an 'equilibrium' condition, where long term
expectations are set equal to current conditions and the spread between expected returns of various asset classes equal the normative assumptions. The
normative approach helps develop broad policy guidelines. Alternatively, the

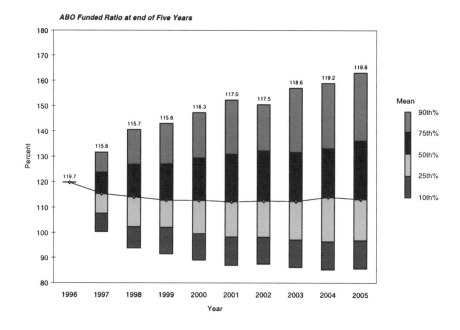

Figure 6: Distribution of ABO funded ratios over the next 10 years. The NEWCO pension plan is likely to be in surplus during the period, given the proposed contribution policy.

assumptions could equal the company's best long-term estimates. This approach is suitable where consideration of strategic shifts in asset allocation are under review. Of course, consideration should be given to whether fund managers will be capable of performing ongoing analysis and adjustments to investments to reflect changing conditions.

Graphs showing the projections of inflation and interest rates (Figures 7, 8, and 9) – the primary liability drivers – are presented below along with the range of returns for each of the asset classes.

After completing this step, we pinpoint a specific concern – funded levels falling below the 90% ratio of assets to the present value of liabilities. In an actual assignment, several elements for further investigation emerge. For example, the time horizon is significant. It is appropriate to investigate both the immediate term (1–3 years) and the longer term (5+ years). The former is critical – due to the nature of evaluating management performance by

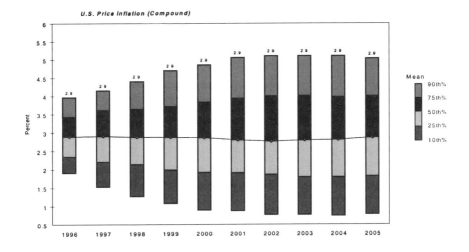

Figure 7: Distribution of estimated price inflation in US. These ranges reflect historical inflation in the US over the past 40 years.

stockholders and other stakeholders. The long term must be considered as well: pension plan sponsors have a fiduciary responsibility to see that the company's promised benefits will be realized. Again a real world case would involve multiple time horizons, perhaps with differing objective elements: a short time horizon for funded status, and a longer one for pension expense levels. In this illustration, we focus on a five year time horizon.

5.2   Stage 2 : Analysis of Risks and Rewards

Efficient frontiers can be developed at various time horizons. Traditional efficient frontiers plot expected investment return versus the volatility (measured by standard deviation) of asset return over a single period – an asset only concept; see Konno *et al.* (1993) and Kroll *et al.* (1984). It is useful to understand the recommendations of the 'standard single period asset only approach' so that they can be compared with a comprehensive ALM approach. To address standard practice, we calculate an asset-only frontier with a five year time horizon (Figure 10). Additional efficient frontiers should reflect differing elements of the objectives identified in Stage 1. Thus, we select average funded ratio at the end of five years as the reward measure and first-order downside risk with a 90% target as the risk measure (Figure 11). First-order downside risk equals the average shortfall below the target averaged over all

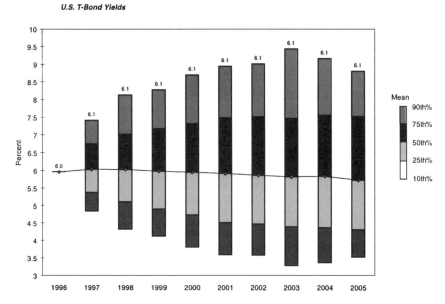

Figure 8: Distribution of estimated US T-bond yields. These results assume that long term interest rates equal the starting interest rates (1 January 1996 in this case).

scenarios (including those where there is no shortfall). For reference, second-order downside risk equals the average squared shortfall. First-order downside risk breaks into the product of two intuitive risk measures: the probability of shortfall and the average shortfall when one exists.

The scope of a risk analysis can be portrayed on a risk ladder (Mulvey 1996) with five rungs: 1) total integrated risk management; 2) dynamic asset-liability; 3) static asset-liability; 4) static asset only (a.k.a. Markowitz mean variance portfolios); and 5) single security risks. Most current risk analyses are conducted at rungs 3 or 4. However, there are several groups that are pushing the analysis to higher, more comprehensive levels on the risk ladder. In this context, several criteria exist for measuring risk, including variance for symmetric outcomes, semi-variance and downside risks for asymmetric outcomes, and von Neumann Morganstern expected utilities. Ultimately, an organization ought to be concerned with risks to its net worth – at the top of the risk ladder.

We might consider several efficient frontier calculations with a variety of risk and reward measures, some which might be a weighted combination of

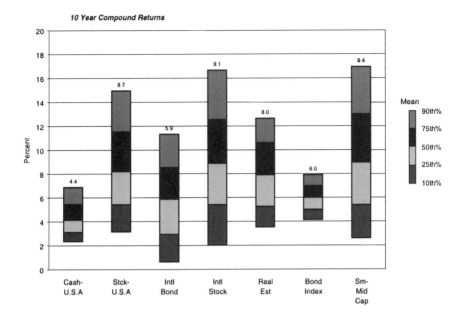

Figure 9: Distribution of ten year compounded returns. Cash becomes dominated by both stock and bond indices as the horizon lengthens.

differing measures. The goal is to find a set of candidate portfolios based on the efficient frontiers which might best meet the competing considerations. In theory, this step could be done by constructing the 'correct' risk and reward measures and performing the efficient frontier calculation on these objectives. In practice, however, the correct measures become known as the analysis progresses. Employing candidate portfolios help draw out this knowledge. In this illustration we limit the discussion to the two frontiers described above. Calculating risk-reward efficient frontiers over multiple time periods requires non-linear stochastic optimization programs; see Mulvey, Armstrong and Rothberg (1995).[5] Additionally, because of the non convexity due to the dynamic aspects of multi-period ALM and the lack of performance guarantees, the investor must be careful to access if a global optimal solution has been found. This consideration argues for the candidate portfolio approach. Based on this analysis, we focus on promising areas on the frontiers. For comparison, portfolios recommended by the asset-only efficient frontier have been plotted on the funded ratio efficient frontier. Differences are quite dramatic

---

[5]Towers Perrin's OPT:Link system solved these efficient frontier calculations.

Illustration E fficient F rontier Asset E fficient F rontier 5 Year Time Horizon

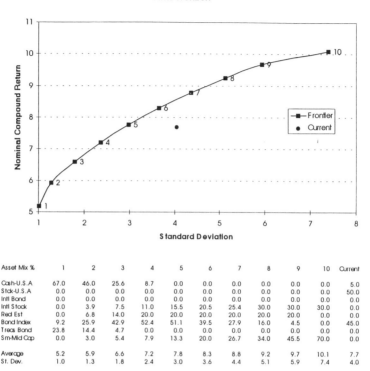

Asset Mix %	1	2	3	4	5	6	7	8	9	10	Current
Cash-U.S.A	67.0	46.0	25.6	8.7	0.0	0.0	0.0	0.0	0.0	0.0	5.0
Stck-U.S.A	0.0	0.0	0.0	0.0	0.0	0.0	0.0	0.0	0.0	0.0	50.0
Intl Bond	0.0	0.0	0.0	0.0	0.0	0.0	0.0	0.0	0.0	0.0	0.0
Intl Stock	0.0	3.9	7.5	11.0	15.5	20.5	25.4	30.0	30.0	30.0	0.0
Red Est	0.0	6.8	14.0	20.0	20.0	20.0	20.0	20.0	20.0	0.0	0.0
Bond Index	9.2	25.9	42.9	52.4	51.1	39.5	27.9	16.0	4.5	0.0	45.0
Treas Bond	23.8	14.4	4.7	0.0	0.0	0.0	0.0	0.0	0.0	0.0	0.0
Sm-Mid Cap	0.0	3.0	5.4	7.9	13.3	20.0	26.7	34.0	45.5	70.0	0.0
Average	5.2	5.9	6.6	7.2	7.8	8.3	8.8	9.2	9.7	10.1	7.7
St. Dev.	1.0	1.3	1.8	2.4	3.0	3.6	4.4	5.1	5.9	7.4	4.0

Figure 10: An asset-only efficient frontier. Cash and bonds are considered the more conservative category for short term investors without liabilities.

– the low-risk portfolios from the asset-only efficient frontier are highly risky from a funded ratio perspective at the five year point (see Figure 11).[6]

By selecting portfolios from the most attractive regions of the efficient frontiers we develop a list of candidate portfolios. NEWCO's efficient frontier in Figure 11 suggests substantial investments in small capitalization and international stocks and real estate. In practice, these recommendation might be limited for liquidity and prudence issues. Constraints on the optimization could be readily applied and the optimization re-run to generate a revised investment strategy.

---

[6]We have discovered numerous similar situations in practice.

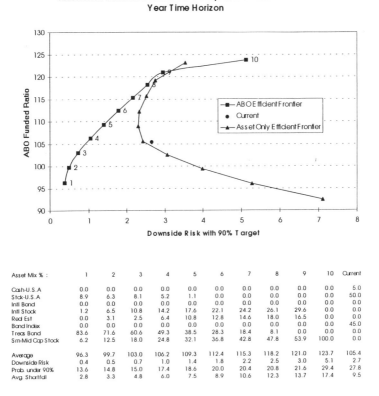

Illustration Efficient Frontier ABO Surplus Efficient Frontier 5
Year Time Horizon

Asset Mix % :	1	2	3	4	5	6	7	8	9	10	Current
Cash-U.S.A	0.0	0.0	0.0	0.0	0.0	0.0	0.0	0.0	0.0	0.0	5.0
Stck-U.S.A	8.9	6.3	8.1	5.2	1.1	0.0	0.0	0.0	0.0	0.0	50.0
Intl Bond	0.0	0.0	0.0	0.0	0.0	0.0	0.0	0.0	0.0	0.0	0.0
Intl Stock	1.2	6.5	10.8	14.2	17.6	22.1	24.2	26.1	29.6	0.0	0.0
Real Est	0.0	3.1	2.5	6.4	10.8	12.8	14.6	18.0	16.5	0.0	0.0
Bond Index	0.0	0.0	0.0	0.0	0.0	0.0	0.0	0.0	0.0	0.0	45.0
Treas Bond	83.6	71.6	60.6	49.3	38.5	28.3	18.4	8.1	0.0	0.0	0.0
Sm-Mid Cap Stock	6.2	12.5	18.0	24.8	32.1	36.8	42.8	47.8	53.9	100.0	0.0
Average	96.3	99.7	103.0	106.2	109.3	112.4	115.3	118.2	121.0	123.7	105.4
Downside Risk	0.4	0.5	0.7	1.0	1.4	1.8	2.2	2.5	3.0	5.1	2.7
Prob. under 90%	13.6	14.8	15.0	17.4	18.6	20.0	20.4	20.8	21.6	29.4	27.8
Avg. Shortfall	2.8	3.3	4.8	6.0	7.5	8.9	10.6	12.3	13.7	17.4	9.5

Figure 11: An alternative efficient frontier with downside ABO
risk. Treasury bonds are considered the least risky asset based on
the investor's liability structure.

5.3   Stage 3 : Decision Making and Implementation

We have chosen two candidate portfolios for comparing the distributions of
contributions, expense, and funded ratio. In addition, it is useful to consider
the differences in potential performance relative to NEWCO's current position. For alternatives 1 and 2, contributions and expenses are reduced, both
on absolute range and as relative improvements over the current allocation
(Figures 12 and 13). In this case, we see that although the first alternative
has the potential for higher absolute levels of expense, the distribution of
improvement from the current portfolio is much more attractive than for the
second. Alternative 1 outperforms the current policy approximately 70% of
the time, while alternative 2 outperforms the current policy about 50% of
the time (Figure 14). From this perspective, alternative 1 appears superior

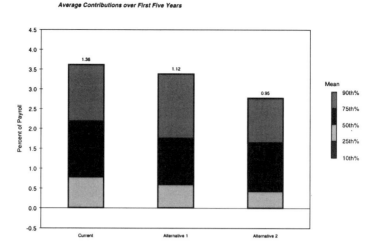

Figure 12: Distribution of contribution over 5 years. Both alternative 1 and 2 reduce contribution over the 5 years.

to the current policy as well as alternative 2. Similarly, alternative 1 has the highest ABO funding ratio on average (Figure 15).

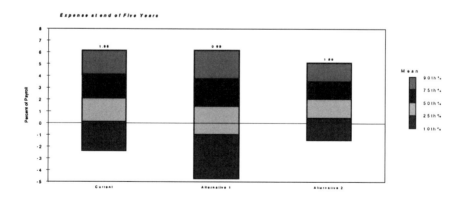

Figure 13: Distribution of expenses at end of 5 years. Alternative 1 provides lower average expenses from the current strategy, while alternative 2 provides lower expenses at the 90% confidence level.

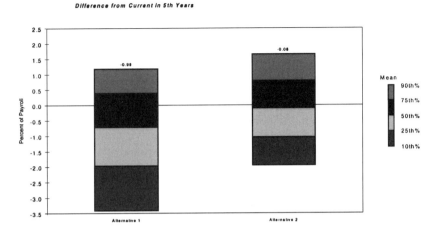

Figure 14: Distribution of expenses as a percentage of payroll as compared with the current policy on a scenario by scenario basis.

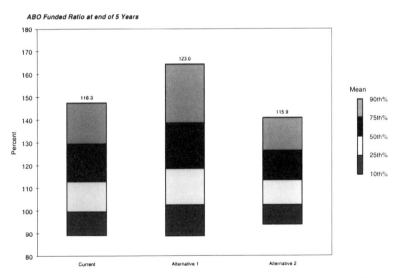

Figure 15: Comparison of ABO ratios for three alternatives. The average ABO is highest on average for alternative 1.

Once an allocation strategy is found, it is essential to carry out a sensitivity analysis. If a higher allocation to equity is suggested over the current asset allocation, for example, we recommend that the investor employ assumptions which reduce this spread. This approach applies to all asset classes targeted for increase in allocation. Observations made at this stage may lead back to an earlier stage. Simulation lends itself to pinpointing the cause of a particular result. Recommendations can be segmented by inflation, interest rates, equity returns, contribution levels, or other proposed causes. The analysis leads to an improved understanding of the investment dynamics. Combining this with other factors, such as liquidity, perceptions, transition problems, and implementation capability produce a decision.

The analysis may also suggest modifying the pension plan design, contribution policy, or asset and liability smoothing in order to improve overall financial performance. In addition, long term ALM systems can address more focused questions, such as whether and how much currency hedging is appropriate, how unexpected inflation would effect the contributions and other measurements, and how short-term funding decisions will impact longer-term risks.

6   Conclusions and Future Directions

Towers Perrin's system provides an internally consistent approach for generating capital market scenarios over long-term horizons. It captures an extensive range of market relationships – making it an effective tool for risk analysis. There are several directions for future research. First, the system easily extents for new asset categories. For example, combining the purchase of equity and selling an off-setting call option on the same stocks – buy-writing – can be modeled as a separate asset category. This strategy is effective in volatile and driftless markets and can be considered on the opposite side of portfolio insurance strategies. The options can be valued in the usual manner via arbitrage free approaches since the risk free rates, volatility and other factors are known within a scenario. Buy-writing may be attractive for risk averse investors since downside risks are considerably reduced over a pure equity category. Another promising asset category involves securitizing catastrophic (CAT) insurance, as implemented by the Chicago Board of Trade. Here, risks are uncorrelated with the returns of other asset categories. Investigating the pros and cons of these and other asset categories is made possible due to the system's flexibility. See Mulvey *et al.* (1997) for further details.

Another extension is to employ the currency exchange rates for hedging strategies. Remember that local returns for countries outside the home country are inaccessible due to uncertain currency movements and the lack of equilibrium with respect to implied forward rates (Brinson 1993). Since the system defines currency as a separate economic factor, we can evaluate cross hedging strategies. See Sorensen *et al.* (1993) for an example application. Next, we can assist corporations in setting organizational goals. For instance, we could evaluate alternative contribution policies. By analyzing policies under stressful conditions such as a recession, we can pinpoint total risks – rather than sub-optimizing the risks for individual components – such as portfolios in a single country.

In conclusion, multi-stage financial models cannot replace an understanding of capital markets. Rather, they aid in analyzing the competing issues of risks and rewards over time. As such, the capital market projections must provide a representative range of plausible scenarios. The global CAP:Link provides this information in a practical and internally consistent fashion.

References

K.I. Amin and R.A. Jarrow, 1991. 'Pricing Foreign Currency Options under Stochastic Interest Rates,' *Journal of International Money and Finance* **10** 310–329.

A.J. Berger and J.M. Mulvey, 1998. 'Asset and Liability Management for Individual Investors.' This volume, 634–665.

F. Black and R. Litterman, 1992. 'Global Portfolio Optimization,' *Financial Analysts Journal*, 28–43, September–October.

G.C.E. Boender, P.C. van Aalst, and F. Heemskerk, 1998. 'Modeling and Management of Assets and Liabilities of Pension Plans in The Netherlands.' This volume, 561–580.

P. Boothe and D. Glassman, 1987. 'The Statistical Distribution of Exchange Rates,' *Journal of International Economics* **22** 297–319.

M.J. Brennan and E.S. Schwartz, 1982. 'An Equilibrium Model of Bond Pricing and a Test of Market Efficiency,' *Journal of Financial and Quantitative Analysis* **17** 75–100, March.

G.P. Brinson, 1993. 'You Can't Access Local-Currency Returns,' *Financial Analysts Journal*, 10, May–June.

D.R. Cariño, T. Kent, D.H. Myers, C. Stacy, M. Sylvanus, A. Turner, K. Watanabe and W.T. Ziemba, 1994. 'The Russell–Yasuda Kasai Model: an Asset Liability Model for a Japanese Insurance Company using Multi-stage Stochastic Programming,' *Interfaces* **24** 24–49, Jan–Feb. Reprinted in this volume, 609–633.

D.R. Cariño and A.L. Turner, 1998. 'Multiperiod Asset Allocation with Derivative Assets.' This volume, 182–204.

K.S. Choie, 1993. 'Currency Exchange Rate Forecast and Interest Rate Differential,' *Journal of Portfolio Management*, 58–64, Winter.

S.J. Cochran and R.H. DeFina, 1995. 'Predictable Components on Exchange Rates,' *The Quarterly Review of Economics and Finance* **35** (1), 1–14, Spring.

Dempster, M A H. and A.M. Ireland, A.M. 1988, 'A financial expert decision support system.' In *Mathematical Models for Decision Support*, B. Mitra (ed.) NATO ASI Series, Vol. F48, Springer-Verlag, 415–440.

M. Hogan, 1993. 'Problems in Certain Two-Factor Term Structure Models', *Annals of Applied Probability* **3** 576–581.

P. Huber, 1995. 'A Review of Wilkie's Stochastic Model,' Actuarial Research Paper No. 70, The City University, London.

H. Konno, S. Pliska and K. Suzuki, 1993. 'Optimal Portfolio with Asymptotic Criteria', *Annals of Operations Research* **45** 187–204.

M. Kritzman, 1992. 'What Practitioners Need to Know About Currencies,' *Financial Analysts Journal*, 27–30, March–April.

M. Kritzman, 1993. 'The Optimal Currency Hedging Policy with Biased Forward Rates,' *Journal of Portfolio Management*, 94–100, Summer.

Y. Kroll, H. Levy and H. Markowitz, 1984. 'Mean Variance Versus Direct Utility Maximization,' *Journal of Finance* **39** 47–62.

R. Mahieu and P. Schotman, 1994. 'Neglected Common Factors in Exchange Rate Volatility,' *Journal of Empirical Finance* **1** 279–311.

J.M. Mulvey, 1994. 'Integrating Assets and Liabilities for Large Financial Organizations.' In *New Directions in Computational Economics,* W.W. Cooper and A.B. Whinston, eds., Kluwer Academic Publishers, 135–150.

J.M. Mulvey, 1996. 'It Always Pays to Look Ahead,' *Balance Sheet* **4** (4) 23–27, Winter.

J. Mulvey, and W. Ziemba, 1995. 'Asset and Liability Allocation in a Global Environment.' In *Finance,* R. Jarrow *et al.,* eds., *Handbooks in Operations Research and Management Science* **9** 435–463, North-Holland.

J.M. Mulvey, J. Armstrong and E. Rothberg, 1995. 'Total Integrative Risk Management,' *Risk Magazine,* June.

Mulvey, J.M., D.P. Rosenbaum and B. Shetty, 1996. 'Parameter Estimation in Stochastic Scenario Generation Systems,' Princeton University Report SOR–96–15.

Mulvey, J.M., S. Correnti and J. Lummis, 1997. 'Total Integrated Risk Management: Insurance Elements,' Princeton University Report, SOR–97–2.

Mulvey, J.M., R. Rush, J.E. Mitchell and T.R. Willemain, 1997. 'Stratified Filtered Sampling in Stochastic Optimization,' Princeton University Report SOR–97–7. To appear in *European Journal of Operations Research.*

A.F. Perold, and W.F. Sharpe, 1988. 'Dynamic Strategies for Asset Allocation,' *Financial Analysts Journal,* 16–27, January.

J.M. Poterba, and L.H. Summers, 1988. 'Mean Reversion in Stock Prices: Evidence and Implications,' *Journal of Financial Economics* **22** 27–59, October.

S. Smith and L. Lasdon, 1992. 'Solving Large Sparse Nonlinear Programs Using GRG,' *ORSA Journal on Computing* **4** 2–15.

E. Sorensen, E. Mezrich, and D. Thadani, 1993. 'Currency Hedging through Portfolio Optimization,' *Journal of Portfolio Management,* 78–85, Spring.

A.D. Wilkie, 1995. 'More on a Stochastic Asset Model for Actuarial Use.' Presented to *Institute of Actuaries and Faculty of Actuaries.*

Part VI
Currency Hedging and Modeling Techniques

An Algorithm for International Portfolio Selection and Optimal Currency Hedging

Markus Rudolf and Heinz Zimmermann

Abstract

The paper investigates the risk return characteristics of constant versus optimized currency hedging strategies with and without investment restrictions. In contrast to the Solnik (1974) model, a portfolio selection approach rather than an equilibrium model is used. In the model introduced here, an efficient set of asset allocations and currency hedges is determined where investors have to choose specific portfolios according to their risk preferences and their expectations. For practical purposes, investment restrictions are imposed. The restricted asset allocation and currency hedging problem is solved by an algorithm based on an extended version of the Markowitz (1956) critical line model. It is shown that optimizing the currency hedge, i.e. including currencies like other asset classes in the asset allocation decision, substantially widens the range of efficient portfolios under traditional investment restrictions. The paper also investigates the stability and out of sample characteristics of minimum variance portfolios.

1   Introduction

International Asset Allocation has become a major investment device for global investors. The merits of international diversification were originally shown by Grubel (1968), Levy and Sarnat (1970), Solnik (1973), and others. Inevitably, international portfolios contain currency risks. Solnik (1974) in his original derivation of the international capital asset pricing model has shown that an optimal internationally diversified portfolios contains long and short positions in currency forward or futures contracts or synthetic positions in the underlying cash market. Thus, currency hedging becomes an integrated part of the overall international asset allocation decision. In practice, however, currency holdings are hardly optimized according to the diversification rules followed in domestic investments. This is particularly surprising since modern capital markets and hedging instruments allow to separate the investments in particular markets from the holding of the respective currencies: currencies can be treated as separate asset classes. Instead, simple ad hoc rules are used (e.g. "50% hedging") mostly without much theoretical justification. In addition, the (optimal) currency hedging decision is implemented with predetermined portfolio positions in the underlying markets. This neglects cross hedging effects between markets and currencies and cannot produce a portfolio which jointly optimizes market and currency holdings.

Fortunately, the problem of optimal currency hedging has been addressed repeatedly in recent papers. Perold and Schulman (1988) advocated the view that currency hedging is a free lunch implying that 100% of foreign currency exposures should be optimally hedged. Eun and Resnick (1988) also find empirical evidence that the performance of international stock portfolios is increased if 100% of the exposure is hedged against currency risk. An opposite view is advocated by Black (1990) who demonstrated (by assuming the validity of Siegel's paradox and homogeneous preferences of investors) that investors worldwide share "universal" currency hedge ratios which he showed to be strictly below 100%. Adler and Prasad (1992) weaken some of the underlying assumptions and generalize Black's result by substituting universal hedges with regression hedges. On the empirical side, however, the effect of currency hedging on the performance of international portfolios is not clear. Only two recent studies reach very different conclusions. While, in a careful study, Glen and Jorion (1993) report that "over the period 1974 to 1990, inclusion of forward contracts results in statistically significant improvements in the performance of unconditional portfolios containing bonds", Levy and Lim (1994) conclude that "almost all the unhedged strategies outperformed the hedged strategies for 1985–88; the opposite held for 1981–84". The results strongly depend on the asset class under investigation (Levy and Lim investigate stock portfolios), the time period, and the biasedness of forward rates in predicting future spot rates.

In this paper no empirical effort is done to resolve the debate over currency hedging. Instead, an algorithm is developed which allows us to investigate a series of methodological questions about optimal international diversification and the associated problem of currency hedging. The model allows for a joint optimization of markets and currency exposures. Hedging is done by using forward contracts; various investment restrictions which limit unconstrained diversification in practice can be included. The algorithm which is used to solve the international asset allocation problem subject to these constraints is based on the critical line method developed by Markowitz (1956) and elaborated by Rudolf (1994). Therefore, the model is able to demonstrate the impact of various restrictions on the efficiency of portfolios as well as the comparative performance of widely used (but inefficient) hedging strategies. A similar paper investigating integrated asset and currency allocation for international bond markets can be found in Lee (1987). Lee's approach is

based on a simplified computation of the return on an international investment which neglects important covariance components in the calculation of
the portfolio variance. Qualitatively, the findings of this study are similar to
those of Lee's paper who also reports a significant improvement in the efficiency of international portfolios by jointly optimizing currency and market
exposures. The study, however, is restricted to bond portfolios. The study of
Levy and Lim (1994) investigates currency hedging effects for investments in
the domestic and a single foreign market. Moreover, they do not investigate
the effects of optimized currency hedge strategies. Their results provide no
consistent outperformance of the hedged portfolios over the unhedged portfolios as measured by the Sharpe ratio. Therefore, in contrast to our results,
the effects of currency hedging is ambiguous. Glen and Jorion (1993) analyze
the effects of currency hedging in the framework of Solnik's (1974) equilibrium model. However, their emphasis is on empirical issues[1] related to the
implementation of various hedging strategies.

The paper is structured as follows. In Section 2, the model is developed.
Section 3 gives a characterization of the data used, and Section 4 presents the
results. The main results are presented in Subsection 4.1, while the results
related to the benchmark currency and to the intertemporal (in)stability of
the efficient frontiers are discussed in subsections 4.2 and 4.3. The main
findings are summarized in Section 5.

2   The Model

Let a_{it} denote the simple return on a foreign asset traded in a foreign currency
i over time period t (for simplicity, assume that there is only one asset in each
currency). c_{it} is the simple return on that currency from the perspective of
a domestic investor. An amount $V_{i,t}$ in domestic currency is invested in the
foreign asset. The end of period value of the investment is

$$V_{i,t+1} = V_{i,t}(1 + a_{it})(1 + c_{it}). \tag{2.1}$$

The currency risk of foreign asset holdings is hedged with forward contracts
which promise to deliver one unit of currency i at the end of the period.
Selling one forward contract at the beginning of each time period, t, generates
a payoff $F_{i,t,t+1} - C_{i,t+1}$ at the end of period, $t + 1$; $F_{i,t,t+1}$ is the forward rate
in t for delivery in $t + 1$, and $C_{i,t+1}$ is the prevailing spot rate at contract
expiration. $V_{i,t}/C_{i,t}$ is the value of the investment in foreign asset i measured
in foreign currency units. It is assumed that a fraction h_i of the investment

---

[1]For bond portfolios, they find that an optimized (unconditional) currency hedge dominates a strategy with unhedged currency hedges which confirms basically our finding.
However, no similar result emerges for stock or mixed stock/bond portfolios. The results
are also ambiguous for unitary and universal hedge strategies.

in country i is hedged by selling the appropriate number of forward contracts (for $h_i < 0$: buying). We will henceforth call h_i the *hedge ratio*. The end of period wealth of the hedged position is therefore

$$V_{i,t+1} = V_{i,t}(1 + a_{it})(1 + c_{it}) + h_i \frac{V_{i,t}}{C_{i,t}}(F_{i,t,t+1} - C_{i,t+1}). \qquad (2.2)$$

However, returns and not payoffs are used in the subsequent analysis. Since the commitment to a forward contract does not generate or use any cash flows, the original investment is still $V_{i,t}$. Define $f_{it} = (F_{i,t,t+1}/C_{i,t}) - 1$ as the percentage deviation of the forward rate from the current spot rate. The return on the hedged position in foreign asset i is then

$$
\begin{aligned}
r_{it} \equiv \frac{V_{i,t+1}}{V_{i,t}} - 1 &= (1 + a_{it})(1 + c_{it}) + h_i \left(\frac{F_{i,t,t+1} - C_{i,t+1}}{C_{i,t}} \right) - 1 \\
&= (1 + a_{it})(1 + c_{it}) + h_i \left((1 + f_{it}) - (1 + c_{it}) \right) - 1 \\
&= a_{it} + c_{it} + a_{it}c_{it} + h_i(f_{it} - c_{it}). \qquad (2.3)
\end{aligned}
$$

In equation (2.3), $f_{it} \equiv \frac{F_{i,t,t+1}}{C_{i,t}} - 1$ is defined as the forward rate return of currency i. It is assumed that the wealth can be invested in n risky foreign assets (each denoted in the respective foreign currency) with returns r_{it}, $i = 1, \ldots, n$, and in one risky domestic asset with return $\tilde{r}_D$. The portfolio weights for the foreign assets are denoted by w_i, and the portfolio weight for the domestic asset is $1 - \sum_i w_i$. The return on the portfolio including $n + 1$ assets, $\tilde{r}_{Pt}$, is then

$$\tilde{r}_{Pt} = \sum_{i=1}^{n} w_i \left(\tilde{a}_{it} + \tilde{c}_{it} + \tilde{a}_{it}\tilde{c}_{it} + h_i(f_{it} - \tilde{c}_{it}) \right) + \left(1 - \sum_{i=1}^{n} w_i \right) \tilde{r}_{Dt} \qquad (2.4)$$

(tildes ($\sim$) denote random variables). Subsequently, the cross product $a_{it}c_{it}$ will be denoted by k_{it}, and the time index t will be dropped for simplicity. The frontier of efficient portfolios is investigated in terms of means and variances of returns (see e.g. Markowitz (1959) and Merton (1972)). Define

$$E \equiv E(\tilde{r}_P) \qquad \sigma^2 \equiv \mathrm{Var}(\tilde{r}_P). \qquad (2.5)$$

The unrestricted efficient frontier is the solution to the minimization problem:

$$\min_x \sigma^2 = x'Vx \quad \text{subject to} \quad x'\mu = E \qquad x'\hat{e} = 1 \qquad (2.6)$$

where x denotes a $(2n + 1)$-dimensional vector whose first n elements are products of the asset weights with the currency hedge ratios ($y_i \equiv h_i w_i$), whose second n elements are the asset weights (w_i), whose last element is the domestic portfolio share, and where $\hat{e}$ is a $(2n + 1)$-dimensional vector with

n zeroes and $(n + 1)$ units:

$$x = \begin{bmatrix} y \\ w \\ 1 - e'w \end{bmatrix} \quad \text{with } \{y_i\} = \{h_i w_i\}, \tag{2.7}$$

where w is a $(n \times 1)$ vector. In (2.6), μ is the $(2n + 1)$-dimensional vector of expected returns

$$\mu = \begin{bmatrix} f - \mu_c \\ \mu_a + \mu_c + \mu_k \\ \mu_d \end{bmatrix} \tag{2.8}$$

where f is the $(n \times 1)$ vector of current forward rate returns, μ_c is the $(n \times 1)$ vector of expected values of c_i, μ_a is the $(n \times 1)$ vector of expected values of a_i, μ_k is the $(n \times 1)$ vector of expected values of k_i and μ_D is the expectation of r_D.

The $(2n + 1) \times (2n + 1)$ covariance matrix V is more complicated to characterize; it can be expressed as

$$V = \begin{bmatrix} V_c & -V_{ca} - V_c - V_{ck} & -V_{Dc} \\ -V_{ac} - V_{kc} - V_c & \begin{matrix} V_a + V_c + V_k + V_{ak} + V_{ca} \\ +V_{ck} + V_{ka} + V_{kc} + V_{ac} \end{matrix} & V_{Da} + V_{Dc} + V_{Dk} \\ -V'_{Dc} & V'_{Da} + V'Dc + V'Dk & V_D \end{bmatrix} \tag{2.9}$$

where the various matrices and vectors are defined as follows:

V_a, V_c, V_k are $(n \times n)$ covariance matrices of asset returns (a_i), currency returns (c_i) and their associated cross product returns (k_i);

V_{ac}, V_{kc}, V_{ka} are $(n \times n)$ covariance matrices between asset returns (a_i), currency returns (c_i), and cross product returns (k_i). V_{ca}, V_{ck}, V_{ak} are the respective transposed matrixes;

V_{Dc}, V_{Da}, V_{Dk} are $(n \times 1)$ covariance vectors of the returns between the domestic asset and currency returns (c_i), international asset returns (a_i), and the cross product returns (k_i);

V_D is the return variance of the domestic asset.

The optimization problem in equation (2.6) gives the unrestricted efficient portfolio frontier which can be derived as

$$\hat{\sigma} = \sqrt{\frac{1}{d}\left(\hat{c}\hat{E}^2 - 2\hat{b}\hat{E} + \hat{a}\right)}, \quad \hat{a} \equiv \mu'V^{-1}\mu, \quad \hat{b} \equiv \mu'V^{-1}\hat{e},$$

$$\hat{c} \equiv \hat{e}'V^{-1}\hat{e}, \quad \hat{d} \equiv \hat{a}\hat{c} - \hat{b}^2 \tag{2.10}$$

in analogy to Roy (1952) and Merton (1972). The optimal asset holdings and, implicitly[2], the hedge ratios, are given by

$$x = \frac{V^{-1}}{\hat{d}}(\hat{c}\mu - \hat{b}\hat{e})E + \frac{V^{-1}}{\hat{d}}(\hat{a}\hat{e} - \hat{b}\mu) = \frac{V^{-1}}{\hat{d}}(\hat{c}E - \hat{b}) - \frac{V^{-1}\hat{e}}{\hat{d}}(\hat{b}E - \hat{a}). \quad (2.11)$$

However, in practice, restrictions in either the asset holdings or the currency exposures must be respected. Three sets of restrictions appear to be relevant:

1. *Non negativity restrictions*: For many investments, negative holdings are impossible (no short sales), expensive, or forbidden. For example, in most countries, mutual funds are not allowed to implement short sales in foreign bond and stock markets. A non negativity restriction also applies for currency hedge ratios; according to our definition, a negative hedge ratio implies a currency exposure in addition to the foreign asset investment (reverse hedge). Many institutional restrictions prevent long positions in currency forward contracts.

2. Restrictions with respect to *currency exposures*: Many institutional investors are not allowed to hedge more than 100% of the value of a foreign investment against currency risks. Therefore, hedge ratios must be lower than 1.

3. Restrictions with respect to *minimum diversification requirements*: Many investors face upper and lower weights according to which they must diversify their assets across markets and currencies.

These restrictions can be added to the optimization problem stated in (2.6). The solution can be obtained by applying the critical line algorithm developed by Markowitz (1956, 1987) and elaborated by Rudolf (1994). The full optimization program can be restated as follows:

$$\min_x x'Vx, \quad \text{such that}$$
$$\begin{aligned}
&\text{(a)} \quad x'\mu = E \\
&\text{(b)} \quad x'\hat{e} = 1 \\
&\text{(c)} \quad y - w \leq 0 \\
&\text{(d)} \quad w \geq b_l \\
&\text{(e)} \quad w \leq b_u \\
&\text{(f)} \quad y, w \geq 0
\end{aligned} \quad (2.12)$$

Restrictions (a) and (b) were discussed earlier. Restriction (c) captures the currency restriction discussed before, which requires $h \leq 1$ or $y \leq w$ or

---

[2]As becomes apparent from the definition of vector x, equation (2.7), the hedge ratios h_i can be calculated from $y_i = h_i w_i$.

Table 1: The investment spectrum

Country	Index
France	CAC 40
German	DAX
Great Britain	FT-SE 100
Italy	BCI
Japan	NIKKEI 225
Sweden	Affarsvarlden
Switzerland	SMI
USA	S&P 100

Reference Currency: US-$
Source: Bloomberg Inc.

$y - w \leq 0$. This guarantees that hedging is strictly below 100% of the investment in the respective market for each currency. Restrictions (d) and (e) formalize the minimum diversification requirements postulated before. Finally, restriction (f) states the non negativity requirement for assets.

3   Data

The empirical results of this paper are based on indexes from seven stock markets outside the US; those indexes are selected where futures contracts are traded in the respective country. An overview on the indexes is provided by Table 1. The study takes the perspective of a US-$ based investor. The *domestic* stock market is characterized by the S&P 100 index. The selection of stock markets reveals a certain preference for investments in Europe.

Spot and forward exchange rates are used to transform local returns to hedged and unhedged portfolio returns measured in US-$, as shown in Section 2. Descriptive statistics of stock market and exchange rate returns are displayed in Tables 2 and 3. Monthly returns from December 1988 to August 1994 are computed from Bloomberg data. The statistics show that all stock markets except Japan exhibit positive average returns in local currency over the investigated time period. Average returns in US-$ are negative for two markets, Japan and Italy. Table 3 (column 1) shows the average performance of the seven currencies; the British Pound, the Lira and the Swedish Krone are weak currencies from the perspective of a US-$ based investor. This affects the hedging results investigated in the next section. One month forward contracts are used for hedging purposes throughout the study. A comparison between spot and forward returns is provided by Table 3.

Table 2: Descriptive statistics of the stock market returns, in %
p.a.

Index	average return in local currency	standard deviation in local currency	average return in $	standard deviation in $
	(1)	(2)	(3)	(4)
S&P 100	9.61	12.34	9.61	12.34
CAC 40	6.47	18.70	8.63	19.20
DAX	10.27	18.57	12.76	20.95
FT-SE 100	11.84	16.71	9.45	19.48
BCI	4.97	23.04	2.15	25.38
NIKKEI	−3.24	26.26	1.71	30.66
Affarsvarlden	9.41	24.89	5.03	23.53
SMI	7.36	16.56	9.59	17.20

Source: monthly closing rates between 12/1988 and 8/1994,
Bloomberg Inc. and personal calculations

The volatility in local currency substantially varies between the markets,
with a minimum observed for the US (12.34%) and a maximum for Japan
(26.73%). The volatility of the markets as measured in US-$ is shown in the

Table 3: Descriptive statistics of currency returns, in % p.a.

Currency	Spot rate, average return	Spot rate, standard deviation	1 month forward rate, average return	1 month forward rate, standard deviation
	(1)	(2)	(3)	(4)
$/FF	2.72	11.28	−3.78	1.64
$/DM	2.72	11.57	−3.08	1.24
$/$	−2.02	12.76	−4.55	1.11
$/ITL	−2.58	12.88	−5.34	1.36
$/Yen	4.36	10.07	0.77	1.12
$/SKR	−3.20	13.48	4.78	10.70
$/Sfr	2.87	12.12	−1.26	1.32

Source: monthly closing rates between 12/1988 and 8/1994,
Bloomberg Inc. and personal calculations

last column of Table 2. A comparison with the figures in column 2 reveals that the exchange rate risk only marginally affects the investment risk of a foreign investor. The marginal impact of currency risk can be compared to the exchange rate volatility in Table 3 (column 2). While the \$/Swiss Franc volatility is more than 12%, the US-\$ investor faces a volatility of the Swiss stock market which is less than 1% higher than the volatility as perceived by Swiss Franc investors. This observation is known from many studies (e.g. Giovannini and Jorion (1989), Adler and Dumas (1983), Knight (1991)) and is due to the low or negative correlations between stock market and currency returns which produces a diversification effect. The correlation matrix for stock market and currency (spot and forward) returns is displayed in Table 4. Surprisingly, the correlation coefficients between markets and currencies are mostly negative (except for Japan). *Direct* market currency correlations (e.g. between the Swiss stock market and the \$/Swiss Franc exchange rate; printed in italics) are particularly negative. With respect to the correlation between spot and forward exchange rates observe that the coefficient for the Swedish Krone is surprisingly low (0.74) which can be only caused by an illiquid forward market for this currency.

4   Results

4.1   General Results

The effect of various currency hedging strategies on the shape of the efficient frontier is shown in Figure 1. The dotted lines represent *constant hedge* strategies. A fixed proportion (0%, 50% or 100%) of the respective (optimized) market exposures is hedged in the respective currency. For example, if the optimal fraction of wealth allocated to Germany is 36%, then 18% of the total portfolio value is hedged with rolling 1 month forward contracts under the 50% constant hedge strategy. The frontier with optimized currency holdings according to equation (2.10) is drawn by the solid line. The Figure immediately and clearly demonstrates that the optimized hedge produces substantially more efficient portfolios than any type of constant hedge. In the range of portfolios shown in the Figure, the standard deviation of the investments can be reduced by 1% to 4% for different levels of expected return. This shows that the covariance matrix of markets and currencies contains important information which must be exploited in international portfolio optimization. But the Figure contains a deeper (although well known) message: The diversification effects are particularly strong in an international setting. Independent of the exact currency hedge strategy, a diversified portfolio is much more efficient than any individual (foreign) market investment. So, the exact nature of currency hedging indeed has a non trivial impact on the

	CAC 40	DAX	FT-SE 100	BCI	NIKKEI 225	Affars-varlden	SMI	$/FF s	$/DM s	$/ s	$/ITL s	$/Yen s	$/SK s	$/SFr s	$/FF f	$/DM f	$/ f	$/ITL f	$/Yen f	$/SK f	$/SFr f
S&P100	0.58	0.44	0.27	0.38	0.47	0.59	0.59	-0.08	-0.05	0.05	-0.03	-0.13	-0.01	-0.14	-0.07	-0.07	0.04	-0.03	-0.11	0.10	-0.13
CAC40		0.74	0.70	0.53	0.40	0.58	0.64	-0.27	-0.25	-0.16	-0.28	-0.12	-0.22	-0.35	-0.27	-0.26	-0.17	-0.27	-0.11	-0.17	-0.34
DAX			0.58	0.62	0.32	0.59	0.59	-0.13	-0.12	-0.02	-0.05	-0.17	-0.14	-0.23	-0.11	-0.11	-0.02	-0.04	-0.14	-0.15	-0.23
FT-SE100				0.58	0.36	0.49	0.69	-0.26	-0.26	-0.17	-0.34	-0.10	-0.21	-0.31	-0.27	-0.27	-0.17	-0.33	-0.10	-0.14	-0.32
BCI					0.43	0.53	0.45	-0.29	-0.29	-0.21	-0.08	-0.10	-0.16	-0.33	-0.29	-0.29	-0.23	-0.08	-0.10	-0.08	-0.35
NIKKEI225						0.45	0.44	0.07	0.11	0.15	0.16	0.21	0.15	0.09	0.06	0.08	0.13	0.15	0.20	0.21	0.07
Affarsvarld.							0.55	-0.32	-0.30	-0.07	-0.18	-0.15	-0.35	-0.31	-0.30	-0.30	-0.09	-0.17	-0.14	-0.26	-0.33
SMI								-0.32	-0.29	-0.16	-0.33	-0.09	-0.16	-0.33	-0.32	-0.31	-0.16	-0.32	-0.09	0.00	-0.33
$/FF s									0.97	0.81	0.81	0.53	0.81	0.92	0.99	0.98	0.81	0.81	0.54	0.58	0.92
$/DM s										0.78	0.80	0.55	0.78	0.92	0.99	0.99	0.80	0.81	0.57	0.61	0.93
$/ s											0.83	0.53	0.74	0.78	0.80	0.78	0.99	0.82	0.53	0.55	0.77
$/ITL s												0.38	0.81	0.75	0.82	0.80	0.82	1.00	0.39	0.61	0.75
$/Yen s													0.41	0.56	0.52	0.54	0.53	0.38	0.99	0.38	0.56
$/SK s														0.79	0.78	0.77	0.74	0.81	0.41	0.74	0.79
$/SFr s															0.90	0.91	0.78	0.74	0.57	0.63	0.99
$/FF f																0.98	0.82	0.83	0.54	0.55	0.92
$/DM f																	0.80	0.81	0.56	0.56	0.93
$/ f																		0.82	0.55	0.54	0.78
$/ITL f																			0.40	0.60	0.76
$/Yen f																				0.37	0.58
$/SK f																					0.61

s denotes spot rates, f denotes forward rates

Table 4: Correlation coefficients

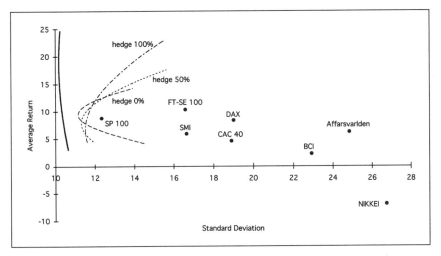

Figure 1: Efficient frontiers - constant and variable currency hedge ratios, optimization problem (2.6), US-$ investor

locus of the efficient frontier of international portfolios, but nevertheless, the far more substantial diversification effect comes from diversification across markets and currencies (and possibly, asset categories). So far, market and currency exposures did not face any restriction. This produces portfolios which are not feasible in reality. For many institutional investors, minimum diversification requirements apply; therefore, upper and lower boundaries for markets and asset classes have to be imposed. For example, in order to avoid *excessive* risk taking it is not allowed to invest more than 30% in a single market, or a minimum investment of 5% in each market should prevent corner solutions and guarantee a minimum diversification effect. The motivation of such restrictions is not analyzed here. It can only be noted that a possible rationale may arise from estimation risk which makes the covariance matrix not perfectly predictable over time and, hence, restricts the benefits from diversification.

The investment restrictions which are applied to our optimization model are displayed in Table 5. On the asset side, we impose lower boundaries of 5% for each foreign stock markets and a non negative holding of the domestic market. I.e. short sales are not allowed for any market. An upper boundary of 30% is selected for the foreign stock markets, while 65% can be allocated in the domestic market. With respect to the currency exposure, the lower boundary is 0%; a hedge ratio of 0% means an unhedged position. A negative hedge ratio (in our setting) implies a reverse hedge i.e. increasing currency exposure. The upper boundary is set at 100% for all currencies which makes

Table 5: Investment Restrictions, percentages (%), US-$ investor

	Assets		Currencies	Hedge Ratios	
Index	lower boundary	upper boundary	Currency	lower boundary	upper boundary
S&P 100	0	65			
CAC 40	5	30	$/FF	0	100
DAX	5	30	$/DM	0	100
FT-SE 100	5	30	$/$	0	100
BCI	5	30	$/ITL	0	100
NIKKEI	5	30	$/Yen	0	100
Affarsvarlden	5	30	$/SKR	0	100
SMI	5	30	$/Sfr	0	100

it impossible to sell a higher percentage of currency forward than is invested in the underlying market. For example, if the optimized portfolio share is 37% for a specific market, then it is impossible to sell more than 37% of the total portfolio value in the respective currency forward. These restrictions, of course, limit the risk return menu of investment opportunities and create inefficiencies. They will be quantified subsequently, and the impact of constant and optimized currency hedging strategies will be compared.

In Figure 2 four efficient frontiers are depicted. The solid lines represent portfolios with optimized currency hedges, while the thin lines are portfolios with a constant hedge ratio of 50%. The first and second frontiers represent unrestricted investments, while the third and fourth frontiers reflect the restrictions as imposed by Table 5. The restrictions substantially shift the frontiers to the right. For the optimized currency hedge portfolios, the volatility increase is at least 2%. The observations have a strong implication: The setting of a priori investment restrictions by law, regulation, charters, investment standards and the like, predetermines the feasible risk return menu in such an extent that the remaining flexibility (with respect to portfolio selection and implementation) is comparatively unimportant. This is supported by several empirical studies showing that the investment style (which basically reflects institutional restrictions) overwhelmingly determines overall portfolio performance (see Sharpe (1992)).

Figure 2 clearly reveals that imposing restrictions produces a very degenerated range of efficient portfolios, particularly under a constant currency hedge policy. For the restricted investment opportunity set, the return potential for taking additional risk is rather limited. In contrast, optimizing the currency hedge offers a much wider risk return menu on the efficient frontier. The mini-

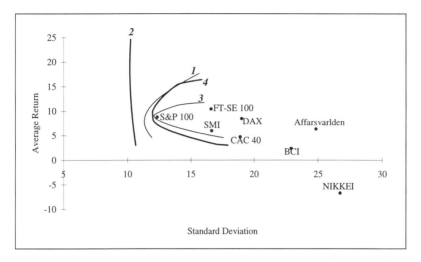

Figure 2: Efficient frontiers–constant hedge ratios 0.5 compared with variable hedge ratios, unrestricted and restricted investments, US-$ investor

mum variance portfolio exhibits a volatility of 12.09% and an expected return of 8.98%, and the risk return parameters of the maximum variance portfolios are 16.04% and 11.74%. Overall, optimizing the currency hedge adds about 5% expected return and widens the risk return menu. It can be concluded that imposing investment restrictions and implementing a constant currency hedge policy narrows the set of efficient portfolios in a rather unattractive way. Of course, the extent of these effects depends on the constant hedge that is selected.

Table 6 illustrates some of the previous observations by examining specific portfolios on the efficient frontiers. For the unrestricted portfolios, the global minimum variance portfolios are examined. Optimizing the currency hedge increases the return by more than 11%[3] compared to the constant hedge strategy, while reducing the standard deviation by approx. 1%. The structure of the portfolios is more interesting. The optimized portfolio contains substantial long positions in the Yen and the Swiss Franc; remember that these are percentages which are related to the market exposures. Since the French stock market share (w) is less than 1, the long currency position ($-h$) of 10298% represents an effective French Franc exposure (y) of roughly 85.47% of the total portfolio value (recall that, according to our definition in Section 2, a short position in a currency exhibits a positive sign). The long position in

---

[3]This figure should be treated with care, since the efficient frontier is almost vertical over a wide range of portfolios.

the Swiss Franc is not surprising since – given the figures in Table 3 – both, the Japanese Yen and the Swiss Franc have been strong currencies from a US-\$ perspective. With respect to the stock market holdings, the optimized as well as the constant hedge minimum variance portfolio exhibit a strong home bias; 82.6% and 83% of the portfolio is invested in the domestic stock market, respectively. Significant negative holdings are observed in the British and the Swedish stock market. Since the performance of the UK stock market was reasonably well over the sample period (see Table 2), this observation is partially due to the correlation structure of the markets: the FT-SE 100 exhibits the highest correlation with the S&P 100 index such that a short position is able to reduce portfolio risk.

The structure of the global minimum variance portfolio subject to the investment restrictions is displayed in the third column of Table 6. Given the observed home bias in the unrestricted portfolio optimization, the domestic equity share is still very high within the imposed boundaries. Besides this, a substantial position is taken in the Swedish stock market. The portfolio with the same average return but with optimized currency holdings is displayed in column 4. The volatility decreases only marginally, however (62 basis points). This is done by a currency hedge strategy where the full exposure in the British, Italian and Swedish stock market is hedged, while the currency exposures in the French Franc and the Swiss Franc remain unhedged. This example also shows that the interaction of covariance structures and restrictions may produce optimal strategies which cannot always be easily defended by pure intuition.

4.2   Benchmark currencies

In the following two sections, several extensions of the previous results are presented. First, the results of Figure 2 and Table 6 are displayed for a different benchmark currency, the Swiss Franc. The investment restrictions which are imposed are displayed in Table 7. The minimum variance frontiers are shown in Figure 3; and they look similar to those for the US-\$ benchmark in Figure 2. The efficient segment of the minimum variance frontier for the most restrictive case (investment restrictions and constant hedge ratio) is reduced to an even smaller *menu*, which leaves not much room for diversifying the portfolio.

The structure and hedging policy of the minimum variance portfolios (mvp) are shown in Table 8. Surprisingly, the market weights of the unrestricted portfolios are quite similar to those of the US-\$ investor. The investor takes substantial short positions in the British and the Swedish stock market, and large long positions in the US, the German and the Swiss stock market. The allocation to the domestic market (SMI) decreases if the optimized currency

Table 6: Efficient asset and currency allocation for specific portfolios – comparison between constant hedge ratios (50%) and optimized hedge ratios, US-$ investor

currency/ country index (*y* value)	no restrictions				restrictions			
	constant hedge ratio (50%)	optimized hedge ratio			constant hedge ratio (50%)	optimized hedge ratio		
	mvp	mvp			mvp	mvp		
	(1)	(2)			(3)	(4)		
S&P 100	82.6	83.0			55.2	59.5		
CAC 40	−2.8	0.8			5.0	5.0		
DAX	12.1	17.3			5.7	5.0		
FT-SE 100	−17.4	−16.2			5.0	5.0		
BCI	12.5	6.8			5.0	5.1		
NIKKEI	−0.4	−0.9			5.0	5.0		
Affarsvarlden	−9.4	−10.9			14.1	5.0		
SMI	23.0	20.1			5.0	10.4		
French Franc*	50	−10298	(−85.47)		50	0	(0)	
DM*	50	866	(150.1)		50	91	(4.56)	
British Pound*	50	−189	(30.5)		50	100	(5.1)	
Italian Lira*	50	−440	(−29.8)		50	100	(5.22)	
Yen*	50	2209	(−20.1)		50	40	(2.01)	
Swedish Krone*	50	−300	(32.8)		50	100	(5.1)	
Swiss Franc*	50	−386	(−77.5)		50	0	(0)	
average return	7.92	19.08			9.90	9.90		
standard dev.	11.32	10.19			12.58	11.96		

mvp: (global) minimum variance portfolio
* *y value* (portfolio exposure to the respective currency)
in percentages in brackets
Source: Personal calculations

hedge is implemented: optimizing the currency hedge decreases overall portfolio risk and thus allows to allocate a larger fraction of wealth to foreign assets. For the restricted case, the Swiss Franc investor is hedged against all the currency exposures, while the US-$ investor has full exposure to the Swiss Franc and the French Franc, i.e. the countries with strong currencies. Optimizing the currency hedge also leads to a improved diversification across the markets, i.e. parts of the Swiss and German market investments is reallocated to the UK and the Italian stock market. Overall, the gains from optimizing the currency hedge in terms of risk and return are slightly more pronounced for the Swiss Franc investor. In the *no restriction* case, the average return

Table 7: Investment Restrictions, percentages (%), Swiss Franc
investor

Assets			Currencies Hedge Ratios		
Index	lower bound	upper bound	Currency	lower bound	upper bound
S&P 100	5	30	US-$	0	100
CAC 40	5	30	French Franc	0	100
DAX	5	30	DM	0	100
FT-SE 100	5	30	British Pound	0	100
BCI	5	30	Italian Lira	0	100
NIKKEI	5	30	Yen	0	100
Affarsvarlden	5	30	Swedish Kr.	0	100
SMI	0	65			

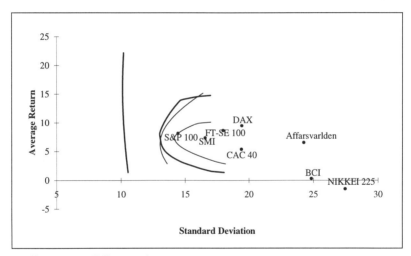

Figure 3: Efficient frontiers – constant hedge ratios 0.5 compared with variable hedge ratios, unrestricted and restricted investments, Swiss Franc investor

more than doubles while the standard deviation decreases by 3%, and in the
restricted case the average return increases by approximately 0.5% while the
standard deviation can be reduced from 14.1% to 13%.

Table 8: Efficient asset and currency allocation for minimum variance portfolios – comparison between constant hedge ratios (50%) and optimized hedge ratios, Swiss Franc investor

currency/ country index (y value)	no restrictions			restrictions		
	constant hedge ratio (50%)	optimized hedge ratio		constant hedge ratio (50%)	optimized hedge ratio	
	mvp	*mvp*		*mvp*	*mvp*	
	(1)	(2)		(3)	(4)	
S&P 100	76.2	84.8		30	30	
CAC 40	−13.1	−0.7		5.0	5.0	
DAX	26.6	17.9		14.0	12.2	
FT-SE 100	−14.2	−17.1		5.0	13.2	
BCI	4.1	6.8		5.0	7.2	
NIKKEI	8.4	−1.3		5.0	5.0	
Affarsvarlden	−13.5	−10.0		5.0	5.0	
SMI	25.5	19.5		31.0	22.5	
US-$*	50	120	(102)	50	100	(30.0)
French Franc*	50	113	(-0.79)	50	100	(5.1)
DM*	50	777	(139)	50	100	(12.3)
British Pound*	50	−164	(28)	50	100	(13.3)
Italian Lira*	50	−361	(−25)	50	100	(7.34)
Yen*	50	1586	(-21)	50	100	(5.1)
Swedish Krone*	50	−281	(28)	50	100	(5.1)
average return	6.33	15.8		7.51	7.91	
standard dev.	13.04	10.1		14.13	13.0	

mvp: (global) minimum variance portfolio
*: *y value* (portfolio exposure to the respective currency) in percentages in brackets
Source: Personal calculations

4.3   Stability and out of sample characteristics

The next and practically important question addressed in this section is the stability of efficient portfolios. It makes little sense to optimize market shares and currency hedges, if the covariance matrix and expected (or average) returns are so unstable over time that an ex ante efficient portfolio proves to be inefficient ex post. In order to investigate this problem, the structure of

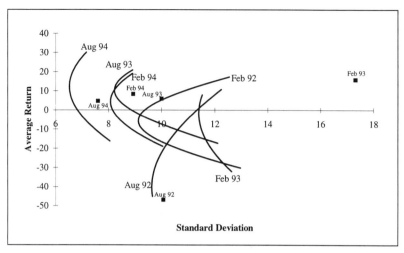

Figure 4a: The changing shape of efficient frontiers – US-$ bench-
mark, variable hedge ratios, no investment restrictions.

minimum variance portfolios[4] is investigated for six subsequent, three year
time periods.

The results are graphically shown in Figures 4a,b and 5a,b. They must be
interpreted as follows: At the end of a specific month (e.g. February 1994),
the minimum variance portfolio is calculated based on the data from the past
three years (i.e. from March 1991 to February 1994). The volatility and return
of this portfolio as well as the minimum variance frontier are recalculated six
months later (i.e. August 1994) using the average return vector and the
covariance matrix based on the most recent three year time interval (i.e.
September 1991 to August 1994). In order to compare the efficiency of the
old minimum variance portfolio with respect to the *new* minimum variance
frontier, both are indicated in the Figures by *Aug 94*[5]. This procedure allows
us to determine whether the asset and currency allocation vectors of the
minimum variance portfolio, determined for a specific sample period, still
represent a minimum variance portfolio after the input data have changed
subsequently.

The Figures reveal that the estimation error has quite a substantial impact
on the stability of the minimum variance portfolio if no investment restrictions
are considered (see Figures 4a/b). However, the position of the minimum

---

[4]Investigating minimum variance portfolios was proposed and justified by Eun and
Resnick (1988) and Jorion (1985). They show that minimum variance portfolios are more
stable over time than certainty equivalent tangency portfolios.

[5]No lagged minimum variance portfolio can be compared to the February 1992 frontier
because the sample starts in March 1989.

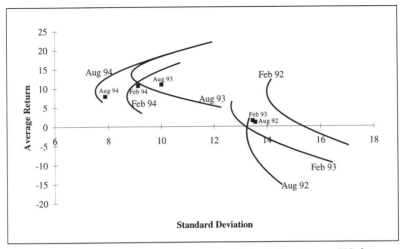

Figure 4b: The changing shape of efficient frontiers – US-\$ investor, constant hedge ratios of 50%, no investment restrictions.

variance portfolio seems to be stable if investment restrictions are imposed (Figures 5a/b). Table 9 provides an explanation for the outlier in Figure 4a showing that the fractions of wealth allocated to the DAX, BCI, SMI and S&P 100 indexes have drastically changed in the six months from August 1992 to February 1993[6]. Therefore, the *Aug 1992* minimum variance portfolio turns out to be very inefficient from the viewpoint of *Feb 1993*.

A general characterization of the portfolio weights and the hedge ratios in the minimum variance portfolios can be found in Table 10. The arithmetic averages of w, h and y and their standard deviation across the six sub intervals are displayed. The results confirm the instability of the minimum variance portfolios for the unrestricted portfolios. There is no apparent difference between the stability of the asset allocation parameters (w) for the constant and the variable optimized hedge strategies. As a practical implication, optimizing the currency exposure of a portfolio does not increase the instability of the asset allocation parameters. For the restricted strategies, the volatility of the hedge ratios seems the to be high. The volatility should be related to the average size of the hedge ratios, which is comparatively lower than in the unrestricted case. In order to quantify the *ex post* inefficiency of the minimum variance portfolio due to estimation risk, the following statistic is

---

[6]The *Feb 93* deviation was investigated in detail. However, no data error was detected.

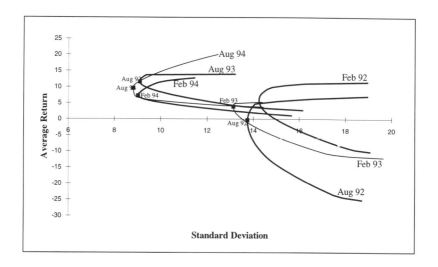

Figure 5a: The changing shape of efficient frontiers – US-$ benchmark, variable hedge ratios, investment restrictions as of Table 5.

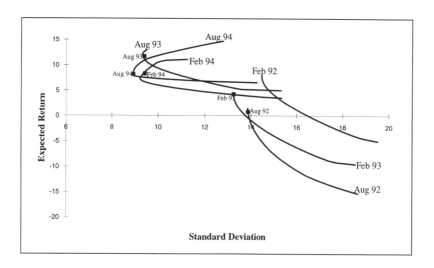

Figure 5b: The changing shape of efficient frontiers – US-$ investor, constant hedge ratios of 50%, investment restrictions as of Table 5.

Table 9: The structure of the minimum variance portfolio in two subsequent periods (variable hedge, no restrictions)

	Aug 92	Feb 93
CAC 40	4.01	9.79
DAX	9.61	20.14
FT-SE 100	−34.3	2.74
BCI	3.4	17.74
NIKKEI 225	9.93	−3.83
Affarsvarlden	−16.42	−4.23
SMI	−6.94	−31.24
S&P 100	130.72	88.89
French Franc	−13597.5	−267.6
DM	2400.8	580.2
British Pound	112.4	961.3
Italian Lira	6916.8	133.1
Yen	−196.2	31.07
Swedish Krone	−1780.8	103.5
Swiss Franc	1979.0	420.6

Source: Personal calculations

calculated[7]:

$$inefficiency =$$
$$((volatility\ out\ of\ sample - volatility)^2 + (ave.\ out\ of\ sample - ave.)^2)^{\frac{1}{2}} \quad (4.1)$$

The statistic takes a value of zero if the structure of the minimum variance portfolio remains constant. The statistic does not discriminate whether the portfolio return increases or decreases compared to the original level, i.e. whether the *move* is in favor or against the investor. This is based on the assumption that any displacement of the location of the minimum variance portfolio reflects a potential estimation risk for the investor. The results in Table 11 show that the average loss of efficiency is substantially higher if no restrictions are imposed in the optimization, and that the loss is substantially smaller for the constant hedge strategy. This implicitly demonstrates that the instability of the parameters exhibits a stronger effect for variable hedging strategies. In Table 12, inefficiency is measured by the volatility difference between out of sample minimum variance portfolios and in sample efficient portfolios with the same mean. In contradiction to the previous findings the results reveal that, in most cases, the inefficiency is considerably low. It

---

[7]Kandel and Stambaugh (1995) propose a different measure for portfolio inefficiency.

Table 10: Structure of the minimum variance portfolios across subperiods – average and standard deviation of market weights and hedge ratios

| | No restrictions | | | | Restrictions | | | |
| | constant hedge | | variable hedge | | constant hedge | | variable hedge | |
	avg	std.	avg	std.	avg	std.	avg	std.
CAC 40	−2.19	9.55	−2.59	8.31	5.00	0.00	5.00	0.00
DAX	13.70	14.40	24.69	12.83	7.49	3.39	8.02	3.07
FT-SE 100	−5.43	7.43	−15.30	12.90	5.00	0.00	5.00	0.00
BCI	13.65	8.46	2.55	9.47	5.44	0.75	5.74	1.12
NIKKEI 225	−0.10	4.73	6.04	4.67	5.00	0.00	5.00	0.00
Affarsvarlden	−2.79	4.90	−9.48	7.48	5.00	0.00	5.00	0.00
SMI	6.36	16.53	1.67	19.31	7.11	3.47	7.10	3.11
S&P 100	6.79	10.61	92.42	25.86	59.96	4.49	59.14	5.24
French Franc	50.00	0.00	−11829	20843	50.00	0.00	32.65	46.24
DM	50.00	0.00	1193.88	1301.36	50.00	0.00	90.39	25.04
British Pound	50.00	0.00	45.99	472.51	50.00	0.00	102.00	0.00
Italian Lira	50.00	0.00	1199.75	2716.07	50.00	0.00	101.26	1.13
Yen	50.00	0.00	78.58	220.16	50.00	0.00	98.25	6.13
Swedish Krone	50.00	0.00	−491.94	836.48	50.00	0.00	82.57	37.31
Swiss Franc	50.00	0.00	271.21	926.77	50.00	0.00	0.00	0.00
y French Franc			−203.86	275.07			2.31	2.42
y DM			148.99	89.54			2.76	5.65
y British Pound			5.64	24.08			0.00	4.08
y Italian Lira			82.88	108.09			1.14	5.09
y Yen			4.52	19.04			0.31	4.14
y Swedish Krone			97.82	137.15			1.87	3.29
y Swiss Franc			−101.30	36.77			0.00	0.00

All figures are percentages.

exceeds 1.5% in only two cases. Again, if investment restrictions are imposed, the bias is essentially non existent. This implies that the minimum variance portfolio remains fairly efficient though time.

5 Conclusions

The paper examines the impact of various currency hedging strategies on the efficiency of internationally diversified portfolios with respect to risk and return. Data from eight international stock markets are analyzed from the

Table 11: In and out of sample characteristics of the minimum variance portfolio

		Feb 92	Aug 92	Feb 93	Aug 93	Feb 94	Aug 94	avg.	std.
	average	−5.85	−40.85	−1.30	10.13	4.45	12.24	−3.53	17.81
No	std.	9.13	9.64	11.40	8.22	8.07	6.52	8.83	1.51
Restrictions	avg. out of	−46.79	15.97	5.98	8.48	4.86	−2.30	22.58	
variable	sample								
hedge	std. out of	10.06	17.32	9.99	8.93	7.59	10.78	3.39	
	sample								
	Inefficiency	5.95	18.25	4.51	4.13	7.46	8.21	5.23	
No	average	9.10	−1.45	5.24	13.47	7.88	9.33	7.26	4.59
Restrictions	std.	14.00	13.24	12.65	8.88	8.71	7.51	10.83	2.53
constant	avg. out of	1.13	1.56	11.06	10.71	7.98	6.49	4.33	
hedge	sample								
(50%)	std. out of	13.56	13.46	10.00	9.12	7.87	10.80	2.31	
	sample								
	Inefficiency	2.60	3.77	2.66	2.85	1.39	2.66	0.76	
	average	5.95	−0.29	3.95	11.63	7.32	9.61	6.36	3.86
Restrictions	std.	14.25	13.74	13.12	9.09	9.00	8.80	11.33	2.39
variable	avg. out of	−2.60	4.08	11.42	8.09	8.87	5.97	4.89	
hedge	sample								
	std. out of	13.78	13.15	9.26	9.17	8.87	10.85	2.15	
	sample								
	Inefficiency	2.30	0.14	0.28	0.79	0.74	0.85	0.77	
	average	7.79	1.42	4.31	12.10	7.39	8.43	6.91	3.35
Restrictions	std.	14.47	13.85	13.23	9.32	9.20	8.91	11.50	2.39
constant	avg. out of	0.81	4.31	11.76	8.32	8.19	6.68	3.76	
hedge	sample								
(50%)	std. out of	13.86	13.23	9.38	9.40	8.91	10.96	2.13	
	sample								
	Inefficiency	0.60	0.00	0.35	0.95	0.24	0.43	0.33	

The months denoted in the first line refer to the final month within a 3–year time interval. All characteristics of the minimum variance portfolios are annualized values.

perspective of a US-\$ based investor. A mean variance model is developed which optimizes stock market and currency exposures with respect to any kind of investment restrictions. The major results of the paper are: Imposing investment restrictions *and* constant currency hedging rules (e.g. 50% of all currency holdings are sold forward) narrows the range of efficient portfolios, so there is not much room left for active portfolio selection. Since institutional or legal investment restrictions with respect to markets or currencies are common, the observation is important that optimizing currency holdings substantially widens the range of efficient portfolios within a given set of restrictions. Therefore, currency optimization within the asset allocation framework is particularly relevant if investors face investment restrictions such as minimum diversification rules etc. A topic which is also addressed in this paper is estimation risk. Optimizing currency hedge ratios implies that the covariance structure of market and currency returns must be predictable. It is possible that part of the advantages of currency optimization presented in this paper lessen if estimation risk (forecasting errors) is explicitly taken into account. Some preliminary results on the (in)stability of the risk return

Table 12: Out of sample minimum variance portfolios – comparison of the standard deviations to in sample efficient portfolios with equal mean

			Aug92	Feb93	Aug93	Feb94	Aug94
no restrictions	variable hedge	average return	−46.8	15.97	5.98	8.48	4.86
		out of sample mvp volatility	10.06	17.32	9.99	8.93	7.59
		in sample efficient portfolio volatility	9.70	12.50	8.35	8.15	6.63
	constant hedge (50%)	average return	1.13	1.56	11.06	10.71	7.98
		out of sample mvp volatility	13.56	13.46	10.00	9.12	7.87
		in sample efficient portfolio volatility	13.25	12.95	9.26	8.94	7.60
restrictions	variable hedge	average return	−2.60	4.08	11.42	8.09	8.87
		out of sample mvp volatility	13.78	13.15	9.26	9.17	8.87
		in sample efficient portfolio volatility	13.78	13.13	9.10	9.08	8.80
	constant hedge (50%)	average return	0.81	4.31	11.76	8.32	8.19
		out of sample mvpvolatility	13.86	13.23	9.38	9.40	8.91
		in sample efficient portfolio volatility	13.86	13.23	9.33	9.35	8.91

The months denoted in the first line refer to the final month within a 3-year time interval. All characteristics of the minimum variance portfolios are annualized values.

characteristics of minimum variance portfolios supporting this hypothesis are presented in this paper. However, whether simple rules in terms of constant hedge policies become more relevant in such a setting, is a topic which needs further empirical investigation.

Acknowledgements

Financial support from the Grundlagenforschungsfond at the University of St. Gallen is gratefully acknowledged. We wish to thank Haim Levy, Harry Markowitz, Nils Tuchschmid, and Rudi Zagst for detailed and valuable comments. We also wish to thank the participants at the fifth workshop of the Studiengruppe für Finanzmarktforschung in Bern, the finance meetings at the Isaac Newton Institute in Cambridge, particularly William Ziemba, and at the INFORMS meeting in Singapore for stimulating discussions. All remaining errors are ours.

References

M. Adler and B. Dumas (1983). 'International Portfolio Choice and Corporate Finance: A Synthesis,' *Journal of Finance* **38** June, 925–984.

M. Adler and B. Prasad (1992). 'On universal currency hedges,' *Journal of Financial and Quantitative Analysis* **27** 19–38.

F. Black (1990). 'Equilibrium exchange rate hedging,' *Journal of Finance* **45** 899–907.

C. Eun and B. Resnick (1988). 'Exchange rate uncertainty, forward contracts, and international portfolio selection,' *Journal of Finance* **43** 197–215.

A. Giovannini and P. Jorion (1989). 'The Time Variation of Risk and Return in the Foreign Exchange and Stock Market,' *Journal of Finance* **44** June, 307–325.

J. Glen and Ph. Jorion (1993). 'Currency hedging for international portfolios,' *Journal of Finance* **48** 1865–1886.

H. Grubel (1968). 'Internationally diversified portfolios. Welfare gains and capital flows,' *American Economic Review* **58** 1299–1314.

Ph. Jorion (1985). 'International Portfolio Diversification with Estimation Risk,' *Journal of Business* **58** 259–278.

S. Kandel and R.F. Stambaugh (1995). 'Portfolio Inefficiency and the Cross Section of Expected Returns,' *Journal of Finance* **50** 157–224.

R.F. Knight (1991). 'Optimal Currency Hedging and International Asset Allocation: An Integration,' *Financial Markets and Portfolio Management* **5** (2) 130–163.

A.F. Lee (1987). 'International Asset and Currency Allocation,' *Journal of Portfolio Management*, Fall, 68–73.

H. Levy and K.C. Lim (1994). 'Forward Exchange Bias, Hedging and the Gains from International Diversification of Investment Portfolios,' *Journal of International Money and Finance* **59** 159–170.

H. Levy and M. Sarnat (1970). 'International diversification of investment portfolios,' *American Economic Review* **60** 668–675.

H.M. Markowitz, (1952). 'Portfolio Selection,' *The Journal of Finance* **7** (1) 77–91.

H.M. Markowitz, (1956). 'The Optimization of a Quadratic Function Subject to Linear Constraints,' *Naval Research Logistics Quarterly* **3** 111-133.

H.M. Markowitz, (1959). *Portfolio Selection: Efficient Diversification of Investment*, Wiley. Reprinted 1991 by Basil Blackwell.

H.M. Markowitz, (1987). *Mean Variance Analysis in Portfolio Choice and Capital Markets*, Basil Blackwell.

A. F. Perold, and E.C. Schulman (1988). 'The free lunch in currency hedging: Implications for investment policy and performance standards,' *Financial Analysts Journal* May–June, 45–50.

M. Rudolf, (1994). *Algorithms for Portfolio Optimization and Portfolio Insurance*, Haupt-Verlag.

W.F. Sharpe, (1992). 'Asset Allocation: Management Style and Performance Measurement,' *The Journal of Portfolio Management* Winter, 7–19.

B. Solnik, (1973),'Why not diversify internationally rather than domestically?,' *Financial Analysts Journal* **30** July-August, 48–54.

B. Solnik, (1974),'An equilibrium model for the international capital market,' *Journal of Economic Theory* **8** 500–524.

Optimal Insurance Asset Allocation in a Multi-Currency Environment

John C. Sweeney, Stephen M. Sonlin, Salvatore Correnti and Amy P. Williams

Abstract

The globalization of the capital markets and the discovery of positive diversification benefits from cross-currency investments have changed the nature of investment management forever. This paper outlines how investment professionals can incorporate multi-currency analysis into their portfolio selection process to capture the available benefits. A simulation/optimization scenario approach to asset-liability management is introduced, which is specific to the insurance industry. Its goal is to maximize an insurer's surplus or equity while minimizing the volatility of the capital markets on both invested assets and liability cash flows. To accomplish this goal, the potential impact of uncertain future macroeconomic environments on an insurance company's assets and liabilities must be evaluated. An integrated, multi-currency stochastic optimization model is described that can be a useful tool for insurers and other financial institutions. The optimization model is an integral element of a comprehensive ALM process that utilizes a dynamically balanced approach to optimal asset allocation. Because many insurers have international operations or investments, three approaches to multi-currency optimization, along with the pros and cons of each alternative, are discussed in detail. A case study of an international insurer illustrates this approach to multi-currency asset-liability modeling and management. The case study demonstrates the use of economic simulation and different global optimization models to identify the optimal asset allocations based on a surplus-driven efficient frontier analyses.

1 Introduction

Over the past 25 years, the focus of the insurance industry has slowly gravitated toward the view that risk management and asset-liability management ('ALM') in particular are central to the health and well being of the industry. ALM is now recognized as playing a prominent role in determining a company's financial flexibility, its liquidity, asset quality, competitive position and ultimately its profitability.

341

Falcon Asset Management, recognizing the importance of integrated risk management, has incorporated ALM into its everyday investment management activities. Falcon Asset Management, a wholly-owned subsidiary of USF&G Corporation, is an international portfolio manager that specializes in ALM based insurance asset management. The firm currently manages over $50 billion of insurance assets, coupling state-of-the-art asset-liability management capabilities with expert fixed income portfolio management skills. Economic simulations and efficient frontier analysis are the core of Falcon's investment management process, enabling its insurance clients to evaluate risk/reward trade-offs under uncertainty and to dynamically select and manage optimal investment strategies. Due to the global nature of insurance, Falcon incorporates multi-currency analysis of both assets and liabilities in its optimization process. This chapter outlines Falcon's approach to multi-currency asset-liability management and presents practical examples within an insurance context.

2   Overview of Asset-Liability Management for Insurance Companies

As outlined in prior chapters, the goal of asset-liability management is to evaluate risk and develop optimal investment strategies that maximize reward at an acceptable level of risk. In general, risk and reward will be dependent on the specific objectives of a company as well as considering the joint affect of both the assets and liabilities on the company's financial results. For insurance companies and other financial institutions, a common objective is to increase economic surplus, defined as the market value of the excess of assets over liabilities. Within an insurance context, surplus is important because it not only represents the level of shareholder wealth, but also determines the level of risk a company can assume. For example, in the property/casualty business, the permissible level of premiums written is generally stated as a multiple of surplus (i.e., 2:1). Consequently, an insurer's marketing power and financial strength are truly a function of its economic surplus. Risk is often defined as the variability of surplus return and can be measured by the standard deviation of surplus or various downside risk measures.

The ALM process must allow management to develop strategies to meet their risk/reward objectives under a wide range of uncertain economic conditions. This is particularly important for the insurance industry where assets and liabilities are inextricably linked to economic conditions.

Within the life insurance industry, interest sensitive products such as single premium deferred annuities (SPDA's), which are basically yield driven investment products, create a critical link between the insurance company cash flows and the economic environment. The economic environment has a direct impact on the capital markets and thus, on SPDA sales, which will influence the asset cash flows of a company. The existence of such ties between the assets and liabilities of a company are the very thing that asset-liability management will attempt to exploit, identifying asset allocation strategies that will form natural hedges against the liabilities, thus, reducing the overall risk of the company.

Property/casualty (P&C) insurers are also faced with liability cash flows that are tied to the capital markets through the economic environment, most importantly interest rates and inflation. In addition, the property/casualty industry is faced with large uncertainty surrounding actual loss payout experience. Although the typical payout patterns for most accident and liability lines are short, business plans and high levels of renewal business make it imperative to consider strategic decisions on a going concern basis. Asset–liability management gives property/casualty insurers a means to determine appropriate liquidity and working capital over a three– to five–year planning period and to find strategic asset allocation policies that will optimize a company's business objectives.

Investment management within the insurance industry becomes even more complex once the unique characteristics of insurance company investing are factored into the ALM process. Accounting, regulatory, tax and operating considerations all impact the investment process and prohibit insurance companies from benchmarking themselves on a simple total return basis. For example, in the USA. Financial Accounting Standard 115 (FAS 115) requires all securities purchased by an insurance company to be designated as 'held to maturity,' 'available for sale' or 'actively traded.' Only those assets classified as actively traded lend themselves to total return management. Unfortunately, by classifying assets as actively traded, all capital gains/losses flow through to a company's earnings and, thus, to its income statement. Since operating constraints are sensitive to earnings and earnings volatility, few, if any, assets are categorized into this basket. Other regulatory constraints such as risk based capital (RBC), statutory solvency requirements and model investment laws further encumber insurance companies from following a pure economic set of objectives. Additionally, the insurance industry is regulated at the state government level. Although many regulations are consistent from state to state, there are numerous state regulations that are different for all 50 states, e.g., many states prohibit international investments, while others prohibit them for Life companies only and allow international securities on a limited basis ($< 10\%$) in P&C companies.

Insurance companies are also subject to operating constraints, such as earnings growth/stability (discussed above), investment income requirements and cash flow needs. The desire for stable earnings and high investment income hinders investment managers from selling higher coupon securities in falling interest rate environments or from realizing capital losses on investments in rising interest rate environments. On the other hand, insurance company cash flow requirements to pay losses or benefits, often force asset managers to sell securities when they might otherwise hold them. Operating constraints such as these make it difficult to match or exceed an unconstrained investment total return benchmark.

Finally, the tax status of a company will have a great impact on the amount of flexibility an asset manager has to buy and sell securities. Since unrealized gains are deferred for tax purposes while realized gains are taxed immediately, a taxable insurer will avoid realizing unnecessary capital gains. To the extent that capital gains have been realized, there may be pressure placed on the asset manager to offset the realized capital gains with realized capital losses. Both the avoidance of realizing capital gains and the intentional realizing of capital losses take much of the discretion attached to the buy/sell decision away from the asset manager. Insurance company unique needs and circumstances, both those mentioned above and others, must be appropriately considered when coming up with strategic investment policy and a performance measurement benchmark.

More and more, insurers have been exploring opportunities to improve both their asset and liability portfolios by looking to the international markets. This introduces yet another level of complexity to the asset-liability management process as the analysis must incorporate the impact of currency fluctuations on the risk/reward objectives of the company. The benefits of international diversification must be weighed against the increased uncertainty implicit in the international markets, and currency hedging strategies will often have to be incorporated into the process. Exposure to international markets requires assumptions concerning appropriate relationships across countries with respect to loss experience, economics and capital markets.

Because the ALM process evaluates all of these various asset, liability, and capital market risks under uncertainty, it must be long term in its vision. As such, it must consider the longer term implications of today's investment decisions in a dynamic and global economic environment. A systematic process that incorporates multiple currencies is therefore needed in order to construct a proper investment strategy for an insurance company with international exposure.

Multi-Currency Asset-liability Model

Falcon's approach toward managing risk for financial institutions deals with these complex issues by weighing the dynamics and interactions of asset, liability and capital market ('A-L-C') factors in both domestic and foreign

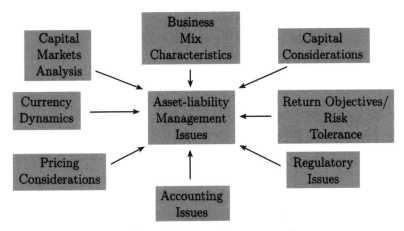

Figure 1: Asset-Liability Management Issues

markets. Assessing the characteristics, relationships, and variability of these A-L-C factors is critical to implementing a global investment strategy. As illustrated in Figure 1, an effective asset-liability management model should incorporate all of these factors.

For example, suppose a company wants to evaluate altering its business mix. First, the company will have to evaluate the impact on its capital in terms of the change in required capital and resulting returns on required capital. In order to assess the capital impact, the company will need to evaluate regulatory and accounting impacts of changing the business mix. These impacts are evaluated with respect to the company's financial objectives and risk tolerance. The cash flow changes arising from the shift in business mix are combined with the simulation of currency and capital markets analysis. Only after assessing the chain reaction of events can a company determine the implication of this single decision to alter its business mix on its overall investment strategy and asset allocation.

The dynamic business environment described above is only one of many complicated situations that can be handled by effective ALM modeling. Another example can be found in the constantly changing world of multi-currency investments. Many securities, such as derivatives, possess contingent cash flows or have embedded options (e.g., callable bonds). Because the value and cash flows of these type of securities change with economic conditions, their behavior must be carefully modeled and monitored in order to implement portfolio strategies. Moreover, the risks inherent in these complex securities should be evaluated at the corporate level to gather a true picture of the total integrated risk of the company – that is, asset decisions blended with liabilities and corporate long range goals. This is the essence of strategic asset-liability management and the critical role it plays in a company's overall financial

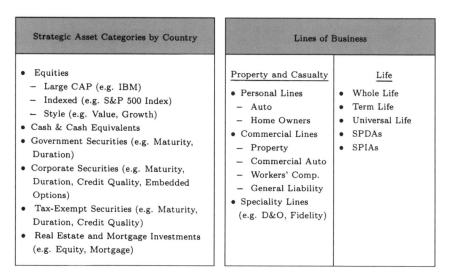

Figure 2: Strategic Asset Categories and Lines-of-Business

management.

The examples above underscore the strategic nature of asset-liability management. The microeconomic world of insurance management in general is becoming more complex and competitive. In order to survive, insurance management must rely on increasingly more sophisticated instruments of analysis to gain market share and determine the direction for their companies. Strategic ALM looks at the asset-liability problem from a macro level over an appropriately long time horizon. As a result, securities with similar characteristics can be grouped together into broad asset categories and individual insurance policies can be aggregated into broader lines of business. In contrast, tactical ALM, is more of a micro analysis, dealing with day-to-day management issues involving individual investment securities and insurance policies. The implementation of the strategic analysis is usually handled within the realm of tactical ALM and can involve the use of various derivative strategies. At Falcon, we perform both strategic and tactical ALM, although our tactical strategies are heavily constrained by our strategic analysis. Figure 2 shows the types of broad asset categories and lines of business that would be used as part of a strategic analysis for an insurance company.

The strategic asset categories are shown in the above left-hand box. Equities are segmented by type, large capitalization, indexed, etc., as is real estate, equity real estate, mortgage loans, etc. Bonds are not only split into types, such as government, corporate, etc., but also by maturity profile and credit quality. Any asset category that can be built based on the core building

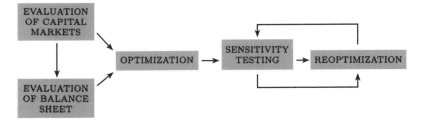

Figure 3: Asset-Liability Management Model

blocks of stock, bonds, and cash can be modeled so as to provide the desired investment profile of the client.

The above box on the right illustrates some of the numerous and varied insurance lines of business. The property and casualty lines cover the standard personal and commercial categories: automobile, homeowner and property insurance, general liability, workers' compensation, and specialty lines, such as director and officers liability (D&O), fidelity bonds and catastrophic risk, such as earthquake and wind coverage. The life business is segmented into traditional core life insurance and term insurance products and annuity products: SPIAs or single premium immediate annuities which begin paying out immediately for a certain or uncertain period. SPDAs, single premium deferred annuities, where the payout is at a later date. Each line of business is modeled based on its unique characteristics.

At the heart of Falcon's approach to insurance asset management is the five-step ALM process (see Correnti and Sweeney, 1994) depicted in Figure 3:

First, an evaluation of the capital market factors is performed employing a stochastic simulation model of the economy. This involves generating numerous simulations of key economic and capital market variables, such as dividend yield, dividend growth rate, inflation and interest rates. Core asset class results, in terms of income return and total return, are derived based on the economic and capital market variables. The result is a robust decision universe to evaluate asset alternatives.

Second, an economic evaluation of the insurance company's balance sheet is done with due consideration given to the ongoing nature of the business. Such a process assesses the critical elements in the balance sheet and business plan. Asset valuation involves the grouping of individual investments into proxy asset classes based on the asset categories simulated in Step One. Liability evaluation is split between known existing liabilities and forecasted new liabilities. The key factors assessed for existing liabilities are the expected magnitude, variability and timing of the liability outflows. Each line of business should be assessed in terms of the expected loss distribution profile, such as normal or lognormal distributions, degree of skewness, etc. For new business, which generates new asset and liability flows, each line of business needs to be evaluated considering expected losses, loss variability, impact of catastrophes, and degree of correlation between lines. These are but some of the factors involved in creating a robust simulation of asset and liability cash flows.

Third, a multi-period non-linear optimization model is utilized to perform surplus optimization in an Efficient Frontier context that explicitly considers the liability cash flows and their unique characteristics. The optimization model dynamically links assets and liability flows to the changing capital markets. This brings together the asset and liability cash flow simulation and the economic and capital market simulation and optimizes the surplus or economic value of the firm over a target time horizon subject to company specific constraints.

Fourth, sensitivity testing of key asset, liability and capital market variables is performed to scrutinize key assumptions and assess the impact of a change in the assumptions on economic surplus and the desired asset allocations. Essentially, this step helps management identify and evaluate the key levers for business and investment strategy decision making.

Fifth, depending on the outcome of the sensitivity tests, step three is performed again with the new assumptions or changed variables. This reoptimization is performed to determine the impact of the new assumptions on investment strategy.

Falcon has found that this process performs well as long as the investing institution limits itself to domestic markets. Many companies, however, employ a global investment philosophy as a result of multi-national operations or simply to diversify their portfolios. Under such circumstances, currency must be taken into account in structuring a global asset-liability model to assess risk and test investment strategies.

Falcon has, therefore, modified the traditional ALM model to explicitly consider currency management issues when evaluating capital market factors and the firm's balance sheet. A multi-currency ALM process would therefore expand the previous five-step process to include 'Currency Factors' as depicted in Figure 4.

In effect, currency factors are fully modeled in the evaluation of the capital markets, step 1 and then carried through to the evaluation of the balance sheet, step 2. This multi-currency ALM approach has been used extensively by Falcon Asset Management to help its international clients assess risk and make investment decisions in a global marketplace.

Modeling currency returns between countries can be a very difficult and complicated task. The multi-currency ALM model must insure arbitrage-free conditions between the spot exchange rates and the forward rates, given the

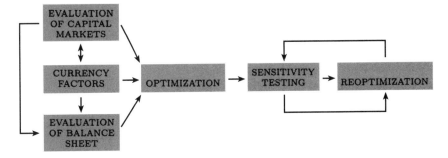

Figure 4: Multi-Currency Asset-Liability Management Model

differential interest rates that prevail. (Additionally, the model must produce scenario results that are symmetrical. That is to say, the distribution of currency returns is the same between country one and country two as are return distributions from country two to country one.) Other important issues related to historical trends, such as purchasing power parity and the level of real interest rates must also be considered in the scenario production process. All of these concepts are not easily modeled but must be evaluated within the structure of the ALM process.

A brief discussion of the various theories behind the functioning of the currency markets, including Purchasing Power Parity (PPP), Interest Rate Parity, the International Fisher Effect and Foreign Exchange Expectations is contained in Appendix A.

Figure 5 illustrates an economic surplus efficient frontier. The efficient frontier shows the possible risk and reward relationship for several points along the curve. In effect, the current portfolio D is well below the efficient frontier and, therefore, suboptimal. By moving to Portfolio A, the insurance company would be selecting a strategy of risk reduction, that is lower volatility of surplus over the higher risk but high returning portfolio B. Obviously, any point on the efficient frontier curve is efficient. It therefore reflects numerous possible asset combinations that increase surplus, reduce risk or, if desired by management, increase risk and increase surplus as in portfolio C.

Figure 6 illustrates the usefulness of the multiple scenario approach to optimization analysis, in that it depicts the probability distribution associated with each portfolio decision. It shows four possible portfolios along with their respective asset mixes and the risk/reward trade-off on a final surplus basis. The results or final surplus for each portfolio are displayed over a probability distribution ranging from 5th to the 95th percentile. Consequently, the size of the distribution depicted as a bar on the chart is a good representation of the level of risk associated with each portfolio decision.

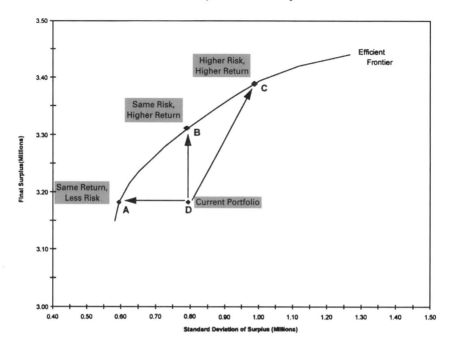

Figure 5: Efficient Frontier – Surplus and Volatility Results

Falcon has favored the use of multi-stage stochastic programs with 'decision rules' for our optimization work, as opposed to traditional stochastic program or stochastic optimal control process. A simple decision rule that we have used for many of our clients is the dynamically balanced strategy. The dynamically balanced strategy rebalances the asset allocation strategy back to a specific allocation at the end of each period over the time horizon being used in the analysis. The dynamically balanced strategy for optimal asset allocation has three distinct advantages over traditional stochastic programming: (1) it is much easier to implement; (2) the quality of its solutions is very good; and, (3) it is much more intuitive. This allows us to both generate the necessary output and communicate the results of our analysis to our clients much more efficiently and effectively. Additionally, our clients also seem to favor the straight forward results that come out of using a decision rule, such as dynamically balanced asset allocations, for determining their strategic investment policy.

An additional feature of our decision rules approach is that it limits the amount of trading turnover and associated commission expense compared to

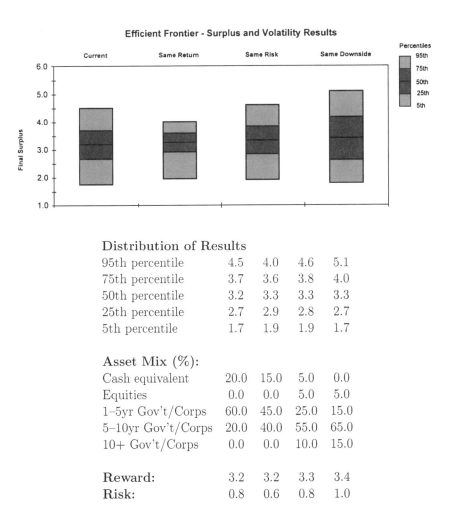

Figure 6: Efficient Frontier – Surplus and Volatility Results

stochastic optimal control methods. Hence, with a turnover in the range of 10%–20% on average, our transactions costs are quite low, which is another feature favored by our clients.

The next section describes various methods for modeling multi-currency exposures and incorporating foreign currency analysis into financial institutions' portfolio optimization and investment processes.

MULTI-CURRENCY OPTIMIZATION OPTIONS

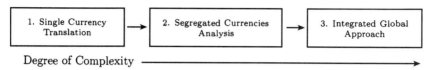

Degree of Complexity ⟶

3 Multi-Currency Analysis Options

Falcon has employed three different approaches to multi-currency analysis
for its clients. The main difference between these three options is the degree
to which cross-currency correlations and the resulting positive diversification
effects are included in the optimization analysis. Which approach is best
for a company will depend upon the complexity and number of its foreign
currency exposures and the degree of analytic rigor required. This section
will describe the various approaches in more detail, as well as outline their
relative advantages and disadvantages.

3.1 Option 1: Single Currency Translation

Translating all foreign currency exposures into a company's home currency
using the exchange rate in effect as of the model valuation date is the easi-
est method for handling currency exposures. It entails converting all foreign
assets and liabilities into the equivalent amount of the company's domestic
currency using a fixed exchange rate and then analyzing the aggregated exposures to determine the appropriate investment strategy. A single, home
currency efficient frontier can then be modeled to identify the optimal asset
allocations. This aggregate investment strategy is proportionately applied
to each local currency, resulting in similarly structured portfolios invested in
each of the local currencies of the foreign countries in which the company
operates.

The appeal of this option is its relative ease, both in terms of modeling
and implementation. Translating currency exposures can be done with a calculator and the net impact incorporated directly into the company's existing
asset-liability management models.

The biggest drawback of this approach, however, is that it can lead to
sub-optimal investment decisions, especially for companies with significant
international operations. This is due to several simplifying implicit assumptions: (1) foreign markets and currency rates move in tandem with those
in the company's domestic country; (2) international liability diversification
effects do not impact the security selection process; (3) liabilities in differ-

ent countries or currencies have similar characteristics, such as duration or inflation sensitivity; and, (4) the optimal result is appropriately measured in terms of the financial statements of the home country. These assumptions are unlikely to be true and oversimplify the company's foreign currency analysis.

In light of these advantages and cautions, this option is most applicable to companies with relatively little foreign currency exposure. Additionally, it can be used as a quick and easy, albeit crude, way to model and assess the potential impact of currency fluctuations on investment strategy.

3.2   Option 2: Segregated Currencies Analysis

For companies with more extensive foreign exposures, Falcon has modeled each major currency separately, with optimal portfolios being developed for each major country. This approach best fits companies that treat each of its international operations as separate business units. This approach entails first identifying the three or four major currencies that comprise the majority (i.e., the 80/20 rule) of a company's foreign currency exposures. Efficient frontier analyses are then performed on each of these major currencies to determine the set of portfolios that generate the highest reward for a given level of risk. As a result, distinct optimal portfolios are selected for each major country of operation, resulting in improved risk/reward profiles compared to Option 1.

The main advantages of this option are clear: companies can drastically reduce the level of currency risks borne while improving their overall financial position. Because each country is modeled separately, asset allocation decisions better reflect the true liability characteristics (e.g., duration) of each specific country. And, the analytic challenge is manageable and less costly than that associated with more sophisticated global models (see Option 3) since only a limited number of currencies are modeled. Additionally, several different portfolio optimization software packages are readily available in the marketplace, enabling many companies to avoid costly and time consuming systems development projects.

As with Option 1, this approach does not, however, incorporate cross-currency correlations and their associated diversification effects. This is because each major currency is modeled and optimized separately. Moreover, it does not generally address all exposures since, typically, only the major currencies would be analyzed. Despite these disadvantages, this option offers considerable benefits to companies whose international exposures are limited to a handful of currencies or for companies who split their international operations into distinct business units.

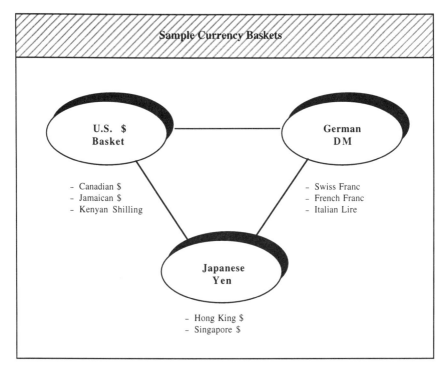

Figure 7: Sample Currency Baskets

Variation of Option 2

For companies whose international exposures are spread across many different currencies, Falcon recommends grouping like currencies into currency 'baskets' and analyzing the various currency 'baskets' in a segregated fashion. These 'baskets' and the home currency can then be modeled and efficient frontiers developed for each. Once the optimal portfolio is determined, an index of currencies comparable to those aggregated into the 'baskets' can be used for hedging/investing purposes. To illustrate this approach, Figure 7 outlines some possible currency baskets. Company-specific currency baskets should be created based on the size, relationship and importance of each currency to the overall financial results of the company.

Because this modeling approach groups multiple currencies into one or two 'baskets', it is inherently inaccurate in its representation of a company's actual currency exposures. The seriousness of this error will largely depend

	US\$/DM	US\$/Yen	US\$/GBP	US\$/AUD
US\$/DM	1.0	—	—	—
US\$/Yen	+0.49	1.0	—	—
US\$/GBP	+0.64	+0.15	1.0	—
US\$/AUD	−0.16	−0.26	+0.06	1.0

Source: First Boston

Figure 8: Multi-Currency Correlation Matrix

on the degree of similarities between the currencies grouped together and the actual number of different currencies included. Simply stated, the more the currencies move in tandem with others in the 'basket,' the more accurate this modeling approach will be.

3.3 Option 3: Integrated Global Approach

Falcon has found that for clients that are truly global, the most accurate approach for analyzing currency exposures is to use an integrated global currency model. A global currency model generates economic simulations which incorporate all major international currencies, as well as their interrelationships with each other. These cross-correlations (see Figure 8) are explicitly included in the portfolio optimization analysis. Unlike the previous model options, this approach integrates the company's domestic and international exposures into one master model, enabling it to understand its global dynamics and the interdependencies within its various operations.

Integrated models such as this require that currencies be treated as a separate asset category and that economic simulations are arbitrage free across currency spot and forward rates. Models of this type allow for hedging strategies that explicitly take into account the fact that local returns are not accessible to international investors. The ultimate objective of an integrated global approach can be optimized either before or after the dynamic conversion into the home currency.

This global optimization approach therefore generates the most accurate modeling results and identifies the highest rewarding investment strategy, selecting an appropriate mix of international securities. It also provides insights as to similarities and/or countervailing forces within a company's business operations that might not otherwise be obvious. Figure 9 summarizes the distinct differences between options 1, 2 and 3 and delineates the associated pros and cons of each approach.

The most generally available models used for this purpose are the standard Markowitz covariance models or variations utilizing GARCH. They are

MULTI-CURRENCY OPTIMIZATION OPTIONS

1. Single Currency Translation	2. Segregated Currencies Analysis	3. Integrated Global Approach
Description: • Convert all exposures into home currency • Develop one efficient frontier for aggregate. translated exposures • Invest in local currencies. as appropriate, to reduce currency fluctuation differences in assets vs. liabilities	• Identify 3–4 currencies that comprise majority of exposures (80/20 rule) • Model separate efficient frontiers for each major currency • Select optimal portfolio for each currency and manage separately	• Build integrated model reflecting major countries' economies and inter-relationships • Incorporate cross-correlations into optimization analysis • Run integrated simulations to generate net total effect
Users: • Companies with relatively small international liability exposures	• Companies that treat each of their international operations as separate business units	• Companies with large diversified international liability exposures who are concerned with the consolidated result of their operations
Pros: • Relatively easy to manage • Provides good picture of aggregate exposures on an equivalent basis	• Addresses majority of exposures in manageable. cost-effective manner • Allows better optimization decisions than Option #1 due to linking same currency assets and liabilities • Less costly to develop and to manage than Option #3	• Incorporates cross-currency correlations. potentially picking up positive diversifications effects • Most accurate and complete view of the business • Conducive to effective hedging since quantifies net position after cross-correlations and offsetting positions
Cons: Sub-optimizes investment decisions because assumes all foreign currencies are perfectly correlated with home currency • Does not incorporate potentially different liability (e.g.. duration) characteristics or capital requirements of various countries	• Does not completely address all exposures • Does not reflect any cross-currency correlations and diversification effects • If basket approach used for minor currencies. may introduce too many inaccuracies	• Global simulation and optimization models of this sophistication are not generally available in the market • Difficult. time consuming. and costly to develop

Figure 9: Multi-Currency Optimization Options

designed to handle single period risk management problems but typically exclude liability considerations. Unfortunately, more sophisticated, global currency software packages that can handle multi-period global asset allocation strategies are not widely available in the marketplace, and would be difficult and costly to build. For companies with extensive international activities, however, it may be worth undertaking such a development project given the potential positive diversification effects in a global portfolio. Falcon uses the Towers Perrin Global CAP:Link model as part of its integrated global ALM approach. Global CAP:Link is one of the few simulation models available for the type of integrated global optimization necessary to implement Option 3.

4 Case Study: Applying Multi-Currency ALM

The increasingly competitive and mature nature of the domestic insurance market in the USA. has many large and to some extent smaller specialty insurers moving their marketing efforts off-shore. They now view their target market as the world. This globalization of insurance premium generation has forced senior management to deal with the complicated and sometimes confusing issue of multi-currency management in all of its many aspects, including marketing and distribution, investments, accounting, finance and, of course, asset-liability management. In the preceding sections, we outlined Falcon's approach for addressing asset-liability management issues, and proposed three possible approaches to analyzing multi-currency ALM extensions. The following case study presents the actual results of applying Falcon's ALM concepts to the investment analysis of a large, international financial institution client.

ABC Corporation is a large multi-line insurance corporation that sells both life and health, as well as general insurance products in over 10 countries worldwide. ABC Corporation separates its operations into five significant markets; its domestic base – the United States, and four international markets. Consequently, ABC Corporation has major investment operations and liability exposures in five different currencies - the US Dollar, the British Pound, the German Deutschmark, the French Franc and the Spanish Peseta. The smaller residual country exposures are all highly correlated to one of the five major currencies and are not modeled separately.

ABC Corporation hired Falcon Asset Management to gather data and develop a series of asset-liability models that would allow their senior management to better understand the financial dynamics of ABC Corporation given their diverse currency exposure. That is to say, they wanted a model that relates the volatility of many different inputs, such as multi-currency assets and liabilities to the volatility of the outputs, income statements, balance sheets and if possible, an optimal investment portfolio. The challenge to Falcon was to optimally allocate assets subject to the company's liability exposures in these diverse capital markets. The optimal allocation of assets would therefore need to consider the risks associated with the company's international liability exposures, each individual country's capital markets, and the corresponding cross-currency exchange rates.

4.1   Falcon's Approach to Modeling ABC Corporation's Assets and Liabilities

There are three main approaches to analyzing multi-currency asset allocation issues such as those faced by ABC Corporation. In the case of ABC, we will compare and contrast these three currency-modeling alternatives, namely single currency translation, segregated currencies analysis and the integrated global approach.

The first, and simplest, alternative for evaluating multi-currency asset

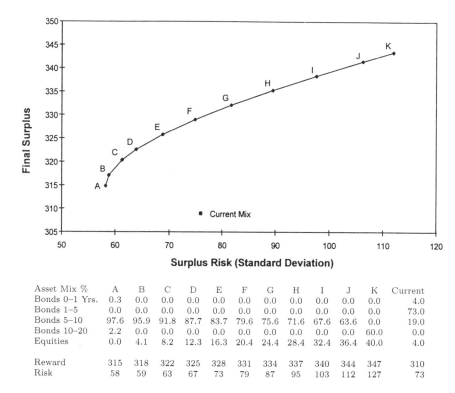

Asset Mix %	A	B	C	D	E	F	G	H	I	J	K	Current
Bonds 0–1 Yrs.	0.3	0.0	0.0	0.0	0.0	0.0	0.0	0.0	0.0	0.0	0.0	4.0
Bonds 1–5	0.0	0.0	0.0	0.0	0.0	0.0	0.0	0.0	0.0	0.0	0.0	73.0
Bonds 5–10	97.6	95.9	91.8	87.7	83.7	79.6	75.6	71.6	67.6	63.6	0.0	19.0
Bonds 10–20	2.2	0.0	0.0	0.0	0.0	0.0	0.0	0.0	0.0	0.0	60.0	0.0
Equities	0.0	4.1	8.2	12.3	16.3	20.4	24.4	28.4	32.4	36.4	40.0	4.0
Reward	315	318	322	325	328	331	334	337	340	344	347	310
Risk	58	59	63	67	73	79	87	95	103	112	127	73

Figure 10: Dollar Based Single Currency Efficient Frontier ($ Millions)

strategy is to limit the analysis to a single currency, in this case the home currency or US dollar. This involves translating all assets and liabilities of the company into US dollar equivalent values using exchange rates in effect as of the valuation date of the analysis. The process for determining an optimal asset allocation strategy thus reduces to a single currency analysis that can be solved using standard asset/liability modeling techniques.

Figure 10 shows the result of such an analysis were Falcon generated an efficient frontier based on stochastic asset and liability cash flow projections. In this analysis, the objective function was designed to maximize economic surplus at the end of a three-year planning horizon subject to minimizing surplus risk, i.e., the standard deviation of the final economic surplus.

Moving to portfolio E, the final economic surplus can be improved by approximately $18 million while maintaining the same level of surplus risk as the current portfolio. To achieve this benefit, the equity exposure would have to be increased from 4% to 16% while the average maturity would have to increase from approximately 4 years to 8 years. Additional analysis evaluating regulatory, operating and tax consequences of moving to such portfolio would be needed before a portfolio repositioning could be approved and implemented. It is clear from this analysis, however, that substantial economic benefits can be achieved by making changes to the current asset strategy.

One problem with single currency analysis is that while it identifies an aggregate portfolio structure for the consolidated assets and liabilities of the company, it leaves open the question of how to optimally manage currency exposures. The company must decide whether to hedge its currency risks or to allocate its international assets in a fashion consistent with the chosen aggregate portfolio identified from the single currency analysis. For companies with significant international operations such as ABC Corporation, the single currency approach leaves too many important questions unaddressed.

A second approach to international ALM attempts to address these questions by generating a separate ALM efficient frontier for each of a company's major currency exposures. This approach requires a company to segment its international operations, treating each segment as a separate operating unit. In our case study, we defined three distinct currency classifications – the US Dollar (31% of total assets), the British Pound (21%) and a European Union Currency Basket (48%). The 'E.U. Currency Basket' category was designed to simplify the analysis from ABC's five major currencies to three by combining the highly linked currencies of the European Union, the German Deutschmark, the French Franc and the Spanish Peseta. The German Deutschmark served as the proxy for the 'E.U. Currency Basket.' Next, the assets and liabilities of the company were segmented according to these three currency classifications. Separate efficient frontiers were then generated for each of the individual currency classifications using standard asset liability modeling techniques. The 'E.U. Currency Basket' classification requires French and Spanish assets and liabilities to be translated into German marks using a process similar to that described for the home currency translation approach.

Figure 11 shows the asset mixes corresponding to the low risk portfolio from each of the three segregated currency efficient frontiers and compares them to the low risk portfolio from the single currency efficient frontier. The maturity and duration of the low risk portfolio from the single currency efficient frontier is approximately equal to the weighted average maturity and duration of the low risk portfolios from the segregated currency efficient frontiers. However, had ABC decided to use the low risk portfolio from the single currency frontier it would have resulted in a shorter portfolio than desired for its US and U.K. operations and a much longer portfolio than desired for its other international business. This illustrates that efficient portfolios determined on an aggregate single currency basis can lead to suboptimal investment strategies when viewed on a country-specific basis.

The segregated currencies approach, though, is not without its drawbacks.

	Single	Segregated Currencies		
	Currencies	US	UK	DM
Bonds 0–1 Yrs.	3.0%	0.0%	0.0%	0.0%
Bonds 1–5 Yrs.	0.0%	0.0%	0.0%	36.8%
Bonds 5-10 Yrs.	97.6%	42.6%	44.0%	63.2%
Bonds 10–20 Yrs.	2.2%	57.4%	56.0%	0.0%
Equities	0.0%	0.0%	0.0%	0.0%
Duration	5.7	7.4	7.3	4.5

Figure 11: Minimum Risk Efficient Portfolios

While it succeeds in answering the question of how to allocate the assets held by each of ABC's separate international business units, it fails to measure the joint impact of these various business units on ABC Corporation's aggregate risk/reward profile. To consider both the optimal mix of investments for an international insurer and the impact such an investment strategy would have on the consolidated results of the company an integrated global approach to multi-currency optimization is needed. An integrated global approach to multi-currency optimization more realistically reflects the various countries' economic situation and inter-relationships, such as relative inflation and interest rates. It incorporates economic cross-correlations into the simulations and as such, into the portfolio optimization process. This allows senior management to get a more accurate and complete view of their international business including the positive effects of diversification and the potential negative effects of unhedged currency positions.

The integrated global approach, similar to the segregated currencies approach, requires that the assets and liabilities be segmented by currency. It differs from the segregated currency analysis however, in that it optimizes the investment strategy for the company as a whole rather than treating each currency of operation as a separate business unit. The resulting efficient frontier thus considers international investment opportunities, the benefits of international asset and liability diversification and the unique characteristics of a company's international liabilities.

Figure 12 shows an efficient frontier that was generated using the integrated global approach. the asset mixes show the breakdown of the investment allocations into US Dollar, British Pound and German Deutschmark asset categories. This approach, while considering the characteristics of ABC's international liabilities, results in the efficient distribution of the companies assets in total. Such a holistic approach to global asset/liability management results in the best risk/reward tradeoff opportunities for the company.

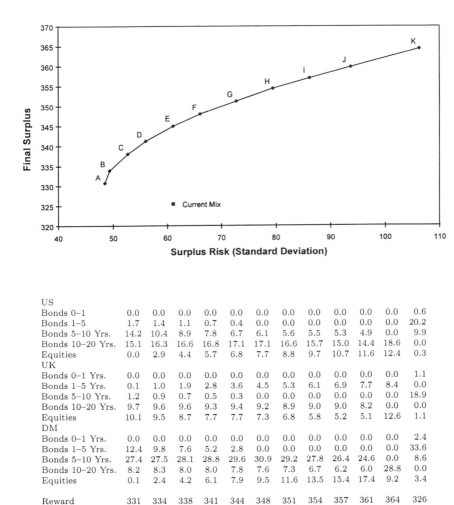

US												
Bonds 0–1	0.0	0.0	0.0	0.0	0.0	0.0	0.0	0.0	0.0	0.0	0.0	0.6
Bonds 1–5	1.7	1.4	1.1	0.7	0.4	0.0	0.0	0.0	0.0	0.0	0.0	20.2
Bonds 5–10 Yrs.	14.2	10.4	8.9	7.8	6.7	6.1	5.6	5.5	5.3	4.9	0.0	9.9
Bonds 10–20 Yrs.	15.1	16.3	16.6	16.8	17.1	17.1	16.6	15.7	15.0	14.4	18.6	0.0
Equities	0.0	2.9	4.4	5.7	6.8	7.7	8.8	9.7	10.7	11.6	12.4	0.3
UK												
Bonds 0–1 Yrs.	0.0	0.0	0.0	0.0	0.0	0.0	0.0	0.0	0.0	0.0	0.0	1.1
Bonds 1–5 Yrs.	0.1	1.0	1.9	2.8	3.6	4.5	5.3	6.1	6.9	7.7	8.4	0.0
Bonds 5–10 Yrs.	1.2	0.9	0.7	0.5	0.3	0.0	0.0	0.0	0.0	0.0	0.0	18.9
Bonds 10–20 Yrs.	9.7	9.6	9.6	9.3	9.4	9.2	8.9	9.0	9.0	8.2	0.0	0.0
Equities	10.1	9.5	8.7	7.7	7.7	7.3	6.8	5.8	5.2	5.1	12.6	1.1
DM												
Bonds 0–1 Yrs.	0.0	0.0	0.0	0.0	0.0	0.0	0.0	0.0	0.0	0.0	0.0	2.4
Bonds 1–5 Yrs.	12.4	9.8	7.6	5.2	2.8	0.0	0.0	0.0	0.0	0.0	0.0	33.6
Bonds 5–10 Yrs.	27.4	27.5	28.1	28.8	29.6	30.9	29.2	27.8	26.4	24.6	0.0	8.6
Bonds 10–20 Yrs.	8.2	8.3	8.0	8.0	7.8	7.6	7.3	6.7	6.2	6.0	28.8	0.0
Equities	0.1	2.4	4.2	6.1	7.9	9.5	11.6	13.5	15.4	17.4	9.2	3.4
Reward	331	334	338	341	344	348	351	354	357	361	364	326
Risk.	49	49	53	56	61	66	73	79	86	94	106	61

Figure 12: Integrated Global Efficient Frontier ($ Millions)

By utilizing the integrated global methodology, the client moved from his current portfolio to the risk neutral portfolio E, which enhanced his expected final surplus by $18 million, the same as in the single currency analysis method. However, due to the better match-up of asset and liability flows by currency exposure, as well as the diversifying effect of multi-currency investments, the risk level was reduced by $12 million from $73 million to $61 million.

5 Conclusion

Asset-liability management is an increasingly useful tool for the management of insurance companies. The use of ALM techniques to help develop an overall risk profile for the company is rapidly emerging as a critical issue for the insurance industry in the 21st century. Concomitantly, the modeling of foreign currency exposures will become an integral part of most insurers' ALM analysis as more and more companies expand their international operations or add foreign securities to their asset portfolios. Multi-currency ALM analysis will enable financial institutions to better assess their international exposures and to take advantage of the positive diversification benefits associated with international investing. Perhaps more importantly, it will help provide a clearer picture of the dynamics within a firm's business operations and the potential financial impact of currency and capital market fluctuations. The world of insurance investment management is becoming more complex and global, requiring sophisticated asset-liability modeling to understand potential risk/reward trade-offs to develop optimal investment strategies. More and more senior management throughout the insurance industry are calling upon their asset-liability teams to pursue issues and challenges that will have a greater and greater impact on the direction of their insurance companies. New investment opportunities are being generated on almost a daily basis from Wall Street. The role of such investments as international bonds and stock, Commercial Mortgage Backed Securities (CMBS), Collateral Loan Obligation (CLO), synthetic equities and Securitized Global Insurance Liabilities must be evaluated both for return and yield potential and also to see if they fit with the emerging liability profile and risk/reward requirements of the company. Additionally, these new investment opportunities must be matched to the pending requirements of the Model Investment Law and its restrictions on foreign investments, not to mention new securities.

Another area of concern for senior insurance managers are the ever changing and increasingly more demanding accounting and regulatory requirements, such as the Financial Accounting Standards Board ruling on certain hedging activities, mortgage-backed security restrictions and the notorious FAS115 rulings on 'marking assets to market.' Finally, the ever increasing competitive pressures of the international insurance market with its fierce price discounting and ever expanding new market entrants presents great new challenges to insurance company management.

There are many different approaches for modeling foreign currency risks. The appropriate choice will depend on the degree of complexity and risk associated with a company's foreign operations and the amount of time and dollars available for modeling purposes. Regardless of the option selected, an investment in modeling a company's foreign currency exposures will likely reap substantial rewards.

These topics can only be evaluated properly using an integrated asset-liability framework. Increasingly, the winners in the insurance industry will be those companies that couple multi-period economic simulation and optimization models to guide strategic decision making under uncertainty.

Appendix. Economic theory of multi-currency markets

For the large financial institutional investor, international markets demand a monetary unit of account for reconciling prices in two different monetary systems. The process of foreign currency exchange ('FX') is just such a system. Understanding the economic theory behind this system and the FX markets is an important first step in modeling the FX effects on assets and liabilities. This section outlines four major theoretical approaches (see Solnik, 1996) to estimating foreign currency exchange rates. Specifically:

- **Purchasing Power Parity (PPP)**, which links spot exchange rates and inflation;

- **Interest Rate Parity**, which links spot exchange rates, forward exchange rates, and interest rates;

- **The International Fisher Effect**, which links interest rates and inflation; and,

- **Foreign Exchange Expectations**, which links forward exchange and expected spot exchange rates.

Purchasing Power Parity

Purchasing Power Parity ('PPP') states that spot exchange rates adjust perfectly to any inflation differences between two countries. PPP suggests that if prices in one country rise relative to those in another country, then the first country's foreign exchange rate must fall in order to maintain the same real price for goods in the two countries. The PPP relationship can be modeled as follows:

$$\frac{E_{S1}}{E_{S0}} = \frac{1 + I_{(F)}}{1 + I_{(D)}} \tag{A.1}$$

where

E_{S0} is the spot exchange rate at the start of the period,

E_{S1} is the spot exchange rate at the end of the period,

I_F is the inflation rate, over the period, in the foreign country,

I_D is the inflation rate, over the period, in the domestic country.

PPP can also be stated as the so-called 'law of one price,' (see Orlin, 1996).

$$P_{(i,t)} = E_{S(t)} P^*_{(i,t)} \quad \text{for any commodity } I, \tag{A.2}$$

where

$E_{S(t)}$ is the current exchange rate,

$P_{(i,t)}$ is the current domestic price of I,

$P^*_{(i,t)}$ is the current foreign price of I,

Interest Rate Parity

A related economic theory to PPP is that of Interest Rate Parity ('IRP'). The IRP relationship states that for free market interest rates, the difference between two countries' interest rates must equal the percentage differential between the forward exchange rate and the spot exchange rate. This is a technical arbitrage as well as a theory. The underlying concept is the law of one price, which as stated before, suggests that a freely traded good or service will sell effectively at one price. Failing this, it is suggested, would induce buying and selling of the good, generating excess economic profits in the process. The IRP relationship is:

$$\frac{F}{E_{S0}} = \frac{1 + r_F}{1 + r_D} \tag{A.3}$$

International Fisher Effect

The International Fisher Effect, named after Irvin Fisher, states that the difference in interest rates between two different countries is equal to the difference in inflation or inflationary expectations in the two countries. This is because 'real' interest rates are assumed to be the same over relatively long periods of time and everywhere. In other words, the International Fisher Effect posits that the difference between the nominal interest rate and the real interest rate in any given country is the expected inflation.

The International Fisher equation is:

$$\frac{1 + r_F}{1 + r_D} = \frac{1 + p_F}{1 + p_D} \times \frac{1 + E_{(I,F)}}{1 + E_{(I,D)}} \tag{A.4}$$

where

r_f is the foreign nominal interest rate,

r_D is the domestic nominal interest rate,

p_F is the foreign real interest rate,

p_D is the domestic real interest rate,

$E_{(I,F)}$ is the foreign expected inflation,

$E_{(I,D)}$ is the domestic expected inflation.

The Fisher model is usually stated in its first-order linear approximation:

$$r_F - r_D = p_F - p_D + E_{(I,F)} - E_{(I,D)}. \qquad (A.5)$$

However, empirical evidence for this assertion is somewhat limited when applied to all currencies. Eurocurrency market studies for six major currencies found limited support for this model (see Kane and Rosenthal 1982). Additionally, several studies have shown that in the period of the 1980s real rates were highly variable over time and differed across countries (see Duxton 1993). Real rates on the US dollar were quite high in the early 1980s and declined to much lower levels than on other currencies by the 1990s.

Other considerations such as political risk perceptions and coordinated intervention by central bankers have further clouded the interpretation of inflation being the sole variable differentiating nominal from real rates.

Foreign Exchange Expectations

The fourth major theory, Foreign Exchange Expectations (FEE), relates the forward exchange rate at time zero for delivery at time one to the expected value of the spot rate at time one. Namely,

$$F = E_{(St+1)} \qquad (A.6)$$

where

F is the forward exchange rate,

$E_{(St+1)}$ is the spot rate at time period one $(t+1)$.

The Foreign Exchange Expectations relationship can also be derived as a function of the current spot exchange rate where the left-hand side of the equation is called the forward discount or premium:

$$\frac{F - E_S}{E_S} = E\left(\frac{S_{t+1} - S_0}{S_0}\right) \qquad (A.7)$$

The FEE relationship states that there is no reward for bearing foreign exchange risk or uncertainty.

5.1   Conclusion

World trade and capital markets do not behave as orderly or precisely as assumed in any of the foreign exchange theories discussed above. These theories do, however, provide important insights for investment professionals as they contemplate asset-liability modeling approaches. Specifically,

- Despite the evidence that real rates vary over time, real interest rate differentials appear to be the primary drivers of foreign currency rates;

- Inflation expectations can differ considerably across countries and directly affect nominal interest rates; and,

- Given the dynamic nature of interest and foreign exchange rates, stochastic modeling of potential movements is critical for developing successful global investment strategies.

With this understanding of the economic theory of exchange rates and market interdependencies, we can evaluate various approaches for incorporating multi-currency analysis into the asset-liability modeling process.

References

M. Adler and B. Dumas (1983), International Portfolio Choice and Corporate Finance: A Synthesis, *Journal of Finance* **38** 925–984

R.T. Baillie and T. Bolleralev (1990), A Multivariarte Generalized ARCH Approach to Modeling Risk Premia in Forward Exchange Rate Markets, *Journal of International Money and Finance* **9** 309–324

A.J. Berger, F. Glover and J. Mulvey (January, 1995), Solving Global Optimization Problems in Long-Term Financial Planning, Statistics and Operations Research Technical Report, Princeton University

J.S. Bhandari (1985), ed, *Exchange Rate Management Under Uncertainty*, MIT Press

F. Black and R. Litterman (September/October, 1992), Global Portfolio Optimization, *Financial Analysts Journal*

F. Black and R. Litterman (1990), Equilibrium Exchange Rate Hedging, *Journal of Finance* **45** 899–908

F. Black and R. Litterman (January/February, 1995), Universal Hedging: Optimizing Currency Risk and Reward in International Equity Portfolios, *Financial Analysts Journal*

Z. Bodie, A. Kane and A. Marcus (1993), *Investments*, Richard D. Irwin, Inc.,

J. Bolsover (1994), Global Asset Allocation. In *Investment IV*, Association for Investment Management and Research

D.R. Cariño, T. Kent, D. Myers, C. Stacy, M. Sylvanus, A. Turner, K. Watanabe and W.T. Ziemba (January/February 1994), The Russell-Yasuda Kasai Model: An Asset Liability Model for a Japanese Insurance Company Using Multi-stage Stochastic Programming, *Interfaces*, **24** (1) 29–49

K. Choie (Winter, 1993), Currency Exchange Rate Forecast and Interest Rate Differential, *The Journal of Portfolio Management* 58–64

S. Correnti and J.C. Sweeney (4th AFIR International Colloquium, 1994), Asset-Liability Management and Asset Allocation for Property and Casualty Companies – The Final Frontier

F.X. Diebold (1988), *Empirical Modeling of Exchange Rate Dynamics*, Springer

R. Dornbush (1976), Expectations and Exchange Rate Dynamics, *Journal of Political Economy* **84** 1161–1176

R. Dornbush (1988), Purchasing Power Parity, in *The New Palgrave Dictionary of Economics*, Macmillan (London)

M.M. Duxton (February ,1993), Real Interest Parity: New Measures and Tests, *Journal of International Money and Finance*

E.F. Fama (1984), Forward and Spot Exchange Rates, *Journal of Monetary Economics* **14** 319–338

J.A. Frankel (1992), Measuring International Capital Mobility: A Review, *American Economic Review* **82** 197–202

J.A. Frankel and J. Aizenman (1982), Aspects of the Optimal Management of Exchange Rates, *Journal of International Economics* **13** 231-256

K.A. Froot and R.H. Thaler (1990), Anomalies: Foreign Exchange, *Journal of Economic Perspectives* **4** 179–192

G. Gastineau (May/June, 1995), The Currency Hedging Decision: A Search for Synthesis in Asset Allocation, *Financial Analysts Journal*

J.O. Grabbe (1996), *International Financial Markets*, Prentice Hall

J. Haizinga (1987), An Empirical Investigation of the Long-Run Behavior of Real Exchange Rates. In *Empirical Studies of Velocity, Real Exchange Rates, Unemployment and Productivity*, K. Brunner and A. Meltzer, eds, Carnegie-Rochester Conference Series on Public Policy **27**, North-Holland 149–214

N. Harris (1994), Currency Management: Issues and Strategies. In *Investment V*, Association for Investment Management and Research

J. Hull (1993), *Options, Futures and Other Derivative Securities*, Prentice-Hall

A.G. Isaac (1989), Exchange Rate Volatility and Currency Substitution, *Journal of International Money and Finance* **8** 277–284

E. Kane and L. Rosenthal (April, 1982), International Interest Rates and Inflationary Expectations, *Journal of International Money and Finance*

D. Karnosky and B. Singer (1994), *Global Asset Management and Performance Attribution*, The Research Foundation of The Institute of Chartered Financial Analysts

C. Koot and J. Tatom (July/August 1988), International Linkages in the Term Structure of Interest Rates. In *Review of the Federal Reserve Bank of St. Louis*

P. Krugman and M. Miller (1991), eds, *Exchange Rate Targets and Currency Bands*, Cambridge University Press

R. Kubarych (1983), *Foreign Exchange Markets in the United States*, New York: Federal Reserve Bank of New York

R.E. Lucas (1982), Interest Rates and Currency Prices in a Two-Country World, *Journal of Monetary Economics* **10** 335–360

J. Maginn and D. Tuttle (1990), *Managing Investment Portfolios*, Warren, Gorham and Lamont

J.P. Morgan (May, 1995), *RiskMetrics - Technical Document*, New York: Morgan Guaranty Trust Company, Global Research

M.G. Morgan and M. Henrion (1990), *Uncertainty*, Cambridge University Press

J.M. Mulvey and E. Thorlacius, The Towers Perrin Global Capital Market Scenario Generation System: CAP:Link. In *World Wide Asset and Liability Modeling*, W.T. Ziemba and J.M. Mulvey, eds, Cambridge University Press, 1998.

J.M. Mulvey and W.T. Ziemba (1995), Asset and Liability Allocation in a Global Environment. In *Finance*, R.A. Jarrow, V. Moksimovic, W.T. Ziemba, eds, North-Holland, 435–463

M. Obstfeld and A.C. Stockman (1985), Exchange-Rate Dynamics, in R.W. Jones and P.B. Kenen, eds, Handbook of International Economics, Volume II, Amsterdam, North-Holland

M. Obstfeld and A.C. Stockman (1990), How Integrated Are World Capital Markets? Some New Tests. In *Debt, Stabilization and Development*, G. Calvo, R. Finelly and J. DeMacedo, eds, Blackwell, 134–155

C. Smith, C. Smithson and D. Wilford (1990), Five Reasons why Companies Should Manage Risk. In *The Handbook of Currency and Interest Rate Risk Management*, R. Schwartz and C. Smith, eds, New York Institute of Finance

B. Solnik (1996), *International Investments*, 3rd Edition, Addison-Wesley

B. Solnik (1994), *Fundamental Considerations in Cross-Border Investment: The European View*, The Research Foundation of The Institute of Chartered Financial Analysts

L.E.O. Sversson (1995), Currency Prices, Terms of Trade and Interest Rates, *Journal of International Economics* **18** 14–44

S. Taylor (1986), *Modeling Financial Time Series*, Wiley

F. Van Der Ploef (1994), ed, *The Handbook of International Macroeconomics*, Blackwell Publishers

A. Wilkie (1994), *A Stochastic Model for Consumer Price Indices and Exchange Rates in Several Countries*, R. Watson and Sons

W.T. Ziemba and R.G. Vickson (1975), eds, *Stochastic Optimization Models in Finance*, Academic Press

Part VII
Dynamic Portfolio Analysis with Assets and Liabilities

Optimal Investment Strategies for University Endowment Funds*

Robert C. Merton

7.1 Introduction

To examine the question of optimal investment strategies for university endowment funds, one must of course address the issue of the objective function by which optimality is to be measured. My impression is that practicing money managers essentially sidestep the issue by focusing on generically efficient risk-return objective functions for investment which are just as applicable to individuals or nonacademic institutions as they are to universities. Perhaps the most common objective of this type is mean-variance efficiency for the portfolio's allocations. Black (1976) provides a deeper approach along those lines that takes account of tax and other institutional factors, including certain types of nonendowment assets held by institutions. The Ford Foundation study of 1969 gave some early practical (if ex post, somewhat untimely) guidance for investment allocations.

Much of the academic literature (which is not copious) seems to focus on appropriate spending policy for endowment, taking as given that the objective for endowment is to provide a perpetual level flow of expected real income (cf. Eisner 1974; Litvack, Malkiel, and Quandt 1974; Nichols 1974; Tobin 1974). Ennis and Williamson (1976) present a history of spending patterns by universities and a discussion of various spending rules adopted. They also discuss the interaction between spending and investment policies. Fama and Jensen (1985) discuss the role of nonprofit institutions as part of a general analysis of organizational forms and investment objective functions, but they do not address the functions of endowment in such institutions.

In contrast, Hansmann (1990) provides a focused and comprehensive re-

---

*Reprinted, with permission, from *Studies of Supply and Demand in Higher Education*, Charles T. Clotfelter and Michael Rothschild, eds., University of Chicago Press, 1993. Copyright 1993, University of Chicago Press.

view of the various possible roles for a university's endowment. Despite the broad coverage of possibilities ranging from tax incentives to promoting intergenerational equity, he is unable to find compelling empirical evidence to support any particular combination of objectives. Indeed, he concludes that "prevailing endowment spending rules seem inconsistent with most of these objectives" (p. 39). Hansmann goes on to assert (pp. 39–40):

> It appears, however, that surprisingly little thought has been devoted to the purposes for which endowments are maintained and that, as a consequence, their rate of accumulation and the pattern of spending from their income have been managed without much attention to the ultimate objectives of the institutions that hold them.

The course taken here to address this question is in the middle range: it does not attempt to specify in detail the objective function for the university, but it does derive optimal investment and expenditure policy for endowment in a context which takes account of overall university objectives and the availability of other sources of revenue besides endowment. In that respect, it follows along lines similar to the discussion in Black (1976, 26–28). In addition, our model takes explicit account of the uncertainties surrounding the costs of university activities. As a result, the analysis reveals another (perhaps somewhat latent) purpose for endowment: namely, hedging against unanticipated changes in those costs. Formal trading rules for implementing this hedging function are derived in sections 7.3 and 7.4. However, the paper neither assesses which costs, as an empirical matter, are more important to hedge nor examines the feasibility of hedging those costs using available traded securities. The interested reader should see Brinkman (1981, 1990), Brovender (1974), Nordhaus (1989), and Snyder (1988), where the various costs of universities are described and modeled, both historically and prospectively.

Grinold, Hopkins, and Massy (1978) develop a budget-planning model which also integrates endowment returns with other revenue and expense flows of the university. However, their model differs significantly from the one presented here, perhaps because their focus is on developing policy guidelines for expenditures instead of optimal intertemporal management of endowment.

In the section to follow, we describe the basic insights provided by our analysis and discuss in a qualitative fashion the prescriptions for endowment policy. The formal mathematical model for optimal expenditures and investment that supports those prescriptions is developed in sections 7.3 and 7.4. It is based on a standard intertemporal consumption and portfolio-selection model. Hence, the formal structure of the optimal demand functions is already widely studied in the literature. It is the application of this model to the management of university endowment which is new. For analytical simplicity and clarity, the model is formulated in continuous time. However, it is evident from the work of Constantinides (1989), Long (1974), and Merton (1977) that a discrete-time version of the model would produce similar results.

7.2 Overview of Basic Insights and Prescriptions for Policy

As indicated at the outset, a standard approach to the management of endowment is to treat it as if it were the only asset of the university. A consequence of this approach is that optimal portfolio strategies are focused exclusively on providing an efficient trade-off between risk and expected return. The most commonly used measure of endowment portfolio risk is the variance (or equivalently, standard deviation) of the portfolio's return. As is well known, the returns on all mean-variance efficient portfolios are perfectly correlated. Thus, a further consequence of treating endowment as the only asset is that the optimal endowment portfolios of different universities should have quite similar risky investment allocations, at least as measured by the correlations of the portfolio returns.

Universities, as we all know, do have other assets, both tangible and intangible, many of which are important sources of cash flow. Examples of such sources are gifts, bequests, university business income, and public- and private-sector grants. Taking explicit account of those assets in the determination of the endowment portfolio can cause the optimal composition of that portfolio to deviate significantly from mean-variance efficiency. That is, two universities with similar objectives and endowments of the same size can nevertheless have very different optimal endowment portfolios if their nonendowment sources of cash flow are different.

A procedure for selecting the investments for the endowment portfolio that takes account of nonendowment assets includes the following steps:

1. Estimate the market value that each of the cash flow sources would have if it were a traded asset. Also determine the investment risk characteristics that each of those assets would have as a traded asset.

2. Compute the *total wealth* or net worth of the university by adding the capitalized values of all the cash flow sources to the value of the endowment.

3. Determine the optimal portfolio allocation among traded assets, using the university's total wealth as a base. That is, treat both endowment and cash flow–source assets as if they could be traded.

4. Using the risk characteristics determined in step 1, estimate the "implicit" investment in each traded-asset category that the university has as the result of owning the nonendowment (cash flow–source) assets. Subtract those implicit investment amounts from the optimal portfolio allocations computed in step 3, to determine the optimal "explicit" investment in each traded asset, which is the actual optimal investment allocation for the endowment portfolio.

As a simple illustration, consider a university with $400 million in endowment assets and a single nonendowment cash flow source. Suppose that the only traded assets are stocks and cash. Suppose further that the university estimates in step 1 that the capitalized value of the cash flow source is $200 million, with risk characteristics equivalent to holding $100 million in stock and $100 million in cash. Thus, the total wealth of the university in step 2 is (400 + 200 =) $600 million. Suppose that from standard portfolio-selection techniques, the optimal fractional allocation in step 3 is .6 in stocks and .4 in cash, or $360 million and $240 million, respectively. From the hypothesized risk characteristics in step 1, the university already has an (implicit) investment of $100 million in stocks from its nonendowment cash flow source. Therefore, we have in step 4 that the optimal amount for the endowment portfolio to invest in stocks is $260 million, the difference between the $360 million optimal total investment in stocks and the $100 million implicit part. Similarly, the optimal amount of endowment invested in cash equals (240 − 100 =) $140 million.

The effect on the composition of the optimal endowment portfolio induced by differences in the size of nonendowment assets can be decomposed into two parts: the wealth effect and the substitution effect. To illustrate the wealth effect, consider two universities with identical preference functions and the same size endowments, but one has nonendowment assets and the other does not. If, as is perhaps reasonable to suppose, the preference function common to each exhibits decreasing absolute risk aversion, then the university with the nonendowment assets (and hence larger net worth) will prefer to have a larger total investment in risky assets. So a university with a $400 million endowment as its only asset would be expected to choose a dollar exposure to stocks that is smaller than the $360 million chosen in our simple example by a university with the same size endowment and a nonendowment asset valued at $200 million. Such behavior is consistent with the belief that wealthier universities can "afford" to take larger risks with their investments. Thus, if the average risk of the nonendowment assets is the same as the risk of the endowment-only university's portfolio, then the university with those assets will optimally invest more of its endowment in risky assets.

The substitution effect on the endowment portfolio is caused by the substitution of nonendowment asset holdings for endowment asset holdings. To illustrate, consider again our simple example of a university with a $400 million endowment and a $200 million nonendowment asset. However, suppose that the risk characteristics of the asset are changed so that it is equivalent to holding $200 million in stocks and no cash. Now, in step 4, the optimal amount for the endowment portfolio to invest in stocks is $160 million, the difference between the $360 million optimal total investment in stocks and the $200 million implicit part represented by the nonendowment asset. The optimal amount of endowment invested in cash rises to (240 − 0 =) $240 million. If instead the risk characteristics of the asset had changed in the other direction to an equivalent holding of $0 in stocks and $200 million in cash, the optimal composition of the endowment portfolio would be (360 − 0 =) $360 million in stocks and (240 − 200 =) $40 million in cash.

Note that the changes in risk characteristics do not change the optimal deployment of *total* net worth ($360 million in stocks and $240 million in cash).

However, the nonendowment assets are not carried in the endowment portfolio. Hence, different risk characteristics for those assets do change the amount of substitution they provide for stocks and cash in the endowment portfolio. Thus, the composition of the endowment portfolio will be affected in both the scale and fractional allocations among assets.

With the basic concept of the substitution effect established, we now apply it in some examples to illustrate its implications for endowment investment policy. Consider a university that on a regular basis receives donations from alums. Clearly, the cash flows from future contributions are an asset of the university, albeit an intangible one. Suppose that the actual amount of gift giving is known to be quite sensitive to the performance of the general stock market. That is, when the market does well, gifts are high; when it does poorly, gifts are low. Through this gift-giving process, the university thus has a "shadow" investment in the stock market. Hence, all else the same, it should hold a smaller portion of its endowment in stocks than would another university with smaller amounts of such market-sensitive gift giving.

The same principle applies to more specific asset classes. If an important part of gifts to a school that specializes in science and engineering comes from entrepreneur alums, then the school de facto has a large investment in venture capital and high-tech companies, and it should therefore invest less of its endowment funds in those areas. Indeed, if a donor is expected to give a large block of a particular stock, then the optimal explicit holding of that stock in the endowment can be negative. Of course, an actual short position may not be truly optimal if such short sales offend the donor. That the school should optimally invest less of its endowment in the science and technology areas where its faculty and students have special expertise may seem a bit paradoxical. But the paradox is resolved by the principle of diversification once the endowment is recognized as representing only a part of the assets of the university.

The same analysis and conclusion apply if alum wealth concentrations are in a different class of assets, such as real estate instead of shares of stock. Moreover, much the same story also applies if we were to change the example by substituting government and corporate grants for donations and gift giving as the sources of cash flows. That is, the magnitudes of such grant support for engineering and applied science may well be positively correlated with the financial performance of companies in high-tech industries. If so, then the prospect of future cash flows to the university from the grants creates a shadow investment in those companies.

The focus of our analysis is on optimal asset allocation for the endowment portfolio. However, the nature and size of a university's nonendowment assets significantly influence optimal policy for spending endowment. As shown in section 7.4, for a given overall expenditure rate as a fraction of the university's total net worth, the optimal spending rate out of endowment will vary, depending on the fraction of net worth represented by nonendowment assets, the expected growth rate of cash flows generated by those assets, and capitalization rates. Hence, neglecting those other assets will generally bias the optimal expenditure policy for endowment.

In addition to taking account of nonendowment assets, our analysis differs from the norm because it takes account of the uncertainty surrounding the costs of the various activities such as education, research, and knowledge storage that define the purpose of the university. The breakdown of activities can of course be considerably more refined. For instance, one activity could be the education of a full-tuition–paying undergraduate, and a second could be the education of an undergraduate who receives financial aid. The unit (net) cost of the former is the unit cost of providing the education less the tuition received, and the unit cost of the latter is this cost plus the financial aid given. As formally demonstrated in section 7.3, an important function of endowment investments is to hedge against unanticipated changes in the costs of university activities.

Consider, for example, the decision as to how much (if any) of the university's endowment to invest in local residential real estate. From a standard mean-variance efficiency analysis, it is unlikely that any material portion of the endowment should be invested in this asset class. However, consider the cost structure faced by the university for providing teaching and research. Perhaps the single largest component is faculty salaries. Universities of the same type and quality compete for faculty from the same pools. To be competitive, they must offer a similar standard of living. Probably the largest part of the differences among universities in the cost of providing this same standard of living is local housing costs. The university that invests in local residential housing hedges itself against this future cost uncertainty by acquiring an asset whose value is higher than expected when the differential cost of faculty salaries is higher than expected. This same asset may also provide a hedge against unanticipated higher costs of off-campus housing for students that would in turn require more financial aid if the university is to compete for the best students. Note that this prescription of targeted investment in very specific real estate assets to hedge against an unanticipated rise in a particular university's costs of faculty salaries and student aid should not be confused with the often-stated (but empirically questionable) assertion that investments in real estate generally are a good hedge against inflation. See Bodie (1976, 1982) for empirical analysis of the optimal assets for hedging against general inflation.

Similar arguments could be used to justify targeted investment of endowment in various commodities such as oil as hedges against unanticipated changes in energy costs. Uncertainty about those costs is especially significant for universities located in extreme climates and for universities with major laboratories and medical facilities that consume large quantities of energy.

The hedging role for endowment can cause optimal investment positions that are in the opposite direction from the position dictated by the substitution effects of nonendowment assets. For example, consider a specialized institute of biology that receives grants from biotech companies and gifts from financially successful alums. As already explained, such an institute has a large shadow investment in biotech stocks, and it should therefore underweight (perhaps to zero) its endowment investments in such stocks. Suppose, however, that the institute believes the cost of keeping top faculty will rise by considerably more than tuition or grants in the event that there is a strong demand for such scientists outside academe. Then it may be optimal to invest a portion of its endowment in biotech stocks to hedge this cost, even though those stocks' returns are highly correlated with alum gifts and industry grants.

As demonstrated in section 7.3, the hedging role for endowment derived here is formally valid as long as there are traded securities with returns that have nonzero correlations with unanticipated changes in the activity costs. However, the practical significance for this role turns on the magnitude of the correlations. As illustrated in Bodie's (1976, 1982) work on hedging against inflation, it is often difficult to construct portfolios (using only standard types of traded securities) that are highly correlated with changes in the prices of specific goods and services. Nevertheless, the enormous strides in financial engineering over the last decade have greatly expanded the opportunities for custom financial contracting at reasonable costs. As we move into the twenty-first century, it will become increasingly more common for the financial services industry to offer its customers private contracts or securities that allow efficient hedging when the return properties of publicly traded securities are inadequate. That is, implementation of the quantitative strategies prescribed in sections 7.3 and 7.4 will become increasingly more practical for universities and other endowment institutions. See Merton (1990b, chap. 14; 1990c, 264–69) for a prospective view on financial innovation and the development of custom financial contracting.

There are of course a variety of issues involving endowment management that have not been addressed but could be within the context of our model. One such issue is the decision whether to invest endowment in specific-purpose real assets such as dormitories and laboratories instead of financial (or general-purpose physical) assets. The returns on those real assets are likely to be strongly correlated with the costs of particular university activities, and thereby the assets form a good hedge against unexpected rises in those costs. However, because the real-asset investments are specialized and largely irreversible, shifting the asset mix toward such investments reduces flexibility for the university. That is, with financial assets, the university has more options as to what it can do in the future. In future research, I plan to analyze this choice problem more formally by using contingent-claims analysis to value the trade-off between greater flexibility in selecting future activities and lower costs in producing a given set of activities.

Another issue not explicitly examined is the impact long-term, fixed liabilities such as faculty tenure contracts have on the management of endowment.

Our formal model of sections 7.3 and 7.4 that uses contingent-claims analysis (CCA) can handle this extension. See McDonald (1974) and Merton (1985) for CCA-type models for valuing tenure and other wage guarantee contracts.

In summary, the paper explores two classes of reasons why optimal endowment investment policy and expenditure policy can vary significantly among universities. The analysis suggests that trustees and others who judge the prudence and performance of policies by comparisons across institutions should take account of differences in both the mix of activities of the institutions and the capitalized values of their nonendowment sources of cash flows.

The overview completed, we now turn to the development of the mathematical model for the process and the derivation of the quantitative rules for implementation.

7.3   The Model

The functions or purposes of the university are assumed to be a collection of activities or outputs such as education, training, research, and storage of knowledge. We further assume that the intensities of those activities can be quantified and that a preference ordering exists for ranking alternative intertemporal programs. In particular, the criterion function for this ranking can be written as

$$(1) \qquad \max E_0\left[\int_0^\infty \bar{U}(Q_1, \ldots, Q_m, t)dt\right],$$

where $Q_j(t)$ denotes the quantity of activity j per unit time undertaken at time $t, j = 1, \ldots, m$; the preference function $\bar{U}$ is assumed to be strictly concave in $(Q_1, \ldots, Q_m)$; and E_t denotes the expectation operator, conditional on knowing all relevant information as of time t. This preference ordering satisfies the classic von Neumann–Morgenstern axioms of choice, exhibits positive risk aversion, and includes survival (of the institution) as a possible objective. The infinite time horizon structure in (1) implies only that there need not be a definite date when the university will liquidate. As shown in Merton (1990b, 149–51, 609–11), $\bar{U}$ can reflect the mortality characteristics of an uncertain liquidation date.

The intertemporally additive and independent preference structure in (1) can be generalized to include nonadditivity, habit formation, and other path-dependent effects on preferences, along the lines of Bergman (1985), Constantinides (1990), Detemple and Zapatero (1989), Duffie and Epstein (1992), Hindy and Huang (1992), Sundaresan (1989), and Svensson (1989). However, as shown in Merton (1990b, 207–9), those more realistic preference functions do not materially affect the optimal portfolio demand functions. Moreover, just as Grossman and Laroque (1990) show for transactions costs in consumption, so it can be shown here that imposing adjustment costs for changing the levels of university activities does not alter the structure of the portfolio demand functions. Hence, because the focus of the paper is on optimal investment (rather than on optimal expenditure) strategies, we assume no adjustment costs for activities and retain the additive independent preference specification to provide analytical simplicity.

Let $S_j(t)$ denote the (net) cost to the university of providing one unit of activity j at time t, $j = 1, \ldots, m$. For example, if $j = 1$ denotes the activity of having full-tuition–paying undergraduates, then S_1 would be the unit cost of providing the education minus the tuition received. If $j = 2$ denotes the activity of having undergraduates who receive financial aid, the unit cost S_2 would equal S_1 plus the financial aid given. In general, all costs and receipts such as tuition that are directly linked to the quantities of specific activities undertaken are put into the activity costs or prices, (S_j). As will be described, fixed costs and sources of positive cash flows to the university that do not depend directly on the activity quantities are handled separately. As in Merton (1990b, 202, 499), we assume that the dynamics for these costs are described by the stochastic differential equations: for $S \equiv (S_1, \ldots, S_m)$,

(2) $$dS_j = f_j(S,t)S_j dt + g_j(S,t)S_j dq_j \, , j = 1, \ldots, m \, ,$$

where f_j is the instantaneous expected rate of growth in S_j, g_j is the instantaneous standard deviation of the growth rate, and dq_j is a Wiener process with the instantaneous correlation coefficient between dq_i and dq_j given by v_{ij}, i, $j = 1, \ldots, m$. f_j and g_j are such that $dS_j \geq 0$ for $S_j = 0$, which ensures that $S_j(t) \geq 0$. Especially since (S) has components that depend on tuition, financial aid, and other variables over which the university has some control, one would expect that the dynamic path for those costs would be at least partially endogenous and controllable by the university, even though competition among universities would limit the degree of controllability. However, as specified, (2) is an exogenous process, not controlled by the university. Alternatively, it can be viewed as the "reduced-form" process for S after optimization over nonportfolio choice variables.

The university is assumed to have N nonendowment sources of cash flows, which we denote by $Y_k(t)dt$ for the kth source at time t. As noted in section 7.2, examples of such sources are gifts, bequests, university business income, and public- and private-sector grants. It can also be used to capture transfer pricing for the use of buildings and other university-specific assets where Y_k is the rental rate and this rental fee appears as an offsetting charge in the (S_j) for the appropriate university activities. The dynamics for these cash flows are modeled by, for $Y \equiv (Y_1, \ldots, Y_N)$,

(3) $$dY_k = \mu_k(Y,S,t)Y_k dt + \delta_k(Y,S,t)Y_k d\varepsilon_k \, ,$$

where μ_k and δ_k depend at most on the current levels of the cash flows and the unit costs of university activities and $d\varepsilon_k$ is a Wiener process, $k = 1, \ldots, N$. Equation (3) can also be used to take account of fixed costs or liabilities of the university such as faculty tenure commitments, by letting $Y_k < 0$ to reflect a cash outflow. However, the focus here is on assets only, and therefore we assume that μ_k and δ_k are such that $dY_k \geq 0$ for $Y_k = 0$, which implies that $Y_k(t) \geq 0$ for all t.

By inspection of (2) and (3), the dynamics for (Y,S) are jointly Markov. A more realistic model would have μ_k and δ_k depend on both current and historical values of $Q_1, \ldots, Q_m$. For example, if a university has undertaken large amounts of research activities in the past, it may attract more grants and gifts in the future. The university may also affect the future expected cash flows from nonendowment sources by investing now in building up those sources. Thus, the dynamic process for Y should be in part controllable by the university. However, again for analytical simplicity, the Y process is taken as exogenous, because that abstraction does not significantly alter the optimal portfolio demand functions.

If for $k = 1, \ldots, N$, $V_k(t)$ denotes the capitalized value at time t of the stream of future cash flows, $Y_k(\tau)$ for $\tau \geq t$, and if $K(t)$ denotes the value of the endowment at time t, then the net worth or wealth of the university, $W(t)$ is given by

$$(4) \qquad W(t) = K(t) + \sum_{1}^{n} V_k(t) .$$

A model for determining the $V_k(t)$ from the posited cash flow dynamics in (3) is developed in section 7.4.

The endowment of the university is assumed to be invested in traded assets. There are n risky assets and a riskless asset. If $P_j(t)$ denotes the price of the jth risky asset at time t, then the return dynamics for the risky assets are given by, for $j = 1, \ldots, n$,

$$(5) \qquad dP_j = \alpha_j P_j dt + \sigma_j P_j dZ_j ,$$

where α_j is the instantaneous expected return on asset j; σ_j is the instantaneous standard deviation of the return; and dZ_j is a Wiener process. The instantaneous correlation coefficients $(\rho_{ij}, \eta_{kj}, \zeta_{lj})$ are defined by, for $j = 1, \ldots, n$,

$$
\begin{aligned}
dZ_i dZ_j &= \rho_{ij} dt \ , \ i = 1, \ldots, n \\
dq_k dZ_j &= \eta_{kj} dt \ , \ k = 1, \ldots, m \\
d\varepsilon_l dZ_j &= \zeta_{lj} dt \ , \ l = 1, \ldots, N \ .
\end{aligned}
$$

(5a)

For computational simplicity and to better isolate the special characteristics of endowment management from general portfolio management, we simplify the return dynamics specification and assume that $(\alpha_j, \sigma_j, \rho_{ij})$ are constants over time, $i, j = 1, \ldots, n$. As shown in Merton (1990b, chaps. 4, 5, 6), this assumption of a constant investment opportunity set implies that $[P_j(t + \tau)/P_j(t)]$, $j = 1, \ldots, n$, for $\tau > 0$ are jointly lognormally distributed. The riskless asset earns the interest rate r, which is also constant over time. Optimal portfolio selection for general return dynamics would follow along the lines of Merton (1990a, sec. 7; 1990b, chaps. 5, 15, 16).

To analyze the optimal intertemporal expenditure and portfolio-selection problem for the university, we begin with a further simplified version of the model in which the university's entire net worth is endowment (i.e., $Y_k[t] = V_k(t) = 0$, $k = 1, \ldots ,N$ and $W(t) = K[t]$). The budget equation dynamics for $W(t)$ are then given by

$$
(6) \quad dW = \{[\sum_1^n w_j(t)(\alpha_j - r) + r] W - \sum_1^m Q_k S_k\}dt + \sum_1^n w_j(t)W\sigma_j dZ_j ,
$$

where $w_j(t) = $ the fraction of the university's wealth allocated to risky asset j at time t, $j = 1, \ldots ,n$; the fraction allocated to the riskless asset is thus $1 - \sum_1^n w_j$. Trustees, donors, and the government are assumed not to impose explicit limitations on investment policy for the endowment, other than general considerations of prudence. In particular, borrowing and short selling are permitted, so the choice for (w_j) is unrestricted. We further posit that spending out of endowment is not restricted, either with respect to overall expenditure or with respect to the specific activities on which it is spent. However, we do impose the feasibility restrictions that total expenditure at time t, $\sum_1^m Q_k S_k$, must be nonnegative and zero wealth is an absorbing state (i.e., $W[t] = 0$ implies $W[t + \tau] = 0$ for $\tau > 0$).

At each time t, the university chooses a quantity of activities $(Q_1, \ldots ,Q_m)$ and a portfolio allocation of its wealth so as to maximize lifetime utility of the university as specified in (1). Just as for the case of multiple consumption goods analyzed in Breeden (1979), Fischer (1975), and Merton (1990b, 205), so the solution for the optimal program here can be decomposed into two parts. First, at each t, solve for the utility-maximizing quantities of individual activities, $(Q_1, \ldots ,Q_m)$, subject to an overall expenditure constraint, $C(t) = \sum_1^m Q_k(t)S_k(t)$. Second, solve for the optimal level of overall expenditures at time t and the optimal portfolio allocation of endowment.

The first part is essentially the static activity-choice problem with no uncertainty

$$
(7) \quad \max_{(Q_1, \ldots ,Q_m)} \bar{U}(Q_1, \ldots ,Q_m,t)
$$

subject to $C(t) = \sum_1^m Q_k S_k(t)$. The first-order conditions for the optimal activity bundle $(Q_1^*, \ldots ,Q_m^*)$ are given by, for $S_k(t) = S_k$,

$$
(8) \quad \frac{\bar{U}_k(Q_1^*, \ldots ,Q_m^*,t)}{S_k} = \frac{\bar{U}_j(Q_1^*, \ldots ,Q_m^*,t)}{S_j} \quad k,j = 1, \ldots ,m
$$

with $C(t) = \sum_1^m Q_k^* S_k$, where subscripts on $\bar{U}$ denote partial derivatives (i.e., $\bar{U}_k \equiv \partial \bar{U}/\partial Q_k$). It follows from (8) that the optimal quantities can be written as $Q_k^* = Q_k^*[C(t),S(t),t]$, $k = 1, \ldots ,m$.

Define the indirect utility function U by $U[C(t),S(t),t] \equiv \bar{U}(Q_1^*, \ldots ,Q_m^*,t)$. By substituting U for $\bar{U}$, we can rewrite (1) as

(9)
$$\max E_0 \left\{ \int_0^\infty U[C(t),S(t),t]dt \right\} ,$$

where the "max" in (9) is over the intertemporal expenditure path $[C(t)]$ and portfolio allocations $[w_j(t)]$. Thus, the original optimization problem is transformed into a single-expenditure choice problem with "state-dependent" utility (where the "states" are the relative costs or prices of the various activities). Once the optimal total expenditure rules, $[C^*(t)]$, are determined, the optimal expenditures on individual activities are determined by (8) with $C^*(t) = \Sigma_1^m Q_k^* S_k$.

The solution of (9) follows by applying stochastic dynamic programming as in Merton (1990b, chaps. 4, 5, 6). Define the Bellman, or derived-utility, function J by

$$J(W,S,t) \equiv \max E_t \left\{ \int_t^\infty U[C(\tau),S(\tau),\tau]d\tau \right\}$$

conditional on $W(t) = W$ and $S(t) = S$. From Merton (1990a, 555; 1990b, 181, 202), J will satisfy

(10)
$$0 = \max_{(C,w)} \left[U(C,S,t) + \lambda C + J_t + J_W \{ [\sum_1^n w_j(\alpha_j - r) + r]W - C \} \right.$$
$$+ \sum_1^m J_i f_i S_i + \frac{1}{2} J_{WW} \sum_1^n \sum_1^n w_i w_j \sigma_{ij} W^2$$
$$+ \left. \sum_1^m \sum_1^n J_{iW} w_j W g_i S_i \sigma_j \eta_{ij} + \frac{1}{2} \sum_1^m \sum_1^m J_{ij} g_i S_i g_j S_j v_{ij} \right]$$

subject to $J(0,S,t) = \int_t^\infty \bar{U}(0, \ldots ,0,\tau)d\tau$, where subscripts on J denote partial derivatives with respect to W, t, and S_i, $i = 1, \ldots ,m$ and $\sigma_{ij} \equiv \rho_{ij}\sigma_i\sigma_j$, the instantaneous covariance between the return on security i and j. λ is a Kuhn-Tucker multiplier reflecting the nonnegativity constraint on C, and at the optimum it will satisfy $\lambda^* C^* = 0$. The first-order conditions derived from (10) are

(11a) $$0 = U_C(C^*,S,t) + \lambda^* - J_W(W,S,t)$$

and

(11b)
$$0 = J_W(\alpha_i - r) + J_{WW} \sum_1^n w_j^* W \sigma_{ij}$$
$$+ \sum_1^m J_{kW} g_k S_k \sigma_i \eta_{ki}, \quad i = 1, \ldots ,n ,$$

where $C^* = C^*(W,S,t)$ and $w_i^* = w_i^*(W,S,t)$ are the optimal expenditure, and portfolio rules expressed as functions of the state variables and subscripts on U denote partial derivatives.

From (11a), the optimal expenditure rule is given by

$$U_c(C^*,S,t) = J_W(W,S,t) \text{ for } C^* > 0$$

(12)
$$\lambda^* = \max [0, J_W(W,S,t) - U_c(0,S,t)]$$

From (11b), the optimal portfolio allocation can be written as

(13)
$$w_i^*W = Ab_i + \sum_1^m H_k h_{ki} , \ i = 1, \ldots, n ,$$

where $b_i \equiv \sum_1^n v_{ij}(\alpha_j - r)$; $h_{ki} \equiv \sum_1^n \sigma_j g_k S_k \eta_{kj} v_{ij}$; v_{ij} is the ij element of the inverse of the instantaneous variance-covariance matrix of returns (σ_{ij}); $A \equiv -J_W/J_{WW}$ (the reciprocal of absolute risk aversion of the derived-utility function); and $H_k \equiv -J_{kW}/J_{WW}$, $k = 1, \ldots, m$. A and H_k depend on the individual university's intertemporal preferences for expenditures and its current net worth. However, b_i and h_{ki} are determined entirely by the dynamic structures for the asset price returns and the unit costs of the various activities undertaken by universities. Hence, those parameters are the same for all universities, independent of their preferences or endowment size.

To provide some economic intuition about the optimal allocation of endowment in (13), consider as a frame of reference the "standard" intertemporal portfolio-selection problem with state-independent utility, $U = U(C(t),t)$. As shown in Merton (1990b, 131–36), given the posited return dynamics in (5), all such investors will hold instantaneously mean-variance efficient portfolios as their optimal portfolios. For $\partial U/\partial S_k \equiv U_k \equiv 0$, $H_k \equiv 0$, $k = 1, \ldots, m$. Hence, in this case, (13) becomes $w_i^*W = Ab_i$, and $w_i^*W/w_j^*W = b_i/b_j$, the same for all investors. This is the well-known result that the relative holdings of risky assets are the same for all mean-variance efficient portfolios. However, the state-dependent preferences for universities induced by the uncertainty surrounding the relative costs of undertaking different desired activities causes the more complex demand structure in (13).

To better understand this differential demand, $w_i^*W - Ab_i = \sum_1^m H_k h_{ki}$, it is useful to examine the special case where for each cost S_k there exists an asset whose instantaneous return is perfectly correlated with changes in S_k. By renumbering securities if necessary, choose the convention that $\eta_{kk} = 1$ in (5a), $k = 1, \ldots, m$ ($m < n$). As shown in Merton (1990b, 203–4), it follows that in this case, $h_{kk} = g_k S_k/\sigma_k$ for $k = 1, \ldots, m$ and $h_{kj} = 0$ for $k \neq j$. Hence, we can rewrite (13) as

$$w_i^*W = Ab_i + \frac{H_i g_i S_i}{\sigma_i} \ i = 1, \ldots, m$$

(14)
$$= Ab_i \qquad i = m + 1, \ldots, n .$$

By the strict concavity of U with respect to C, J is strictly concave in W. Hence, $J_{WW} < 0$ and $H_i = -J_{iW}/J_{WW}$ is positively proportional to J_{iW}. Thus, relative to a "normal" investor with state-independent preferences (i.e., $H_i \equiv 0$, $i = 1, \ldots, m$) but the same current level of absolute risk aversion (i.e., $-J_{WW}/J_W$), the university will optimally hold more of asset i if $J_{iW} > 0$ and less if $J_{iW} < 0$, $i = 1, \ldots, m$.

If $J_{iW} > 0$, then at least locally the university's marginal utility (or "need") for wealth or endowment becomes larger if the cost of undertaking activity i increases, and it becomes smaller if this cost decreases. Because the return on asset i is perfectly positively correlated with the cost of activity i, a greater than expected increase in S_i will coincide with a greater than expected return on asset i. By holding more of asset i than a "normal" investor would, the university thus assures itself of a relatively larger endowment in the event that S_i increases and the need for wealth becomes more important. The university, of course, pays for this by accepting a relatively smaller endowment in the event that S_i decreases and wealth is less important. The behavioral description for $J_{iW} < 0$ is just the reverse, because the need for endowment decreases if the cost of activity i increases.

To perhaps help in developing further insights, we use (12) to interpret the differential demand component in (14) in terms of the indirect utility and optimal expenditure functions. By differentiating (12), we have that, for $C^*(W,S,t) > 0$,

(15)
$$J_{WW} = U_{CC}(C^*,S,t) \frac{\partial C^*}{\partial W}$$

$$J_{kW} = U_{CC}(C^*,S,t) \frac{\partial C^*}{\partial S_k} + U_{Ck}(C^*,S,t)$$

$$A = \frac{-U_C(C^*,S,t)}{U_{CC}(C^*,S,t) \dfrac{\partial C^*}{\partial W}}$$

$$H_k = \frac{-\dfrac{\partial C^*}{\partial S_k}}{\dfrac{\partial C^*}{\partial W}} + \frac{U_{Ck}(C^*,S,t)}{-U_{CC}(C^*,S,t) \dfrac{\partial C^*}{\partial W}}$$

for $k = 1, \ldots, m$. Because $U_{CC} < 0$ and $\partial C^*/\partial W > 0$ for $C^* > 0$, we see that the sign of H_k is determined by the impact of a change in the cost of activity k on two items: the optimal level of total current expenditure and the marginal utility of expenditure. So, for example, if an increase in S_k would cause both a decrease in optimal expenditure ($\partial C^*/\partial S_k < 0$) and an increase in the marginal utility of expenditure ($U_{Ck} > 0$), then, from (15), $H_k > 0$ and the university will optimally hold more of asset k than the corresponding investor with a mean-variance efficient portfolio.

Following (14) causes the university's optimal portfolio to be mean-variance inefficient, and therefore the return on the endowment will have greater volatility than other feasible portfolios with the same expected return. However, the value of the endowment or net worth of the university is not the "end" objective. Instead, it is the "means" by which the ends of a preferred expenditure policy can be implemented. Viewed in terms of the volatility of the time path of *expenditure* (or more precisely, the *marginal utility of expenditure*), the optimal strategy given in (14) is mean-variance efficient (cf. Breeden 1979; Merton 1990b, 487–88). That is, because $\partial C^*/\partial W > 0$, the additional increment in wealth that, by portfolio construction, occurs precisely when S_k increases will tend to offset the negative impact on C^* caused by that increase. There is thus a dampening of the unanticipated fluctuations in expenditure over time. In sum, we see that in addition to investing in assets to achieve an efficient risk-return trade-off in wealth, universities should optimally use their endowment to hedge against unanticipated and unfavorable changes in the costs of the various activities that enter into their direct utility functions.

In closing this section, we note that the interpretation of the demand functions in the general case of (13) follows along the same lines as for the special case of perfect correlation leading to (14). As shown for the general case in Merton (1990a, 558–59; 1990b, 501–2), the differential demands for assets reflect attempts to create portfolios with the maximal feasible correlations between their returns and unanticipated changes in the S_k, $k = 1, \ldots, m$. These maximally correlated portfolios perform the same hedging function as assets $1, \ldots, m$ in the limiting case of perfect correlation analyzed in (14). Furthermore, if other state variables besides the various activities' costs (e.g., changes in the investment opportunity set) enter a university's derived utility function, then a similar structure of differential asset demands to hedge against the unanticipated changes in these variables will also obtain.

7.4 Optimal Endowment Management with Other Sources of Income

In the previous section, we identified hedging of the costs of university activities as a reason for optimally deviating from "efficient" portfolio allocations when endowment is the only means for financing those activities. In this section, we extend the analysis to allow other sources of cash flow to support the activities. To simplify the analysis, we make two additional assumptions. First, we posit that μ_k and δ_k in (3) are constants, which implies that $Y_k(t)/Y_k(0)$ is lognormally distributed, $k = 1, \ldots, N$. Second, we assume that for each k there exists a traded security whose return is instantaneously perfectly correlated with the unanticipated change in Y_k, $k = 1, \ldots, N$. By renumbering if necessary, we use the convention that traded security k is instantaneously perfectly correlated with Y_k. Hence, it follows that $\zeta_{kk} = 1$ in (5a) and

$$(16) \qquad d\varepsilon_k = dZ_k, \quad k = 1, \ldots, N .$$

These two assumptions permit us to derive a closed-form solution for the capitalized values of the cash flows, $[V_k(t)]$, using contingent-claims analysis. As will be shown, those valuation functions are independent of the university's preferences or wealth level.

From (3), (5), and (16) with μ_k and δ_k constant, we have that the cash flows can be written as a function of the traded asset prices as follows, for $k = 1$, . . . ,N,

$$(17) \qquad\qquad Y_k(t) = Y_k(0)\exp(-\phi_k t) \, [P_k(t)/P_k(0)]^{\beta_k} \, ,$$

where $\phi_k \equiv \beta_k (\alpha_k - \sigma_k^2/2) - (\mu_k - \delta_k^2/2)$ and $\beta_k \equiv \delta_k/\sigma_k$. That (17) obtains can be checked by applying Itô's Lemma. We now derive the capitalized value for Y_k, following Merton (1990a, 562–63; 1990b, 415–19).

Let $F^k (P_k, t)$ be the solution to the partial differential equation, for $0 \leq t \leq T_k$,

$$(18) \qquad\qquad 0 = 1/2 \, \sigma_k^2 P_k^2 F_{11}^k + r P_k F_1^k - r F^k + F_2^k + Y_k$$

subject to the boundary conditions:

$$(19a) \qquad\qquad F^k(0,t) \quad = 0$$
$$(19b) \qquad\qquad F^k/(P_k)^{\beta_k} \text{ bounded as } P_k \to \infty$$
$$(19c) \qquad\qquad F^k(P_k,T_k) = 0 \, ,$$

where subscripts on F^k in (18) denote partial derivatives with respect to its arguments P_k and t; Y_k is given by (17); and T_k is the last date at which the university receives the cash flows from source k, $k = 1, \ldots ,N$. It is a mathematical result that a solution exists to (18)–(19) and that it is unique. Moreover, for $Y_k \geq 0$, $F^k \geq 0$ for all P_k and t.

Consider a dynamic portfolio strategy in which $F_1^k[P_k(t),t]P_k(t)$ is allocated to traded asset k at time t and $V(t) - F_1^k[P_k(t),t]P_k(t)$ is allocated to the riskless asset, where $V(t)$ is the value of the portfolio at time t. Furthermore, let the portfolio distribute cash (by selling securities if necessary) according to the flow-rate rule

$$(20) \qquad\qquad D_2(P_k,t) = Y_k(t)$$

as given by (17). Then the dynamics of the portfolio can be written as, for $P_k(t) = P_k$ and $V(t) = V$,

$$(21) \qquad dV = F_1^k(P_k,t)dP_k + \{[V - F_1^k(P_k,t)P_k]r - D_2(P_k,t)\}dt \, .$$

Since F^k satisfies (18), it is a twice continuously differentiable function and therefore, by Itô's Lemma, we can write the dynamics for F^k as

$$(22) \qquad\qquad dF^k = 1/2 \, \sigma_k^2 \, P_k^2 \, F_{11}^k + F_2^k \, dt + F_1^k dP_k \, .$$

But F^k satisfies (18) and hence, $1/2 \, \sigma_k^2 \, P_k^2 \, F_{11}^k + F_2^k = r F^k - r P_k F_1^k - Y_k$. Substituting into (22), we can rewrite it as

(23) $$dF^k = F^k_1 dP_k + (rF^k - rP_kF^k_1 - Y_k)dt .$$

From (21) and (23), we have that

(24) $$dV - dF^k = (rV - rP_kF^k_1 - D_2 - rF^k + rP_kF^k_1 + Y_k)dt$$
$$= r(V - F^k)dt$$

because $D_2 = Y_k$. By inspection, (24) is an ordinary differential equation with solution

(25) $$V(t) - F^k[P_k(t),t] = \{V(0) - F^k[P_k(0),0]\}\exp(rt) .$$

Thus, if the initial investment in the portfolio is chosen so that $V(0) = F^k[P_k(0),0]$, then for all t and $P_k(t)$, we have that

(26) $$V(t) = F^k[P_k(t),t] .$$

To ensure that the proposed portfolio strategy is feasible, we must show that its value is always nonnegative for every possible sample path for the price P_k and all t, $0 \le t \le T_k$. Because F^k is the solution to (18) and $Y_k \ge 0$, $F^k \ge 0$ for all P_k and t. It follows from (26) that $V(t) \ge 0$ for all P_k and t. We have therefore constructed a feasible dynamic portfolio strategy in traded asset k and the riskless asset that produces the stream of cash flows $Y_k(t)dt$ for $0 \le t \le T_k$ and has zero residual value $[V(T_k) = 0]$ at T_k.

Because the derived strategy exactly replicates the stream of cash flows generated by source k, it is economically equivalent to owning the cash flows $Y_k(t)$ for $t \le T_k$. It follows that the capitalized value of these cash flows satisfies

(27) $$V_k(t) = F^k[P_k(t),t]$$

for $k = 1, \ldots ,N$. Note that by inspection of (18)–(19), F^k, and hence $V_k(t)$, does not depend on either the university's preferences or its net worth. The valuation for source k is thus the same for all universities.

Armed with (27), we now turn to the optimal policy for managing endowment when the university has N nonendowment sources of cash flows. The procedure is the one outlined in section 7.2. To derive the optimal policy, note first that even if those nonendowment sources cannot actually be sold by the university for legal, ethical, moral hazard, or asymmetric information reasons, the university can achieve the *economic* equivalent of a sale by following the "mirror image," or reverse, of the replicating strategy. That is, by (short selling or) taking a $-F_1^k[P_k(t),t]P_k$ position in asset k and borrowing $(F^k - F_1^kP_k)$ of the riskless asset at each t, the portfolio will generate a positive amount of cash, $F^k(P_k,t)$, available for investment in other assets at time t. The entire liability generated by shorting this portfolio is exactly the negative cash flows, $(-Y_kdt)$, for $t \le T_k$, because $V_k(T_k) = F^k(P_k, T_k) = 0$. But, since the university receives Y_kdt for $t \le T_k$ from source k, this short-portfolio liability is entirely offset. Hence, to undertake this strategy beginning at time

t is the economic equivalent of selling cash flow source k for a price of $V_k(t)$ $= F^k[P_k(t),t]$.

As discussed more generally in Merton (1990b, sec. 14.5, esp. 465–67), the optimal portfolio strategy will be as if all N nonendowment assets were sold and the proceeds, together with endowment, invested in the n risky traded assets and the riskless asset. This result obtains because it is feasible to sell (in the economic sense) the nonendowment assets and because all the economic benefits from those assets can be replicated by dynamic trading strategies in the traded assets. Hence, there is neither an economic advantage nor a disadvantage to retaining the nonendowment assets. It follows that the optimal demand for the traded risky assets is given by (13) and the demand for the riskless asset is given by $(1 - \Sigma_1^n w_i^*)W(t)$, where from (4) and (27)

$$(28) \qquad W(t) = K(t) + \sum_1^N F^k[P_k(t),t] \ .$$

Because, however, the university has not actually sold the nonendowment assets, the optimal demands given by (13) and (28) include both *implicit* and *explicit* holdings of the traded assets. That is, the university's ownership of nonendowment cash flow source k at time t is equivalent to having an additional net worth of $F^k[P_k(t),t]$, as reflected in (28), *and* to having $F_1^k[P_k(t),t]P_k(t)$ invested in traded asset k and $\{F^k[P_k(t),t] - F_1^k[P_k(t),t]P_k(t)\}$ invested in the riskless asset. Thus, ownership of source k causes implicit investments in traded asset k and the riskless asset. Optimal explicit investment in each traded asset is the position actually observed in the endowment portfolio, and it is equal to the optimal demand given by (13) and (28) minus the implicit investment in that asset resulting from ownership of nonendowment assets. Let $D_i^*(t)$ denote the optimal explicit investment in traded asset i by the university at time t. It follows from (13) that

$$(29) \qquad \begin{aligned} D_i^*(t) &= Ab_i + \sum_1^m H_k h_{ki} - F_1^i[P_i(t),t]P_i(t) \ , \quad i = 1, \ldots ,N \\ &= Ab_i + \sum_1^m H_k h_{ki} \qquad\qquad\qquad , \quad i = N + 1, \ldots ,n \ , \end{aligned}$$

where $W(t)$ used in the evaluation of A and H_k is given by (28). If we number the riskless asset by "$n + 1$," then explicit investment in the riskless asset can be written as

$$(30) \qquad \begin{aligned} D^*_{n+1}(t) &= [1 - \sum_1^n w_j^*(t)]W(t) - \sum_1^N \{F^i[P_i(t),t] - F_1^i[P_i(t),t]P_i(t)\} \\ &= K(t) - \sum_1^n D_j^*(t) \ . \end{aligned}$$

By inspection of (29), it is apparent that, in addition to the hedging of activity costs, the existence of nonendowment sources of cash flow will cause further differences between the observed holdings of assets in the optimal endowment portfolio and the mean-variance efficient portfolio of a "standard" investor. Similarly, from (30), the observed mix between risky assets and the riskless asset will differ from the true economic mix.

To explore further the effects of those nonendowment sources of cash flows, we solve the optimal expenditure and portfolio-selection problem for a specific utility function, $\tilde{U}$. However, in preparation for that analysis, we first derive explicit formulas for the capitalized values of those sources when $Y_k(t)$ is given by (17). As already noted, there exists a unique solution to (18) and (19). Hence, it is sufficient to simply find a solution. As can be verified by direct substitution into (18), the value of cash flow source k is given by, for $k = 1, \ldots, N$,

$$(31) \quad F^k[P_k(t),t] = Y_k(0)\exp(-\phi_k t)\{1 - \exp[-\theta_k(T_k - t)]\} \frac{\left(\frac{P_k(t)}{P_k(0)}\right)^{\beta_k}}{\theta_k},$$

where β_k, ϕ_k are as defined in (17) and

$$(31a) \qquad \theta_k \equiv r + \beta_k(\alpha_k - r) - \mu_k .$$

It follows from (31) that, for $k = 1, \ldots, N$,

$$(32) \qquad F_1^k[P_k(t),t]P_k(t) = \beta_k F^k[P_k(t),t] ,$$

which implies that the capitalized value of source k has a constant elasticity with respect to the price of traded asset k. Equation (32) also implies that the replicating portfolio strategy is a constant-proportion or rebalancing strategy which allocates fraction β_k of the portfolio to traded asset k and fraction $(1 - \beta_k)$ to the riskless asset. In the case when positive fractions are allocated to both assets (i.e., $[1 - \beta_k] > 0$ and $\beta_k > 0$), then F^k is a strictly concave function of P_k. If $\beta_k > 1$, then F^k is a strictly convex function of P_k, and the replicating portfolio holds traded asset k leveraged by borrowing. In the watershed case of $\beta_k = 1$, F^k is a linear function of P_k, and the replicating portfolio holds traded asset k only.

Using (17) and (27), we can rewrite (31) to express the capitalized value of source k in terms of the current cash flow it generates:

$$(33) \qquad V_k(t) = Y_k(t) \frac{1 - \exp[-\theta_k(T_k - t)]}{\theta_k}, \quad k = 1, \ldots, N .$$

From (17), (31a), and (32), it is a straightforward application of Itô's Lemma to show that the total expected rate of return for holding source k from t to $t + dt$ is given by

$$(34) \qquad \frac{E_t[Y_k(t)dt + dV_k]}{V_k(t)} = (\mu_k + \theta_k)dt$$

$$= [r + \beta_k(\alpha_k - r)]dt .$$

Thus, if the rights to the cash flows Y_k between t and T_k were sold in the marketplace, the expected rate of return that would be required by investors to bear the risk of these flows is $r + \beta_k(\alpha_k - r)$. Therefore, θ_k equals the required expected rate of return (the capitalization rate) minus the expected rate of growth of the cash flows, μ_k. By inspection of (33), $V_k(t)$ can be expressed by the classic present-value formula for assets with exponentially growing cash flows. For $\theta_k > 0$, the perpetual ($T_k = \infty$) value is $Y_k(t)/\theta_k$, and the limiting "earnings-to-price" ratio, $Y_k(t)/V_k(t)$, is θ_k, a constant. Applying the closed-form solution for F^k, we can by substitution from (27) and (32) into (29) and (30) rewrite the optimal demand functions as

$$D_i^*(t) = Ab_i + \sum_1^m H_k h_{ki} - \beta_i V_i(t) , i = 1, \ldots, N$$

(35a)

$$= Ab_i + \sum_1^m H_k h_{ki} \qquad , i = N + 1, \ldots, n$$

and

(35b)
$$D^*_{n+1}(t) = [1 - \sum_1^n w_j^*(t)]W(t) - \sum_1^N (1 - \beta_i)V_i(t)$$

$$= K(t) - A\sum_1^n b_j - \sum_1^m \sum_1^n H_k h_{kj} + \sum_1^N \beta_i V_i(t) .$$

Having derived explicit formulas for the values of nonendowment assets, we turn now to the solution of the optimal portfolio and expenditure problem in the special case where the university's objective function is given by

(36) $$\bar{U}(Q_1, \ldots, Q_m, t) = \exp(-\rho t)\sum_1^m \Gamma_j \log Q_j ,$$

with $\rho > 0$ and $\Gamma_j \geq 0, j = 1, \ldots, m$. Without loss of generality, we assume that $\sum_1^m \Gamma_i = 1$. From (8), the optimal Q_j satisfy

(37) $$Q_j^*(t) = \frac{\Gamma_j C(t)}{S_j(t)}, j = 1, \ldots, m .$$

From (36) and (37), the indirect utility function can be written as

(38) $$U(C,S,t) = \exp(-\rho t)\{\log C - \sum_1^m \Gamma_j[\log S_j - \log (\Gamma_j)]\} .$$

It follows from (11a) that the optimal expenditure rule is

(39) $$C^*(t) = \exp(-\rho t)\frac{1}{J_W(W,S,t)} .$$

It is straightforward to verify by substitution into (10), (11a), and (11b) that

(40) $$J(W,S,t) = \frac{1}{\rho} \exp(-\rho t)\log W + I(S,t)$$

for some function $I(S,t)$. By the verification theorem of dynamic programming, satisfaction of (10), (11a), and (11b) is sufficient to ensure that J in (40) is the optimum.

It follows from (40) that $J_{kW} = 0$ and hence that $H_k = 0$ in (13) and (35), $k = 1, \ldots, m$. Therefore, for the log utility specified in (36), there are no differential hedging demands for assets to protect against unanticipated changes in the costs of university activities. The optimal allocation of the university's total net worth is thus instantaneously mean-variance efficient. Noting that $A = -J_W/J_{WW} = W$, we have that (35) can be written in this special case as

$$
(41a) \qquad
\begin{aligned}
D_i^*(t) &= b_i W - \beta_i V_i(t) \, , \, i = 1, \ldots, N \\
&= b_i W \qquad\qquad , \, i = N + 1, \ldots, n
\end{aligned}
$$

and

$$
(41b) \qquad D_{n+1}^*(t) = \Big(1 - \sum_1^n b_j\Big)W - \sum_1^N (1 - \beta_i)V_i(t) \, .
$$

By inspection of (41), in the absence of nonendowment assets, the fraction of endowment allocated to risky asset i in the university's optimal portfolio is b_i, $i = 1, \ldots, n$, and the fraction allocated to the riskless asset is $(1 - \sum_1^n b_j)$, independent of the level of endowment. If $\chi_i^* \equiv D_i^*(t)/K(t)$ is the optimal fraction of endowment invested in asset i, then from (41) the difference in fractional allocations caused by the nonendowment assets is

$$
(42a) \qquad
\begin{aligned}
\chi_i^*(t) - b_i &= R(b_i - \beta_i \lambda_i) \, , \, i = 1, \ldots, N \\
&= Rb_i \qquad\qquad , \, i = N + 1, \ldots, n
\end{aligned}
$$

and

$$
(42b) \qquad \chi_{n+1}^*(t) - \Big(1 - \sum_1^n b_j\Big) = -R\Big(\sum_1^n b_k - \sum_1^N \beta_k \lambda_k\Big) \, ,
$$

where $\lambda_k \equiv V_k(t)/\sum_1^N V_i(t)$ is the fraction of the capitalized value of the university's total nonendowment assets contributed by cash flow source k at time t, $k = 1, \ldots, N$, and $R \equiv \sum_1^N V_i(t)/K(t)$ is the ratio of the values of the university's nonendowment assets to its endowment assets at time t.

As discussed in section 7.2, the differences in (42) are the result of two effects: (1) the "wealth" effect caused by the difference between the net worth and the endowment of the university and (2) the "substitution" effect caused by the substitution of nonendowment asset holdings for traded asset holdings. Suppose, for concreteness, that the expected returns, variances, and covariances are such that a positive amount of each traded risky asset is held in mean-variance efficient portfolios. Then, $b_i > 0$, $i = 1, \ldots, n$. It follows that the impact of the wealth effect in (42a) and (42b), (Rb_i), is unambiguous: it causes a larger fraction of the optimal endowment portfolio to be allocated to each risky asset and therefore a smaller percentage allocation to the riskless asset. Because $\beta_i \geq 0$ and $\lambda_i > 0$, $i = 1, \ldots, N$, we have that the impact of the substitution effect in (42a) and (42b), $(R\beta_i\lambda_i)$, is also unambiguous: for those traded assets $1, \ldots, N$ for which the nonendowment assets are substitutes, the fractional allocation is smaller; for the traded assets $N + 1, \ldots, n$, the fractional allocation is unchanged; and the allocation to the riskless asset thus increases.

Because the wealth and substitution effects are in opposite directions for $b_k > 0$, whether the optimal endowment portfolio allocates an incrementally larger or smaller fraction to traded asset k depends on whether $b_k > \beta_k\lambda_k$ or $b_k < \beta_k\lambda_k$. $\beta_k\lambda_k$ is the fraction of the total increment to net worth (from nonendowment assets) that is implicitly invested in asset k as the result of owning cash flow source k. If that fraction exceeds the optimal one for total wealth, b_k, then the optimal endowment portfolio will hold less than the mean-variance efficient allocation. Indeed, if $\lambda_k > (1 + R)b_k/(R\beta_k)$, then $\chi_k^*(t) < 0$ and the university would optimally short sell traded asset k in its portfolio. This is more likely to occur when R is large (i.e., nonendowment assets are a large part of university net worth) and λ_k is large (i.e., cash flow source k is a large part of the value of nonendowment assets).

The implications of (42a) and (42b) for optimal endowment fit the intuitions discussed at length in section 7.2. For instance, if a significant amount of gift giving to a particular university depends on the performance of the general stock market, then in effect that university has a "shadow" investment in that market. Hence, all else the same, it should hold a smaller portion of its endowment in stocks than another university with smaller amounts of such market-sensitive gift giving. As noted in section 7.2, much the same substitution-effect story applies to concentrations in other assets, including real estate. The same analysis also follows where grants from firms or the government are likely to be strongly correlated with the financial performance of stocks in the related industries. However, the underweightings in those assets for substitution-effect reasons can be offset by sufficiently strong demands to hedge against costs, as is illustrated by the biotech example in section 7.2.

The analysis leading to (29) and (30) requires that there exist traded securities which are instantaneously perfectly correlated with the changes in Y_1, $\ldots, Y_N$. If this "complete market" assumption is relaxed, then the capitalized values of those nonendowment cash flow sources will no longer be independent of the university's preferences and endowment. However, the impact on endowment investments will be qualitatively similar. This more general case of nonreplicable assets can be analyzed along the lines of Svensson (1988).

We can use our model to examine the impact of nonendowment cash flow sources on optimal expenditure policy. From (39) and (40), we have that the optimal expenditure rule is the constant-proportion-of-net-worth policy

(43) $$C^*(t) = \rho W(t) .$$

However, current expenditure *from endowment* will not follow a constant proportion strategy. Optimal expenditure from endowment at time t is $[C^*(t) - \sum_1^N Y_k(t)]dt$, which can be either positive or negative (implying net saving from nonendowment cash flow sources). If $s^*(t)$ denotes the optimal expenditure rate as a fraction of endowment ($\equiv [C^*(t) - \sum_1^N Y_k(t)]/K(t))$, then from (4) and (43),

(44) $$s^*(t) = \rho + R(t) [\rho - y(t)] ,$$

where $R(t)$ is as defined in (42a) and (42b) and $y(t) \equiv [\sum_1^N Y_k(t)]/[\sum_1^N V_k(t)]$ is the current yield on the capitalized value of the nonendowment sources of cash flow. In the special case of (33), where the cash flows are all perpetuities (i.e., $T_k = \infty$ and $\theta_k > 0$, $k = 1, \ldots, N$), $V_k(t) = Y_k(t)/\theta_k$ and the current yield on source k is constant and equal to θ_k. In that case, $y(t) = \sum_1^N \lambda_k \theta_k$, the value-weighted current yield. From (31a), θ_k will tend to be smaller for assets with higher expected growth rates of cash flow, (μ_k). If on average the current yield on nonendowment assets is less than ρ, then the current spending rate out of endowment will exceed ρ. If the current yield is high so that $y(t) > \rho$, then $s^*(t) < \rho$. Indeed, if $y(t) > \rho(1 + R)/R$, then $s^*(t) < 0$ and optimal total expenditure is less than current cash flow generated by nonendowment sources. Because both $R(t)$ and $\lambda_k(t)$ change over time, we have from (44) that the optimal current expenditure rate from endowment is not a constant, even when expected returns on assets, the interest rate, and the expected rate of growth of nonendowment cash flows are constants.

We can also analyze the dynamics of the mix of the university's net worth between endowment and nonendowment assets. If $\alpha \equiv r + \sum_1^n b_i(\alpha_i - r)$ denotes the instantaneous expected rate of return on the growth-optimum, mean-variance efficient portfolio, then, as shown in Merton (1990b, 169–71), the resulting distribution for that portfolio is lognormal with instantaneous expected return $\alpha(> r)$ and instantaneous variance rate equal to $(\alpha - r)$. It follows from (6), (41), and (43) that the dynamics for the university's net worth are such that $W(t)/W(0)$ is lognormally distributed with

$$E_0[W(t)] = W(0)\exp[(\alpha - \rho)t]$$

(45) $$E_0[\log\frac{W(t)}{W(0)}] = \left(\frac{\alpha + r}{2} - \rho\right)t$$

$$\text{Var}[\log\frac{W(t)}{W(0)}] = (\alpha - r)t .$$

If $X_k(t) \equiv V_k(t)/W(t)$ denotes the fraction of net worth represented by nonendowment cash flow source k, then, because V_k and W are each lognormally distributed, $X_k(t)$ is lognormally distributed, and from (33) and (45)

$$E_0[X_k(t)] = X_k(0)\exp[(\rho - \theta_k)t]$$

(46)
$$E_0\left[\log\frac{X_k(t)}{X_k(0)}\right] = \left[\mu_k + \rho - \frac{(\alpha + r + \delta_k^2)}{2}\right]t$$

$$\text{Var}[\log\frac{X_k(t)}{X_k(0)}] = [\delta_k^2 + \alpha + r - 2(\mu_k + \theta_k)]t$$

for $k = 1, \ldots, N$.

From (46), the fraction of total net worth represented by all sources of non-endowment cash flow, $X(t) \equiv \Sigma_1^N X_k(t) = R(t)/[1 + R(t)]$, is expected to grow or decline depending on whether $\rho > \theta_{\min}$ or $\rho < \theta_{\min}$ where $\theta_{\min} \equiv \min(\theta_k)$, $k = 1, \ldots, N$. In effect, a university with either a high rate of time preference or at least one (perpetual) high-growth nonendowment asset (i.e., with $\rho > \theta_{\min}$) is expected to "eat" its endowment. Indeed, it may even go to a "negative" endowment by borrowing against the future cash flows of its nonendowment assets. Whether this expected growth in $X(t)$ is the result of declining expected net worth or rising asset values can be determined from (45). Because $\alpha > r$, if $\rho \leq r$, then both the arithmetic and geometric expected rates of growth for net worth are positive. For $\rho < \theta_{\min}$, it follows that $E_0[X(t)] \to 0$ as $t \to \infty$. Hence, in the long run of this case, endowment is expected to become the dominant component of the university's net worth. Of course, these "razor's edge" results on growth or decline reflect the perpetual, constant-growth assumptions embedded in nonendowment cash flow behavior. However, this special case does capture the essential elements affecting optimal portfolio allocation and expenditure policies (cf. Tobin 1974).

The formal analysis here assumes that endowment is fungible for other assets and that neither spending nor investment policy are restricted. Such restrictions on endowment could be incorporated, using the same Kuhn-Tucker type analysis used in section 7.3 to take account of the constraint that total expenditure at each point in time is nonnegative. The magnitudes of the Kuhn-Tucker multipliers at the optimum would provide a quantitative assessment of the cost of each such restriction. However, including those restrictions is not likely to materially change the basic insights about hedging and diversification derived in the unrestricted case. The model can also be integrated into a broader one for overall university financial planning. Such integration would permit the evaluation of other nonendowment financial policies such as whether the university should sell forward contracts for tuition.

References

Bergman, Yakov. 1985. Time preference and capital asset pricing models. *Journal of Financial Economics* 14(March):145–59.

Black, Fischer. 1976. The investment policy spectrum: Individuals, endowment funds and pension funds. *Financial Analysts Journal* 32(January–February):23–31.

Bodie, Zvi. 1976. Common stocks as a hedge against inflation. *Journal of Finance* 31(May):459–70.

———. 1982. Investment strategy in an inflationary environment. In *The changing roles of debt and equity in financing U.S. capital formation,* ed. Benjamin Friedman, 47–64. Chicago: University of Chicago Press.

Breeden, Douglas T. 1979. An intertemporal asset pricing model with stochastic consumption and investment opportunities. *Journal of Financial Economics* 7(September):265–96.

Brinkman, Paul T. 1981. Factors affecting instructional costs at major research universities. *Journal of Higher Education* 52(May–June):265–79.

———. 1990. Higher education cost functions. In *The economics of American universities,* ed. Stephen Hoenack and Eileen Collins, 107–28. Albany: State University of New York.

Brovender, Shlomo. 1974. On the economics of a university: Toward the determination of marginal cost of teaching services. *Journal of Political Economy* 82(May–June):657–64.

Constantinides, George. 1989. Theory of valuation: Overview and recent developments. In *Theory of valuation: Frontiers of modern finance theory,* ed. Sudipto Bhattacharya and George Constantinides. Vol. 1, 1–23. Totowa, N.J.: Rowman and Littlefield.

———. 1990. Habit formation: A resolution of the equity premium puzzle. *Journal of Political Economy* 98(June):519–43.

Detemple, Jerome B., and Fernando Zapatero. 1989. Optimal consumption-portfolio policies with habit formation. Working Paper. New York: Graduate School of Business, Columbia University.

Duffie, Darrell, and Larry Epstein. 1992. Asset pricing with stochastic differential utility. *Review of Financial Studies* 5:411–36.

Eisner, Robert. 1974. Endowment income, capital gains and inflation accounting: Discussion. *American Economic Review* 64(May):438–41.

Ennis, Richard, and J. Peter Williamson. 1976. Spending policy for educational endowments. Research and Publication Project. New York: Common Fund, January.

Fama, Eugene, and Michael C. Jensen. 1985. Organizational forms and investment decisions. *Journal of Financial Economics* 14(March):101–19.

Fischer, Stanley. 1975. The demand for index bonds. *Journal of Political Economy* 83(June):509–34.

Ford Foundation Advisory Committee on Endowment Management. 1969. Managing educational endowments: Report to the Ford Foundation. New York: Ford Foundation.

Grinold, Richard, David Hopkins, and William Massy. 1978. A model for long-range university budget planning under uncertainty. *Bell Journal of Economics* 9(Autumn):396–420.

Grossman, Sanford, and Guy Laroque. 1990. Asset pricing and optimal portfolio choice in the presence of illiquid consumption goods. *Econometrica* 58(January):25–52.

Hansmann, Henry. 1990. Why do universities have endowments? *Journal of Legal Studies* 19(January):3–42.

Hindy, Ayman, and Chi-fu Huang. 1992. Intertemporal preferences for uncertain consumption: A continuous time approach. *Econometrica* 60(July):781–801.

Hoenack, Stephen, and Eileen Collins, eds. 1990. *The economics of American universities.* Albany: State University of New York Press.

Litvack, James, Burton Malkiel, and Richard Quandt. 1974. A plan for the definition of endowment income. *American Economic Review* 64(May):433–42.

Long, John B. 1974. Stock prices, inflation, and the term structure of interest rates. *Journal of Financial Economics* 1(July):131–70.

McDonald, John. 1974. Faculty tenure as a put option: An economic interpretation. *Social Science Quarterly* 55(September):362–71.

Merton, Robert C. 1977. A reexamination of the capital asset pricing model. In *Risk and return in finance,* ed. Irwin Friend and James Bicksler, Vol. 1, 141–60. Cambridge, Mass.: Ballinger.

———. 1985. Comment: Insurance aspects of pensions. In *Pensions, labor and individual choice,* ed. David A. Wise, 343–56. Chicago: University of Chicago Press.

———. 1990a. Capital market theory and the pricing of financial securities. In *Handbook of monetary economics,* ed. Benjamin Friedman and Frank Hahn, 498–581. Amsterdam: North-Holland.

———. 1990b. *Continuous-time finance.* Oxford: Basil Blackwell.

———. 1990c. The financial system and economic performance. *Journal of Financial Services Research* 4(December):263–300.

Nichols, Donald. 1974. The investment income formula of the American Economic Association. *American Economic Review* 64(May):420–26.

Nordhaus, William. 1989. Risk analysis in economics: An application to university finances. New Haven, Conn.: Cowles Foundation, Yale University, May. Unpublished paper.

Snyder, Thomas. 1988. Recent trends in higher education finance: 1976–77 to 1985–86. In *Higher education administrative costs: Continuing the study,* ed. Thomas Snyder and Eva Galambos, 3–23. Washington, D.C.: Office of Educational Research and Improvement, Department of Education.

Sundaresan, Suresh. 1989. Intertemporally dependent preferences and the volatility of consumption and wealth. *Review of Financial Studies* 2:73–89.

Svensson, Lars. 1988. Portfolio choice and asset pricing with nontraded assets. NBER Working Paper no. 2774. Cambridge, Mass.: National Bureau of Economic Research, November.

———. 1989. Portfolio choice with non-expected utility in continuous time. *Economic Letters* 39(October):313–17.

Tobin, James. 1974. What is permanent endowment income? *American Economic Review* 64(May):427–32.

Optimal Consumption-Investment Decisions Allowing for Bankruptcy: A Survey

S.P. Sethi

Abstract

This paper surveys the research on optimal consumption and investment problem of an agent who is subject to bankruptcy that has a specified utility (reward or penalty). The bankruptcy utility, modeled by a parameter, may be the result of welfare subsidies, the agent's innate ability to recover from bankruptcy, psychic costs associated with bankruptcy, etc. Models with non-negative consumption, positive subsistence consumption, risky assets modeled by geometric Brownian motions or semimartingales are discussed. The paper concludes with suggestions for open research problems.

Acknowledgement

This research is supported in part by SSHRC Grant 410-93-042.

1 Introduction

This paper surveys the research on the optimal consumption-investment problem facing a utility maximizing agent (an individual or a household) that is subject to bankruptcy, the utility being associated with consumption and with bankruptcy; for an in depth study of the problem, see Sethi (1997). The problem has its beginning in the classical works of Phelps (1962), Hakansson (1969), Samuelson (1969), and Merton (1969). In a finite-horizon discrete-time framework, Samuelson (1969) showed that for isoelastic marginal utility functions (i.e., $U'(c) = c^{\delta-1}, \delta < 1$), the optimal portfolio decision is independent of wealth in each period and independent of the consumption decision. More specifically, the portfolio is rebalanced at each period so that the fraction of wealth invested in the risky asset remains a constant. Merton (1969) confirmed the result in the continuous-time infinite horizon case.

A significant plateau was reached by Merton (1971), who formulated many interesting problems in continuous time with geometric Brownian motions to model the uncertainties in the prices of risky assets. He chose the utility of consumption $U(c)$ to belong to the HARA (**H**yperbolic **A**bsolute **R**isk **A**version) class and obtained explicit solution in the case when the marginal utility at zero consumption is infinite (i.e., $U'(0) = \infty$). Among the important findings was the statement of the so-called *mutual fund theorem* that allows, under certain conditions, efficient separation of the decision to invest in the individual assets from the more macro allocational choices among classes of assets. This result represents a multiperiod generalization of the well-known Markowitz-Tobin mean-variance portfolio rules.

Merton's (1971, 1973) analysis was erroneous in the case of HARA utility functions with $U'(0) < \infty$ as identified years later by Sethi and Taksar (1988). What Merton had done was to formally write the dynamic programming equation for the value function of the problem and provided an explicit solution of the equation. In the absence of a verification theorem, however, there is no guarantee that the solution obtained is the value function. Indeed, when $U'(0) < \infty$, not only his solution is not the value function, but if it were it would also imply negative consumption levels at some times. Missing in Merton's formulation were an all important boundary condition that the value function should satisfy and the requirement that consumption be nonnegative. Without a boundary condition, it is not possible to obtain a verification theorem and without the nonnegativity requirement, negative consumption may occur.

A simple boundary condition specifies the value function at zero wealth. In addition to being mathematically expedient, the value function at zero wealth signifies the reward or penalty, or more generally utility, associated with bankruptcy. The value of the reward or penalty associated with bankruptcy will have consequences on agent's decisions. Lippman, McCall and Winston (1980) underscore the importance of bankruptcy when they write 'Valid inferences concerning an agent's neutrality or aversion to risk must necessarily emanate from a highly robust model. Failure to include a constraint such as bankruptcy might very well produce the *maximally incorrect inference* (italics supplied)'.

The specific value of the utility at bankruptcy depends on what is assumed to happen in its wake. In most modern societies, the agent can count on welfare if and when he goes bankrupt. In this case, the value may represent the discounted expected utility of future consumption stream provided by the government. In addition, as Gordon and Sethi (1997) indicate, bankruptcy may carry with it negative or positive psychic income, the former to the extent that shame attaches to going bankrupt or living on the dole and the latter to the extent that poverty may be a blessing to devoutly religious people. Mason (1981) considered the case in which the agent might be reendowed and allowed to restart the decision problem. Sethi and Taksar (1992) consider delayed recovery model of bankruptcy. Whatever the case, it is sufficient for mathematical purposes to assign a utility P to bankruptcy, and include P as a parameter of the problem. Karatzas, Lehoczky, Sethi, and Shreve (1986) (KLSS hereafter) do this in their comprehensive treatment of the consumption-investment problem with nonnegative consumption requirement and bankruptcy.

We begin our survey with the discussion of the KLSS model in the next section. We also indicate how it generalizes the existing results, and discuss its implication for the agent's risk-aversion behavior as studied in Presman and Sethi (1991). In Section 3, we discuss models that require a subsistence or a minimum *positive* consumption rate, and the impact of this requirement on the risk-aversion behavior of the agent. Section 4 extends the models in the previous two sections to cases in which the asset prices are modeled with random coefficients. Section 5 examines the influence of imposing borrowing and shortselling constraints. The constraints can give rise to more complicated value functions than the concave ones obtained earlier. Section 6 concludes with a brief discussion of related research and open research problems.

2 Constant Market Coefficients with Nonnegative Consumption

In this section, we shall review models that assume constant interest rate, constant average mean rates of return on risky assets, and a variance-covariance matrix of constants. The models require nonnegative consumption rates. Models with consumption rates not less than a minimum required positive subsistence level will be reviewed in the next section. All models reviewed here allow explicitly for bankruptcy.

2.1 The KLSS Model

KLSS consider a single agent attempting to maximize total discounted utility from consumption over an infinite horizon. The agent begins with an initial wealth x and makes consumption and investment decisions over time, which is assumed to be continuous. The agent has his wealth in $N+1$ distinct assets available to him. One is riskless (deterministic) with a rate of return $r > 0$, whereas the others are risky and are modeled by geometric Brownian motions. More specifically, the price dynamics of the available assets are given by

$$\frac{dP_0(t)}{P_0(t)} = rdt, \tag{2.1}$$

$$\frac{dP_i(t)}{P_i(t)} = \alpha_i dt + \mathbf{e}_i \mathbf{D} d\mathbf{w}_t^T, \tag{2.2}$$

where $P_0(t)$ is the price of the riskless asset and $\mathbf{P}(t) = (P_1(t), P_2, \ldots, P_N(t))$ is the vector of prices of N risky assets at time t, with given initial prices $P_0(0)$ and $\mathbf{P}(0)$. Furthermore, $\{\mathbf{w}_t, t \geq 0\}$ is an N-dimensional standard Wiener process given on the probability space $(\Omega, \mathcal{F}, \mathcal{P})$, $\mathbf{e}_i$ is the unit row vector with a 1 in the i-th position, α_i is the average rate of return on the i-th asset, the volatility matrix $\mathbf{D}$ is an $N \times N$ matrix with $\Sigma = \mathbf{D}\mathbf{D}^T$, a positive definite variance-covariance matrix, and $(^T)$ denotes the transpose operation.

The agent specifies a consumption rate c_t, $t \geq 0$, and an investment policy $\boldsymbol{\pi}_t = (\pi_1(t), \ldots, \pi_N(t))$, $t \geq 0$, where $\pi_i(t)$ denotes the fraction of wealth invested in the i-th investment at time t. The remaining fraction $\pi_0(t) = 1 - (\pi_1(t) + \pi_2(t) + \ldots + \pi_N(t))$ is invested in the riskless asset. The vector $\boldsymbol{\pi}_t$ is unconstrained, implying that unlimited borrowing and shortselling are allowed. We assume no transaction costs for buying and selling assets. The consumption rate must be nonnegative, i.e.,

$$c_t \geq 0, \text{ a.s. } \omega, \text{ a.e. t.} \tag{2.3}$$

Both $C \triangleq \{c_t, t \geq 0\}$ and $\Pi \triangleq \{\boldsymbol{\pi}_t, t \geq 0\}$ must depend on the price vector $\{\mathbf{P}(t), t \geq 0\}$ in a nonanticipative way.

Given C and Π, it can be shown that the dynamics of the agent's wealth x_t, $t \geq 0$, satisfy the Itô stochastic differential equation

$$dx_t = (\boldsymbol{\alpha} - r\mathbf{1})\boldsymbol{\pi}_t^T x_t dt + (rx_t - c_t)dt + x_t \boldsymbol{\pi}_t \mathbf{D}d\mathbf{w}_t^T, \; x_0 = x, \tag{2.4}$$

where $\boldsymbol{\alpha} = (\alpha_1, \alpha_2, \ldots, \alpha_N)$ and $\mathbf{1} = (1, 1, \ldots, 1)$.

A complete formulation of the model requires some assumption concerning the options available to the agent if and when his wealth reaches zero, since further consumption would result in negative wealth. One possible, and quite general, treatment is to assign a value $P \in (-\infty, \infty)$ to bankruptcy and include it as a parameter of the model.

To define the agent's objective function, one needs to specify his utility function of consumption. This function U defined on $(0, \infty)$ is assumed to be strictly increasing, strictly concave, and *thrice* continuously differentiable. Extend U to $[0, \infty)$ by defining $U(0) = \lim_{c \downarrow 0} U(c)$. The agent chooses C and Π in order to maximize

$$V_{C,\Pi}(x) \triangleq E_x \left[\int_0^{T_x} e^{-\beta t} U(c_t)dt + Pe^{-\beta T_x} \right], \tag{2.5}$$

where

$$T_x = \inf\{t | x(t) = 0\} \tag{2.6}$$

is the stopping time of bankruptcy when the initial wealth is x and $\beta > 0$ is the agent's discount rate. $P = U(0)/\beta$ is equivalent to continuing the problem indefinitely after bankruptcy with only zero consumption, and is termed the *natural payment.*

The value function is defined as

$$V(x) = \begin{cases} \sup_{C,\Pi} V_{C,\Pi}(x), & \text{if } x > 0, \\ P, & \text{if } x = 0. \end{cases} \tag{2.7}$$

Define the nonnegative constant

$$\gamma = (\frac{1}{2})(\boldsymbol{\alpha} - r\mathbf{1})\Sigma^{-1}(\boldsymbol{\alpha} - r\mathbf{1})^T \tag{2.8}$$

and consider the quadratic equation

$$\gamma\lambda^2 - (r - \beta - \gamma)\lambda - r = 0 \tag{2.9}$$

with the solutions $\lambda_- < -1$ and $\lambda_+ > 0$ when $\gamma > 0$. When $\boldsymbol{\alpha} = r\mathbf{1}$ and $\beta < r$, define $\lambda_- = -r/(r - \beta)$. It is shown that $V(x)$ is finite for every $x > 0$ if

$$\int_c^\infty \frac{d\theta}{U'(\theta)^{\lambda_-}} < \infty, \ \forall c > 0. \tag{2.10}$$

Moreover, if (2.10) does not hold, then $V(x)$ could become infinite.

Presman and Sethi (1997c) show that if the agent had an exponentially distributed random lifespan with the mortality rate λ, his problem could be reduced to the KLSS problem of an infinite horizon agent whose discount rate is $\beta + \lambda$.

2.2 The Mutual Fund Theorem and the Reduced Model

In order to simplify the problem, choose any α and $\sigma > 0$ so that

$$\frac{(\alpha - r)^2}{2\sigma^2} = \gamma, \tag{2.11}$$

and consider the 'reduced' problem with a single risky asset with drift α and variance σ^2 and the riskless asset with the rate of return r. The term $(\alpha - r)$ is known as the risk premium and $(\beta + \gamma)$ the risk-adjusted discount rate.

The mutual fund theorem states that, at any point in time, the agent will be indifferent between choosing from a linear combination of the above two assets or a linear combination of the original $(N + 1)$ assets. It is termed the mutual fund theorem, because the single risky asset can be thought of as a mutual fund. If one constructs a mutual fund which trades continuously using a self-financing strategy to maintain the proportions of the riskless and N risky assets given by the $(N + 1)$-dimensional vector

$$(1 - (\boldsymbol{\alpha} - r\mathbf{1})\sum\nolimits^{-1}\mathbf{1}^T, \ (\boldsymbol{\alpha} - r\mathbf{1})\sum\nolimits^{-1}),$$

then the mutual fund has mean return $\alpha = r + 2\gamma$ and variance $\sigma^2 = 2\gamma$, which satisfy (2.11). Moreover, if $(\boldsymbol{\alpha} - r\mathbf{1})\Sigma^{-1}\mathbf{1}^T \neq 0$, then the mutual fund consisting only of risky stocks held with proportions $(\boldsymbol{\alpha} - r\mathbf{1})\Sigma^{-1}/(\boldsymbol{\alpha} - r\mathbf{1})\Sigma^{-1}\mathbf{1}^T$ also satisfies (2.11).

This important theorem was first stated by Merton (1971) for the dynamic consumption-investment problem without bankruptcy considerations and without a rigorous proof. The rigorous proof is supplied by KLSS for all values of P. The theorem generalizes the Markowitz-Tobin separation theorem to multiple periods. Moreover, in the special case when $(\boldsymbol{\alpha} - r\mathbf{1})\Sigma^{-1}\mathbf{1}^T \neq 0$, the derived optimal portfolio policy has the same structure as that prescribed in the mean-variance model.

The mutual fund theorem is based on the strict concavity of the value function $V(x)$, which, in turn, is brought about by the assumption that the investment vector $\boldsymbol{\pi}_t$ is unconstrained.

In view of the mutual fund theorem, it suffices to consider the reduced problem with the modified wealth dynamics

$$dx_t = (\alpha - r)\pi_t x_t dt + (rx_t - c_t)dt + x_t \pi_t \sigma dw_t, \quad x_0 = x, \qquad (2.12)$$

in place of (2.4), where $\{w_t, t \geq 0\}$ is a standard Wiener process and π_t denotes the fraction of the wealth invested in the risky asset.

2.3 The HJB Equation and the Solution of the Problem

From the theory of stochastic optimal control, it is known that the value function $V(x)$ must satisfy the HJB (Hamilton-Jacobi-Bellman) equation

$$\left.\begin{aligned}
\beta V(x) &= \max_{c \geq 0, \pi}[(\alpha - r)\pi x V'(x) + (rx - c)V'(x) \\
&\qquad\qquad + \tfrac{1}{2}\pi^2\sigma^2 x^2 V''(x) + U(c)], \\
x &> 0, \\
V(0) &= P.
\end{aligned}\right\} \qquad (2.13)$$

Assume $\alpha \neq r$; see Lehoczky, Sethi and Shreve (1983) or Section 5.2 for the special case $\alpha = r$.

The optimal feedback policies for investment and consumption, respectively, are

$$\pi(x) = -\frac{(\alpha - r)V'(x)}{\sigma^2 x V''(x)} \qquad (2.14)$$

and

$$c(x) = \max\{U'^{-1}(V'(x)), 0\}. \qquad (2.15)$$

When (2.14) and (2.15) are substituted in (2.13), it results in a highly non-linear differential equation, which appears to be very difficult to solve at first sight. However, KLSS discovered a change of variable that allowed them to convert the nonlinear equation into a linear second-order differential equation in a variable that represents the inverse of the marginal (indirect or derived) utility of wealth given by the first derivative $V'(x)$ of the value function. Since the resulting equation has many solutions depending on the constants of integration, one needs to identify the values of the constants that would yield the value function. Furthermore, when the candidate feedback policies are expressed in terms of the solution of the linear differential equation involving the constants, KLSS discovered surprisingly that the candidate marginal utility of wealth over time can be written as a process satisfying a linear Itô's stochastic differential equation. It is then a simple matter to evaluate the objective function value associated with the candidate policies and identify the one satisfying the HJB equation. The procedure yields the value function in view of the additional fact that any function satisfying (2.13) majorizes the value function as shown in KLSS.

Solutions for the general consumption utility functions have been obtained in KLSS. Because of the space limitation, we characterize the results in Table 1. In this table, q denotes the probability of bankruptcy under the optimal policy, and P^*, $\bar{x}$, and a are given as:

$$P^* = \frac{1}{\beta}U(0) - \frac{U'(0)^{1+\lambda_-}}{\beta\lambda_-}\int_0^\infty \frac{d\theta}{U'(\theta)^{\lambda_-}}, \tag{2.16}$$

$$\bar{x} = \frac{[\beta P - U(0)]^{\frac{\lambda_- - \lambda_+}{1+\lambda_-}} - \left[\int_0^\infty \frac{d\theta}{U'(\theta)^{\lambda_-}}\right]^{\frac{1+\lambda_+}{1+\lambda_-}}}{\gamma(\lambda_+ - \lambda_-)(-\lambda_-)^{\frac{1+\lambda_+}{1+\lambda_-}}}U'(0)^{\lambda_+}$$

$$- \frac{\int_0^\infty \frac{d\theta}{U'(\theta)^{\lambda_-}}}{\gamma\lambda_-(\lambda_+ - \lambda_-)}U'(0)^{\lambda_-}, \tag{2.17}$$

and a is given by the unique positive solution for c in the equation

$$-\frac{U'(c)^{1+\lambda_-}}{\gamma\lambda_-(1+\lambda_-)}\int_c^\infty \frac{d\theta}{U'(\theta)^{\lambda_-}} - \frac{1+\lambda_+}{\beta}U(c) + \frac{\lambda_+}{r}cU'(c) = -(1+\lambda_+)P. \tag{2.18}$$

Formulas for the value function $V(x)$, modulo some transcendental equations, are derived in KLSS. Given $V(x)$, the optimal feedback policies can be obtained from (2.14) and (2.15).

2.4 Solutions for the HARA Utility Class

The HARA utility functions on $(0, \infty)$ have the form

	$P \leq \frac{1}{\beta}U(0)$	$\frac{1}{\beta}U(0) < P \leq P^*$	$P^* < P < \frac{1}{\beta}U(\infty).$	$\frac{1}{\beta}U(\infty) \leq P$
$U'(0)$ $= \infty$	$c_t > 0,$ $q = 0.$	$c_t > a > 0,$ $0 < q < 1$ if $\beta < r+\gamma,$ $q = 1$ if $\beta \geq r+\gamma.$		Consume quickly to bankruptcy.
$U'(0)$ $< \infty$	$c_t = 0$ if $x_t \in (0,\bar{x}],$ $c_t > 0$ if $x_t \in (\bar{x},\infty),$ $q = 0.$	$c_t = 0$ if $x_t \in (0,\bar{x}],$ $c_t > 0$ if $x_t \in (\bar{x},\infty),$ ($\bar{x}$ when $P = P^*$.) $0 < q < 1$ if $\beta < r+\gamma,$ $q = 1$ if $\beta \geq r+\gamma.$	$c_t > a > 0,$ $0 < q < 1$ if $\beta < r+\gamma,$ $q = 1$ if $\beta \geq r+\gamma.$	No optimal policy. $V(x) = P,$ $x \geq 0$

Table 1: Characterization of optimal consumption and bankruptcy probability

$$U = (1/\delta)(c+\eta)^\delta, \quad \delta < 1, \quad \delta \neq 0, \quad \eta \geq 0, \qquad (2.19)$$

$$U = \log(c+\eta), \quad \eta \geq 0. \qquad (2.20)$$

The log utility function (2.20) is referred to as the HARA function with $\delta = 0$. In these cases, the growth condition (2.10) specializes to $\beta > r\delta + \gamma\delta/(1-\delta)$, which is weaker than $\beta > r\delta + \gamma(2-\delta)/(1-\delta)$ imposed by Merton [1969, (condition (41))].

Merton [1971, 1973] provides explicit solutions for $V(x)$ in these cases. His solutions, however, are correct *only* for $\eta = 0$, i.e., when $U'(0) = \infty$. For $\eta = 0$, these solutions are

$$V_\delta(x) = \frac{1}{\delta}\left[\frac{1-\delta}{\beta - r\delta - \gamma\delta/(1-\delta)}\right]^{1-\delta} x^\delta, \quad x \geq 0, \qquad (2.21)$$

$$V_0(x) = (1/\beta)\log\beta x + \frac{(r-\beta+\gamma)}{\beta^2}, \quad x \geq 0. \qquad (2.22)$$

for utility functions (2.19) and (2.20), respectively. By (2.14) and (2.15), we have the optimal investment and consumption policies

$$\pi(x) = \frac{\alpha - r}{(1-\delta)\sigma^2} \quad \text{and} \quad c(x) = \frac{1}{1-\delta}\left(\beta - r\delta - \frac{\gamma\delta}{1-\delta}\right)x. \qquad (2.23)$$

2.5 Bankruptcy with Delayed Recovery

Sethi and Taksar (1992) introduced a model of nonterminal bankruptcy that is equivalent to the KLSS model. In this model, an agent, upon going bankrupt, may recover from it after a temporary but random sojourn in bankruptcy.

Such recovery may be brought about in a number of ways, e.g., the individual may generate an innovative idea having commercial value. The rate of such recovery reflects essentially his innate ability or resourcefulness. However, such a recovery is not instantaneous. The individual must stay in the bankruptcy state for a positive amount of time and during this time, his consumption rate must be zero. This type of bankruptcy can be modeled by a continuous diffusion process with a delayed reflection.

The wealth equation changes to

$$
\left.
\begin{aligned}
dx(t) &= [(\alpha - r)\pi(t)x(t) + rx(t) - c(t)]1_{x(t)>0}dt \\
&\quad + \mu 1_{x(t)=0}dt + x(t)\pi(t)\sigma dw(t), \\
x(0) &= x
\end{aligned}
\right\}
\qquad (2.24)
$$

The equation shows that the recovery rate μ can be viewed as the rate of wealth accumulation during the time when $x(t) = 0$; this permits the investor to leave the bankruptcy state (Gihman and Skorohod, 1972, Section 24).

Sethi and Taksar (1992) show that for every recovery rate μ, there is a bankruptcy utility P that makes their model equivalent to the KLSS model, and vice versa.

In addition to providing an alternative model of bankruptcy, the nonterminal bankruptcy may be a way toward an eventual development of an equilibrium model that incorporates bankruptcy. Further discussion in this regard is deferred to Section 6.

2.6 Analysis of the Risk-Aversion Behavior

While KLSS had obtained an explicit solution of the problem, the specification of the value function was still too complicated to examine the implied risk-aversion behavior in detail. The analysis was made possible by yet another change of variable introduced by Presman and Sethi (1991). They defined a variable equal to the logarithm of the inverse of the marginal utility of wealth. This allowed them to obtain a linear second-order differential equation in wealth as a function of the new variable, and whose solution can be obtained in a parametric form with the parameter standing for the utility of bankruptcy. In other words, given the bankruptcy utility P, there is a unique choice of this parameter that makes the solution of the differential equation correspond exactly to the value function. Thus, unlike in KLSS, it unifies the cases in which optimal solution may or may not involve consumption at the boundary. Furthermore, it extends the KLSS analysis to utility functions that need only to be continuously differentiable rather than thrice so as assumed in KLSS.

Presman and Sethi (1991) studied the Pratt-Arrow risk-aversion measures, namely the coefficient of the absolute risk aversion

$$l_V(x) = -\frac{d\ln V'(x)}{dx} = -\frac{V''(x)}{V'(x)} \tag{2.25}$$

and the coefficient of the relative or proportional risk aversion

$$L_V(x) = -\frac{d\ln V'(x)}{d\ln x} = xl_V(x). \tag{2.26}$$

with respect to the value function $V(x)$ denoting the derived utility associated with the wealth level x. Note for later discussion purposes that (2.25) also defines the coefficient $l_U(c)$ associated with the consumption utility $U(c)$.

Merton (1971) obtained some results relating the nature of the value function to the nature of the utility function for consumption assumed to be of HARA type. When $\eta = 0$ (i.e., when $U'(0) = \infty$) and $P \le U(0)/\beta$, the value function of the problem is also of HARA type with the same parameter as the one for the HARA utility of consumption used in the problem. Thus, the coefficient of absolute risk aversion decreases with wealth, while that of relative risk aversion is constant of value $(1 - \delta)$.

Merton's results obtained for the HARA case are not correct for $\eta > 0$ or $P > U(0)/\beta$. In these cases, Presman and Sethi (1991) show that the agent's value function is no longer of HARA type; while Merton (1990) recognizes the errors in Merton (1971) as pointed out by Sethi and Taksar (1988), he does not update the risk-aversion implications of the corrected solutions.

With regards to an agent's relative risk aversion, first we note that $L'_U > 0$ for $U(c)$ specified in (2.19) and (2.20) with $\eta > 0$. The agent's relative risk aversion increases with wealth provided $\eta > 0$ or $P > U(0)/\beta$. In other words, while not of HARA type, the value function inherits the qualitative behavior from the HARA utility of consumption used in the problem. However, for $\eta > 0$ and $P \le U(0)/\beta$, the inheritance holds only at higher wealth levels, while at lower wealth levels, the agent's relative risk aversion remains constant.

The agent's absolute risk aversion behavior is more complicated for $\eta > 0$ or $P > U(0)/\beta$. If δ is sufficiently large, for which it is necessary that $\beta + \gamma - r > 0$, then absolute risk aversion decreases with wealth. For smaller values of δ and $\beta + \gamma - r \ge 0$, however, absolute risk aversion decreases with wealth if bankruptcy payment is sufficiently low; otherwise risk aversion increases at lower levels of wealth, while it decreases at higher levels of wealth. Furthermore, if $\beta + \gamma - r < 0$, then for every δ and every $P > U(0)/\beta$, the absolute risk aversion increases at lower levels of wealth, while it decreases at higher levels of wealth.

From the above discussion, one may draw the following general conclusion regarding the risk aversion behavior in the HARA case with $\eta > 0$. At higher wealth levels, the agent's absolute (relative) risk-aversion decreases (increases) with wealth. This qualitative behavior at high wealth levels is inherited from the agent's HARA type consumption utility, as the agents seems quite immune from the bankruptcy payment parameter P. Of course, what is considered to be a high wealth level itself may depend on P.

At lower wealth levels, the agent is no longer immune from the amount of payment at bankruptcy. His behavior at these wealth levels is somewhat complicated, and it results from the interaction of his consumption utility, the bankruptcy payment, and the relationship of his risk-adjusted discount and the risk-free rate of return; see Presman and Sethi (1991) for details.

To describe the risk-aversion behavior with general concave utility functions, the situation is far more complex. The most surprising observation is that while the sign of the derivative of the coefficient of local risk-aversion depends on the entire utility function, it is nevertheless explicitly independent of U'' and U''' or even their existence. Both absolute and relative risk aversions decrease as bankruptcy payment P increases. Also derived for all values of P are some necessary and sufficient conditions for absolute risk aversion to be decreasing and relative risk aversion to be increasing as wealth increases. Furthermore, the relative risk aversion increases with wealth in the neighborhood of zero wealth. Moreover, if there exists an interval of zero consumption (which happens when $U(0)/\beta \leq P < P^*$), then the relative risk aversion increases with wealth in this interval. In the neighborhood of zero wealth and in the interval of zero consumption, the absolute risk aversion increases (decreases) with wealth according as $\beta + \gamma - r < 0 (> 0)$ for $P < P^*$.

Presman and Sethi (1991) also show that if $\beta + \gamma - r \leq 0$, then either the absolute risk aversion increases with wealth for all P or for each wealth level there exists a bankruptcy payment $P(x)$ such that at x the risk aversion is decreasing for payments smaller than $P(x)$ and increasing for payments larger than $P(x)$.

Finally, contrary to the intuitive belief that the absolute risk aversion is nonincreasing as wealth approaches infinity, the limiting behavior at infinity is much more complex.

3 Constant Market Parameters and Positive Subsistence Consumption

Sethi, Taksar and Presman (1992) provided an explicit specification of the optimal policies in a general consumption- investment problem of a single agent with subsistence consumption and bankruptcy. In doing so, they used the methods developed in KLSS and Presman and Sethi (1991). Their model and results are discussed in the next two subsections. In Section 3.3, we characterize the risk-aversion behavior of the agent studied by Presman and Sethi (1997a). Not reported here is the agent's consumption behavior obtained in Presman and Sethi (1997b). Explicit formulas for the bankruptcy probability were obtained by Presman and Sethi (1996), and are reported in Section 3.4.

3.1   The Sethi–Taksar–Presman Model and the HJB Equation

A number $s \geq 0$, termed the subsistence consumption rate, and a concave utility function $U(c)$ of consumption are given such that $U = -\infty$ for $c < s$, $U(c)$ is strictly increasing and continuously differentiable for $c > s$, $U(s) = \lim_{c \downarrow s} U(c)$, and $U(c)$ is sublinear, i.e., $U'(\infty) = \lim_{c \to \infty} U'(c) = 0$.

The goal of the agent is to maximize the objective defined in (2.5) over all policies, i.e., over all random processes (C, Π) defined in Section 2.1 except that the condition (2.3) is replaced by

$$c_t \geq s, \text{ a.s. } \omega, \text{ a.e. } t. \tag{3.1}$$

The value function, now denoted as $V(x; P)$ to show its dependence on P explicitly, can be defined as before in (2.7).

Let

$$\bar{P} = U(\infty)/\beta, \qquad \tilde{P} = \begin{cases} -\infty & \text{if } s > 0, \\ U(0)/\beta & \text{if } s = 0. \end{cases} \tag{3.2}$$

For $P \geq \bar{P}$, one would like to consume instantaneously to bankruptcy and the value function $V(x; P)$ is then identical to P. Since an instantaneous lump-sum consumption is not admissible, strictly speaking there exists no optimal strategy in this case. However, one can approach instantaneous bankruptcy arbitrarily closely, and attain an objective function value arbitrarily close to P. For $P \leq \tilde{P}, V(x; P) = V(x; \tilde{P}), x > 0$. When $s = 0$, the value $P = U(0)/\beta$, called the *natural payment* in KLSS, is the value of consuming nothing over the entire future (a feasible consumption policy) following bankruptcy. When $s > 0$, the value $U(s)/\beta$ represents the value of consuming at the subsistence level over the entire future following the time of bankruptcy. Therefore, we shall assume $P \in [\tilde{P}, \bar{P})$ in the rest of the paper.

Suppose that for some $c > s$, the inequality (2.10) holds; otherwise there are cases when $V(x; P) = \infty$. The value function $V(x; P)$ satisfies the following HJB equation:

$$\left.\begin{aligned} \beta V(x; P) &= \max_{c \geq s, \pi}\{(\alpha - r)\pi x V'(x; P) + (rx - c)V'(x; P) \\ &\qquad + \tfrac{1}{2}(\sigma \pi x)^2 V''(x; P) + U(c)\}, \\ x &> 0, \end{aligned}\right\} \tag{3.3}$$

with

$$V(0; P) = P \text{ for } \tilde{P} < P < \bar{P} \qquad \text{and} \qquad \lim_{x \to s/r} V(x; \tilde{P}) = U(s)/\beta. \quad (3.4)$$

Equation (3.3) has the same form as the HJB equation in (2.13); we have put in the parameter P as an argument of the value function to emphasize its dependence on P.

3.2 Results and Interpretation

The optimal solution (x_t^*, c_t^*, π_t^*), $t \geq 0$, of the agent's problem exists and is unique, and the value function is strictly concave in wealth x. Moreover, it is possible to express the optimal rate of consumption and the fraction of the wealth invested in the risky asset as $c_t^* = c(x_t^*; P)$ and $\pi_t^* = \pi(x_t^*; P)$, respectively, where $c(x; P)$ and $\pi(x; P)$ are appropriate functions that are determined as part of the solution. More precisely, the following properties hold:

(i) If $(\Pi^*, C^*) = (\pi_t^*, c_t^*)_{t \geq 0}$ denotes the optimal investment/consumption strategy given the initial wealth x, and x_t^* denotes the corresponding wealth process $(x_0^* = x)$, then $\pi_t^* = \pi(x_t^*; P)$ and $c_t^* = c(x_t^*; P)$, where

$$\pi(x; P)x = -\frac{\alpha - r}{\sigma^2} \frac{V'(x; P)}{V''(x; P)}, \qquad (3.5)$$

$$c(x; P) = \begin{cases} (U')^{-1}(V'(x; P)) & \text{if } 0 < V'(x; P) \leq U'(s), \\ s & \text{if } V'(x; P) \geq U'(s). \end{cases} \qquad (3.6)$$

(ii) Let $P \in (\tilde{P}, \bar{P})$. The optimal wealth process x_t^* leads to bankruptcy with positive probability, which equals one iff $\beta \geq r + \gamma$. The optimal behavior of consumption depends on the values $U'(s)$ and P. When $U'(s) = \infty$, the subsistence is not binding for all values of P, i.e. optimal consumption is strictly above subsistence $(c(x; P) > s$ for all $x > 0)$. Moreover, $c(x; P)$ is bounded away from s and is strictly increasing in x for all $x > 0$. The same property is true in case $0 < U'(s) < \infty$ only for big values of P (i.e. for $P \geq P^*(s)$, where $P^*(s)$ for a given s is some number in the interval $(\tilde{P}, \bar{P})$ that can be determined as part of the solution), but for $P = P^*(s)$ the optimal consumption is not bounded away from s. When $0 < U'(s) < \infty$ and $P < P^*(s)$, there exists an nonempty interval of wealth near zero (we denote this interval by $[0, \bar{x}(P, s)]$), where consumption is at subsistence. In this case $c(x; P) = s$ for $0 \leq x \leq \bar{x}(P, s)$ and $c(x; P)$ is strictly increasing on x for $x > \bar{x}(P)$.

(iii) Let $P = \tilde{P}$ and $x > s/r$. Then the optimal wealth process is always bounded below by s/r, and so there is no bankruptcy. The wealth never falls to the level s/r if $U(s) = -\infty$. If $U'(s) = \infty$, then the optimal consumption is strictly above subsistence, but not bounded away from s. However, if $U'(s) < \infty$, then $c(x; P) = s$ for $s/r < x \le \bar{x}(s) = \lim_{P \downarrow \tilde{P}} \bar{x}(P, s)$, $c(x; P) > s$ and is strictly increasing on x for $x > \bar{x}(s)$.

(iv) Let $P = \tilde{P}$ and $0 < x \le s/r$ (in this case $s > 0$ and $\tilde{P} = -\infty$).

(a) Consider $x = s/r$. Then for $U(s) > -\infty$, the optimal consumption is at subsistence, $V(s/r; P) = U(s)/\beta$, the optimal wealth process identically equals to s/r, and there is no bankruptcy. For $U(s) = -\infty$, the value function $V(s/r; P) = -\infty$ and, therefore, consumption is arbitrary and the probability of bankruptcy can take any value in $[0, 1]$ depending on the particular consumption chosen by the agent.

(b) If $0 < x < s/r$, then bankruptcy cannot be avoided with certainty. Therefore, $V(x; P) = -\infty$ and every consumption policy is optimal. The probability of bankruptcy takes any value in $(0, 1]$ depending on the selected consumption policy.

Now that we have summarized the results, let us provide their economic interpretations.

In (i), explicit formulas (3.5) and (3.6) are provided for the optimal investment and consumption policies, respectively. In (ii)-(iv), consumption policies are qualitatively characterized for $P \in [\tilde{P}, \bar{P})$. For any $P \le \tilde{P}$, the problem has the same solution as that for $P = \tilde{P}$. For $P \ge \bar{P} = U(\infty)/\beta$, it is obvious that the bankruptcy utility is the best there can be and, if it were admissible, a quick consumption to bankruptcy would be optimal. Thus, technically, there does not exist any optimal solution in this case.

Equation (3.6) states that if equating marginal utility of consumption with the marginal value of wealth yields a feasible consumption level, then it is the optimal consumption level. On the other hand, if it yields a consumption level below the subsistence level, then it is optimal to consume at the subsistence level. That is, while the marginal utility of the subsistence consumption is lower than the marginal value of wealth, the agent has no choice but to consume at the subsistence level because of the lower bound constraint imposed on the consumption level. Finally, from the strict concavity of the value function in wealth x and of the utility function in consumption c, it follows from (3.6) that consumption rate increases with the wealth level x. Moreover, it increases strictly whenever the consumption rate is above the subsistence level.

In Case (ii) when $P \in (\tilde{P}, \bar{P})$, bankruptcy occurs with probability one if and only if the discount rate is not less than the rate of interest plus

γ. This implies that the higher the discount rate, or the lower the interest rate, or the lower the desirability of the risky investment, then the higher the propensity to consume and the higher the probability of bankruptcy. When $\beta < r + \gamma$, on the other hand, the bankruptcy happens with a positive probability less than one, i.e., it happens depending on the luck of the draw; see Section 3.4 for an explicit formula for the probability of bankruptcy. If $U'(s) = \infty$, i.e., if the marginal utility at the subsistence level is infinity, then the agent obtains a tremendous incremental utility by consuming at a slightly higher than the subsistence level, and that is exactly what the agent does; he consumes at the rate strictly above the subsistence rate as long as he is not bankrupt. If $0 < U'(s) < \infty$ and P is sufficiently large, then the consumption behavior is the same as the above. For $P \leq P^\star(s)$, however, the optimal consumption rate is not bounded away from the subsistence level. In fact, there exists a wealth level $\overline{x}(P, s)$ with $\overline{x}(P^\star(s), s) = 0$ such that for wealth levels below $\overline{x}(P, s)$, consumption is at subsistence. Here, the bankruptcy value is sufficiently unpleasant inducing the agent to consume as little as possible at small wealth levels to reduce his chance of going bankrupt. For higher wealth levels, the agent can afford to indulge and consume at levels higher than the subsistence level.

Cases (iii) and (iv) are limiting cases of (ii) when P approaches $\tilde{P}$, a very unpleasant bankruptcy penalty. The peculiarity of these cases is that for $s > 0$ we have no continuity in the limit, namely, the value s/r appears in the characterization of the optimal policy, a value which plays no role when $P > \tilde{P} = -\infty$. We begin with Case (iv)a when $P = \tilde{P}$ and $x = s/r$. If $U(s) > -\infty$, then the agent can invest all his wealth in the riskless asset and consume only the interest and keep his principal intact at the level s/r. The interest earned is just enough for him to consume at the subsistence level forever. If $U(s) = -\infty$, it does not really matter what he does, since his value function $V(s/r; P) = -\infty$ regardless. In Case (iv)b when $P = \tilde{P}$ and $0 < x < s/r$, we must have $s > 0$ and, therefore, $\tilde{P} = -\infty$. It is obvious that bankruptcy cannot be avoided with certainty if the agent must consume at least at the subsistence rate. Once again $V(x; P) = -\infty$ regardless of what the agent does.

Finally, we examine Case (iii) when $P = \tilde{P}$ and $x > s/r$. From the discussion of Case (iv), it is obvious that the agent must keep his wealth strictly above s/r if $U(s) = -\infty$, and must not allow his wealth to fall below s/r if $U(s) > -\infty$. Furthermore, if $U'(s) = \infty$, the optimal consumption is strictly above subsistence as in Case (ii) when $P > \tilde{P}$ and $U'(s) = \infty$. But unlike in Case (ii) with $P > \tilde{P}$ and $U'(s) = \infty$, the optimal consumption with $P = \tilde{P}$ and $U'(s) = \infty$ is not bounded away from the subsistence level. In other words, the unpleasantness of the bankruptcy utility $\tilde{P}$ forces the individual to consume at a lower rate so that it is no longer bounded away from s. If $U'(s) < \infty$, on the other hand, then the agent can resort to consuming at the subsistence level for low wealth levels $x \in (s/r, \bar{x}]$, in order to decrease his probability of going bankrupt. At wealth levels $x > \bar{x}$, the agent can afford to consume more than the subsistence level.

We have now explained the consumption behavior of the agent. Next we turn to describing his investment behavior implicit in equation (3.5). The optimal investment strategy in (3.5) can be expressed in terms of the risk-aversion measures defined in (2.25) and (2.26) as follows:

$$x_t^* \pi_t^* = \frac{\alpha - r}{\sigma^2 l_V(x_t^*; P)} \quad \text{or} \quad \pi_t^* = \frac{\alpha - r}{\sigma^2 L_V(x_t^*; P)}. \tag{3.7}$$

Thus, the optimal fraction of wealth invested in the risky asset is inversely proportional to the coefficient of relative risk aversion, while the optimal amount of wealth invested in the same asset is inversely proportional to the coefficient of absolute risk aversion. Moreover, the constant of proportionality $(\alpha - r)/(\sigma^2)$ means that the higher the risk premium available on the risky asset, the more is invested in it, and the higher the variance σ^2 of the risky returns, the less is invested in the risky asset. Therefore, the study of the optimal investment strategy with respect to the parameters of the agent's problem and his current wealth is tantamount to analyzing the coefficients (2.25) and (2.26) of risk aversion. This analysis was carried out by Presman and Sethi (1997a), and is discussed in the next section.

3.3   Analysis of the Risk Aversion Behavior

In a way similar to their earlier paper discussed in Section 2.6, Presman and Sethi (1997a) have examined the risk-aversion behavior implied by the optimal solution in the Sethi–Taksar–Presman model reported in Sections 3.1 and 3.2.

In the analysis of the coefficient of absolute risk aversion $l_V(x; P)$ given in (2.25), an interesting question is how it relates to the corresponding coefficient $l_U(c)$ associated with the utility of consumption, which can be formally defined also by (2.25). In the HARA case, i.e., when the utility of consumption for $\eta \in (-\infty, +\infty)$ is given by

$$U = \begin{cases} U_{\delta,\eta}(c) & \text{for} \quad c \geq s, \\ -\infty & \text{for} \quad 0 \leq c < s, \end{cases} \tag{3.8}$$

where

$$U_{\delta,\eta}(c) = \begin{cases} |\delta|(c+\eta)^\delta/\delta & \text{for} \quad 0 < \delta < 1,\ c \geq -\eta,\ \text{or}\ \delta < 0,\ c > -\eta, \\ \ln(c+\eta) & \text{for} \quad \delta = 0,\ c > -\eta, \\ -\infty & \text{for} \quad \delta \leq 0,\ c = -\eta, \end{cases}$$

$$\tag{3.9}$$

it is shown in Sethi, Taksar and Presman (1992) that the value function is also of HARA type for $s = -\eta$, $P = \tilde{P}$. Thus, the coefficient of absolute risk aversion in these cases decreases with wealth. In all other cases, the agent's value function is no longer of HARA type. Nevertheless, as it will be shown below, for $P = \tilde{P}$ and $s > -\eta$, the agent is once again decreasingly absolute risk averse.

The absolute risk-aversion coefficient of the agent does not always inherit the monotonicity property of the corresponding coefficient associated with the given HARA utility function. It was shown in Presman and Sethi (1991) that there are two possibilities for the absolute risk-aversion coefficient of the agent in the case $s = 0$. For some values of the parameters, the property of decreasing absolute risk aversion is inherited for all levels of wealth, where as for other values of the parameters the property holds only if the agent is rich enough, i.e., when his wealth exceeds some critical level denoted as $\tilde{x}(P, 0)$. For wealth levels less than $\tilde{x}(P, 0)$, the agent is increasingly absolute risk averse. In both cases there exists such a level of wealth $\tilde{x}(P, 0) \in [0, \infty)$ that $l'(x; P) > 0$ for $x < \tilde{x}(P, 0)$ and $l'(x; P) < 0$ for $x > \tilde{x}(P, 0)$, i.e., according to (3.7), the amount of wealth invested in the risky asset decreases for small values of wealth ($x < \tilde{x}(P, 0)$) and increases for large values of wealth ($x > \tilde{x}(P, 0)$). A similar result holds in the case $s > 0$ with some critical level $\tilde{x}(P; s)$ for wealth, i.e., there exists a critical wealth level $\tilde{x}(P, s) \in [0, \infty)$ for any $P \in (\tilde{P}, \bar{P})$, such that $l'(x; P) > 0$ for $x < \tilde{x}(P, s)$ and $l'(x; P) < 0$ for $x > \tilde{x}(P, s)$. Furthermore, $\tilde{x}(P, s)$ is continuous in both its arguments. For the bankruptcy value $\tilde{P}$, $l(x; \tilde{P})$ is defined only for $x > s/r$, and $l'(x; \tilde{P}) < 0$ for $x > s/r$.

The nature of the dependence of $\tilde{x}(P, s)$ on the bankruptcy value $P \in (\tilde{P}, \bar{P})$ and the subsistence consumption $s \geq 0$ can be quite different for different values of the parameters of the problem; see Presman and Sethi (1997a) for details.

Let us consider the proportional risk-aversion behavior of the agent in the HARA case. The relative risk-aversion coefficient of the HARA utility function increases for $\eta > 0$, decreases for $\eta < 0$, and is constant for $\eta = 0$.

It was shown in Presman and Sethi (1991) that for $s = 0$ (in this case $\eta \geq 0$) and $P > \tilde{P}$, the agent is increasingly proportionally risk averse for all values of wealth, i.e., the relative risk-aversion coefficient of the derived utility of wealth inherits the monotonicity property from the HARA consumption utility function in the case $\eta > 0$. For $s = 0, \eta > 0$, and $P = \tilde{P} = U(0)/\beta$ (in this case $\bar{x}(\tilde{P}, 0) > 0$), it inherits this property only for $x > \bar{x}(\tilde{P}, 0)$, and the corresponding relative risk-aversion coefficient remains constant for $0 < x < \bar{x}(\tilde{P}, 0)$. For $s = \eta = 0$ (i.e., when $U'(0) = \infty$) and $P \leq \tilde{P} = U(0)/\beta$, the value function of the problem is also of HARA type with the same parameter as the one for the HARA utility of consumption used in the problem. When $s > 0$, the situation is more complicated and it differs depending on $\eta < 0, \eta > 0$, or $\eta = 0$. When $\eta < 0$ and $P = \tilde{P}$, the agent inherits from the HARA consumption utility function the property of decreasing proportional risk aversion, while for $\eta < 0$ and $P > \tilde{P}$, this is true only in case he is sufficiently wealthy, i.e., when his wealth exceeds some value denoted by $x_1(P, s)$. For wealth levels below $x_1(P, s)$, the agent is increasingly proportionally risk averse, and $x_1(P, s) > 0$.

When $\eta > 0$ with P exceeding some value denoted by $P(s)$, the agent inherits from the HARA consumption utility function, the property of increasing proportional risk aversion for all values of wealth, while for small values of $P(P < P(s))$, the agent is increasingly proportionally risk averse only when he is either sufficiently rich or sufficiently poor. For intermediate wealth levels, the agent is decreasingly proportionally risk averse. When $\eta = 0$, the agent inherits from the HARA consumption utility function, the property of relative risk-aversion coefficient to be constant only for $P = P(s)$ and only in the case when his wealth exceeds some value denoted by $x_1(s)$. He is increasingly proportionally risk averse in the case $P > P(s)$ for all values of wealth, and in the case $P < P(s)$ (resp. $P = P(s)$) when his wealth is less than $x_1(P, s)$ (resp. $x_1(s)$). The agent is decreasingly proportionally risk averse in the case $P < P(s)$ when his wealth exceeds $x_1(P, s)$.

From the above results, one may draw the following general conclusions regarding the risk-aversion behavior in the HARA case. At higher wealth levels, the agent's absolute risk aversion decreases with wealth, while his relative risk aversion increases with wealth for $\eta > 0$ and decreases for $\eta < 0$. This qualitative behavior is inherited from the agent's HARA-type consumption utility, as the agent seems quite immune from the bankruptcy payment parameter P when he is sufficiently wealthy.

Of course, what is considered to be a high enough wealth level itself may depend on P and s. At lower wealth levels, the agent is no longer immune from the utility of bankruptcy, except when $\eta > 0$ and the bankruptcy utility is sufficiently high. In this exceptional case, the high bankruptcy utility allows the individual to not worry about bankruptcy and lets him inherit the increasing relative risk aversion property of his HARA-type consumption utility also at low wealth levels. On the other hand, when bankruptcy is not too pleasant, the behavior is quite complicated because of the nature of his consumption utility and the utility of bankruptcy. In the case $P \in [\tilde{P}, P(s))$ in particular, the agent shoots for the moon by putting a high fraction of his wealth at very low levels. As the wealth increases, the fraction decreases up to a certain level, at which point the fraction starts increasing. Here, the increasing wealth provides a greater cushion against an unpleasant bankruptcy. This behavior continues until a sufficiently high level of wealth is reached. Beyond this level, the agent reverts to his elements by behaving according to the increasing relative risk-aversion property of his HARA type consumption utility.

The description of the risk-aversion behavior with a general concave utility function is far more complex. An intuitive result states that the agent is more risk averse, both absolutely and relatively, if the state of bankruptcy is more unpleasant. The other results are much more complicated to describe; see Presman and Sethi (1997a) for details.

3.4 Probability Distribution of the Bankruptcy Time

Presman and Sethi (1996) provide an explicit formula for $q(x, t; P)$ which represents the probability that the agent, endowed with an initial wealth x, exposed to a bankruptcy value $P \in (\tilde{P}, \bar{P})$, and behaving in the optimal fashion, will become bankrupt in the interval $[0, t)$ for $t \leq \infty$. The case $P = \tilde{P}$ has already been dealt with in Section 3.2(iv). See KLSS for the probability of an eventual bankruptcy $q(x, \infty; P)$ in the case $s = 0$.

The distribution function of the time of the bankruptcy under optimal behavior with $P \in (\tilde{P}, \bar{P})$ is given by

$$q(x, t; P) = \mathcal{P}[T_x^* < t]$$
$$= \frac{1}{\sqrt{2\pi}} \int_0^{t/b^2} \exp\left[-\left(\frac{\sigma}{\alpha - r}\{r + \gamma - \beta\}by + 1\right)^2 / 2y\right] y^{-3/2} dy, \quad (3.10)$$

where

$$b = \frac{\sigma}{\alpha - r} \ln \frac{V'(0; P)}{V'(x; P)} > 0. \quad (3.11)$$

Moreover, the probability of an eventual bankruptcy is

$$\mathcal{P}[T_x^* < \infty] = q(x; P) = q(x, \infty; P)$$
$$= \begin{cases} 1 & \text{if } r + \gamma - \beta \leq 0, \\ (V'(x; P)/V'(0; P))^{(r+\gamma-\beta)/\gamma} & \text{if } r + \gamma - \beta > 0. \end{cases} \quad (3.12)$$

We conclude with some observations that follow immediately from the formula. When $\beta \geq r + \gamma$, it implies that the future is severely discounted and current enjoyment is highly valued. Therefore, not surprisingly, the agent's optimal consumption-investment policy leads to a certain bankruptcy at some finite time. However, when $\beta < r + \gamma$, the future is quite important and there is a positive probability of no bankruptcy altogether except when $P = \bar{P}$, in which case the agent will choose to go bankrupt immediately. Moreover, as expected, $q(x, t; P)$ decreases as the initial wealth x increases for any given t and for $t = \infty$.

4   Random Market Parameters and Positive Subsistence Consumption

In this section we describe the model and the results obtained by Cadenillas and Sethi (1997), who introduce random market parameters in the models of Sections 2.1 and 3.1. Thus, their model also extends the models of Karatzas, Lehoczky and Shreve (1987) and Cox and Huang (1989) to allow for explicit consideration of bankruptcy.

In Section 2.1, the asset dynamics (2.1) and (2.2) involve constant parameters r, $\boldsymbol{\alpha}$, and $\mathbf{D}$ denoting the interest rate, the vector of mean rates of return on the risky assets, and the volatility matrix, respectively. More generally, these parameters are stochastic processes denoted as r_t, $\boldsymbol{\alpha}_t$ and $\mathbf{D}_t$ respectively. Also, the constant discount rate β can be generalized to a stochastic process β_t, $t \geq 0$. An agent facing the non- Markovian asset-price dynamics (2.1) and (2.2) with $r, \boldsymbol{\alpha}, \mathbf{D}$, replaced by $r_t, \boldsymbol{\alpha}_t, \mathbf{D}_t$ respectively, cannot use dynamic programming to solve his consumption-investment problem. Therefore, Karatzas, Lehoczky and Shreve (1987) use martingale methods to perform a beautiful analysis of the problem. While recognizing bankruptcy, they simply let the agent consume at the zero level following his bankruptcy. Such a bankruptcy does not allow for a positive subsistence consumption either. Cox and Huang (1989) use martingale methods as well, with the asset prices modeled as (Markov) diffusion processes and consumption utility not stochastically discounted, and when dynamic programming theory does apply.

In what follows, we discuss the model developed by Cadenillas and Sethi (1997), who use random market coefficients as in Karatzas, Lehoczky and Shreve (1987), constrain consumption to be not below a subsistence level as in Sethi, Taksar and Presman (1992), allow explicitly for bankruptcy as in KLSS (1986), and permit generalized consumption policies.

4.1   The Cadenillas–Sethi Model

Cadenillas and Sethi (1997) use random market parameters r_t, $\boldsymbol{\alpha}_t$ and $\mathbf{D}_t$ such that for some $\varepsilon > 0$, $\xi \mathbf{D}_t \mathbf{D}_t^T \xi \geq \varepsilon |\xi|^2$ for all $\xi \in R^n$ and for all $t \geq 0$. This makes the variance-covariance matrix $\Sigma_t = \mathbf{D}_t \mathbf{D}_t^T$ invertible; see Karatzas, Lehoczky and Shreve (1987) for details. Furthermore, the random discount rate β_t is assumed to be bounded below by a nonnegative constant.

Using the price dynamics (2.1) and (2.2) with random parameters and allowing for instantaneous lumpsum consumption, it is possible to proceed from the price dynamics and modify the wealth equation (2.4) as follows:

$$dx_t = (\boldsymbol{\alpha} - r_t \mathbf{1})\boldsymbol{\pi}_t^T x_t dt + r_t x_t dt - dC_t + \boldsymbol{\pi}_t x_t \mathbf{D}_t d\mathbf{w}_t^T, \quad x_0 = x, \quad (4.1)$$

where C_t denotes the cumulative consumption by time t with $C_0 = 0$, and is of bounded variation. By defining $\Pi_t = \boldsymbol{\pi}_t x_t$, we can rewrite (4.1) as

$$dx_t = \{r_t x_t + \Pi_t(\boldsymbol{\alpha}_t - r_t \mathbf{1})\}dt - dC_t + \Pi_t \mathbf{D}_t d\mathbf{w}^T(t), \quad x_0 = x, \qquad (4.2)$$

The objective function is

$$V_{\{C_t, \Pi_t\}}(x) = E\left[\int_0^{T_x} e^{-\int_0^t \beta_\tau d\tau} U(c_t)dt + B(T_x)e^{-\int_0^{T_x} \beta_\tau d\tau}\right], \qquad (4.3)$$

where

$$T_x = \inf\{0 \le t \le T | x_t = 0\}. \qquad (4.4)$$

Here T denotes the length of the finite horizon and the continuous process $B : [0, T] \times \Omega \to R$, $B(T) = 0$ a.s., represents the utility of bankruptcy. Note that (4.3) does not include the part of consumption that is not absolutely continuous because of the assumption that $U'(\infty) = 0$. The objective function is to be maximized over a class of suitable admissible policies $\{c_t, \pi_t\}$. These are defined by progressively measurable portfolio processes $\Pi : [0, T] \times \Omega \to R^n$ satisfying

$$P\left\{\int_0^T |\pi_t|^2 dt < \infty\right\} = 1 \qquad (4.5)$$

and progressively measurable cumulative consumption processes $C : [0, T] \times \Omega \to [0, \infty)$ that are nondecreasing with $C_0 = 0$ and are left-continuous with right limits. Furthermore,

$$P\left\{\int_0^T c_t dt < \infty\right\} = 1 \qquad (4.6)$$

where C_t is the absolutely continuous part of c_t and satisfies

$$P\{\forall 0 \le t \le T_x : c_t \ge s\} = 1. \qquad (4.7)$$

Cadenillas and Sethi solve the problem in two steps. In the first step they show that for any given stopping time θ almost surely in $[0, T]$, the martingale method of Karatzas, Lehoczky and Shreve (1987) can be extended to obtain a candidate solution that has the agent going bankrupt at θ. Using the candidate policies in (4.2) and (4.3), one can thus obtain the objective function value associated with the solution. This yields a functional of the stopping time θ. Moreover, it can be shown that if this functional could be maximized over the class of stopping times to obtain an optimal θ^*, then the candidate policy associated with θ^* would be optimal.

Since the theory of solving the second step is not yet available, Cadenillas and Sethi settle for an ε-optimal stopping time and show that the candidate consumption-investment policy associated with that stopping time is ε-optimal. This ε-optimality is not a real restriction in practice.

In two important special cases, Cadenillas and Sethi are able to obtain optimal solutions. These extend some cases of high initial wealth treated by Bardhan (1994) and the case of zero subsistence consumption (i.e., $s = 0$) treated by Karatzas, Lehoczky and Shreve (1987). In both cases, the terminal time T can be used as the optimal stopping time θ^*.

5 Borrowing and Shortselling Constraints

We have described models with bankruptcy that allow unlimited borrowing and shortselling. In this section we discuss models with constrained borrowing and shortselling. These constraints give rise to value functions that may not be concave. Observe that in regions where value functions are convex, the agent will put all his investment in the riskiest asset available.

In Section 5.1 we briefly discuss a discrete-time model developed by Sethi, Gordon and Ingham (1979). While only limited results are obtained, the model is discussed because it is to our knowledge the first model that considers subsistence consumption and bankruptcy in a multiperiod consumption-investment problem. Next in Section 5.2, we discuss a model due to Lehoczky, Sethi and Shreve (1983), which is the first rigorous analysis of the continuous-time consumption-investment problem allowing for subsistence consumption, bankruptcy, and borrowing and shortselling constraints. Since the authors' focus was to examine presence of convex portions in the value function, it was sufficient to consider the special case in which there was only one risky asset with a mean rate of return equal to the interest rate. Lehoczky, Sethi and Shreve (1985) further generalize their 1983 model to allow for the risky asset price to be modeled by a semimartingale and for generalized consumption as in Section 4.1.

5.1 An Early Discrete-Time Model with Bankruptcy and Welfare

Sethi, Gordon and Ingham (1979) consider an agent who can count on a government consumption maintenance program (welfare) when his wealth falls below a predetermined poverty level. Without loss of generality, let the poverty level be one. The agent goes on welfare when his wealth is less than one, and his consumption from then on is equal to the consumption level of welfare recipients, which is also set at one. When not on welfare, the agent allocates his wealth (no borrowing is allowed) among consumption, a risky asset, and a risk-free asset. The latter provides a return of r, and the return on the risky asset is the random variable

$$z = \begin{cases} y > 1 & \text{w.p. } p_1, \ 0 < p_1 < 1 \\ 0 & \text{w.p. } 1 - p_1. \end{cases} \tag{5.1}$$

For simplicity, the fair game condition $p_1 x = (1 + r)$ is assumed.

The consumption utility is $U(c) = \log c$. The discount rate is set at zero. The problem of the agent with a finite life of T periods is:

$$V(x, \tau) = \max_{\{c_t, \pi_t\}} E \sum_{t=\tau}^{T} f(x_t, c_t), \qquad (5.2)$$

where

$$f(x_t, c_t) = \begin{cases} 0 & \text{if} \quad x_t < 1 \\ \log c_t & \text{if} \quad x_t \geq 1 \end{cases}, \qquad (5.3)$$

$$x_{t+1} = \begin{cases} 0 & \text{if} \quad x_t < 1 \\ (x_t - c_t)[(1 + r)(1 - \pi_t) + \pi_t z] & \text{if} \quad x_t \geq 1 \end{cases}, \quad x_1 = x, \qquad (5.4)$$

and

$$c_t \geq 0, \qquad 0 \leq \pi_t \leq 1. \qquad (5.5)$$

Using stochastic dynamic programming, the problem is solved for $T = 3$. The agent with two periods remaining consumes all his wealth when the level of his wealth is just above the welfare level. As his wealth increases, there comes a point when he switches to consuming a constant fraction which is more than half of his wealth and investing the remainder in the risky asset. This continues until a certain wealth level beyond which the agent consumes half of his wealth and invests the remainder in the risk-free asset. As this level of wealth, the individual can be considered 'wealthy' since the welfare program does not distort his decisions.

With three periods remaining, the agent's investment behavior has no systematic pattern. The agent, as he becomes wealthier, can switch back and forth to risky, risk-free, and intermediate-risk investments. Moreover, there are levels and ranges of wealth where the agent's investment policy is not unique. In particular cases, it is optimal even to have both the extreme policies of investing entirely in the risky asset or entirely in the risk-free asset.

Consumption on the other hand increases systematically as wealth increases, whereas the fraction of wealth consumed decreases to one-third with various intermediate stages.

At wealth levels just over the welfare-cutoff level, the agent consumes all his wealth up to a certain point. Above that his consumption is $(1 + p_1)/(1 + r)$, which can fall below the welfare level if $p_1 < r$. In this case, the risky asset is more of a lottery, and the agent is tempted to consume below the welfare level in the hope of luxuriant consumption next period.

The results lend theoretical support for the findings of Gordon, Paradis and Rorke (1972) obtained from an experimental game where consumption and portfolio decisions were made over time with the knowledge that a fixed periodic income would be received when wealth fell below a certain level. As one might intuitively expect, risk aversion fell sharply as wealth fell towards the welfare level. Recognizing that agents in general and proprietors in particular may change their state upon bankruptcy provides an alternative and perhaps superior explanation of risk preference than previous efforts by Friedman and Savage (1948) and others.

The computations involved in Sethi, Gordon and Ingham are quite difficult and yield limited answers. Besides, with more risky alternatives including borrowing, the agent's wealth in a discrete-time model may fall below zero, a situation which should not be allowed and is difficult to deal with. All these point to the development of a continuous-time model, listed as an open problem by Sethi, Gordon and Ingham. The development was carried out by Lehoczky, Sethi and Shreve (1983), and is discussed in the next section.

5.2   The Lehoczky–Sethi–Shreve Model

The model developed by Lehoczky, Sethi and Shreve (1985) can be related to the model developed in Section 3.1 as follows. Impose an additional constraint that disallows shortselling, i.e.,

$$0 \leq \pi_t \leq 1, \tag{5.6}$$

and set the risky asset's mean rate of return

$$\alpha = r. \tag{5.7}$$

While (5.6) appears to permit no borrowing, we shall see how a reformulation of the problem allows us to convert a special case of this model into a model that allows unlimited borrowing. Furthermore, (5.7) is imposed to simplify the solution and to focus entirely on the distortions caused by consumption constraints and bankruptcy, and thus eliminate other factors which might induce risk-taking behavior. The HJB equation (3.3) and (3.4) can be modified for the present model as

$$\beta V(x) = \max_{c \geq s, 0 \leq \pi \leq 1} \left\{ (rx - c)V'(x) + \frac{1}{2}(\sigma \pi x)^2 V''(x) + U(c) \right\},$$
$$x > 0, \ V(0) = P. \tag{5.8}$$

The wealth dynamics is given as

$$dx_t = (rx_t - c_t)dt + \sigma \pi_t x_t dw_t, \quad x(0) = x. \tag{5.9}$$

Letting $\nu = \pi \sigma$, we can reformulate the HJB equation (5.8) as

$$\beta V(x) = \sup_{c \geq s, 0 \leq \nu \leq \sigma} \left\{ (rx - c)V'(x) + \frac{1}{2}\nu^2 x^2 V''(x) \right\}, \quad x > 0, V(0) = P.$$
$$\tag{5.10}$$

By setting $\sigma = \infty$, ν becomes unconstrained. This means unlimited borrowing is allowed at rate r for investing in the risky asset.

Lehoczky, Sethi and Shreve (1985) obtain an explicit solution of the problem. When $s = 0$, the results are conveniently summarized in Table 2 of KLSS. Therefore, we shall only focus on the positive subsistence case, i.e., when $s > 0$. It is no loss of generality to set $s = 1$. The results are summarized next in three distinct cases.

When $\beta < r$, the agent exhibits risk-taking behavior at low wealth levels. For $x < \frac{1}{r}$, the agent must reach bankruptcy unless he selects the risky investment. For all values of P, the value function will be convex at low wealth levels that extend beyond $1/r$, the point at which the agent need no longer become bankrupt. As for consumption, it is at subsistence up to some wealth level and beyond that it increases with increasing wealth if $U(1) > U'(1) + \beta P$. When not, the consumption begins at a level higher than subsistence and *decreases* as wealth increases, reaches the subsistence level at some wealth level, and thereafter increases with increasing wealth.

When $\beta = r$, the value function consists of a strictly concave function or of a linear segment attached to a concave function. On the linear segment the agent is indifferent between the two assets, thus the value functions is independent of σ.

When $\beta > r$, the value function will be concave at all wealth levels indicating risk-aversion behavior. The agent always selects the riskless investment and thus the value function does not depend on σ. When $P < U(0)/\beta$, the value function has a discontinuity at 0. When $U(1) \leq U'(1) + \beta P$, consumption begins at a level higher than subsistence and increases with increasing wealth. When $U(1) > U'(1) + \beta P$, the agent consumes at the subsistence level at wealth $x \leq (1/r)[1 - \{U'(1)/U(1) - \beta P\}^{r/(\beta-r)}]$. Thereafter, consumption increases as wealth increases.

Lehoczky, Sethi and Shreve (1983) have provided computational plots of the value functions with $U(c) = \log c$, $s = 1$, $P = 0$, and $\beta < r$, and $\sigma = 0, \infty$ and an intermediate value. Lehoczky, Sethi and Shreve (1985) have generalized their model presented in this section to include risky assets modeled as semimartingales. This is discussed in the next section.

5.3 A Martingale Formulation of the Problem

To see how Lehoczky, Sethi and Shreve (1985) have generalized their earlier model, let us begin with the wealth dynamics (5.9) expressed in the integral form as follows:

$$x_t = x + \int_0^t r x_\tau d\tau + \int_0^t \sigma \pi_\tau x_\tau dw_\tau - \int_0^t c_\tau d\tau. \qquad (5.11)$$

They generalize the diffusion term to a martingale M_t and the consumption integral to a cumulative consumption process C_t of bounded variation as in Section 4.1. Thus,

$$x_t = x + \int_0^t rx_\tau d\tau + M_t - C_t, \tag{5.12}$$

where M_t is a right-continuous martingale with $M_0- = 0$ interpreted as cumulative winnings from the investment in the risky asset over and above the cumulative interest on the investment. The process C_t representing cumulative consumption by time t is adapted, nondecreasing, and satisfies $C_0- = 0$ and $EC_t < \infty$, $t > 0^-$. More compactly, we can write (5.12) as

$$x_t = x + \int_0^t rx_\tau d\tau + S_t, \tag{5.13}$$

where S_t is a supermartingale with $S_0- = 0$, $-1 \leq t < 0$ and satisfying conditions DL, which allow us to decompose it into M_t and C_t; see Karatzas and Shreve (1988, pp. 24–25).

As in Section 4.1, denote the absolutely continuous part of C_t as c_t, and define the bankruptcy time as

$$T_x = \inf\{t \geq 0^- | x_t \leq 0\}. \tag{5.14}$$

In view of (5.14), we define an admissible supermartingale to be that which satisfies the DL condition and for which

$$x_{T_x} = 0 \text{ on } \{T_x < \infty\}. \tag{5.15}$$

The optimization problem is to

$$\max E\left[\int_0^{T_x} e^{-\beta t} U(c_t)dt + Pe^{-\beta T_x}\right] \tag{5.16}$$

over the class of admissible supermartingales. For consumption utilities $U(c)$ satisfying the assumptions in Section 3.1 or Section 4.1, only c_t, the absolutely continuous part of C_t, appears in (5.16).

Lehoczky, Sethi and Shreve (1985) show that the results of their earlier model reported in the previous section generally carry over to this fairly general formulation. Furthermore, they are able to solve explicitly for risky investments such as impulse or jump gambles or those involving Poisson processes, and HARA consumption utilities. These give rise to interesting but complicated value functions.

6   Open Research Problems and Concluding Remarks

We have reviewed the literature on consumption-investment problems that explicitly incorporate bankruptcy. This concluding section briefly refers to related research on consumption-investment problems that does not deal with bankruptcy issue. This suggests some open research problems; see also Sethi (1997).

In all the papers discussed in this survey, there is no cost of buying and selling assets. Davis and Norman (1990), Shreve, Soner and Xu (1991), and Shreve and Soner (1995) have considered proportional transition costs in consumption-investment models with two assets and nonnegative consumption constraint. It would be interesting to incorporate positive subsistence level and bankruptcy utility in these models. Another extension would be to include fixed transaction costs; such a cost has not been considered in the consumption-investment context.

Karatzas, Lehoczky, Shreve and Xu (1991) and He and Pearson (1991) have considered incomplete markets. One would like to incorporate such markets in consumption-investment models with bankruptcy and subsistence requirement.

Finally, Karatzas, Lehoczky, Shreve and Xu (1990, 1991) have developed equilibrium models with many agents consuming and trading securities with one another over time. In these models, consumption utilities are chosen so that agents do not go bankrupt. This way if one begins with n agents, one stays with n agents throughout the horizon. Thus, there is no easy way to see how these models can be extended to allow for bankruptcy. Sethi and Taksar (1992) introduced a concept of nonterminal bankruptcy as discussed in Section 2.3. This allows agents to stay in the system and may facilitate the eventual development of an equilibrium model with bankruptcy. Several important open research problems flow from these considerations.

The Sethi–Taksar nonterminal bankruptcy model needs to be extended to allow for random coefficients and subsistence consumption as in Section 4. It is not clear how to prove the equivalence between the terminal and nonterminal bankruptcies in the more general setup.

The Cadenillas–Sethi model treats an almost surely finite horizon agent. In addition, it deals with only nearly optimal policies. One needs to extend the model to allow for infinite horizon and to obtain optimal policies. If this problem is solved, and if it can be shown to be equivalent to a model with nonterminal bankruptcy as mentioned above, then we would have a single agent model as a starting point in the development of an equilibrium model with bankruptcy.

Another important consideration is how to provide for the bankruptcy value P if it consists of welfare or the subsistence consumption while in the state of bankruptcy. This would call for a different kind of agent, called the government, who must collect taxes and provide for welfare to agents who are in the bankruptcy state.

We hope that work will be carried out in addressing the open research problems described above, and that a suitable equilibrium model that allow for bankruptcy and subsistence consumption will eventually be developed.

References

I. Bardhan (1994), 'Consumption and Investment under Constraints,' *J. Econ. Dyn. and Control* **18**, 909–929.

A. Cadenillas and S.P. Sethi (1997), 'The Consumption Investment Problem with Subsistence Consumption, Bankruptcy, and Random Market Coefficients,' *J. Opt. Theory & Appl.* **93** (2).

J.C. Cox and C.F. Huang (1989), 'Optimal Consumption and Portfolio Policies when Asset Prices Follow a Diffusion Process,' *J. Econ. Theory* **49**, 33–83.

M.H.A. Davis and A. Norman (1990), 'Portfolio Selection with Transaction Costs.' *Math. Oper. Research* **15**, 676–713.

I.I. Gihman and A.V. Skorohod (1972), *Stochastic Differential Equations.* Springer-Verlag.

M.J. Gordon, G. Paradis and C. Rorke (1972), 'Experimental Evidence on Alternative Portfolio Decision Rules,' *Amer. Econ. Review* **62**, 107–118.

M.J. Gordon and S.P. Sethi (1997), 'A Contribution to the Micro Foundation for Keynesian Macroeconomic Models,' Chapter 11 in Sethi (1997), 217–244.

N.H. Hakansson (1969), 'Optimal Investment and Consumption Strategies Under Risk, An Uncertain Lifetime, and Insurance,' *International Econ. Review* **10**, 443–466.

H. He and N. Pearson (1991), 'Consumption and Portfolio Policies with Incomplete Markets and Short-Sale Constraints: The Infinite Dimensional Case.' *J. Econ. Theory* **54**, 259–304.

I. Karatzas (1989), 'Optimization Problems in the Theory of Continuous Trading.' *SIAM J. Control Optim.* **27**, 1221–1259.

I. Karatzas, J. Lehoczky, S.P. Sethi and S. Shreve (1986), 'Explicit Solution of a General Consumption/Investment Problem,' *Math. Oper. Research* **11** (2), 261–294; reprinted as Chapter 2 in Sethi (1997).

I. Karatzas, J. Lehoczky and S. Shreve (1987), 'Optimal Portfolio and Consumption Decisions for a Small Investor on a Finite Horizon.' *SIAM J. Control and Optim* **25**, 1557–1586.

I. Karatzas, J. Lehoczky and S. Shreve (1990), 'Existence and Uniqueness of Multi-Agent Equilibrium in a Stochastic, Dynamic Consumption/ Investment Model.' *Math. Oper. Research* **15**, 80–128.

I. Karatzas, J. Lehoczky and S. Shreve (1991), 'Equilibrium Models with Singular Asset Prices.' *Math. Finance* **1**, 11–29.

I. Karatzas, J. Lehoczky, S. Shreve and G. Xu (1991), 'Martingale and Duality Methods for Utility Maximization in an Incomplete Market.' *SIAM J. Control Optim.* **29**, 702–730.

I. Karatzas and S. Shreve (1988), *Brownian Motion and Stochastic Calculus.* Springer-Verlag.

J. Lehoczky, S.P. Sethi and S. Shreve (1983), 'Optimal Consumption and Investment Policies Allowing Consumption Constraints and Bankruptcy,' *Math. Oper. Research* **8**, 613–636; an unabridged version as Chapter 14 in Sethi (1997), 303–378.

J. Lehoczky, S.P. Sethi and S. Shreve (1985), 'A Martingale Formulation for Optimal Consumption/Investment Decision Making,' *Optimal Control Theory and Economic Analysis 2*, G. Feichtinger, editor, North-Holland; an unabridged version as Chapter 15 in Sethi (1997), 379–406.

S.A. Lippman, J.J. McCall, W.L. Winston (1980), 'Constant Absolute Risk Aversion, Bankruptcy, and Wealth-Dependent Decisions,' *J. of Business* **53** (3), Part 1, 285–296.

S.P. Mason (1981), 'Consumption and Investment Incentives Associated with Welfare Programs,' Working Paper, No. 79–34, Graduate School of Business Administration, Harvard University.

R.C. Merton (1969), 'Lifetime Portfolio Selection Under Uncertainty: The Continuous-Time Case,' *Rev. Econ. Statist.* **51**, 247–257.

R.C. Merton (1971), 'Optimum Consumption and Portfolio Rules in a Continuous Time Model,' *J. Econ. Theory* **3**, 373–413.

R.C. Merton (1973), Erratum, *J. Econ. Theory* **6**, 213–214.

R.C. Merton (1990), *Continuous-Time Finance*, Basil Blackwell.

E. Presman and S.P. Sethi (1991), 'Risk-Aversion Behavior in Consumption/Investment Problem,' *Mathematical Finance* **1** (1), 100–124; Erratum, **1**, p. 86.

E. Presman and S.P. Sethi (1996), 'Distribution of Bankruptcy Time in a Consumption/Portfolio Problem,' *J. Econ. Dyn. and Control* **20**, 471–477.

E. Presman and S.P. Sethi (1997a), 'Risk-Aversion Behavior in Consumption/Investment Problems with Subsistence Consumption,' Chapter 8 in Sethi (1997), 155–184.

E. Presman and S.P. Sethi (1997b), 'Consumption Behavior in Investment/ Consumption Problems,' Chapter 9 in Sethi (1997), 185–205.

E. Presman and S.P. Sethi (1997c), 'Equivalence of Objective Functionals in Infinite Horizon and Random Horizon Problems,' Chapter 10 in Sethi (1997), 207–216.

P.A. Samuelson (1969), 'Lifetime Portfolio Selection by Dynamic Stochastic Programming,' *Rev. Econ. Statist.* **51**, 239–246.

S.P. Sethi (1997), *Optimal Consumption and Investment with Bankruptcy*, Kluwer.

S.P. Sethi, M.J. Gordon and B. Ingham (1979), 'Optimal Dynamic Consumption and Portfolio Planning in a Welfare State,' *Portfolio Theory, 25 Years After*, E.J. Elton and M.J. Gruber (eds.), *TIMS Studies in the Management Sciences* **II**, 179–196; reprinted as Chapter 13 in Sethi (1997).

S.P. Sethi and M. Taksar (1988), A Note on Merton's 'Optimum Consumption and Portfolio Rules in a Continuous Time Model,' *J. Econ. Theory* **46**, 395–401; reprinted as Chapter 3 in Sethi (1997).

S.P. Sethi and M. Taksar (1992), 'Optimal Consumption and Investment Policies Modeled by a Diffusion Process with Delayed Reflections,' *J. Opt. Theory and Appl.* **74**, 333–346.

S. Shreve and M. Soner (1994), 'Optimal Investment and Consumption with Transaction Costs.' *Annals of Appl. Probab.* **4**, 609–692.

S.P. Sethi, M. Taksar and E. Presman (1992), 'Explicit Solution of a General Consumption/Portfolio Problem with Subsistence Consumption and Bankruptcy,' *J. Econ. Dyn. and Control* **16** 747–768; Erratum, **19**, (1995), 1297–1298.

S. Shreve, M. Soner and G. Xu , (1991), 'Optimal Investment and Consumption with Two Bonds and Transaction Costs.' *Math. Finance* **1**, 53–84.

W. Ziemba and R. Vickson (1975, Eds.), *Stochastic Optimization Models in Finance*, Academic Press.

Solving Stochastic Programming Models for Asset/Liability Management using Iterative Disaggregation

Pieter Klaassen

1 Introduction

This paper describes a novel approach to solving multiperiod stochastic programming models for practical portfolio investment problems, called the iterative disaggregation algorithm. A well-known complication of using stochastic programming models in practice is that only a limited amount of uncertainty can be included, due to the numerical optimization methods which have to be used. An important question is therefore how to choose such a limited description of the uncertainty. On the one hand, the description should be representative of the true uncertainty, while on the other hand one would like to exclude uncertainty which does not affect optimal decisions. Furthermore, one would like to know how sensitive the optimal solution is to the specific choice made. The iterative disaggregation algorithm has been developed with these issues in mind.

The most common method for obtaining an approximate description of the true uncertainty is to randomly sample scenarios from an underlying distribution. Hiller and Eckstein [14], Zenios [26] and Golub *et al.* [10] use sampled interest-rate scenarios in their models for fixed-income portfolio optimization, and exploit the structure of the resulting models by using parallel optimization methods to obtain solutions. Cariño *et al.* [6], [7], Dert [9] and Mulvey and Thorlacius [23] (see also elsewhere in this volume) describe scenario generators for large sets of economic variables, and employ these in multiperiod models for integrated asset and liability management for pension funds and insurance companies. Cariño *et al.* and Dert combine sampling techniques with event trees to model the uncertainty. Many other papers on stochastic programming models for dynamic portfolio investment problems focus on the development of efficient optimization methods, and do not explicitly address the question of how to obtain an approximate description of the true uncertainty. Examples are Bradley and Crane [4], Kusy and Ziemba [22] and Mulvey and Vladimirou [24].

Important properties of any description of the uncertainty in future asset prices and returns are that it is free of arbitrage opportunities and consistent with current market prices. However, these properties often seem to be neglected when stochastic programming models for portfolio investment problems are formulated. As is shown in Klaassen [20], a violation of these properties may lead to optimal portfolios in stochastic programming models which are severely biased towards spurious profit opportunities.

Central to the iterative disaggregation algorithm which we will present are aggregation methods which can be applied to condense a description of the asset-price uncertainty by combining states and time periods in such a manner that the condensed description does not contain arbitrage opportunities or inconsistencies with current market prices if this is true for the original description. Given a detailed description of the uncertainty which is arbitrage-free and consistent with market prices, these aggregation methods can thus be used to arrive at a concise but still arbitrage-free description of the uncertainty on which a stochastic programming model can be based.

The iterative disaggregation algorithm starts with the solution to a small, and therefore relatively easy to solve, stochastic programming model with such an aggregated but arbitrage-free description of the asset-price uncertainty. In each iteration of the algorithm, the description of the uncertainty in the model is refined by reversing one or more of the aggregations that were applied to arrive at the initial model (*disaggregations*), and the model re-optimized. To choose which aggregations to reverse in an iteration, an estimate is made of what additional uncertainty will have the largest impact on the optimal solution, where use is made of optimal solutions found in previous iterations.

In this way, uncertainty is only added to the model in places where it seems to be relevant for the optimal solution. Moreover, the sequence of optimal solutions provides direct insight into the sensitivity of the optimal solution to increases in the level of uncertainty. These are clear advantages over the usual way of solving a stochastic programming model only once, and with a description of the uncertainty that one has to decide on ex-ante.

We describe the type of multiperiod asset/liability management problems which we are considering in section 2, and formulate it as a multistage stochastic program. In section 3 we discuss why it is both reasonable and important to require that the description of the asset-price uncertainty in this formulation is arbitrage-free, and provide a useful characterization of arbitrage-free asset prices. The aggregation methods are described in section 4, as well as their effect on the stochastic programming formulation. Section 5 discusses the steps in the iterative disaggregation algorithm in more detail, and in section 6 we present the results of the application of the algorithm to a small portfolio insurance problem. Section 7 contains conclusions.

2   Problem Formulation

In this section we describe a multiperiod asset/liability management (ALM) problem, and formulate it as a multistage stochastic programming model.

2.1   Multiperiod Asset/Liability Management

We consider an investor who wants to determine a portfolio investment strategy over time to meet a sequence of liability payments in the future. We assume that the investor can only rebalance his portfolio at a finite number of points in time (*trading dates*) within a planning horizon of fixed length. Security prices at the initial date are known, but prices and returns at future trading dates are unknown. We approximate this uncertainty by assuming that at each future trading date only one of a finite number of *states of the world* can occur. We can then depict the uncertainty in future security prices and returns in the form of an *event tree*.

As an example, Figure 1 shows a recombining binomial event tree with six periods. The nodes in the tree represent states of the world, and the arcs transitions with positive probability. It is a binomial tree because two states can occur at the end of a period for each given state at the beginning of a period. It is a recombining tree (also called a *lattice*) because states in the inner part of the tree can be visited by multiple paths. This is often assumed to limit the number of different states at each trading date.[1] As the numerical example in section 6 employs a recombining binomial event tree, we will use it for illustrative purposes throughout this chapter. However, our analysis can be extended directly to trees which are not binomial and/or do not recombine.

The investor faces a trade-off between the initial cost of the trading strategy which must enable him to meet his liabilities, and the value of the portfolio which is left at the model horizon (*surplus*). We assume that he can borrow money at intermediate trading dates, but require that the surplus must be nonnegative in all cases. In executing a trading strategy, we furthermore assume that the investor has to pay transaction costs, is not allowed to short sell assets, and faces constraints on the amount that he can borrow as well as a spread between the interest rate for borrowing and (risk-free) lending.

The investor's optimal portfolio may be composed differently in different states at a given trading date. Moreover, in the presence of transaction costs, the optimal portfolio composition in a state at a future trading date does in

---

[1]It is not possible to use a recombining event tree if securities with path-dependent payoffs (and thus prices) such as mortgage-backed securities are included in the analysis, as each state at a trading date can in that case only correspond to one path in the tree up to that date.

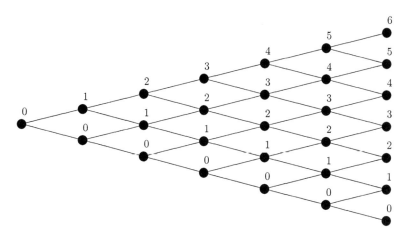

Figure 1: A recombining binomial event tree with six periods.

general not only depend on the state itself, but also on the composition of the portfolio that is carried over from the previous period. That is, optimal trading strategies in the event tree are generally *path-dependent*. However, the trading strategies are not allowed to depend on knowledge about the actual course of events in the future. Thus, when two paths in the event tree share the same history up to a certain trading date t, the optimal trading strategy up to time t must be identical on both paths. The trading strategies are then said to be *non-anticipative*.

2.2 Notation

The trading dates are denoted by the index t. The initial date is $t = 0$, the terminal date $t = T$, and intermediate trading dates are $t = 1, \ldots, T - 1$. States of the world are referred to by the index n. Given a state n at trading date $t < T$, each state which can occur with positive probability at time $t + 1$ is called a *successor* of state n, and will as such be referred to by the index n^+. For any state n at trading date $t > 0$, there is at least one state at time $t - 1$ which has state n as its successor. Such a state is called a *predecessor* of state n, and will be referred to by the index n^-. In the recombining binomial tree of Figure 1, each state at time $t < T$ has two successors, every state in the interior of the tree has two predecessors, while each state on the boundary has one predecessor. The (subjective) conditional probability of a transition from state n at time t to its successor state n^+ at time $t + 1$ is denoted by $\eta_{t/t+1}^{n/n^+}$, and the corresponding probability measure on the event tree by η.

All data in the problem are assumed to be functions of the nodes in the event tree. L_t^n denotes the liability which is due at the end of period t if state n occurs, $D_{i,t}^n$ the dividend payment on security i at the end of period t

if state n occurs, and $S_{i,t}^n$ its ex-dividend price in state n at time t. Let I be the total number of securities that is considered by the investor.

The riskless one-period interest rate (continuously compounded and annualized) in state n at time t is r_t^n, and the corresponding discount factor $P_t^n \equiv \exp(-r_t^n \Delta)$, where Δ is the length of a time period in the event tree. P_t^n can be interpreted as the price in state n at time t of a riskless one-period zero-coupon bond that pays one dollar at time $t+1$. The interest-rate spread (continuously compounded and annualized) between the investor's one-period borrowing rate and r_t^n is assumed to be constant through time and denoted by κ. The upper bound on one-period borrowing for the investor in state n at time $t < T$ is written as $\bar{Z}_t^n$.

A *scenario* s at time t is a path in the event tree between time 0 and time t, and the set of all possible scenarios at time t is denoted by $\mathcal{S}_t$. In the recombining binomial tree of Figure 1, the set $\mathcal{S}_t$ consists of 2^t scenarios. The unconditional probability of scenario s at time t as implied by the probability measure η is denoted by η_t^s.

A scenario $s \in \mathcal{S}_t$ visits one node in the event tree at each trading date between time 0 and time t, and such a node will be referred to by the index $n(s)$. For each scenario s at time t there is exactly one scenario at each time $\tau < t$ which follows the same path in the event tree between time 0 and time τ. This scenario is called the *predecessor* at time τ of scenario s at time t, and will as such be denoted by s^-. Furthermore, each scenario s at time t is the predecessor of one or more scenarios at time $t + 1$, which are called *successors* of scenario s. They will be referred to by the index s^+.

Let $\bar{x}_{i,0}$ denote the current holding of security i in the portfolio of the investor (a known number). The variables $xb_{i,t}^s$ and $xs_{i,t}^s$ denote the units of security i that are bought and sold, respectively, in scenario s at time t, and $xh_{i,t}^s$ the holding of security i in scenario s at time t *after* portfolio rebalancing (i.e., the portfolio holdings during period $t + 1$). The holding in the riskless one-period security during period $t + 1$ if scenario $s \in \mathcal{S}_t$ occurs is denoted by the separate variable y_t^s, while y_T^s denotes the portfolio surplus in scenario $s \in \mathcal{S}_T$. z_t^s represents the amount which has to be repaid by the investor at time $t + 1$ due to borrowing in scenario s at time t.

Transaction costs are assumed to be a fraction c of the dollar value traded and apply to both purchases and sales of securities, but not to investment in the risk-free security or borrowing.

2.3   Stochastic Programming Formulation

We capture the investor's trade-off between the initial investment and the value of the portfolio surplus in the objective function: the initial portfolio investment is minimized, but any positive final portfolio value is credited to the objective using a concave utility function $\mathcal{U}(\cdot)$. To be able to make a sensible trade-off, the utility of a surplus must be measured in units of current investment. We assume that the utility function satisfies the expected utility property (see Varian [25, chapter 11]).

The asset/liability management problem can now be formulated as the following multistage stochastic program (each stage corresponds to a time period):

minimize

$$(1+c)\sum_{i=1}^{I} S_{i,0} x b_{i,0} - (1-c)\sum_{i=1}^{I} S_{i,0} x s_{i,0} + P_0 y_0 - e^{-\kappa\Delta} P_0 z_0 - \sum_{s\in\mathcal{S}_T} \eta_T^s \mathcal{U}(y_T^s)$$

subject to

$$- x s_{i,0} + x b_{i,0} - x h_{i,0} = -\bar{x}_{i,0} \qquad\qquad \forall\, i = 1,\dots,I \quad (2.1)$$

$$x h_{i,t-1}^{s^-} - x s_{i,t}^s + x b_{i,t}^s - x h_{i,t}^s = 0 \;\; \forall\, i = 1,..,I, s\in\mathcal{S}_t, t = 1,..,T-1 \quad (2.2)$$

$$\sum_{i=1}^{I} D_{i,t}^{n(s)} x h_{i,t-1}^{s^-} + y_{t-1}^{s^-} - z_{t-1}^{s^-} + (1-c)\sum_{i=1}^{I} S_{i,t}^{n(s)} x s_{i,t}^s$$

$$- (1+c)\sum_{i=1}^{I} S_{i,t}^{n(s)} x b_{i,t}^s - P_t^{n(s)} y_t^s + e^{-\kappa\Delta} P_t^{n(s)} z_t^s = L_t^{n(s)}$$

$$\forall\; s\in\mathcal{S}_t, t = 1,\dots,T-1 \qquad\qquad\qquad (2.3)$$

$$\sum_{i=1}^{I}\left(D_{i,T}^{n(s)} + S_{i,T}^{n(s)}\right) x h_{i,T-1}^{s^-} + y_{T-1}^{s^-} - z_{T-1}^{s^-} - y_T^s = L_T^{n(s)} \quad \forall\, s\in\mathcal{S}_T \quad (2.4)$$

$$x s_{i,t}^s, x b_{i,t}^s, x h_{i,t}^s \geq 0 \qquad\qquad \forall\, i = 1,\dots,I,\; s\in\mathcal{S}_t,\; t = 0,\dots,T-1 \quad (2.5)$$

$$y_t^s \geq 0 \qquad\qquad\qquad\qquad\qquad \forall\, s\in\mathcal{S}_t,\; t = 0,\dots,T \quad (2.6)$$

$$0 \leq z_t^s \leq \bar{Z}_t^{n(s)} \qquad\qquad\qquad \forall\, s\in\mathcal{S}_t,\; t = 0,\dots,T-1 \quad (2.7)$$

The first four terms in the objective function represent the net cost of *additional* investments at time 0. These additional investments consist of asset purchases (including transaction costs) and investment in the riskless one-period security, while revenues from the sale of assets (net of transaction costs) and borrowing are subtracted. The last term in the objective is the expected utility of a final portfolio surplus.

There are three types of constraints in the model: *portfolio-balance* constraints, *cash-balance* constraints and *borrowing* constraints. The portfolio-balance constraints link portfolio holdings between successive periods (i.e., before and after rebalancing) for each scenario and each asset. The portfolio-balance constraints are given by (2.1) for all assets at time 0, and by (2.2) for all assets in each scenario after time 0.

The cash-balance constraints make sure that sufficient cash is generated to meet the liability payment in each scenario at each trading date. For each scenario at time $t < T$, this constraint is given by (2.3). At the end of a period, the investor receives dividend payments on his asset holdings and the return on his investment in the one-period riskless security, but has to repay the amount borrowed in the previous period plus interest. This is what the first three terms on the left-hand side of (2.3) represent. The next two terms reflect rebalancing of the portfolio: revenues are generated by selling assets, and money can be invested by buying assets, where both are adjusted for transaction costs. The final two terms on the left-hand side are the investment in the riskless one-period security and the amount borrowed, respectively, during the next period.

The cash-balance constraints (2.4) define the portfolio surplus in each scenario at the terminal date T. The first three terms on the left-hand side determine the final portfolio value before meeting the liability: the portfolio holdings are converted at the appropriate market prices, the return on the investment in the riskless one-period security is added, and the amount due because of borrowing is subtracted. The difference between this portfolio value and the liability payment in a scenario $s \in \mathcal{S}_T$ is the portfolio surplus y_T^s.

The nonnegativity restrictions on $xh_{i,t}^s$ and y_T^s preclude short sales of assets and a negative surplus, respectively, while equation (2.7) limits the amounts which can be borrowed.

3 Arbitrage-free Asset Prices

The formulation of the ALM problem as a stochastic program was based on a description of the uncertainty about future asset prices and returns in the form of an event tree. In this section we discuss why it is both reasonable and important that asset prices in such a description are arbitrage-free, and indicate how it may be constructed using financial asset-pricing models.

3.1 Arbitrage Opportunities and Arbitrage-Free Asset Prices

We speak of an *arbitrage opportunity* if it is possible to construct a self-financing trading strategy (a trading strategy is self-financing if no investments are required after time 0) with payoffs that are nonnegative everywhere and strictly positive in at least one state in the event tree, and for which the initial investment is nonpositive. With such a trading strategy it is thus possible to create something from nothing, and many investors will try to take advantage of that. In fact, there is a large group of investors in today's financial markets, called arbitrageurs, whose main objective is to look for and exploit arbitrage opportunities. This will influence the prices of the securities involved and lead to the elimination of the arbitrage opportunity. When no arbitrage opportunities exist, asset prices are said to be *arbitrage-free*.[2]

Although arbitrage opportunities may occasionally exist in reality, they are usually short-lived due to the presence of arbitrageurs. Moreover, even if one is able to detect and exploit arbitrage opportunities today, it is preposterous to assume that anyone would be able to foresee their occurrence at future points in time. As the primary objective in asset/liability management is to determine a portfolio investment strategy which forms a robust hedge against the future uncertainty, it furthermore seems imprudent to base such a strategy on the presence of arbitrage opportunities.

In the financial literature, arbitrage opportunities are usually defined for a frictionless world (i.e., a world without transaction costs and taxes, and in which securities are infinitely divisible, interest rates for borrowing and lending are the same, and short sales of assets with full use of proceeds are allowed). In reality, an investor may not be able to exploit such arbitrage opportunities directly because of market imperfections and trading restrictions. Nonetheless, Klaassen [20] illustrates that their presence in a portfolio optimization model with realistic market imperfections and trading restrictions may still significantly bias its optimal solution in unrealistic ways. We therefore require in the sequel that security prices in our ALM model are such that they do not admit arbitrage opportunities in a world without frictions.

Harrison and Kreps [11] have derived an important characterization of arbitrage-free security prices in an event tree under the assumption of a frictionless world, which is contained in the following theorem.

Theorem 1 (Harrison and Kreps) *Security prices in an event tree are arbitrage-free if and only if there exists a positive probability measure on the event tree such that in each given state the expected one-period return with respect to this probability measure is identical for all assets.*

The theorem states that there are no arbitrage opportunities in the event tree if and only if there exists a probability measure π such that

$$\frac{\sum_{n^+} \pi_{t/t+1}^{n/n^+} \left(S_{i,t+1}^{n^+} + D_{i,t+1}^{n^+} \right)}{S_{i,t}^n} = \frac{\sum_{n^+} \pi_{t/t+1}^{n/n^+} \left(S_{j,t+1}^{n^+} + D_{j,t+1}^{n^+} \right)}{S_{j,t}^n} \qquad (3.1)$$

for all assets i, j and in every state n at each trading date $t = 0, \ldots, T - 1$ in the event tree. The summations in (3.1) are over all successor states n^+ of state n.

---

[2] A concept related to arbitrage opportunities is that of "locks" in racetrack betting; see Hausch and Ziemba [12].

Because we have assumed that a riskless one-period security exists in every state in the event tree, this characterization can be restated as

$$S_{i,t}^n = P_t^n \sum_{n^+} \pi_{t/t+1}^{n/n^+} \left(S_{i,t+1}^{n^+} + D_{i,t+1}^{n^+} \right) \tag{3.2}$$

for all assets i in every state n at each trading date $t = 0, \ldots, T - 1$. This equation states that the price of each security i in state n at time t can, under the risk-neutral probability measure π, be written as the expected value of its payoffs at time $t + 1$, discounted by the riskless interest rate.

Because all assets have the same one-period expected return under the probability measure π, this measure is often called *risk neutral*. We emphasize that this risk-neutral probability measure is only a theoretical construct (namely, a necessary and sufficient condition for the absence of arbitrage opportunities), and that it should not be viewed as representing either actual probabilities or subjective probability beliefs of an investor. In other words, whether asset prices in an event tree are arbitrage-free is *independent* of the probabilities assigned to the states in the tree. Thus, an investor may very well believe that the expected return on one asset is higher than that on other assets, which will be reflected by the fact that he assigns probabilities to states (and therefore scenarios) in the tree which differ from risk-neutral probabilities. However, as long as a risk-neutral probability measure can be found he will not able to construct a portfolio which yields a riskless return in excess of the risk-free rate. For more background on the theory of arbitrage-free asset prices, see Huang and Litzenberger [16] and Hull [17].

3.2   Using Financial Asset-Pricing Models

The presence of arbitrage opportunities in portfolio optimization models may stem from several sources. One possible source is that the description of the uncertainty in the model is inconsistent with current market prices, which are usually taken as the prices at which the investor can trade at the initial date. As current market prices of securities are based on expectations of market participants about future prices and cashflows, these expectations should be properly reflected in the event tree which is used as description of the uncertainty in a portfolio optimization model. Another possible source is that prices in the event tree itself are not arbitrage-free. This will be the case, for example, if randomly sampled scenarios from a larger (and possibly arbitrage-free) model are used to construct the event tree.

To construct an event tree in which security prices are arbitrage-free and consistent with market prices, a natural starting point is one of the arbitrage-free asset-pricing models in discrete time which have been proposed in the financial literature. These models are primarily used to calculate prices of derivative securities, and they do so by assuming that arbitrage opportunities do not exist in financial markets. Examples are the binomial stock-price model of Cox, Ross and Rubinstein [8] and the term-structure models of Black, Derman and Toy [3], Brennan and Schwartz [5], Heath, Jarrow and Morton [13], Ho and Lee [15] and Hull and White [18].

These asset-pricing models, or discretized versions thereof, describe the uncertainty in the underlying variable(s) in the form of a discrete-time, discrete-state event tree. The arbitrage-free value of a derivative security is computed in a recursive manner, starting from the maturity date on which the cash flows (as a function of the value(s) of the underlying variable(s)) are known, and proceeding backwards through time in the tree until the arbitrage-free value at time 0 is found (this recursion employs relation (3.2)). In this way, the security's arbitrage-free value is calculated for every state in the event tree at or before its maturity date.

Although prices of securities in the event tree are therefore arbitrage-free when calculated in this way, there is no guarantee that their arbitrage-free value at time 0 equal current market prices. To accomplish this (at least approximately), one typically selects a set of benchmark securities, and chooses values for the parameters in the model (representing, for example, the volatility of the underlying variable(s)) so that the calculated model prices match the market prices as closely as possible. These parameter values can vary significantly with the number of time steps in the event tree, but they converge to certain limit values when the number of time steps increases.

However, the number of time steps which is necessary to obtain satisfactory convergence is usually much larger than the 10 or at most 20 stages (corresponding to time periods of possibly different lengths) which can be included in a stochastic program[3]. If the number of time steps in the event tree would be restricted to such a small number, the obtained parameter values are typically very sensitive to an increase or decrease in the number of time steps. This implies that the stochastic nature of the underlying variable(s) will not be modelled correctly. Furthermore, the calculated model price of a security which is not included in the benchmark set may vary significantly as a function of the chosen number of steps.

Thus, an event tree that is derived from a financial asset-pricing model in which security prices are arbitrage-free and consistent with current market pices, and which captures the stochastic nature of the underlying uncertainties in a satisfactory way, is typically much too large to include fully in a stochastic program. In the next section we describe aggregation methods which enable a reduction in the size of the event tree without sacrificing its desired properties.

---

[3]With 10 securities and a binomial event tree with 10 time periods, the ALM model would already have 33,740 variables and 12,266 constraints, and these numbers would approximately double with each additional time period.

4 Aggregation of the Uncertainty

We will combine both states and time periods in a large event tree to reduce its size, and refer to this as *state aggregation* and *time aggregation*, respectively. To illustrate the mechanics of state and time aggregation, we consider the two-period binomial tree of Figure 2a. The vectors at each node in the tree represent the price of a one-period discount bond with a face value of one dollar (P_t^n), the dividend payment on security i $(D_{i,t}^n)$ and the ex-dividend price of security i $(S_{i,t}^n)$. We assume that the prices in the tree are arbitrage-free so that a risk-neutral probability measure π exists on the tree. Let π_t^n denote the conditional risk-neutral probability of an upward transition in the tree from state n at time t. These conditional probabilities are written next to the arcs in Figure 2a. We will reduce this two-period binomial tree through the application of state and time aggregation to a binomial "tree" with only one period. See Klaassen [21] for a complete discussion of the aggregation methods.

4.1 State Aggregation

We say that *state aggregation* is performed in state n at time t if all successor states of state n in the event tree are combined into one state, the *aggregate successor*. Obviously, the conditional probability of this aggregate successor state is one. Figure 2b shows the resulting event tree when state aggregation is performed in both states at time 1 in the event tree of Figure 2a. The dividends and ex-dividend prices in the aggregate successor states are defined as follows $(n = 0, 1)$:

$$\bar{D}_{i,2}^n \equiv \pi_1^n D_{i,2}^{n+1} + (1 - \pi_1^n)D_{i,2}^n \tag{4.1}$$
$$\bar{S}_{i,2}^n \equiv \pi_1^n S_{i,2}^{n+1} + (1 - \pi_1^n)S_{i,2}^n \tag{4.2}$$
$$\bar{P}_2^n \equiv \pi_1^n P_2^{n+1} + (1 - \pi_1^n)P_2^n. \tag{4.3}$$

Using the no-arbitrage relation (3.2), it follows that

$$\begin{aligned} S_{i,1}^n &= P_1^n \left[\pi_1^n \left(D_{i,2}^{n+1} + S_{i,2}^{n+1} \right) + (1 - \pi_1^n) \left(D_{i,2}^n + S_{i,2}^n \right) \right] \\ &= P_1^n \left(\bar{D}_{i,2}^n + \bar{S}_{i,2}^n \right) \end{aligned} \tag{4.4}$$

which shows that the prices in the aggregated tree of Figure 2b are arbitrage-free.

To obtain Figure 2c from 2b, we perform state aggregation at $t = 0$. Define:

$$\bar{D}_{i,1}^0 \equiv \pi_0 D_{i,1}^1 + (1 - \pi_0)D_{i,1}^0 \tag{4.5}$$
$$\bar{S}_{i,1}^0 \equiv \pi_0 S_{i,1}^1 + (1 - \pi_0)S_{i,1}^0 \tag{4.6}$$
$$\bar{P}_1^0 \equiv \pi_0 P_1^1 + (1 - \pi_0)P_1^0 \tag{4.7}$$
$$\bar{\pi}_1^0 \equiv \left(\pi_0 P_1^1 \right)/\bar{P}_1^0. \tag{4.8}$$

(a) The original tree:

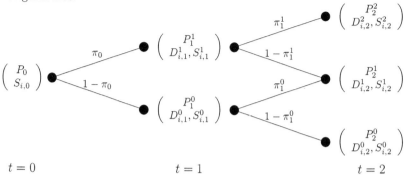

$t = 0$ $t = 1$ $t = 2$

(b) After state aggregation in the nodes at $t = 1$:

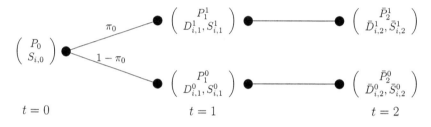

$t = 0$ $t = 1$ $t = 2$

(c) After state aggregation at $t = 0$:

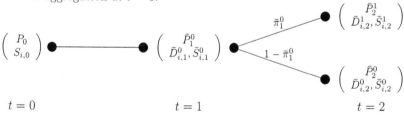

$t = 0$ $t = 1$ $t = 2$

(b) After time aggregation at $t = 0$:

$$\left(\begin{array}{c} P_{0 \to 2} \\ S_{i,0} - D_{i,0 \to 2} \end{array} \right)$$

$\pi_{0 \to 2}$

$1 - \pi_{0 \to 2}$

$$\left(\begin{array}{c} \bar{P}_2^1 \\ \bar{D}_{i,2}^1, \bar{S}_{i,2}^1 \end{array} \right)$$

$$\left(\begin{array}{c} \bar{P}_2^0 \\ \bar{D}_{i,2}^0, \bar{S}_{i,2}^0 \end{array} \right)$$

$t = 0$ $t = 2$

Figure 2: Illustration of state and time aggregation.

Using again the no-arbitrage relation (3.2) yields

$$S_{i,0} = P_0 \left(\bar{D}_{i,1}^0 + \bar{S}_{i,1}^0 \right) \tag{4.9}$$

$$\bar{S}_{i,1}^0 = \bar{P}_1^0 \left[\bar{\pi}_1^0 \left(\bar{D}_{i,2}^1 + \bar{S}_{i,2}^1 \right) + (1 - \bar{\pi}_1^0) \left(\bar{D}_{i,2}^0 + \bar{S}_{i,2}^0 \right) \right]. \tag{4.10}$$

This shows that prices in the aggregated tree of Figure 2c are arbitrage-free, with $\bar{\pi}_1^0$ acting as risk-neutral probability.

Although state aggregations thus preserve arbitrage-free prices in an event tree, we note that certain relationships between securities with nonlinear payoffs in the original event tree may be violated in an aggregated event tree. For example, suppose the event tree models the evolution of a stock price, and consider a call option on the stock. At its maturity date the option pays the difference between the stock price and its exercise price, if positive, and nothing otherwise. Although this relation will hold exactly in the original event tree, it may be violated in some nodes in an aggregated tree due to the nonlinearity of the option payoff as function of the stock price. Such deviations are unavoidable to maintain arbitrage-free prices for general assets in an aggregated event tree as well as consistency between each asset's payoffs in the tree and its market price.

4.2   Time Aggregation

Time aggregation involves the merging of successive time steps in the event tree. Specifically, time aggregation is performed in state n at time t in an event tree if we replace the transitions from this state to its successors by direct transitions to the successors of its successors. Figure 2d depicts the change in event tree when time aggregation is performed at time 0 in the event tree of Figure 2c.

Time aggregation in Figure 2d eliminates trading date $t = 1$ from the event tree. This affects the definition of a riskless one-period security in the tree as well as the dividend payments. Obviously, dividends which are paid on the trading date that is eliminated cannot be ignored. We will take them into account by assuming that *their arbitrage-free value* is prepaid at the last trading date in the event tree before the true payment date, i.e., at $t = 0$. The arbitrage-free value (at time 0) of all dividends that are paid on security i between time 0 and time 2, not including the dividends at time 0 and time 2 itself, is denoted by $D_{i,0 \to 2}$. Furthermore, the riskless one-period security at time 0 in the event tree is redefined as a security which pays one dollar at time 2, and nothing at other times. Its price is denoted as $P_{0 \to 2}$.

Define

$$D_{i,0 \to 2} \equiv P_0 \bar{D}_{i,1}^0 \tag{4.11}$$

$$P_{0\to2} \equiv P_0 \bar{P}_1^0 \tag{4.12}$$

$$\pi_{0\to2} \equiv \bar{\pi}_1^0 \tag{4.13}$$

and combine this with equations (4.9) and (4.10) to get

$$S_{i,0} = D_{i,0\to2} + P_{0\to2}\left[\pi_{0\to2}\left(\bar{D}_{i,2}^1 + \bar{S}_{i,2}^1\right) + (1 - \pi_{0\to2})\left(\bar{D}_{i,2}^0 + \bar{S}_{i,2}^0\right)\right]. \tag{4.14}$$

This relation shows that the price of security i at time 0 can be written as the sum of the present (arbitrage-free) value of its dividend at time 1 and the expected present value of dividend plus ex-dividend price at time 2. As the discounting is performed using the riskless interest rate ($P_{0\to2}$), the probability that is used to take the expected value ($\pi_{0\to2}$) is a risk-neutral probability. Because $\pi_{0\to2}$ is independent of i, the prices in the aggregated event tree of Figure 2d are arbitrage-free.

4.3   The Aggregated ALM model

In the formulation of the ALM model we assumed that in between trading dates no dividends were paid on securities and no liabilities were due. Clearly, we can not assume this any longer if we perform time aggregations in the underlying event tree. We will therefore indicate the changes which have to be made to the formulation in order to take prepayment of the arbitrage-free value of intermediate dividends as well as liabilities into account. This adjusted formulation, which results after performing state and time aggregations in the event tree which underlies the original ALM model, will be called the *aggregated* ALM model. For its exact mathematical formulation and the precise definition of all aggregated quantities, see Klaassen [21].

Let the trading dates in the aggregated event tree be $t_0, t_1, \ldots, t_T$, with $t_0 = 0$ and $t_T = T$. The period between two successive trading dates in the aggregated tree thus comprises one or more periods in the original tree. In the aggregated ALM model, an extra term is subtracted from the objective function which represents the prepayment of the arbitrage-free value of dividends which are in reality paid between the trading dates t_0 and t_1, not including the dividends on these trading dates itself. These dividends are prepaid over the portfolio holdings of the investor between dates t_0 and t_1, i.e., *after* rebalancing the initial portfolio at time t_0. Furthermore, an extra term is added to the objective function, which represents the arbitrage-free value of the liability payments which are due at dates in between times t_0 and t_1.

The cash-balance constraint for each scenario at times $t_j \in \{t_1, t_2, \ldots, t_{T-1}\}$ is modified as follows. To the left-hand side of the constraint an extra term is added which represents the prepayment of the arbitrage-free value of dividends which are in reality paid between the trading dates t_j and t_{j+1}, conditional on the specific scenario at time t_j. This arbitrage-free value does not include the dividends on the trading dates t_j and t_{j+1}. These intermediate dividends are prepaid over the portfolio holdings of the investor between dates t_j and t_{j+1} in the specific scenario under consideration, i.e., after rebalancing at time t_j. To the liability on the right-hand side of the cash-balance constraint is added the arbitrage-free value of the liability payments which are due at dates in between times t_j and t_{j+1}, conditional on the scenario at time t_j.

An important property of the aggregated ALM model is that it is neither a restriction nor a relaxation of the original ALM model. This property follows from the fact that the constraint set of the ALM model after an aggregation can be obtained from the constraint set of the model before the aggregation by performing both elementary row *and* column operations. We will list the sequence of these operations for both state and time aggregation below. A proof of these results can be found in Klaassen [19].

The fact that the aggregated ALM model is neither a restriction nor a relaxation of the original model indicates that the state and time aggregation methods do not discard part of the uncertainty in the original event tree, but instead condense its description. This contrasts with stochastic programming models in which only a (sampled) subset of scenarios from a large event tree is used as description of the uncertainty.

4.3.1 State Aggregation as Row and Column Operations in the ALM Model

We consider state aggregation in state n at trading date t_j in the event tree. This state aggregation corresponds to the following row and column operations in the ALM model for each scenario s at trading date t_j which visits state n:

- For each successor scenario s^+ at trading date t_{j+1}, multiply its cash-balance constraint, the portfolio-balance constraint of each asset, and the borrowing constraint with the conditional risk-neutral probability $\pi_{t_j/t_{j+1}}^{n(s)/n(s^+)}$. Subsequently, sum the cash-balance constraints of *all* successor scenarios s^+ at time t_{j+1}, and do the same for the borrowing constraints and each of the portfolio-balance constraints.

- Add the columns which correspond to the portfolio sales variables $xs_{t_{j+1}}^{s^+}$ over all successor scenarios s^+ at trading date t_{j+1} (that is, sum the column coefficients of $xs_{t_{j+1}}^{s^+}$ over all s^+ in each constraint). Repeat this for the variables $xb_{t_{j+1}}^{s^+}$, the variables $xh_{t_{j+1}}^{s^+}$, the variables $y_{t_{j+1}}^{s^+}$, and the variables $z_{t_{j+1}}^{s^+}$.

When $t_{j+1} = t_T$ only those row and column aggregations are performed which are applicable, as there are no portfolio-balance and borrowing constraints in the model for scenarios at time t_T, nor are there variables $xs^s_{t_T}$, $xb^s_{t_T}$, $xh^s_{t_T}$, and $z^s_{t_T}$.

Aggregation of the columns which correspond to variables $y^{s^+}_T$ when $t_{j+1} = t_T$ affects the evaluation of the surplus in the objective function. If the probabilities η^s_T in the objective function are equal to the risk-neutral probabilities, and if the utility function $\mathcal{U}$ has the form

$$\mathcal{U}(y^s_T) = \lambda \left(\prod_{j=0}^{T-1} P^{n(s^-)}_{t_j} \right) y^s_T \tag{4.15}$$

with $\lambda \leq 1$ a scalar, then this aggregation is uniquely defined. In this case the expected utility of a surplus is equal to its expected present value, weighted by the scalar λ, with discounting taking place against the one-period interest rates along the scenario path, and the expectation being calculated under the risk-neutral probability measure. This is also the case that will be considered in the numerical example of section 6, and we will therefore focus our discussion on this case. In other cases, however, one must decide how to define the utility of an aggregated surplus in the objective function (see Klaassen [21]).

If the underlying event tree is recombining, then some scenarios which are distinguishable before the state aggregation may become duplicates of each other after the aggregation. This will be the case for scenarios which visit node n at time t_j, and follow the same path through the tree before the aggregation *except* for the node at trading date t_{j+1}. Obviously, one can combine such duplicate scenarios in the ALM model after the aggregation, and sum their respective probabilities. (This is equivalent to multiplying the duplicated constraints of these scenarios by their relative risk-neutral probabilities, summing the corresponding constraints, and adding the columns of corresponding duplicated variables.)

4.3.2 Time Aggregation as Row and Column Operations in the ALM Model

Consider now time aggregation in state n at trading date t_j. We assume that this state has only one successor state at the next trading date t_{j+1} (compare with Figures 2c and d). Otherwise, we can always first perform a state aggregation in this state to obtain this situation. The following row and column operations are performed in the ALM model for each scenario s in state n at trading date t_j:

- The time aggregation implies that no securities can be bought and sold in the successor scenario s^+ at time t_{j+1}, and thus we essentially add the constraints $xs^{s^+}_{i,t_{j+1}} = 0$, $xb^{s^+}_{i,t_{j+1}} = 0$ and $xh^{s^+}_{i,t_{j+1}} = xh^s_{i,t_j}$ for all assets i. Instead of adding them, however, we use these constraints to remove the variables $xs^{s^+}_{i,t_{j+1}}$, $xb^{s^+}_{i,t_{j+1}}$ and $xh^{s^+}_{i,t_{j+1}}$ from the ALM model. This renders the portfolio-balance constraints for the successor scenario s^+ at time t_{j+1} vacuous, and they are therefore removed from the formulation.

- Multiply the cash-balance constraint of the successor scenario s^+ at trading date t_{j+1} by the one-period discount bond price $P^n_{t_j \to t_{j+1}}$ in state n at time t_j. If $t_j = 0$, add this constraint to the objective function. Otherwise, add it to the cash-balance constraint of scenario s at time t_j.

- If the borrowing-lending spread κ is positive, multiply the column that corresponds to the borrowing variable $z^s_{t_j}$ by $\left(e^{-\kappa(t_{j+2} - t_{j+1})} P^{n(s^+)}_{t_{j+1} \to t_{j+2}} \right)$ and add it to the column that corresponds to the borrowing variable $z^{s^+}_{t_{j+1}}$.

5   The Iterative Disaggregation Algorithm

We now describe the iterative disaggregation algorithm in more detail. Assume that at the start of the algorithm one has an event tree as description of the uncertainty in asset prices and returns which is arbitrage-free and consistent with current market prices, for example by using arbitrage-free financial asset-pricing models. Such an event tree will most likely be much too large to include in a stochastic programming model for ALM problems. The aggregation methods from the previous section can then be used to reduce the size of the event tree without introducing arbitrage opportunities or losing consistency with current market prices. In this manner one can obtain a small aggregated ALM model which can be solved relatively easily.

In the extreme, one could continue to perform state and time aggregations in the event tree until nothing more than one "expected-value" scenario remains. Obviously, this is not what one wants. In contrast, it is important to include in the ALM model as much of the relevant uncertainty as possible so that its optimal solution is a good approximation to the optimal solution that would result if one were able to incorporate all uncertainty in the model. By starting with a relatively coarse approximation of the true uncertainty in the ALM model, and iteratively refining this approximation based on the information one gathers from the sequence of solutions, this is exactly what the iterative disaggregation algorithm aims to accomplish.

Each iteration of the iterative disaggregation algorithm consists of the following steps:

1. Disaggregate in the aggregated event tree; i.e., reverse one or more aggregations.

2. Find a feasible solution to the disaggregated ALM model, based on the optimal solution from the previous iteration, and re-optimize the disaggregated model.

We will discuss both steps in the remainder of this section.

5.1 Construction of a Feasible Solution after a Disaggregation

From the analysis in section 4 it should be clear that the ALM model after a disaggregation is performed in the underlying event tree will contain both more variables and more constraints than the ALM model before the disaggregation. Hence, the ALM model after a disaggregation is not just a relaxation of the model before the disaggregation, which implies that the optimal solution to the ALM model before the disaggregation may not define a feasible solution in the ALM model after the disaggregation. Here we will be concerned with the question how to construct a feasible solution for the ALM model after a disaggregation, given an optimal solution to the ALM model before the disaggregation. This analysis will be useful when we consider the question in the next subsection how to decide where to disaggregate in the event tree.

Our starting point for the derivation of a feasible solution is the *fixed-weight solution* that is defined by Zipkin [27], [28] in the context of row and column (dis)aggregations in linear programs, and extended by Birge [1] to stochastic linear programs. This fixed-weight solution is defined for both the primal variables (i.e., corresponding to sales and purchases of assets, portfolio holdings, borrowing and lending) and dual variables (shadow prices on the cash-balance, portfolio-balance and borrowing constraints), and relates directly to the column, respectively row, operations of the corresponding aggregation in the ALM model. We will summarize the results for state and time disaggregation in turn. More details can be found in Klaassen [19].

5.1.1 Fixed-Weight State Disaggregation

We consider the reversal of the state aggregation which was described in section 4.3.1. The fixed-weight solution after the state disaggregation in state n at trading date t_j is defined as follows for each scenario s in this state:

- Primal variables:

  The optimal values for the single successor scenario at trading date t_{j+1} before the disaggregation are taken as initial values for each of the successor scenarios at trading date t_{j+1} after the disaggregation. If the original event tree was not recombining, then no scenarios and thus variables are added at other places in the event tree.

If the original event tree did recombine, however, single scenarios at trading dates beyond t_{j+1} in the event tree before the disaggregation may have to be splitted into multiple scenarios after the disaggregation (see the discussion in section 4.3.1). Initial values for the variables of these new scenarios are set equal to the optimal values of the variables of the single scenario from which they originate.

- Dual variables:

  Initial values for the shadow prices of the constraints for each of the successor scenarios at trading date t_{j+1} after the disaggregation are set equal to the optimal shadow prices for the single successor before the disaggregation, multiplied by the conditional risk-neutral probability of the corresponding state. When the original event tree was a recombining one, and the state disaggregation leads to the splitting of scenarios at later trading dates in the tree, the shadow prices of the constraints for these new scenarios are initialized at the optimal shadow prices of the constraints for the scenario from which they originate, multiplied by their relative risk-neutral probabilities.

The fixed-weight solution for the primal variables will satisfy all portfolio-balance constraints, but it may violate the cash-balance and borrowing constraints in some scenarios. However, the violated constraints are satisfied *on average*. Moreover, if the upper bound on borrowing is state-independent, the borrowing constraints will be satisfied by the fixed-weight primal solution in all scenarios.

5.1.2   Fixed-Weight Time Disaggregation

We consider the reversal of the time aggregation in state n at trading date t_j which was described in section 4.3.2. For each scenario s in this state, the fixed-weight solution after the time disaggregation is defined as follows:

- Primal variables:

  The variables $xh^{s^+}_{i,t_{j+1}}$, $y^{s^+}_{t_{j+1}}$ and $z^{s^+}_{t_j}$ for the new successor scenario s^+ at trading date t_{j+1} are initialized at the optimal values of these variables for scenario s at trading date t_j. Furthermore, the variables $xs^s_{i,t_{j+1}}$ and $xb^{s^+}_{i,t_{j+1}}$ are initialized at zero. The optimal values of the lending and borrowing variables for scenario s at trading date t_j are modified as follows:

$$y^s_{t_j} = P^n_{t_j \to t_{j+1}} \hat{y}^s_{t_j}$$
$$z^s_{t_j} = \left(e^{-\kappa(t_{j+1}-t_j)} P^n_{t_j \to t_{j+1}} \right) \hat{z}^s_{t_j}$$

where $\hat{y}_{t_j}^s$ and $\hat{z}_{t_j}^s$ are the optimal values of the one-period lending and borrowing variables before the disaggregation.

- Dual variables:

  Initial values for the shadow prices of the cash-balance and portfolio-balance constraints for the new successor scenario at trading date t_{j+1} are set equal to the optimal shadow prices for scenario s at trading date t_j before the disaggregation, multiplied by the discount bond price $P_{t_j \to t_{j+1}}^n$. For scenario s at trading date t_j itself, the initial shadow prices of these constraints are set equal to the optimal shadow prices before the disaggregation.

  For the borrowing constraints of these scenarios, the initial shadow prices ξ are chosen as follows:

  $$\xi_{t_j}^s = \begin{cases} \dfrac{\hat{\xi}_{t_j}^s}{e^{-\kappa(t_{j+2}-t_{j+1})} P_{t_{j+1} \to t_{j+2}}^{n(s^+)}} & \text{if } \bar{Z}_{t_j \to t_{j+2}}^n = \dfrac{\bar{Z}_{t_j \to t_{j+1}}^n}{e^{-\kappa(t_{j+2}-t_{j+1})} P_{t_{j+1} \to t_{j+2}}^{n(s^+)}} \\ 0 & \text{otherwise} \end{cases}$$

  $$\xi_{t_{j+1}}^{s^+} = \begin{cases} \hat{\xi}_{t_j}^s & \text{if } \bar{Z}_{t_j \to t_{j+2}}^n = \bar{Z}_{t_{j+1} \to t_{j+2}}^{n(s^+)} \\ 0 & \text{otherwise} \end{cases}$$

  where $\hat{\xi}_{t_j}^s$ is the optimal shadow price on the borrowing constraint of scenario s at trading date t_j before the disaggregation.

The only constraints which may be violated by the fixed-weight primal solution are the cash-balance constraints of each scenario s in state n at trading date t_j and its successor scenario s^+ at the added trading date t_{j+1}. (The upper bounds on one-period borrowing in an aggregated ALM model are defined in such a way that fixed-weight time disaggregation will never result in a violation of the borrowing constraint.) Furthermore, it is easy to see that these cash-balance constraints can only be violated if prepayment of dividends and/or liabilities occurs in state n at trading date t_j before the time disaggregation.

5.1.3 Modifying Fixed-Weight Solutions to Obtain Feasibility

We have indicated that fixed-weight disaggregation may result in violations of the cash-balance constraints for some scenarios in the ALM model, and also of the borrowing constraints if state disaggregation is performed. If the cash-balance constraint for a scenario s at time t is violated after a fixed-weight disaggregation, an obvious way to make it feasible is by increasing the amount of short-term lending in case of a cash surplus, and the amount of short-term borrowing in case of a cash deficit. In the last case, however, this may lead to a violation of the borrowing constraint in the scenario. Furthermore, additional short-term lending or borrowing in scenario s at time t also increases the amounts of short-term lending, respectively borrowing, in its descendant scenarios if we correct for violations in the cash-balance constraints in these scenarios in the same manner. Thus, even if additional borrowing in scenario s at time t does not violate the borrowing constraint at that time, it may cause a violation of the borrowing constraint in one of its successors.

The upper bounds on short-term borrowing may therefore hinder the construction of a feasible solution from the fixed-weight solution through adjustments of the one-period borrowing and lending amounts. To circumvent this problem, we can relax the ALM model by including for all scenarios at each trading date an additional variable for short-term borrowing on which no upper bound is imposed. A high borrowing-lending spread will be associated with these additional borrowing variables, however. Furthermore, an extra variable is added for each scenario at the terminal date to take care of a negative final portfolio value. The expected present value of such a final portfolio deficit is added to the objective function, and penalized using a (large) penalty parameter.

As is shown in Klaassen [19, section 4.3], it is possible to derive lower bounds on the borrowing-lending spread and the penalty parameter so that the optimal solution will involve neither borrowing nor final portfolio deficits for any pair of values above these lower bounds. Thus, for any pair of values above these lower bounds it does not matter whether one solves the true ALM model, or its relaxation as just defined. The lower bounds are shown to be positively correlated to the variability of asset returns, and negatively to the value of λ when the utility function is defined according to equation (4.15). For example, if $\lambda = 1$ in equation (4.15), no borrowing will take place in the optimal solution whenever the borrowing-lending spread is positive.

5.2   Choosing Disaggregations

In each iteration of the iterative disaggregation algorithm one has to select one or more states in the event tree in which to perform state and/or time disaggregation. The method we will discuss for doing this is based on the infeasibilities of fixed-weight disaggregation, and estimates the effect of these infeasibilities on the objective function. This method is used in the implementation of the iterative disaggregation algorithm in the next section.

Suppose we perform state disaggregation in state n at trading date t_j in the event tree. For each scenario s in this state the fixed-weight primal solution may violate the cash-balance and the borrowing constraints for its new successor scenarios s^+ at trading date t_{j+1}. Let $U_{t_{j+1}}^{s^+}$ denote the shortfall in the cash-balance constraint for successor scenario s^+ at trading date t_{j+1}, i.e., the difference (if positive) between the liabilities in state $n(s^+)$ and the cash flows from the portfolio as implied by the fixed-weight primal solution. Similarly, let $V_{t_{j+1}}^{s^+}$ denote the amount of one-period borrowing in this scenario which exceeds its upper bound.

We estimate the effect of these infeasibilities on the objective function[4] by the following quantity ε:

$$\varepsilon_{t_j}^n \equiv \sum_{s \in \mathcal{S}_{t_j}^n} \sum_{s^+} \left(\tilde{\varphi}_{t_{j+1}}^{s^+} U_{t_{j+1}}^{s^+} + \tilde{\xi}_{t_{j+1}}^{s^+} V_{t_{j+1}}^{s^+} \right) \tag{5.1}$$

where $\tilde{\varphi}_{t_{j+1}}^{s^+}$ and $\tilde{\xi}_{t_{j+1}}^{s^+}$ denote the shadow prices on the cash-balance and borrowing constraint, respectively, as defined by the fixed-weight dual solution, while $\mathcal{S}_{t_j}^n$ denotes the set of all scenarios which visit state n at trading date t_j.

A similar measure can be defined for time disaggregation in state n at trading date t_j. Fixed-weight time disaggregation may lead to a violation of the cash-balance constraint for a scenario s in this state as well as for its successor scenario s^+ at the newly added trading date (which we denote as t_{j+1} for simplicity). With U_t^s again denoting the shortfall in the cash-balance constraint for scenario s at trading date t, we estimate the effect on the objective function of the infeasibilities due to fixed-weight time disaggregation as:

$$\zeta_{t_j}^n \equiv \sum_{s \in \mathcal{S}_{t_j}^n} \left(\tilde{\varphi}_{t_j}^s U_{t_j}^s + \tilde{\varphi}_{t_{j+1}}^{s^+} U_{t_{j+1}}^{s^+} \right) \tag{5.2}$$

where $\tilde{\varphi}_{t_j}^s$ and $\tilde{\varphi}_{t_{j+1}}^{s^+}$ denote the shadow prices on the cash-balance constraints as defined by the fixed-weight dual solution.

We can calculate the quantities ε and ζ for each state in the event tree in order to decide where to perform a state or time disaggregation. Clearly, the higher a quantity is in a state, the larger the estimated effect of the corresponding disaggregation is on the optimal objective value, and thus the more important this disaggregation is estimated to be.

Another approach to deciding where to perform disaggregations in the event tree is to calculate *bounds* on the possible change in the objective function. Zipkin [28] shows how such bounds can be derived for (dis)aggregations in general linear programs if (generalized) upper bounds on the primal and dual variables are known. For the ALM problem, Klaassen [19] shows how upper bounds on the primal and dual variables can be derived if an upper bound on the investment at time 0 is known.

The difference between the arbitrage-free value of all liabilities and the value of the investor's initial portfolio always forms a lower bound on the

---

[4]Strictly speaking, we consider infeasibilities in the formulation of the ALM problem in which the cash-balance constraints are written as greater-than-or-equal-to constraints, as we only take shortfalls and not surpluses into account.

optimum objective value. Furthermore, the fixed-weight solution in the relaxation of the ALM model after one or more disaggregations always defines
an upper bound. One hopes that the bounds which follow from Zipkin's
method will be tighter than these general bounds.

5.3   Termination of the Algorithm

For the decision when to terminate the iterative disaggregation algorithm, we
would like to have a measure of how close the current solution is to the true
optimal solution, i.e., the solution to the ALM model when all uncertainty
is considered. It is not obvious, however, how to construct such a measure.
Furthermore, because an aggregated ALM model is neither a restriction nor
a relaxation of the unaggregated model, it is impossible to tell precisely how
the optimal solution to an aggregated model relates to the solution in the
unaggregated model.

One possibility would be to calculate the bounds of Zipkin on the change
in objective value, which were discussed earlier. However, these bounds are
most likely too weak to be meaningful if one would translate the optimal
solution of an aggregated ALM model to a solution for the unaggregated
model by fixed-weight disaggregation, because the unaggregated model is
typically very much larger than the aggregated models that are solved in the
algorithm. Moreover, the sheer size of the unaggregated ALM model may
make it practically impossible to perform this calculation.

A more feasible approach is to base the decision to terminate the algorithm
on the results in past iterations. In practical applications, an investor will
primarily be interested in the optimal portfolio decisions at time 0. One may
therefore decide to terminate the algorithm if these portfolio decisions have
remained sufficiently stable in recent iterations.

6   A Numerical Example

We now present the results from the implementation of the iterative disaggregation algorithm for a small asset/liability management problem.

6.1   Problem Statement

Consider an investor with interest-rate exposure in his investment portfolio,
who wants to limit the downside risk. More specifically, we assume that the
investor owns a zero-coupon treasury bond with a maturity of two years from
the current date. Furthermore, he has to make a payment after one year, for
which he will have to sell the bond. As he expects that interest rates will fall

in the coming year, he wants to hold the two-year bond during the first year, but he also wants to be guaranteed that he can fulfill his obligation after one year. He could realize this guarantee if he could buy a one-year put option on the two-year zero-coupon bond.

Assume that all traded options have a maturity of at most four months. The problem therefore becomes to construct a dynamic trading strategy, involving the traded options, with payoffs that replicate the payoffs of the desired one-year put option. The liabilities in the ALM formulation are thus equal to the difference between the scheduled payment and the value of the discount bond, if positive, and zero otherwise. The investor does not accept shortfalls at the payment date.

The face value of the two-year zero-coupon bond is $1000. Assume that the current zero-coupon yield curve is flat with a (continuously compounded) yield of 8% for all maturities. The current price of the bond is therefore $852.14, and the forward price of the bond for delivery after one year is $923.12. Assume that the scheduled payment equals $932.35 (101% of the forward price), irrespective of the state of the world. That is, the exercise price of the one-year put option that he wants to replicate is $932.35.

Assume that traded option contracts on the two-year bond are initiated at the beginning of every two-month period, and that the options have an initial maturity of four months. For every option maturity, three call options and three put options are traded, which differ only in their exercise prices (respectively 99.5%, 100% and 100.5% of the forward bond price on the maturity date of the options). Thus, at every point in time the investor can trade in six different put options and six different call options on the two-year bond.

The term-structure model of Ho and Lee [15] is used to model the interest-rate uncertainty in the stochastic programming formulation of this problem[5]. This term-structure model describes the uncertainty in the future term structure of interest rates by means of a binomial lattice (see Figure 1). It is a one-factor model as the evolution of the short-term (one-period) interest rate completely determines the evolution of the whole term structure. As input parameters, the Ho and Lee model requires the number of time steps in the binomial lattice for a horizon of given length (in this case one year), the volatility of the one-period interest rate, and the conditional risk-neutral probability of an upward movement in the lattice. This probability is assumed to be the same in every node in the lattice.

Prices of interest-rate derivative securities (such as bonds and options on bonds) in the event tree can be calculated in the recursive manner which is described in section 3.2. For ordinary bonds, the price at the start of the tree

---

[5] Although the simple Ho and Lee model suffices to illustrate the iterative disaggregation algorithm in this example, one may want to use a more sophisticated term-structure model in real-world applications (see the references in section 3.2).

| Option | Initiation | Expiration | | Price at time 0: | |
number	date	date	Strike price	put	call
10		20	$859.26 (99.5%)	$0.40	$4.66
11		20	$863.58 (100.0%)	$1.76	$1.76
12		20	$867.90 (100.5%)	$4.66	$0.40
20	0	40	$870.80 (99.5%)	$0.76	$5.03
21	0	40	$875.17 (100.0%)	$2.28	$2.28
22	0	40	$879.55 (100.5%)	$5.03	$0.77
30	20	60	$882.49 (99.5%)	-	-
31	20	60	$886.92 (100.0%)	-	-
32	20	60	$891.36 (100.5%)	-	-
40	40	80	$894.33 (99.5%)	-	-
41	40	80	$898.83 (100.0%)	-	-
42	40	80	$903.32 (100.5%)	-	-
50	60	100	$906.34 (99.5%)	-	-
51	60	100	$910.89 (100.0%)	-	-
52	60	100	$915.44 (100.5%)	-	-
60	80	120	$918.50 (99.5%)	-	-
61	80	120	$923.12 (100.0%)	-	-
62	80	120	$927.73 (100.5%)	-	-

Table 1: Data for traded options on the two-year zero-coupon bond.

which is derived in this recursive manner is guaranteed to be equal to the current market price in the Ho and Lee model, irrespective of the values of its input parameters. For options and other interest-rate derivative securities, however, this is not the case.

Assume that the market prices of the traded options are consistent with a version of the Ho and Lee model which incorporates 120 time steps, in which the risk-neutral binomial probability is 1/2, and the volatility of the short-term interest rate 0.7% per year. This number of time steps is large enough so that the calculated option values have converged to at least two-decimal precision with the given parameter values, and the chosen volatility level prevents negative interest rates in the model at any point in time. The theoretical value of the replicated put option at time 0 according to this model is $8.73. The data for the traded options, including their price at time 0 if applicable, are listed in Table 1. The initiation and expiration dates are specified in terms of the time steps in the Ho and Lee model. Because options 10, 11 and 12 were initiated before time 0, no initiation date is specified for them. The exercise price of each option is both given as absolute number and as percentage of the forward bond price (between brackets).

6.2   Disaggregation Strategy

To start the iterative disaggregation algorithm, we have aggregated states and time steps in the Ho and Lee model of the previous section to obtain the aggregated event tree of iteration 0 in Figure 4. This initial event tree has only four different scenarios at $t = 120$, and the corresponding ALM model is thus small and easy to solve. The interest rate ranges between 8.128% per year in the lowest state to 7.872% in the highest state, and the corresponding liability (payoffs on the replicated put option) between $10.55 and $8.38. Because trading dates are included in this aggregated event tree for all points in time at which dividends are paid (i.e., options expire) and liabilities are due ($t = 120$), no adjustment for the prepayment of dividends and liabilities is necessary.

To prevent that only state and no time disaggregations are performed in this tree, we have slightly modified the use of the sensitivity measures ε and ζ that were defined in section 5.2. First, we have imposed that a state in an aggregated event tree can never have more than two successor states, and that its successor states occur at the same point in time.

Second, we only calculate the sensitivity measure ε in states that have successor states in which liabilities are specified. In this problem these are the states with successors at $t = 120$. For each of these states we identify a *critical scenario*, which is the scenario with the highest contribution to the sensitivity measure ε in that state. If a state is selected for a state disaggregation based on its value for ε, then a disaggregation is performed along the path in the event tree that corresponds to the critical scenario in that state. If possible, a state disaggregation is performed somewhere along this critical path. If there are multiple possibilities, then the state disaggregation is performed at the earliest point in time at which it is possible. If no state disaggregation is possible, then a time disaggregation is performed in the state at the beginning of the longest period on the critical path (i.e., comprising the largest number of time steps of the Ho and Lee model with 120 time steps). If there is more than one possibility, then the time disaggregation is performed as early as possible in the tree.

We have also used the critical scenario to define the way in which new states after a state disaggregation are connected to the existing event tree. After a state disaggregation is performed in some state along the critical path, then the state disaggregation is basically pushed forward along the path until all new states are connected to the existing tree, or the end of the tree is reached. This is illustrated in Figure 3 for the state disaggregation in iteration 31 of the iterative disaggregation algorithm. The critical path in the event tree of iteration 30 is indicated by the fat line. In iteration 31, a state disaggregation is performed in the state on the path at time 30. The new arcs in the event tree of iteration 31 after the state disaggregation are indicated in bold, and

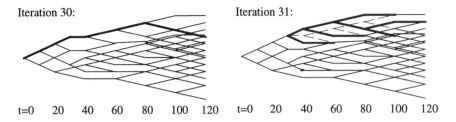

Iteration 30:   Iteration 31:

t=0   20   40   60   80   100   120   t=0   20   40   60   80   100   120

Figure 3: Change in event tree after a state disaggregation on the critical path.

the ones that have disappeared from the event tree are drawn as dashed lines.

A time disaggregation in our implementation consists of both the time and the state disaggregation which were described in sections 5.1.2 and 5.1.1: first a time disaggregation adds a new trading date and a single successor state, and then a state disaggregation in the same state splits the single successor in two successor states (this corresponds to the change from situation d to situation b in Figure 2). The new trading date is added in the middle between two existing trading dates in the tree.

6.3   Computational Results

We have coded the iterative disaggregation algorithm for this problem on a Sun 10 workstation with 32 MB of internal memory (RAM) in the C programming language. The ALM model has been re-optimized in each iteration as a large linear program, using the CPLEX callable library, where the optimal basis from the previous iteration is used to define a starting basis. The optimal basis columns from one iteration typically do not define a full basis for the model in the next iteration, but CPLEX allows the specification of an incomplete basis, and will complement it with additional columns to construct an initial basis.

The utility function of equation (4.15) with $\lambda = 0.9$ is used to value a portfolio surplus in the objective function, and the scenario probabilities are equal to the risk-neutral probabilities. Assume a transaction cost rate of 1%. The investor can borrow up to $10 in each state at the riskless one-period interest rate plus one basis point (0.01 percent), and he faces a 1% borrowing spread for any amount in excess of that. Assume that the investor is only allowed to buy options.

Figure 4 depicts the development of the aggregated event tree in the course of the algorithm. Time disaggregations are performed in iterations 17, 26, 30 and 38, and state disaggregations in all other iterations. In the event tree of iteration 40, the interest rate at time 120 decreases from 8.384% at the

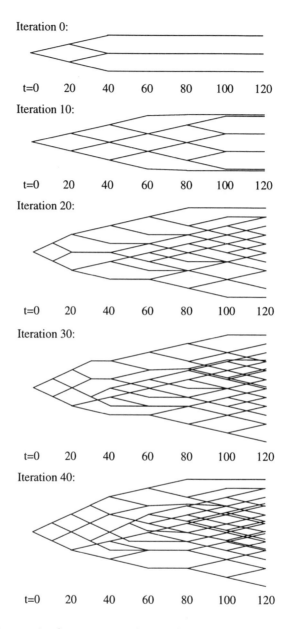

Figure 4: Changes in the event tree during the iterative disaggregation algorithm.

bottom of the tree to 7.617% at the top. The corresponding payoffs from the replicated put option (the liabilities) range from $12.82 to $6.33.

The number of states and scenarios in the event tree increases from 18 and 23, respectively, in iteration 0 to 68 and 342 in iteration 40. The corresponding ALM model has 131 constraints, 485 variables, and 1007 nonzeros in iteration 0 and 1572 constraints, 5120 variables, and 11770 nonzeros in iteration 40. The complete run of 40 iterations only takes a few minutes in real time. The number of simplex pivots required to re-optimize the ALM model in each iteration varies between less than ten and a few hundred. We could not continue the algorithm for many more than 40 iterations because at that point CPLEX required more memory than was available on our computer to perform the re-optimizations.

The optimal value of the objective function in the course of the algorithm is depicted in Figure 5, together with the cost (including transaction costs) of the optimal portfolio at time 0. Both seem to converge in the course of the algorithm. The difference between the two lines represents the value of the portfolio surplus at the terminal date which is credited to the objective. The average excess in a scenario at time 120 is \$0.11 with a standard deviation of \$0.28. The standard deviation is relatively high because surpluses almost exclusively occur in the upper part of the event tree (corresponding to low interest rates, and therefore low liabilities).

Figure 6 compares the value of the sensitivity measure with the actual change in the optimum objective value of the ALM model in each iteration. Their correlation is high in early iterations, but decreases later on. A study of the sensitivity measure across different states in each iteration shows that the number of scenarios in a state (and correspondingly, its probability of occurrence) is an important determinant of its value. Our disaggregation strategy thus exhibits a bias towards states in the center of the event tree. As the variability in the number of scenarios per state increases with the growth of the event tree, this bias becomes stronger in the course of the algorithm. This implies that the critical scenario in the state with the largest value of the sensitivity measure plays an increasingly important role for the actual disaggregation that is performed in the event tree.

Typically, interest rates along the critical path decrease initially, and increase after a certain point in time. In the event tree, this corresponds to paths that move upward in the event tree at first, and downward later on. That is, the critical path is initially favorable for the investor, but turns unfavorable after some time. This explains why a majority of the disaggregations are performed in the upper half of the event tree.

The optimal portfolio at time 0 only involves short-term lending and investments in the put options 10 and 20 in every iteration. The selected options are the ones that are most out-of-the-money (i.e., with the lowest strike price). These options provide the investor with the largest *relative* difference in payoff in different states of the world per option bought. Figure 7 depicts both the optimal amount of short-term lending and the optimal number of options

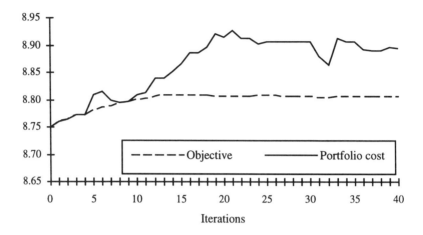

Figure 5: Optimum objective value and initial portfolio cost during the iterative disaggregation algorithm.

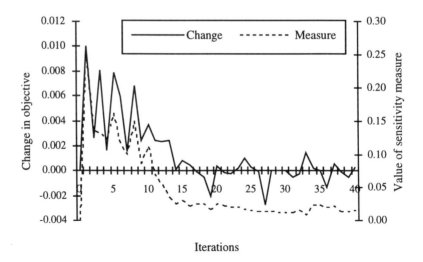

Figure 6: Change in the optimum objective value and the value of the sensitivity measure during the iterative disaggregation algorithm.

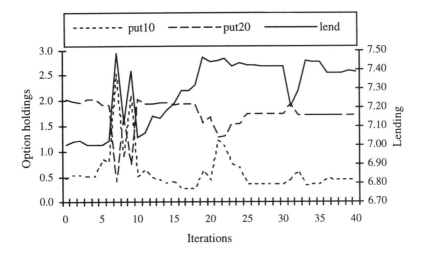

Figure 7: Optimum portfolio composition during the iterative disaggregation algorithm.

bought in each iteration of the algorithm. Although the option holdings appear somewhat volatile, it should be noted that the dollar amount invested in the options as fraction of the portfolio cost remains fairly constant at a little below 20%. Comparison with Figure 5 shows that the objective value is insensitive to shifts in the option holdings, but that they result in a different trade-off between the initial portfolio cost and the final portfolio value.

6.3.1   Variations in the Transaction Cost Rate

Table 2 shows the optimal portfolio composition at time 0 and the corresponding objective value when the transaction cost rate c varies between 0.1% and 2% (in each case after 40 iterations in the iterative disaggregation algorithm). With higher transaction costs, more of the initial portfolio is invested in short-term lending (which involves no transaction costs) and less in the options. A consequence of this change in portfolio composition is that the liabilities are matched less precisely when transaction costs incease.

The transaction cost rate has a significant impact on portfolio rebalancing after time 0. If the transaction cost rate is 1.0%, additional investments in (out-of-the-money) put options are made after every upward move in the event tree. In contrast, when the transaction cost rate is 2.0%, investments in new options are only made at the expiration date of options in the portfolio, and then only in some of the states of the world.

c =	Objective	Portfolio cost time 0	Expected final surplus	Portfolio composition		
				put 10	put 20	lending
0.1%	$8.74	$8.74	$0.00	0.872	1.713	$7.08
1.0%	$8.81	$8.89	$0.11	0.450	1.718	$7.39
2.0%	$8.85	$9.18	$0.40	0.118	1.711	$7.80

Table 2: Optimal solution for different transaction cost rates.

λ =	Objective	Portfolio cost time 0	Expected final surplus	Portfolio composition		
				put 10	put 20	lending
0.80	$8.81	$8.81	$0.00	0.000	2.423	$6.94
0.90	$8.81	$8.89	$0.11	0.450	1.718	$7.39
0.98	$8.77	$10.25	$1.64	0.000	0.084	$10.19

Table 3: Optimal solution for different values of the final-portfolio weight.

6.3.2 Variations in the Final-Portfolio Weight

Table 3 displays the optimization results when the final portfolio weight λ varies between 0.8 and 0.98, each time with a transaction cost rate of 1%. When λ increases, it becomes less important to match the liabilities exactly. As a consequence, the transaction cost rate increases in relative importance, and less portfolio rebalancing takes place. This results in an increase in short-term lending in the initial portfolio, and a significantly higher expected value of the portfolio surplus.

7   Conclusions

There are at least two advantages of using the iterative disaggregation algorithm as described in this paper for the solution of stochastic programming models for asset/liability management problems. First, it is based on state and time aggregation methods which can be used to condense a description of the uncertainty in asset prices and returns to any desired level, while being guaranteed that asset prices in this condensed description are arbitrage-free and consistent with current market prices if this was true for the original description. We have discussed why it is both reasonable and important that this description in stochastic programming models satisfies these properties, and indicated how financial asset-pricing models may be employed to construct one.

Second, the algorithm relieves one of the task of having to specify all relevant uncertainty in a stochastic programming model *ex-ante*. Instead, one can start to solve a formulation with a very coarse description of the uncertainty – in the extreme, one expected-value scenario – and iteratively refine this description based on information that one obtains from solutions in the course of the algorithm. Besides, this sequence of solutions also provides useful insight in the sensitivity of the optimal solution to additional uncertainty in the model. We have illustrated these advantages of the iterative disaggregation algorithm in the numerical example of section 6.

Our description and analysis of the iterative disaggregation algorithm is based on a (arbitrage-free) description of the uncertainty about future asset prices and liabilities in the form of an event tree, but does not depend on assumptions about the nature of the assets or the specific economic factors which influence asset prices and liabilities. In practical applications, however, one usually writes asset returns and liabilities as a function of the development in a set of economic variables. One therefore first needs a description of the uncertainty in the relevant economic variables before one is able to derive a sensible description of the uncertainty in asset prices and liabilities. This process was illustrated in section 6 with interest rates as the only economic variables and interest-rate derivative securities as the only possible investments. In a real-world setting, more variables will need to be considered (e.g., inflation) so that more asset classes can be included (e.g, inflation-linked bonds and stocks). An important area of research is therefore how a proper and practical description of the joint uncertainty in all relevant economic variables can be obtained, and how a usable, arbitrage-free description of the uncertainty in asset prices and liabilities can be derived from it. Cariño *et al.* [6], [7], Dert [9] and Mulvey and Thorlacius [23] describe different approaches to this problem.

Several other directions for future research remain. In the implementation of the iterative disaggregation algorithm in section 6, we imposed several restrictions on the disaggregation strategy, concerning for example the number of successor states per state and the trade-off between state and time disaggregations. Furthermore, our disaggregation strategy exhibited a preference for disaggregations in states with a high probability of occurrence, whereas in some situations one may be especially interested in hedging against extreme and unlikely events. Experimentation with different disaggregation strategies, and application to a variety of problems, should provide more insight in the sensitivity of the results to the choice of disaggregation strategy.

In the example of section 6 we re-optimized the ALM model in each iteration as a large linear program, using optimal basis information from the previous iteration. Although this re-optimization was fast, we encountered memory problems due to the size of the stochastic program after a certain number of iterations. This suggests the use of decomposition methods for the re-optimizations. A widely used and generally efficient decomposition method for stochastic programs is Benders' decomposition (see Birge [2]). However, direct application of this decomposition method in the iterative disaggregation algorithm is not very attractive as cuts which are generated in one iteration of the algorithm may not be feasible in the next iteration, due to the fact that the ALM model after a disaggregation is neither a relaxation nor a restriction of the model before the disaggregation (see section 5). Thus, Benders' decomposition is not able to exploit solution information from one iteration to the next. Another decomposition method which does have the ability to exploit such information is the primal-dual decomposition method as described in Klaassen [19]. A drawback of this method is, however, that each subproblem must be solved to full optimality, whereas primal-dual methods are known to converge quite slowly to the optimum in general.

The description of the iterative disaggregation algorithm in this paper is based on the formulation of the general ALM problem in section 2, which only includes cash-balance, portfolio-balance and borrowing constraints. In practical applications, one often needs to include additional constraints which reflect policy or regulatory restrictions. The analysis in sections 4.3 and 5 must then be extended to include these additional constraints. For example, to obtain a feasible solution after a disaggregation when additional constraints are present, one may need to introduce extra slack variables for these constraints and penalize non-zero values of these variables in the objective function, similar to what was done for the borrowing constraints in section 5.1.3. Depending on the specific problem instance, one may also be able to prove that the slack variables are always zero in optimum solutions if the penalties are chosen large enough.

Another area of research is to consider more general utility functions in the formulation of the ALM model to evaluate the portfolio surplus in the objective function. One can preserve linearity of the model by approximating a nonlinear utility function with a piecewise linear function. Alternatively, a generalization of the iterative disaggregation algorithm to convex stochastic programming models must be developed.

The idea of iterative disaggregation of the uncertainty in stochastic programs may be an attractive solution approach for application areas other than asset/liability management as well. In other areas, the restriction of no-arbitrage may not be applicable, which can provide more freedom in how to perform (dis)aggregations. Furthermore, if the stochasticity in the stochastic program is restricted to the right-hand-side vector, Benders' decomposition method can be used efficiently to perform the re-optimizations in the course of the algorithm, as the Benders' cuts from one iteration remain valid for the next iteration.

References

[1] J.R. Birge. Aggregation bounds in stochastic linear programming. *Mathematical Programming*, 31:25–41, 1985.

[2] J.R. Birge. Decomposition and partitioning methods for multi-stage stochastic linear programs. *Operations Research*, 33(5):989–1007, September–October 1985.

[3] F. Black, E. Derman, and W. Toy. A one-factor model of interest rates and its application to treasury bond options. *Financial Analysts Journal*, pages 33–39, January–February 1990.

[4] S.P. Bradley and D.B. Crane. A dynamic model for bond portfolio management. *Management Science*, 19(2):139–151, October 1972.

[5] M.J. Brennan and E.S. Schwartz. An equilibrium model of bond pricing and a test of market efficiency. *Journal of Financial and Quantitative Analysis*, 17:301–330, September 1982.

[6] D.R. Cariño, T. Kent, D.H. Myers, C. Stacy, M. Sylvanus, A.L. Turner, K. Watanabe, and W.T. Ziemba. The Russell–Yasuda Kasai model: an asset/liability model for a Japanese insurance company using multistage stochastic programming. *Interfaces*, 24(1):29–49, January–February 1994. Reprinted in this volume, 609–633.

[7] D.R. Cariño, D.H. Myers, and W.T. Ziemba. Concepts, technical issues, and uses of the Russell–Yasuda Kasai financial planning model. Technical report, Frank Russell Company, 1996. To appear in *Operations Research*.

[8] J.C. Cox, S.A. Ross, and M. Rubinstein. Option pricing: A simplified approach. *Journal of Financial Economics*, 7:229–263, September 1979.

[9] C.L. Dert. *Asset-Liability Management for Pension Funds*. PhD thesis, Erasmus University Rotterdam, September 1995. See also this volume, 501–536..

[10] B. Golub, M. Holmer, R. McKendall, L. Pohlman, and S.A. Zenios. A stochastic programming model for money management. *European Journal of Operational Research*, 85:282–296, 1995.

[11] M.J. Harrison and D.M. Kreps. Martingales and arbitrage in multiperiod securities markets. *Journal of Economic Theory*, 20:381–408, 1979.

[12] D.B. Hausch and W.T. Ziemba. Locks at the racetrack. *Interfaces*, 20(3):41–48, May–June 1990.

[13] D. Heath, R. Jarrow, and A. Morton. Bond pricing and the term structure of interest rates: A discrete time approximation. *Journal of Financial and Quantitative Analysis*, 25(4):259–304, December 1990.

[14] R.S. Hiller and J. Eckstein. Stochastic dedication: Designing fixed income portfolios using massively parallel benders decomposition. *Management Science*, 39(11):1422–1438, November 1993.

[15] T. Ho and S.-B. Lee. Term structure movements and pricing interest rate contingent claims. *Journal of Finance*, 41(5):1011–1029, December 1986.

[16] C.-F. Huang and R.H. Litzenberger. *Foundations for Financial Economics*. North-Holland, 1988.

[17] J. Hull. *Options, Futures and Other Derivative Securities*. Prentice-Hall, 2nd edition, 1993.

[18] J. Hull and A. White. Valuing derivative securities using the explicit finite difference method. *Journal of Financial and Quantitative Analysis*, 25(1):87–99, March 1990.

[19] P. Klaassen. *Stochastic Programming Models for Interest-Rate Risk Management*. PhD thesis, Sloan School of Management, MIT, Cambridge, Massachusetts, May 1994. Published as IFSRC Discussion Paper 282–94.

[20] P. Klaassen. Discretized reality and spurious profits in stochastic programming models for asset/liability management. *European Journal of Operations Research*, 101(2):374–392, September 1997.

[21] P. Klaassen. Financial asset pricing theory and stochastic programming models for asset/liability management: A synthesis. *Management Science*, 44(1):31–48, January 1998.

[22] M.I. Kusy and W.T. Ziemba. A bank asset and liability management model. *Operations Research*, 34(3):356–376, May-June 1986.

[23] J.M. Mulvey and A.E. Thorlacius. The Towers Perrin global capital market scenario generation system. This volume, 286–312.

[24] J.M. Mulvey and H. Vladimirou. Stochastic network programming for financial planning problems. *Management Science*, 38(11):1642–1664, November 1992.

[25] H.R. Varian. *Microeconomic Analysis*. Norton, 3rd edition, 1992.

[24] J.M. Mulvey and H. Vladimirou. Stochastic network programming for financial planning problems. *Management Science*, 38(11):1642–1664, November 1992.

[25] H.R. Varian. *Microeconomic Analysis*. Norton, 3rd edition, 1992.

[26] S.A. Zenios. A model for portfolio management with mortgage-backed securities. *Annals of Operations Research*, 43:337–356, 1993.

[27] P. Zipkin. Bounds on the effect of aggregating variables in linear programs. *Operations Research*, 28(2):403–418, March–April 1980.

[28] P. Zipkin. Bounds for row-aggregation in linear programs. *Operations Research*, 28(4):903–916, July–August 1980.

The CALM Stochastic Programming Model for Dynamic Asset-Liability Management*

*G. Consigli** and M.A.H. Dempster*

Abstract

Multistage stochastic programming – in contrast to stochastic control – has found wide application in the formulation and solution of financial problems characterized by a large number of state variables and a generally low number of possible decision stages. The literature on the use of multistage recourse modelling to formalize complex portfolio optimization problems dates back to the early seventies, when the technique was first adopted to solve a fixed income security portfolio problem. We present here the CALM model which has been designed to deal with uncertainty affecting both assets (in either the portfolio or the market) and liabilities (in the form of scenario dependent payments or borrowing costs). We consider as an instance a pension fund problem in which portfolio rebalancing is allowed over a long-term horizon at discrete time points and where liabilities refer to five different classes of pension contracts. The portfolio manager, given an initial wealth, seeks the maximization of terminal wealth at the horizon, with investment returns modelled as discrete state random vectors. Decision vectors represent possible investments in the market and holding or selling assets in the portfolio, as well as borrowing decisions from a credit line or deposits with a bank. Computational results are presented for a set of 10-stage portfolio problems using different solution methods and libraries (OSL, CPLEX, OB1).The portfolio problem with an underlying vector data process which allows up to 2688 realizations at the 10 year horizon is solved on an IBM RS6000/590 for a set of twenty four large scale test problems using the simplex and barrier methods provided by CPLEX (the latter for either linear or quadratic objective), the predictor/corrector interior point method provided in OB1, the simplex method of OSL, the MSLiP-OSL code instantiating nested Benders decomposition with subproblem solution using OSL simplex and the current version of MSLiP.

---

*Reprinted, with permission, from *Annals of OR*.

** This work was completed while the first author was at the Judge Institute, University of Cambridge.

1 Introduction

In this paper we develop a generic model for the integrated dynamic management of financial assets and liabilities – the *computer-aided asset/liability management* (CALM) model – which we feel certain will provide a basis for ubiquitous model-based corporate financial planning within a decade or so. Such planning will be brought about by the realistic nature of models of this type, rapid advances in software tools and computer hardware capability, and a universal requirement to model and more tightly control financial risks in modern global corporations, from banks to manufacturers, employing a wide variety of technologies from low to high.

The CALM model has been developed from dynamic stochastic programming research which began over two decades ago with the seminal work of Bradley and Crane (1972) and Lane and Hutchinson (1980). A partial list of subsequent related works includes Cariño *et al.* (1994), Dempster and Ireland (1991), Kusy and Ziemba (1986), Mulvey and Vladimirou (1989) and Zenios (1992), some of which will be treated in more detail in Section 3.

Perhaps the most important of these papers are those of Bradley and Crane (1972) – who introduced the *inventory approach* to modelling financial decisions in which each asset or liability in the model has an initiate, hold and terminate variable in each period, an invaluable aid to realistic detailed modelling – Dempster and Ireland (1991) – who explored the modern information system context required for such models – and Cariño *et al.* (1994) – who developed the first genuine *commercial* application of dynamic stochastic programming in spite of the fact that most of their predecessors were involved in prototype applications with financial institutions.

To indicate the current state-of-the-art we present computational results for an instance of the CALM model – the *Watson model* – which is specialized to a pension fund manager's dynamic integrated asset/liability management problem in a single currency. Comparisons are made of current approaches to the numerical solution of such problems utilizing a contemporary high end workstation and the question of adequate representation of financial uncertainties in the models is addressed.

In the next section (Section 2) we review the multistage recourse formulation of dynamic portfolio management problems. Section 3 introduces the CALM and Watson integrated asset/liability management models. (A detailed presentation of the CALM model is given in the appendix.) Multistage recourse models are very complex large-scale linearly constrained problems whose constraint structure grows in size linearly with the number of data paths or *scenarios* representing the uncertainties facing the decision maker. Hence practical models must be handled with software tools which generalize – and utilize – linear programming (LP) modelling languages such as

GAMS, AMPL and MODLER (used here). This is the topic of Section 4, which describes briefly the STOCHGEN subroutine library (Corvera Poiré (1995a)) developed by our research group. Section 5 reviews current LP and QP solution solution techniques and software – of simplex, interior point and decomposition type – with emphasis on their features relevant to the solution of large scale financial planning models. Our comparative computational results – showing the clear overall superiority of the nested Benders decomposition technique for these large scale portfolio management problems – are presented in Section 6. Section 7 contains conclusions and directions for further work.

2 Multistage recourse formulation of dynamic portfolio management

We consider a stochastic programming problem in the form of a multistage recourse problem (cf. Dempster (1980a), Ermoliev and Wets (1988a)) (here boldface denotes random):

$$\min_{x_1 \in \mathbb{R}^{n_1}} \{ f_1(x_1) + \mathbb{E}_{\boldsymbol{\omega}_2}[\min_{\mathbf{x}_2}(\mathbf{f}_2(\mathbf{x}_2) + \ldots + \mathbb{E}_{\boldsymbol{\omega}_T|\boldsymbol{\omega}^{T-1}}[\min_{\mathbf{x}_T} \mathbf{f}_T(\mathbf{x}_T)])]\}$$

such that

$$
\begin{array}{rcll}
A_1 x_1 & & & = & b_1 \\
\mathbf{B}_2 x_1 & + & \mathbf{A}_2 \mathbf{x}_2 & & = & \mathbf{b}_2 & a.s. \\
& & \mathbf{B}_3 \mathbf{x}_2 & + & \mathbf{A}_3 \mathbf{x}_3 & = & \mathbf{b}_3 & a.s. \\
& & & \ddots & & \vdots \\
& & & \mathbf{B}_T \mathbf{x}_{T-1} & + & \mathbf{A}_T \mathbf{x}_T & = & \mathbf{b}_T & a.s.
\end{array}
$$

$$(2.1)$$

$$l_1 \le x_1 \le u_1$$

$$\mathbf{l}_t \le \mathbf{x}_t \le \mathbf{u}_t \quad a.s., \quad t = 2, \ldots, T .$$

In (2.1) the separable **objective** is defined by the period functionals f_t, for $t = 1$ up to the *horizon T*; $A_1 \in \mathbb{R}^{m_1 n_1}$ and $b_1 \in \mathbb{R}^{m_1}$ define deterministic constraints on the first stage decision x_1, while, for $t = 2, \ldots, T$, $A_t : \Omega \to \mathbb{R}^{m_t n_t}$, $B_t : \Omega \to \mathbb{R}^{m_{t-1} n_{t-1}}$ and $b_t : \Omega \to \mathbb{R}^{m_t}$ define stochastic constraint regions for the recourse decisions $\mathbf{x}_2, \ldots, \mathbf{x}_T$; and $\mathbb{E}_{\boldsymbol{\omega}_t | \boldsymbol{\omega}^{t-1}}$ denotes conditional expectation of the state $\boldsymbol{\omega}_t$ of the *data process* $\boldsymbol{\omega}$ at time t with respect to the *history* $\boldsymbol{\omega}^{t-1}$ of the process up to time t, where the data process $\boldsymbol{\omega}$ may be regarded as a random vector defined in a canonical probability space $(\Omega, \mathcal{F}, P)$, with $\boldsymbol{\omega}_t := (\boldsymbol{\xi}_t, \mathbf{A}_t, \mathbf{B}_t, \mathbf{b}_t)$ in which $\boldsymbol{\xi}_t$ is the random parameter in the objective functional given by $f_t(\xi_t, x_t)$.

The recourse formulation (2.1) shows explicitly the dependence of the optimal policy or *decision process* $\mathbf{x}^0 := (\mathbf{x}_1^0, \mathbf{x}_2^0, \ldots, \mathbf{x}_T^0)$ on the realizations of the vector data process $\boldsymbol{\omega} := (\boldsymbol{\omega}_2, \ldots, \boldsymbol{\omega}_T)$ in $(\Omega, \mathcal{F}, P)$, with the sample space defined as $\Omega := \Omega_2 \times \Omega_3 \times \ldots \times \Omega_T$ and the filtration $\mathcal{F}_1 := \{0, \Omega\} \subset \mathcal{F}_2 \subset \mathcal{F}_3 \subset \ldots \subset \mathcal{F}_T := \mathcal{F}$, where $\mathcal{F}_t := \sigma\{\boldsymbol{\omega}^t\}$ is the σ-field generated by the history $\boldsymbol{\omega}^t$ of the data process $\boldsymbol{\omega}$ for $t = 2, \ldots, T$ and P is a probability measure on this space.

This model embodies the main features of a decision problem under uncertainty:-

- The objective in this nested representation formalizes a sequence of optimization problems corresponding to different stages: at time 1 the decision maker has to select a decision whose outcome completely depends on the future realizations of the underlying multidimensional stochastic data process. Solution of this problem is sometimes referred to as the *here and now* problem.

- Thereafter, for each realization of the history $\boldsymbol{\omega}^t$ of the data process up to time t, a recourse problem is considered in which decisions are allowed to be a function of the observed realization (x^{t-1}, ω^t) only.

  At each stage previous decisions affect current problems through the stochastic matrices $\mathbf{B}_t$, $t = 2, \ldots, T$, with the decision sequence:

$$\begin{array}{ccccccccc} \text{decide} & & \text{observe} & & \text{decide} & & \text{observe} & & \text{decide} \\ x_1 & \rightsquigarrow & \omega_2 & \rightsquigarrow & x_2 & \rightsquigarrow \cdots \rightsquigarrow & \omega_T & \rightsquigarrow & x_T. \end{array}$$

  The recourse decisions depend on the current state of the system as determined by previous decisions and by random events. Each decision thus is required to be *adapted* to the σ-field generated by the data process implying *nonanticipativity* in these decisions.

- The data process $\boldsymbol{\omega}$ in $(\Omega, \mathcal{F}, P)$ is defined as a discrete-time, possibly autocorrelated, vector stochastic process. A finite sample of its paths is conveniently represented as a *scenario tree*: each scenario corresponding to a trajectory of the process $\boldsymbol{\omega}^T := (\boldsymbol{\omega}_1, \boldsymbol{\omega}_2, \ldots, \boldsymbol{\omega}_T)$ at the horizon T. It is this branching structure of data paths that ensures that all decisions are made in the face of uncertainty.

- In (1) – as is generally the case in financial planning problems – both the RHS and LHS of the constraint set are scenario-dependent. The generation of the sample data paths for the problem represents a crucial step for a valid specification of the stochastic optimization problem. (We will consider briefly in Section 3 some of the implications of the so-called investment model (Wilkie (1995a)) that has been adopted for this purpose in the WATSON portfolio problems.)

Dynamic portfolio problems are easily formulated as *dynamic recourse problems* (DRP). This approach to model-based portfolio management was first adopted by Bradley and Crane in 1973 for a portfolio problem restricted to fixed income securities. Other applications of DRP for financial planning are reported in Lane and Hutchinson (1980), Kallberg, White and Ziemba (1982), Kusy and Ziemba (1986), Dempster and Ireland (1991), Mulvey and Vladimirou (1989), Dantzig and Infanger (1993), Zenios (1992), Cariño *et al.* (1994, 1995).

In many of these examples uncertainty takes the form of unknown future rates of return on market investments and funding sources, as well as cash flow misalignments, and the objective function is typically characterized in terms of the expected value of a linear or nonlinear utility function of wealth at the horizon and sometimes beyond.

Problem (2.1) may be given a more compact *dynamic programming* representation which takes advantage of the structure exhibited by the set of constraints (here Markovian or staircase, but more generally lower triangular). For each $t = 1, \ldots, T - 1$, we have

$$\min_{x_t}[f_t(x_t) + v_{t+1}(\omega^t, x^t)] \tag{2.2}$$

such that

$$B_t x_{t-1} + A_t x_t = b_t,$$

where v_{t+1} expresses the optimal expected cost for the stages from $t+1$ to T, given the decision history $x^t := (x_1, \ldots, x_t)$ and the realized history of the random process $\omega^t := (\omega_1, \ldots, \omega_t)$. Specifically,

$$v_{t+1}(\omega^t, x^t) := \mathsf{E}_{\boldsymbol{\omega}_{t+1}|\boldsymbol{\omega}^t}[\min_{\mathbf{x}_{t+1}}(f_{t+1}(\mathbf{x}_{t+1}) + \ldots + \mathsf{E}_{\boldsymbol{\omega}_T|\boldsymbol{\omega}^{T-1}} \min_{\mathbf{x}_T}(f_T(\mathbf{x}_T))], \tag{2.3}$$

where the minimizations are taken subject to the appropriate constraints which will be discussed further in Section 5.

Correspondingly, at the end of each time period and on the basis of the current information a portfolio manager selects an optimal decision in the face of the uncertainty that (s)he is now facing. This decision needs to be feasible with respect to the constraints induced by the future values of the random data process and is influenced by the current composition of the portfolio.

Due to the convexity of $v_{t+1}(\omega^t, x^t)$ with respect to x^t, problem (2.2) can be also formulated in the form:

$$\min_{x_t} f_t(x_t) + \theta_t \tag{2.4}$$

such that

$$B_t x_{t-1} + A_t x_t = b_t$$
$$v_{t+1}(\omega^t, x^t) \le \theta_t .$$

The adoption of a cutting-plane algorithm for the solution of (2.1) is based on the simple decomposition (2.4) by approximating the value function at $t+1$ by a set of cuts of the form

$$d'x_t \leq \theta_t ,$$

(see Birge (1985), Dempster and Thompson (1996a), Ermoliev and Wets (1988a)). Cuts are also used to enforce constraints on current decisions which ensure that later recourse actions are feasible.

3 The CALM and Watson models

Multistage stochastic programming techniques, unlike typical stochastic control formulations, allow the representation and solution of financial problems characterized by a large number of state variables and a generally low number of possible decision stages.

Recent applications of multistage stochastic programming techniques for financial planning confirm the possibility of solving large problems with complex structure. The Russell–Yasuda Kasai model and the Towers Perrin investment system for pension plans (Mulvey and Thorlacius (1998)) are important examples.

Later we discuss the formulation and solution of a long-term portfolio problem with uncertainty affecting price processes, interest rates and liability payments. The CALM model (Dempster 1993, see the appendix) represents a general formulation of an asset-liability model as a linearly constrained problem with an objective function explicitly taking into account the risk attitude of the portfolio manager. It is based on well-established financial theory and addresses a multicurrency portfolio problem with randomness affecting rates of return and liabilities in domestic and foreign terms.

The general CALM model assumes a portfolio manager possessing a utility function from a general class which expresses his or her attitude to possible distributions of the terminal wealth $\mathbf{W}^T$. This class of isoelastic utility functions u characterizes a risk-averse portfolio manager possessing an Arrow-Pratt relative risk measure which is inverse affine. Logarithmic utility functions, as well as certain exponential and polynomial utilities, belong to this class and a well-established theory is available with theoretical results for the case of portfolio problems formulated as dynamic programming problems under uncertainty (see e.g. Bertsekas (1987)).

A practical dynamic portfolio management theory requires, as well as an accurate representation of the risk attitude of the portfolio manager, fast and accurate solution techniques for a decision problem involving large numbers of decision variables – so far impossible to handle in a dynamic programming framework, but currently practical for the scenario-based multistage stochastic programming DRP formulation.

For this reason and given the current state of algorithm development for large-scale nonlinear problems, we consider an objective in linear or quadratic form and we shall restrict ourselves to these two cases in the numerical experiments that follow, although piecewise linear/quadratic forms are also computationally tractable (King (1994)). Specifically:

$$\mathbb{E}u(\mathbf{W}^T) := \begin{cases} \mathbb{E}\mathbf{W}^T & \text{in the linear case} \\ \mathbb{E}(\mathbf{W}^T - \tilde{W})^2 & \text{in the quadratic case,} \end{cases} \tag{3.1}$$

where $\tilde{W}$ denotes a *target* determined *a priori* by the portfolio manager and the expectation is taken with respect to the probability measure assigned at the horizon to scenarios.

The *Watson model* – named after Watson and Sons Consulting Actuaries who have provided the set of financial data for the parameter specification of the problem – represents a particular instance of the CALM model for the case of pension fund management with uncertainty affecting assets, either in the portfolio or in the market, and liabilities, in the form of scenario dependent pension payments or borrowing costs.

We consider two possible ways of generating scenarios for the test-problems which are both based on the *Wilkie investment model* (Wilkie (1995a)) regarded as a complex nonlinear predictor of real and nominal returns for five asset classes and pension payments on an equal number of pension schemes.

In particular, either:-

1. We take the simulations as provided by Watson and Sons, a set of 10 year independently generated sample paths and generate a scenario tree by 'imposing' a tree-structure (as explained in Section 4) which largely destroys the temporal stochastic properties of the underlying data process, or

2. By modifying the generator as originally supplied, we generate data paths in conditional mode, correctly representing paths from Wilkie investment model in the form of a scenario tree.

Our computational results show that the two procedures have different implications for the nature of the optimization problem and the performance of a solution method based on nested Benders decomposition.

Wilkie's model is based on actual data from the U.K. for the period 1924–1991 and is formulated as a set of simultaneous autoregressive equations of up to the third order in Wold recursive form, all dependent on an underlying inflation process. It generates data paths for annual returns in the UK market for (see Figure 1):

Marginal densities - 5 investment opportunities

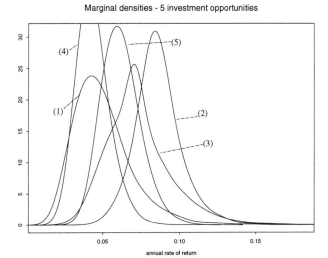

Figure 1: Return distributions - 1,000 independent simulations

1. ordinary shares

2. fixed-interest irredeemable bonds

3. bank deposits

4. index-linked securities

5. real estate,

together with predictions of annual pension payments and an estimated *reass-urance-to-close* representing future payment liabilities discounted to the horizon. All these are used to calculate the required random coefficients for the WATSON problems studied in the numerical experiments.

In order to characterize the risk faced by the portfolio manager, we show in Figures 1 and 2 the marginal empirical probability densities for the different returns at the horizon resulting respectively from the simulation of 1 000 independent data paths and 1,024 conditional scenarios. A comparison shows the impacts of the two procedures on the marginal distributions at the horizon; with a clear path-dependent 'humpy' effect in the case of the conditionally generated scenarios.

The Watson model is a dynamic A/L management optimization problem with linear constraints with general financial and accounting considerations as follows.

a. *Cash Balance Constraint*

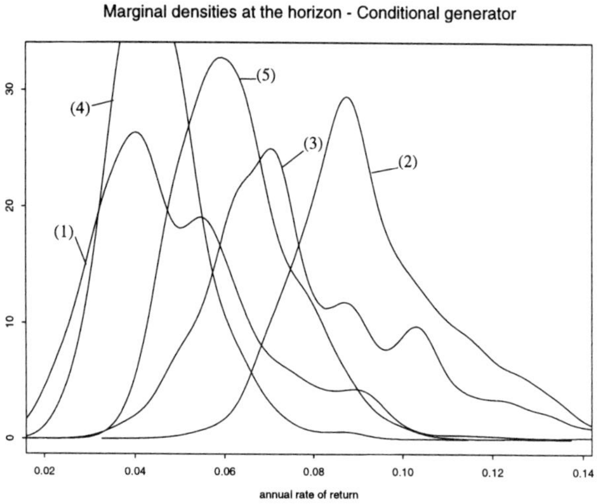

Marginal densities at the horizon - Conditional generator

Figure 2: Return distributions - 1,024 conditional scenarios

In each period cash inflows, due either to borrowing or selling decisions, must be equal to cash outflows, due to pension payments, debt reimboursements or other sources. We consider five pension funds, with an estimated payment also required at the horizon to compensate for all future liabilities.

b. *Inventory Balance Constraints*

Each decision of the portfolio manager with an impact on the portfolio composition needs to be accounted for. In the Watson model – following Bradley and Crane (1972) and unlike most of the other models presented in the past – we specify at each stage the composition of the portfolio for each asset category according to the period in which the investment has taken place. This is the mechanism by which both tax and regulatory constraints are easily added to the model.

c. *Other Conditions*

All decisions are constrained by upper or lower bounds on investment classes, lines of credit and specific corporate and regulatory constraints. In general these conditions have to be modelled as scenario dependent; in the first instance, however, we keep them fixed over the planning horizon.

d. *Terminal Wealth Definition*

Finally, the specification of the terminal wealth that enters the objective function represents an important aspect of the portfolio problem. In

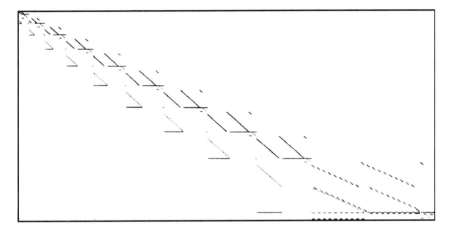

Figure 3: Non-zero entries in the WATSON coefficient matrix

the Watson model, we include in the definition the market values at the horizon of all assets in the portfolio, together with the position with the bank, net of the terminal reassurance-to-close payments for the five pension funds.

Figure 3 shows the sparsity structure of the constraint matrix for the Watson problem on a single scenario.

The constraint matrix exhibits a block-bidiagonal structure as a result of the Markovian nature of the A/L management decision problem and suggests scope for decomposition methods.

Given the model formulation and a data specification defining its random coefficients, the resulting stochastic programming problem must be generated in standard input format for numerical solution. A complete specification of the problem also requires the path probabilities for the different scenarios – here equally likely since they have been generated by simulation.

4   Model generation

A stochastic programming problem specified for numerical solution is conventionally in SMPS format (Birge *et al.* (1987)) described in terms of three

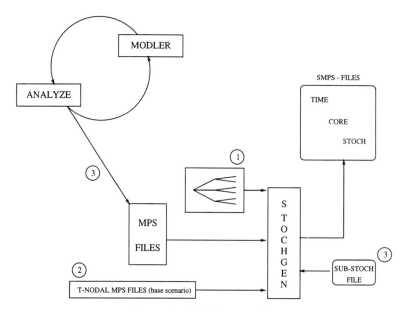

Figure 4: Model generation

input files as:- a *time file* defining the dynamic structure of the problem, a *core file* supplying all the data along one scenario path (generally regarded as the *base* scenario of the problem) and a *stoch file* containing the stochastic information of the problem, in which the scenario probabilities are also specified. The profile and dimension of the stoch file depends on the number of stages in the model as well as on the number of realizations allowed at each stage.

The data specification represents an essential step in the formulation of a stochastic problem. In a one-scenario, i.e. deterministic, problem we remain in the area of linear or quadratic programming. An MPS file generator for deterministic problems (MODLER, AMPL, MIMI, etc.) can be usefully incorporated into the design of an SMPS generator for the specification of a stochastic version of the problem. This is the approach taken in the STOCHGEN subroutine library (SSL) recently created by Corvera Poiré (1995a). Its structure is shown in Figure 4.

STOCHGEN takes as inputs a sequence of mps files generated for each scenario by the modelling language, here MODLER (Greenberg (1992a)), together with the specification of the tree structure for the data process and a set of nodal MPS files for the base scenario, and provides the required SMPS files as output. The SSL library expands the utilization of current modelling facilities for linear programming to the stochastic case and provides insight into the understanding of dynamic stochastic programs. It is constructed to encourage the LP user to approach the field of stochastic programming and allows for the combination of trees and subtrees in an error-checked non-redundant manner.

5 Standard solution techniques

The specification of the probability measure P as discrete for the measurable sample space $(\Omega, \mathcal{F})$ allows an easy derivation of the deterministic equivalent of a stochastic program with recourse for numerical solution.

We consider two possible solution procedures for the deterministic equivalent of the dynamic portfolio problem treated here:

- Direct solution of the deterministic equivalent problem in which conditional probabilities (or, equivalently, path probabilities) are used to define expectations and constraints, are replicated for all data paths at each stage. Very large linear or quadratic programs are created that can be solved using respectively standard LP algorithms (i.e. simplex or interior point methods) or QP algorithms (pivoting, iterative or interior point based algorithms).

- Alternatively, the original problem may be decomposed into a sequence of subproblems which are computationally closely related across scenarios and nested Benders decomposition applied. This algorithm passes solution information from subproblem to subproblem forward in time, and feasibility and optimality information in the form of cuts backward in time.

Assuming for simplicity equal numbers of branches descending from each node of the decision tree in a time period, together with the specification of the corresponding conditional probabilities as $p_t^{k_2,k_3,\ldots,k_t}$, $t = 2,\ldots,T$, allows the formulation of the *deterministic equivalent* of (2.1) in the form

$$\min\Big\{f_1(x_1) + \sum_{k_2=1}^{K_2} [p_2^{k_2} f_2(x_2^{k_2}) + \sum_{k_3=1}^{K_3} [p_3^{k_2,k_3} f_3(x_3^{k_2,k_3}) + \ldots$$
$$+ \sum_{k_T=1}^{K_T} [p_T^{k_2,\ldots,k_T} f_T(x_T^{k_2,\ldots,k_T})]]]\Big\}$$

such that

$$
\begin{aligned}
A_1 x_1 &= b_1 \\
B_2^{k_2} x_1 + A_2^{k_2} x_2^{k_2} &= b_2^{k_2} \\
&\quad k_2 = 1, \ldots, K_2 \\
B_3^{k_2,k_3} x_2^{k_2} + A_3^{k_2,k_3} x_3^{k_2,k_3} &= b_3^{k_2,k_3} \\
&\quad k_t = 1, \ldots, K_t \\
&\quad t = 2, 3 \\[6pt]
\vdots \qquad \vdots &\qquad \vdots \\[6pt]
B_T^{k_2,\ldots,k_T} x_{T-1}^{k_2,\ldots,k_{T-1}} + A_T^{k_2,\ldots,k_T} x_T^{k_2,\ldots,k_T} &= b_T^{k_2,\ldots,k_T} \\
&\quad k_t = 1, \ldots, K_t, \\
&\quad t = 2, \ldots, T
\end{aligned}
$$

$$
l_1 \leq x_1 \leq u_1, \qquad l_t \leq x_t^{k_2,\ldots,k_T} \leq u_t, \; t = 2, \ldots, T, \tag{5.1}
$$

in which a vector of decision variables corresponds to each node in the decision tree.

In concise matrix notation the multistage linear problem is

$$
\min_{\hat{x}} \hat{c}' \hat{x} \quad \text{such that } \hat{A}\hat{x} = \hat{b}, \; \hat{x} \geq 0, \tag{5.2}
$$

where

$$
\hat{c} := \left(c_1', (p_2^1 c_2^1)', \ldots, (p_2^{K_2} c_2^{K_2})', (p_3^{1,1} c_3^{1,1})', \ldots, (p_3^{K_2,K_3} c_3^{K_2,K_3})', \ldots \right.
$$
$$
\left. \ldots, (p_T^{1,\ldots,1} c_T^{1,\ldots,1})', \ldots, (p_T^{K_2,\ldots,K_T} c_T^{K_2,\ldots,K_T})' \right)'
$$
$$
\hat{x} := (x_1', x_2^{1'}, \ldots, x_2^{K_2'}, \ldots, x_T^{1,\ldots,1'}, \ldots, x_T^{K_2,\ldots,K_T'})'
$$
$$
\hat{b} := (b_1', b_2^{1'}, \ldots, b_2^{K_2'}, \ldots, b_T^{1,\ldots,1'}, \ldots, b_T^{K_2,\ldots,K_T'})'
$$

$$
\hat{A} := \begin{pmatrix}
A_1 & 0 & \cdots & & \cdots & & & 0 \\
B_2^1 & A_2^1 & 0 & \cdots & & \cdots & & 0 \\
\vdots & & \ddots & & & \cdots & & \vdots \\
B_2^{K_2} & 0 & \cdots & A_2^{K_2} & 0 & & \cdots & 0 \\
0 & B_3^{1,1} & & A_3^{1,1} & 0 & \cdots & & 0 \\
\vdots & \vdots & & & \ddots & & & 0 \\
0 & B_3^{1,K_3} & & & & & \cdots & 0 \\
& & B_3^{K_2,K_3} & \cdots & & A_3^{K_2,K_3} & \cdots & 0 \\
\vdots & \vdots & & \vdots & & \vdots & & \vdots \\
& & & & B_4^{K_2,\ldots,K_4} & & & \\
& & \cdots & & \vdots & & \ddots & 0 \\
0 & 0 & \cdots & & & B_T^{K_2,\ldots,K_T} & \cdots & A_T^{K_2,\ldots,K_T}
\end{pmatrix}
$$

In a T-stage problem with $\Pi_{s=2}^t K_s$ nodal problems in stage t with associated $(m_t \times n_t)$ submatrix dimension, this results in $\hat{c} \in \mathbb{R}^{\hat{n}}$, $\hat{x} \in \mathbb{R}^{\hat{n}}$, $\hat{b} \in \mathbb{R}^{\hat{m}}$,

$\hat{A} \in \mathbb{R}^{\hat{m}\hat{n}}$ (see Figure 3), where $\hat{n} := n_1 + K_2(n_2 + K_3(n_3 + \cdots + K_T n_T))$ and $\hat{m} := m_1 + K_2(m_2 + K_3(m_3 + \cdots + K_T m_T))$, generating eventually *very large* problems as the number of scenarios increases.

The (LP)

$$\min_{\hat{x}}\{\hat{c}'\hat{x} \,|\, \hat{A}\hat{x} = \hat{b}, \ \hat{x} \geq 0\} \tag{5.3}$$

and its dual (DLP)

$$\max_{\hat{\pi}}\{\hat{b}'\hat{\pi} \,|\, \hat{\pi}'\hat{A} \leq \hat{c}'\} \tag{5.4}$$

are those considered in the solution with simplex or interior point methods.

The derivation of a complete certainty equivalent linear, or with a simple extension quadratic, program from the SMPS files generated by STOCHGEN is achieved with the STD2MPS utility created by Gassmann (1987).

The main problems in solving either (5.3) or (5.4) are related to the extremely large dimensions that such an LP may attain. In the sequel we consider different algorithms for the solution of large scale programs and compare the results from *state-of-the-art* commercially available solvers with the performance of two versions of the MSLiP implementation of the nested Benders decomposition algorithm.

5.1 Simplex

Modern implementation of the revised simplex method, such as in IBM's OSL Release 2 (1992) and CPLEX simplex (1994), incorporate many of the theoretical results of research in this area, making commercially available two excellent versions of this technique for the solution of LP problems.

The performance of the simplex method, recognized as very efficient for the solution of relatively small dense problems, may be considered to depend on:

- the implementation of a *scaling* procedure, before phase 1 of the algorithm takes place, and an efficient *basis factorization* and update after the dimension of the original matrix has been reduced by *preprocessing*,

- the determination of a convenient advanced starting basis through a *crash* procedure,

- the adoption of an efficient *pricing strategy* during the solution phase, possibly with multiple pricing, in order to determine the pivot direction from the current solution point.

The scaling procedure generally results in substantial improvements in terms of the pivot path and numerical stability. The OSL and CPLEX libraries both provide a subroutine for scaling the original entries of the constraint matrix. In the solution of the WATSON problems we always make use of the scaling option.

The presolution of LP problems reduces the dimension of the original matrix, say $\hat{A}$ of (5.2), by eliminating redundant rows and columns; both OSL and CPLEX have quite effective presolvers. However CPLEX, by implementing a two-phase method with a presolver-aggregator, effects substantial reductions of the original matrix. This is the distinguishing feature of the CPLEX library which markedly improves the overall solution time for large problems – for both simplex and interior point algorithms.

Once the matrix dimension has been reduced and the columns for the basic and non-basic variables identified, the basis submatrix is factorized and inverted. Complete refactorization takes place regularly thereafter during the solution of the problem. In this respect the performance of the algorithm may be affected by the number of factor update iterations allowed between refactorizations of the basis matrix. Levkovitz and Mitra (1993) report that in order to avoid numerical instability, the *Markowitz merit count* approach, which guides pivot selection during the basis factorization, can improve the numerical properties of the solution. The CPLEX library provides a subroutine that enables the user to select the number of iterations between refactorizations and to fix the Markowitz tolerance. Similarly, OSL allows the selection of a parameter for the frequency of the basis refactorization, but no use is made of the Markowitz approach to pivot selection during basis factorization.

In order to reduce the number of iterations required to achieve the optimum a good starting basis is essential; this is usually realized by adopting the crash procedure before the start of the solution phase. In all the test problems we have made use of the crash subroutines provided by OSL and CPLEX.

Finally, the efficiency of the simplex algorithm depends on the pricing strategy that has been adopted in order to select the direction of the step from one vertex to the next. Both OSL and CPLEX provide a parameter for the choice of a convenient pricing method for the problem under consideration. Particularly for large sparse problems, DEVEX pricing (Harris (1973)) turns out to be quite effective and can be used with both CPLEX and OSL; its idea being that the computation of the reduced cost vector for the complete set of basic variables is appropriately and efficiently scaled at each iteration. Examining the complete set of columns with any pricing method can result in unnecessary extra work when the coefficient matrix is large, so that heuristics limiting the reduced cost computations determining the pivot column may be much more efficient in this case.

In the implementation of the simplex method with OSL we have used a pricing method based on pricing a random subset of basic columns with an automatic switch to DEVEX on the full set. Using CPLEX, the best results have been achieved by selecting DEVEX pricing for very large problems and allowing the hybrid ordinary reduced cost-DEVEX pricing method for small and medium problems. Utilizing the work of Goldfarb and Reid (1977),

CPLEX has also implemented a steepest edge crossing procedure which can be thought of as a logical extension of the heuristics introduced by DEVEX pricing. However in our different tests, we could not find evidence of improved solution times utilizing this method.

5.2 Interior point method

Interior point methods have become increasingly popular after Karmarkar's 1984 contribution. We utilize a primal/dual barrier method with predictor-corrector. Following Mehrotra's study (Mehrotra (1992)), the algorithm is based on approximation of primal-dual trajectories in the interior of the feasible region, as was first implemented in the OB1 code and then incorporated, with some improvements, in CPLEX. The general theory of primal/dual interior point (P/D IP) methods is due to Megiddo (1989).

The two implementations of the algorithm we used differ in their coding language – Fortran for OB1 and C for CPLEX – and significantly in the power of the preprocessing subroutines, with a very effective presolver/aggregator implemented by CPLEX.

The search direction method adopted by both libraries (see Lustig *et al.* (1992)) may be regarded as an attempt to generate steps in the feasible region according to first-order information, with a following 'centralizing' step as a correction.

The primal/dual IP method is based on the simultaneous solution of the primal and dual problems (5.3) and (5.4) with a progressive reduction of the duality gap at each iteration. A good property of P/D IP methods is that the convergence rate for these algorithms in general is seen to be independent of the problem size.

In requiring disk storage for the primal and the dual problem simultaneously, however, current implementations of P/D IP methods are memory-intensive and this may seriously compromise the solution of very large problems. In this respect, unlike OB1, CPLEX has an efficient dynamic memory allocation system and is able to handle very large problems – although *not* the largest in this study.

The P/D IP method requires the computation at each iteration of the Cholesky factorization of a matrix of the form ADA', where D is a diagonal matrix whose nature is determined by the choice of the algorithm. This factorization absorbs most of the computational effort and the time spent on each iteration is substantially affected by the method used, which needs to be fast and memory requirement minimizing. These properties are influenced by the matrix ordering procedure before the Cholesky factor is actually determined. On the IBM/RS6000 platform used in our computations, the *multiple minimum degree* procedure (Liu (1985)) has so far proved to be preferable (cf.

also Lustig, Shanno and Marsten (1992)) to the *minimum local fill ordering* (Duff *et al.* (1986)) which is also available in both the OB1 and CPLEX systems. In this respect, Berger *et al.* (1995) have claimed that Liu's procedure – giving rise to considerable fill-in in the Cholesky factor – is in general quite inefficient, particularly for multistage stochastic programs in split-variable formulation (i.e. with explicit nonanticipative constraints on the decision variables). This evidence lead to the development of an alternative ordering heuristic termed *tree-dissection*. Comparing the above-mentioned methods, however, we have so far had empirical confirmation of the superiority for large sparse matrices of multiple minimum degree, due largely to the fact that solution times for problems formulated with explicit nonanticipative constraints are in general extremely slow relative to the so-called *compact* formulation of (5.1).

This conclusion also holds in the case of the quadratic problems solved by CPLEX interior point QP solver which was beta-tested on the Watson test problems. In this case, despite the increased density of the Cholesky factor and a significant increase in the fill-in for the different test-problems, the ordering procedure appears to be very effective and the solution times found in Table 3 are competitive with the performance of the best LP solvers.

In IP algorithms the determination of a starting point again follows the reduction of the original coefficient matrix by elimination of redundant rows and columns. As opposed to the presolver-aggregator implemented by CPLEX, OB1 bases the preprocessing of the problem on simple inferences; the reductions achieved by OB1 for the different problems are generally considerably less than those achieved by the OSL and the CPLEX presolvers.

Levkovitz and Mitra (1993) report results of experiments on the determination of an efficient starting point by selecting different starting points in the dual space, but this remains in general an open problem.

5.3   Nested Benders Decomposition

The nested Benders decomposition method implemented by MSLiP (Gassmann (1987, 1990, 1992)) specifically applies to the solution of multistage stochastic linear programs in the DRP form. Unlike the methods considered above, it does not require the simultaneous determination of a solution for the complete deterministic equivalent problem. Instead it involves an efficient decomposition procedure by extending to the multistage case the Benders decomposition method established for the two stage case.

A good introduction to the application of Benders' decomposition to stochastic linear programs can be found for two or three stage problems in respectively Van Slyke and Wets (1969) and Birge (1985) and for the multistage case in Gassmann (1990).

The idea behind nested Benders decomposition is to express at each stage t, as the impact of stages $t + 1, \ldots, T$, the expected future costs in terms of a variable θ_t and by 'cuts' – linear necessary conditions for both *feasibility* and *optimality* – expressed only in terms of the current stage variables. It is thus a special type of cutting plane algorithm in which the convex constraint $\theta_t \geq v_{t+1}$ in (4) is replaced by a polyhedral description which is updated as more information becomes available from the dual problems at the descendants of current nodes. In the MSLiP implementation each nodal subproblem is solved with a (1974) simplex subroutine created by Pfefferkorn and Tomlin (1976).

The efficiency of nested Benders decomposition depends on the number of subproblems to be solved – each one corresponding to a node in the scenario tree, the performance in solving each subproblem and the order in which the nodes of the tree are visited.

The number of subproblems to solve depends in general on the number of cuts generated by dual information. A high number of feasibility cuts is likely to cause numerical instability and eventually jeopardizes the convergence of the implemented algorithm (see below). We show in Section 6 that the performance of the nested Benders cutting plane method depends critically on the stochastic properties of the problem in terms of the number of feasibility cuts generated. Scenarios produced in conditional mode lead to a lower level of stochasticity as measured by the *expected value of perfect information* (EVPI) value at the root node of a problem (Dempster and Gassmann (1991), Dempster and Thompson (1996b)) and imply a significant decrease in the number of cuts generated and the overall solution times.

In order to improve the numerical properties of the algorithm, the OSL Release 2 simplex solver has been implemented in the MSLiP code, introducing the possibility of preprocessing the subproblems and selecting adequate crash procedures (Thompson 1997). As demonstrated in the following section, this has improved the stability of the method without affecting solution times. If stages are aggregated to create larger subproblems more efficiently solved by OSL simplex considerable speed-ups result with MSLiP-OSL (see Dempster and Thompson (1996b)).

Three alternative approaches may be adopted to selecting the sequence of subproblems to solve. In the context of deterministic dynamic problems Wittrock (1985) suggested that the fastest way to pass information between nodal subproblems would be to change direction as little as possible and to move along the chain until a direction is stopped either because the horizon has been reached or because an infeasible problem has been found. This method applied to trees is referred to as the *fast forward fast back* (FFFB) approach. Alternatively, a 'forward first' or 'backward first' approach as described by Gassmann (1990) may be considered. The FFFB protocol has been shown to give the best results and has been adopted in our experiments.

6   Computational Results

Results are presented for different solution methods applied to two sets of 10-stage WATSON problems with 'independent' and conditionally generated scenarios (see Section 3). All problems were run in double precision on an IBM RS6000/590 with 128 MB of main memory and 2.5 GB of disk running under AIX 3.2.5.

By expanding the sample space $\Omega := \Pi_{t=1}^{10} \Omega_t$ of the multidimensional random process, we can generate successive problems of increasing complexity and derive a comparative evaluation of the algorithms discussed in Section 5. A check on the actual randomness, or *stochasticity*, of the problems is made by computing the relative EVPI (as a ratio with respect to the objective value) at the root node of each problem (see Table 2).

In Table 1 we give details of the the two sets, of 11 and 13 test problems respectively, that have been generated by increasing the number of scenarios – both 'independently' and conditionally generated. These 24 problems represent the basis of the comparative evaluation of the solution algorithms for our long-term portfolio problem. Statistics are reported regarding the original dimension of the complete deterministic equivalent problem and the reduction achieved by preprocessing using the different libraries.

Thanks to a very rich set of financial data and the reliability of the SSL problem generator (given the memory and hard-disk constraints of the IBM platform) we have been able to generate 10 stage problems with up to 2688 possible realizations of the process at the horizon. Additional sample paths give increasing confidence in the representation of the portfolio problem, for which we eventually wish to study the sensitivity of the optimal policy to very small variations in the parameters underlying market rates of return and borrowing conditions.

In Figure 5 we depict the almost linear relationship between the dimension of the WATSON problems (expressed in terms of the number of variables) and the number of scenarios.

We have tested the performance of the preprocessing subroutines provided by OSL, OB1, and CPLEX for the different problems. Both OSL and CPLEX achieve a significative reduction in the dimension of the original problems, but the presolver/aggregator of CPLEX proves to be extremely effective, with a remarkable impact on solution times.

From a numerical viewpoint, long term portfolio decision problems such as the WATSON problems result in stochastic programs whose general form is characterized by:-

- A very sparse coefficient matrix with entries representing financial returns and cash inflows and outflows per unit of value.

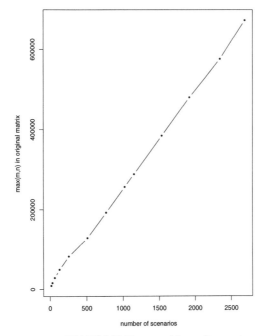

Figure 5: WATSON Test problem dimensions

- A generally small feasible region, with most of the constraints specified by upper and lower bounds on borrowing and investment decisions binding at each stage, effectively reducing the size of the set of feasible policies to be a direct function of the initial wealth of the decision maker.

- At each stage the information provided by the possible subsequent realizations of the random data process strongly affects the state of the system (in the Watson case identified with the current portfolio value process).

Interior point methods are particularly efficient in this setting. For the different test problems, CPLEX barrier converges to the solution in a number of iterations which is independent of the size of the problem. This property – generally attributed to P/D IP methods – is also possessed by MSLiP and MSLiP-OSL; in both cases the number of iterations is computed on the basis of the number of times the *master* (root-node) problem is solved in the course of the solution procedure.

Table 1: Definition of WATSON Test Problems

Number of scenarios	16	32	64	128
Tree branching structure	$2^4 1^5$	$2^5 1^4$	$2^6 1^3$	$2^7 1^2$
Number of nodes	111	191	319	511
Original matrix	$4\,573 \times 8\,401$	$8\,413 \times 15\,553$	$15\,101 \times 28\,097$	$26\,237 \times 49\,153$
Entries	21\,368	39\,848	72\,648	128\,648
Density	0.0556%	0.0305%	0.0171%	0.00997%
OSL reduced matrix	$2\,886 \times 5\,961$	$5\,318 \times 10\,953$	$9\,478 \times 19\,465$	$16\,134 \times 33\,033$
Entries	15\,359	28\,511	51\,359	88\,863
Density	0.0893%	0.0489%	0.0278%	0.01667%
OB1 reduced matrix	$3\,970 \times 7\,062$	$7\,362 \times 8\,430$	$13\,314 \times 23\,366$	$23\,298 \times 40\,326$
Entries	19\,525	36\,437	66\,389	117\,269
Density	0.06964%	0.05871%	0.02134%	0.01248%
CPLEX reduced matrix	$837 \times 3\,897$	$1\,630 \times 7\,230$	$3\,101 \times 13\,025$	$5\,981 \times 22\,753$
Entries	16\,669	30\,799	55\,099	94\,255
Density	0.5119%	0.2613%	0.1341%	0.0693%

Number of scenarios	256	512	768	—
Tree branching structure	$2^8 1^2$	2^9	$3^1 2^8$	—
Number of Nodes	767	1023	1534	—
Original matrix	$43\,517 \times 82\,177$	$67\,069 \times 128\,001$	$100\,598 \times 191\,994$	—
Entries	218\,888	351\,228	516\,784	—
Density	0.00612%	0.00406%	0.00267%	—
OSL reduced matrix	$25\,606 \times 52\,233$	$35\,846 \times 72\,713$	$53\,768 \times 109\,067$	—
Entries	144\,927	244\,765	367\,140	—
Density	0.01083%	0.00939%	0.00626%	—
OB1 reduced matrix	$38\,914 \times 65\,798$	$60\,418 \times 97\,798$	$90\,624 \times 146\,692$	—
Entries	198\,421	314\,389	471\,572	—
Density	0.00775%	0.00532%	0.00354%	—
CPLEX reduced matrix	$11\,517 \times 37\,889$	$21\,757 \times 58\,113$	$32\,636 \times 87\,167$	—
Entries	154\,47	239\,403	359\,106	—
Density	0.0354%	0.0189%	0.0126%	—

Table 1 ctd.

Number of scenarios	1 024	1 152	1 536
Tree branching structure	$4^1 2^8$	$3^2 2^7$	$4^1 3^1 2^7$
Number of Nodes	2 045	2 299	3 065
Original matrix	134 127 × 255 987	150 869 × 287 949	201 155 × 383 927
Entries	689 156	775 288	1 033 268
Density	0.002007%	0.001784%	0.001338%
OSL reduced matrix	71 690 × 145 421	80 642 × 163 577	107 522 × 218 101
Entries	489 515	550 653	734 199
Density	0.00469%	0.004174%	0.003131%
OB1 reduced matrix	120 830 × 195 586	135 918 × 220 006	181 222 × 293 338
Entries	628 755	707 285	943 039
Density	0.00266%	0.002365%	0.001774%
CPLEX reduced matrix	43 513 × 116 221	48 950 × 130 733	65 266 × 174 309
Entries	478 801	538 650	718 193
Density	0.00941%	0.00842%	0.00631%
Number of scenarios	1 920	2 304	2 688
Tree branching structure	$5^1 3^1 2^7$	$6^1 3^1 2^7$	$7^1 3^1 2^7$
Number of Nodes	3 831	4 597	5 363
Original matrix	251 441 × 479 905	301 828 × 575 983	352 114 × 671 961
Entries	1 292 258	1 578 528	1 841 528
Density	0.001071%	0.0009073%	0.000778%
OSL reduced matrix	134 402 × 272 626	out of memory	out of memory
Entries	917 745	—	—
Density	0.002504%	—	—
OB1 reduced matrix	226 526 × 366 652	out of memory	out of memory
Entries	1 178 193	—	—
Density	0.001418%	—	—
CPLEX reduced matrix	81 582 × 217 886	out of memory	out of memory
Entries	897 750	—	—
Density	0.00505%	—	—

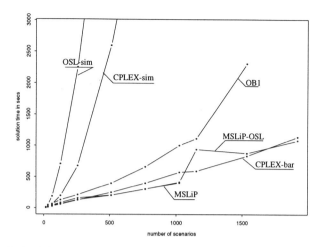

Figure 6: WATSON problem comparative solution times – independent scenarios

A major advantage of nested Benders decomposition relative to P/D IP methods is that a negligible amount of memory needs to be allocated during the solution procedure. The WATSON.10.1920 problem, for instance, has not been solved by OB1 because of lack of memory space, whereas CPLEX barrier, thanks to its dynamic memory allocation, has succeeded in finding an optimum although requiring roughly 85% of the total memory (128 MB) available. On the other hand, MSLiP and MSLiP-OSL needed for the same problem only about 21% of the memory to be allocated.

OSL simplex has been tested up to the solution of the 512-scenario problem and CPLEX simplex up to 768 scenarios; at these points the sizes of the problem made simplex already very inefficient. All other problems, up to system sustainability, have been tried with MSLiP, MSLiP-OSL, CPLEX barrier and OB1.

The numerical results shown in Table 2 refer to the complete set of 'independent' problems generated with the original Watson simulator. Solution accuracy agreement of the various methods is displayed in the number of decimal places shown for each problem in the objective value (at least 2 decimal places) (see also Table 3).

All solution methods have been tested on the 'independent' problems shown in Table 2, while only the more efficient solvers are considered in the case of problems shown in Table 3 with scenarios generated in conditional mode.

There are evident variations in the values of both the objective and the relative EVPI which appear to be independent of the number – but clearly not the nature – of randomly generated scenarios. *A priori* we expect more

Table 2: WATSON numerical results – Independent scenario generation

Number of scenarios	16	32	64	128	256	512
Obj (£m)	2158.7519	1866.398	2310.51	1637.8125	2000.838	1959.635
Relative EVPI	12.64%	28.31%	30.43%	42.75%	36.63%	44.63%
OSL simplex	11.6″	47″	188″	707″	2253″	6510″
n.iterations	1141	2696	5721	11747	23089	45457
OB1 pred./corr.	12.1″	27.1″	63.4″	128.2″	210.7″	394.9″
n.iterations	41	52	68	80	73	85
CPLEX simplex	6.37″	17.6″	56.4″	196.1″	674.7″	2600.2″
n.iterations	2257	4623	10623	20126	36315	50512
CPLEX barrier	5.77″	11.79″	26.27″	58.9″	125.8″	248.8″
n.iterations	23	26	35	44	50	63
CPLEX QP barrier	5.1″	16.3″	28.6″	74.3″	113.3″	400.4″
n.iterations	19	31	28	46	34	67
MSLiP-OSL	9.87″	20.03″	43.5″	83.0″	157.9″	204.6″
n.iterations	6	7	9	11	12	12
MSLiP	9.93″	19.5″	37.6″	65.2″	131.2″	202.9″
n.iterations	6	8	11	12	22	12

Number of scenarios	768	1024	1152	1536	1920
Obj (£m)	1813.105	1926.787	1687.92	1790.276	1778.954
Relative EVPI	45.33%	46.53%	63.45%	61.86%	62.13%
OB1 pred./corr.	658.6″	1002.2″	1111.1″	2280.7″	n.a.
n.iterations	96	111	108	123	—
CPLEX simplex	5380.5″	—	—	—	—
n.iterations	72206	—	—	—	—
CPLEX barrier	397.5″	573.6″	591.5″	833.4″	1137.6″
n.iterations	66	70	63	67	73
CPLEX QP barrier	510.1″	735.7″	926″	2135.8″	n.a.
n.iterations	55	60	67	90	—
MSLiP-OSL	303.1″	413.8″	937.9″	876.25″	1079.74″
n.iterations	10	12	7	15	18
MSLiP	304.4″	402.4″	unsolved	unsolved	unsolved
n.iterations	7	15	—	—	—

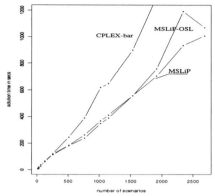

Figure 7: WATSON problem comparative solution times – conditional scenarios

conservative strategies and lower terminal wealths as a result of increasing number of scenarios! Decomposition methods, unlike solution methods based on the complete certainty equivalent problem, are affected by the stochastic properties of the underlying random process.

Tables 2 and 3 show the impact of the relative EVPI values at the root node on the different procedures: reduced randomness of the conditional problems improves speed of the solution and stability of the decomposition method making MSLiP much faster then CPLEX barrier. Very large problems – previously unsolved by MSLiP – can be solved with conditionally generated scenarios since the number of feasibility cuts needed decreases.

The IP–QP solver of CPLEX (which was made available to us by CPLEX for beta-testing) appears to be competitive with respect to the other solvers (cf. Table 2) and becomes relatively slow only for extremely large problems. The WATSON.10.1920 QP problem could not be solved again because of lack of memory. The target terminal wealth $\widetilde{W}$, for a utility function of the form $(W - \widetilde{W})^2$ has been defined for all test problems to be slightly higher then the corresponding linear optimal value, in order to make sensible comparisons between solution times. The solution times reported in Table 2 thus refer to a set of problems that have been run in order to have explicit comparisons with LP solution times. In general, for problems with a quadratic objective and the same number of scenarios as with a linear objective, a reduction of the target value $\widetilde{W}$ implies a decrease in solution times and in some cases an IP–QP solution time even lower than the corresponding LP solution time by the CPLEX IP–LP solver.

Table 3: WATSON numerical results - Conditional scenario generation

Number of scenarios	16	32	64	128	256
Obj (£m)	2 401.875	2 167.624	2 027.4293	1 879.432	1 841.2985
Relative EVPI	10.47%	13.85%	19.51%	24.85%	28.19%
CPLEX barrier	5.89″	10.94″	25.71″	60.37″	120.34″
n.iterations	28	27	33	45	48
MSLiP	10.23″	18.28″	35.92″	66.44″	118.08″
n.iterations	9	8	9	12	11
MSLiP-OSL	12.48″	21.42″	37.51″	64.58″	113.28″
n.iterations	8	8	9	8	9
Number of scenarios	512	768	1 024	1 152	—
Obj (£m)	1 787.1977	1 780.126	1 791.176	1 725.53	—
Relative EVPI	32.87%	30.26%	39.20%	33.83%	—
CPLEX barrier	245.52″	389.55″	618.20″	647.71″	—
n.iterations	62	63	75	74	—
MSLiP	184.47″	261.93″	367.94″	411.30″	—
n.iterations	10	10	11	11	—
MSliP-OSL	182.14″	238.01″	347.85″	389.23″	—
n.iterations	11	9	10	10	—
Number of scenarios	1 536	1 920	2 304	2 688	—
Obj (£m)	1 693.474	1 769.45	1 664.66	1 679.921	—
Relative EVPI	45.37%	37.04%	47.48%	45.00%	—
CPLEX barrier	898.54″	1 305.74″	—	—	—
n.iterations	74	82	—	—	—
MSLiP	557.71″	704.03″	933.69″	1 006.59″	—
n.iterations	12	12	16	15	—
MSLiP-OSL	557.41″	758.85″	1 192.18″	1 070.84″	—
n.iterations	11	12	17	13	—

Table 4: Performance of MSLiP and MSLiP-OSL on selected test problems.

Independent vs Dependent Scenario Generation.			
Number of scenarios	128	256	512
MSLiP			
solution time:			
- Independent	65.16″	131.07″	202.88″
- Conditional	66.44″	118.18″	184.47″
no. iterations:			
- Independent	12	22	12
- Conditional	12	11	10
optimality cuts/total:			
- Independent	1 697/1 699	2 842/2 887	3 762/3 795
- Conditional	1 549/1 549	2 403/2 405	2 973/2 977
no. of subproblems:			
- Independent	7 398	19 696	15 605
- Conditional	7 805	10 090	12 503
MSLiP-OSL			
solution time:			
- Independent	82.99″	157.9″	204.64″
- Conditional	64.58″	113.3″	182.14″
no. iterations:			
- Independent	11	13	12
- Conditional	8	9	11
optimality cuts/total:			
- Independent	1 474/1 474	2 262/2 273	2 855/285
- Conditional	1 365/1 365	2 043/2 043	2 577/2 577
no. of subproblems:			
- Independent	6 708	11 700	13 977
- Conditional	5 104	8 290	12 741

of nested Benders decomposition in the two cases of scenarios conditionally or unconditionally generated.

Summarizing, we see that for nested Benders decomposition a reduced level of stochasticity, regardless of the implementation considered, results in:

(i) a lower number of feasibility cuts generated throughout the solution procedure,

(ii) a significant reduction of the number of subproblems solved,

Table 4 ctd.
Independent vs Dependent Scenario Generation.

Number of scenarios	1 024	1 536	1 920
MSLiP			
solution time:			
- Independent	402.42″	unsolved	unsolved
- Conditional	367.94″	557.71″	704.03″
no. iterations:			
- Independent	15	—	—
- Conditional	11	12	12
optimality cuts/total:			
- Independent	7 375/7 469	—	—
- Conditional	5 922/5 922	8 652/8 652	10 756/10 763
no. of subproblems:			
- Independent	35 247	—	—
- Conditional	25 988	38 134	51 646
MSLiP-OSL			
solution time:			
- Independent	413.76″	876.25″	1 079.74″
- Conditional	347.85″	557.41″	758.85″
no. iterations:			
- Independent	12	15	18
- Conditional	10	11	12
optimality cuts/total:			
- Independent	5 723/5 729	8 726/8 880	10 942/11 153
- Conditional	5 330/5 330	7 595/7 595	9 734/9 734
no. of subproblems:			
- Independent	27 936	59 570	77 657
- Conditional	23 570	41 874	52 299

(iii) faster CPU times with an almost linear relationship between solution time and problem dimension.

7   Conclusions

We have shown how a dynamic portfolio problem can be conveniently represented as a discrete time linearly constrained stochastic optimization problem which can be formulated as a dynamic (multistage) stochastic program with recourse in which the principal decisions correspond to portfolio rebalancing over time.

The main steps in the formulation and solution of such decision problems have been described, with an emphasis on both the financial assumptions implied by the representation adopted and the extended numerical work needed for correct model generation and solution of the resulting stochastic optimization problem. Many sophisticated tools are available for generating, solving, analyzing and simulating such models, eventually resulting in representing the full corporate and regulatory environment in them (as, for example, is the case in the Russell–Yasuda Kasai model (Cariño *et al.* (1994)).

We conclude with some directions for future research under active pursuit by our group at the time of writing:-

- The EVPI value computed at each node of the scenario tree (Dempster and Gassmann (1991), Dempster and Thompson (1996b)) provides a useful measure of the randomness embedded in the remaining stochastic decision problem at each stage. An EVPI-based sampling procedure interfaced with the STOCHGEN subroutine library has been designed by Dempster and Corvera Poiré (Consigli and Dempster (1996), Corvera Poiré (1995b)) which can be effective in the solution of financial planning problems typically characterized as highly stochastic systems. A marginal EVPI-based (Dempster (1988)) sampling procedure is also under development.

- The two implementations of nested Benders decomposition (MSLiP and MSLiP-OSL) tested provide very efficient solvers in the case of linear problems. An extension of the method to handle nonlinear objectives has been developed which can significantly extend the applicability of the method to general linearly constrained decision problems with convex objective functions representing arbitrary utility functions – i.e. attitudes to risk – in financial planning problems.

- The solution of these very large and complex problems needs to be followed by a detailed computer-based analysis of the results in order to supply conveniently represented information to the decision maker. This is a necessary step towards the implementation of a user-friendly decision support system and can be expected to make extensive use of visualization and other sophisticated software tools (Dempster and Ireland (1991)).

- The general implications and advantages of the DRP formulation studied here relative to other portfolio management paradigms – such as the well-established static Markowitz representation which still dominates the financial services industry or the dynamic stochastic control formulation with continuous rebalancing (Brennan *et al.* (1995)) – also represents an area of increasing interest. We shall shortly be in a position to report on such a comparative study conducted under the auspices of the Frank Russell Company.

Acknowledgements

The research reported here represents one output of a long term research project conducted by the Finance Research Group, first at Dalhousie University, subsequently at the University of Essex and now at the University of Cambridge. The second author is grateful for considerable financial support over the years from these universities and from the Natural Science and Engineering Research Council of Canada and the Engineering and Physical Sciences Research Council of the UK. Our colleagues, Horand Gassmann, Xavier Corvera Poiré and Robin Thompson have provided both crucial software development support for the results reported here and lively discussion regarding their presentation. We owe a sincere debt of gratitude to Professor David Wilkie of Watson Wyatt Limited who suggested the pension fund managers' problem resulting in the Watson version of the CALM model and the WATSON problems and who generously gave us his time, simulation data from the Wilkie investment model and finally his Fortran code for its generation. We also wish to acknowledge with pleasure the supply and tuning of the OB1 Fortran code by Professor David Shanno of Rutgers University.

Appendix: The CALM Model

Notation

T	horizon
$s, t = 1, \ldots, T+1$	time periods
$i = 1, \ldots, I$	asset type
$j = 1, \ldots, J$	liability type
$k = 1, \ldots, K$	riskless instrument type
$\omega = 1, \ldots, |\Omega|$	data paths (scenarios)
Ω^t	all distinct paths at time t ($\Omega := \Omega^T$).
ω^t	data path history to the beginning of period t.

Decision variables

$x_{it}^+(\omega^t)$	amount purchased of asset i in period t.
$x_{ist}(\omega^t)$	amount held of asset i in period t which was purchased in period $s \leq t$.
$x_{ist}^-(\omega^t)$	amount sold of asset i in period t which was purchased in period $s < t$.
$y_{jt}^+(\omega^t)$	amount incurred of liability j in period t.
$y_{jst}(\omega^t)$	amount held of liability j in period t which was incurred in period $s \leq t$.
$y_{jst}^-(\omega^t)$	amount discharged of liability j in period t which was incurred in period $s < t$.
$z_{kt}^+(\omega^t)$	amount held of riskless asset k in period t.
$z_{kt}^-(\omega^t)$	amount owed of riskless asset k in period t.
$\delta_{x_{it}^+}(\omega^t), \delta_{x_{ist}^-}(\omega^t)$	binary action of buying (selling) asset i in period $t(s < t)$.
$\delta_{y_{jt}^+}(\omega^t), \delta_{y_{jst}^-}(\omega^t)$	binary action of incurring (discharging) liability j in period $t(s < t)$.
w	net wealth at the beginning of period $T+1$.

Parameters

$r_{ist}(\omega^t)$	cash return in period t on asset i purchased in period $s < t$ and held in period $t - 1$.
$s_{jst}(\omega^t)$	unit cost in period t of liability j incurred in period $s < t$ and held in period $t - 1$.
$r_{kt}^+(\omega^t)$	return in period t on riskless asset k held in period $t - 1$.
$r_{kt}^-(\omega^t)$	unit cost of borrowing riskless asset k in period $t - 1$.

$e_{it}(\omega^t)$, $e_{jt}(\omega^t)$ lump sum transaction cost of purchasing (incurring) asset i (liability j) in period t.

$f_{it}(\omega^t)$, $f_{jt}(\omega^t)$ unit cash outflow (inflow) upon purchasing (incurring) asset i (liability j) in period t.

$g_{ist}(\omega^t)$, $g_{jst}(\omega^t)$ unit cash inflow (outflow) upon selling (discharging) in period t asset i (liability j) bought (incurred) in period $s < t$.

$h_{ist}(\omega^t)$, $h_{jst}(\omega^t)$ lump sum transaction cost of selling (discharging) in period t asset i (liability j) bought (incurred) in period $s < t$.

$\rho_{it}(\omega^t)$, $\rho_{jt}(\omega^t)$, $\rho_{kt}(\omega^t)$ exchange rate appropriate to asset i (liability j) (riskless asset k) held in period $t-1$.

$v_{is}(\omega)$, $v_{js}(\omega)$ market value at the horizon $(T+1)$ of asset i (liability j) purchased (incurred) in period $s \le T$.

$\underline{X_{it}}$, $\overline{X_{it}}$ limits on investment on asset i in period t.

$\underline{Y_{jt}}$, $\overline{Y_{jt}}$ limits on incurring liability j in period t.

$\overline{Z_{kt}}$ short position limit on riskless asset k in period t.

$\overline{X_t^+}$, $\overline{Y_t^+}$ maximum new investment (liabilities) in period t.

$\overline{Y_t}$ maximum liability in period t.

Objective

$\max \mathsf{E} u(\mathbf{w})$ (expected utility of terminal wealth)

$u \in C^2(\mathbb{R})$ $\dfrac{u'(\mathbf{w})}{-u''(\mathbf{w})} := a\mathbf{w} + b.$

Constraints

Terminal wealth:

$$\sum_{s=1}^{T} \left[\sum_{i=1}^{I} v_{is}\mathbf{x}_{isT} - \sum_{j=1}^{J} v_{js}\mathbf{y}_{jsT} \right] + \sum_{k=1}^{K} \left[(1 + \mathbf{r}_{k(T+1)}^+)\mathbf{z}_{kT}^+ - (1 + \mathbf{r}_{k(T+1)}^-)\mathbf{z}_{kT}^- \right]$$

$$= \mathbf{w}$$

Cash balance $t = 1, \ldots, T+1$:

$$\sum_{i=1}^{I} \rho_{it} \left[-e_{it}\boldsymbol{\delta}_{x_{it}^+} - f_{it}\mathbf{x}_{it}^+ + \sum_{s=1}^{t-1}(\mathbf{r}_{ist}\mathbf{x}_{is(t-1)} + \mathbf{g}_{ist}\mathbf{x}_{ist}^- - \mathbf{h}_{ist}\boldsymbol{\delta}_{x_{ist}^-}) \right]$$

$$- \sum_{j=1}^{J} \rho_{jt} \left[e_{jt}\boldsymbol{\delta}_{y_{jt}^+} - f_{jt}\mathbf{y}_{jt}^+ + \sum_{s=1}^{t-1}(\mathbf{s}_{jst}\mathbf{y}_{js(t-1)} + \mathbf{g}_{jst}\mathbf{y}_{jst}^- + \mathbf{h}_{jst}\boldsymbol{\delta}_{y_{jst}^-}) \right]$$

$$+ \sum_{k=1}^{K} \rho_{kt}[(1 + \mathbf{r}_{kt}^+)\mathbf{z}_{k(t-1)}^+ - (1 + \mathbf{r}_{kt}^-)\mathbf{z}_{k(t-1)}^- - \mathbf{z}_{kt}^+ + \mathbf{z}_{kt}^-] = 0$$

$$\mathbf{x}_{itt} - \mathbf{x}_{it}^{+} = 0 \qquad \text{asset purchase inventory balance}$$
$$\mathbf{y}_{jtt} - \mathbf{y}_{jt}^{+} = 0 \qquad \text{liability incurrence inventory balance}$$
$$i = 1, \ldots, I \quad j = 1, \ldots, J \quad t = 1, \ldots, T.$$

$$\mathbf{x}_{ist} - \mathbf{x}_{is(t-1)} + \mathbf{x}_{ist}^{-} = 0 \qquad \text{asset sale inventor } y \text{ balance}$$
$$\mathbf{y}_{jst} - \mathbf{y}_{js(t-1)} + \mathbf{y}_{jst}^{-} = 0 \qquad \text{liability discharge inventory balance}$$
$$i = 1, \ldots, I \quad j = 1, \ldots, J \quad s = 1, \ldots, t-1 \quad t = 1, \ldots, T+1.$$

$$\mathbf{x}_{i(T+1)}^{+} = 0 \quad \mathbf{y}_{j(T+1)}^{+} = 0 \qquad \text{no horizon decisions}$$
$$\mathbf{x}_{is(T+1)}^{-} = 0 \quad \mathbf{y}_{js(T+1)}^{-} = 0$$
$$i = 1, \ldots, I \quad j = 1, \ldots, J \quad s = 1, \ldots, T.$$

$$\underline{X_{it}}\boldsymbol{\delta}_{x_{it}^{+}} \leq \mathbf{x}_{it}^{+} \leq \overline{X_{it}}\boldsymbol{\delta}_{x_{it}^{+}} \qquad \text{investment limits by type}$$
$$\underline{Y_{jt}}\boldsymbol{\delta}_{y_{jt}^{+}} \leq \mathbf{y}_{jt}^{+} \leq \overline{Y_{jt}}\boldsymbol{\delta}_{y_{jt}^{+}} \qquad \text{liability limits by type}$$
$$i = 1, \ldots, I \quad j = 1, \ldots, Jt = 1, \ldots, T.$$

$$0 \leq \mathbf{z}_{kt}^{+}$$
$$0 \leq \mathbf{z}_{kt}^{-} \leq \overline{Z_{kt}^{-}} \qquad \text{short position limit by type}$$
$$k = 1, \ldots, Kt = 1, \ldots, T.$$

$$\sum_{i=1}^{I} \mathbf{x}_{it}^{+} \leq \overline{X_{t}^{+}} \qquad \text{maximum new investments}$$
$$\sum_{j=1}^{J} \mathbf{y}_{jt}^{+} \leq \overline{Y_{t}^{+}} \qquad \text{maximum new liabilities}$$
$$\sum_{j=1}^{J} \sum_{s=1}^{t} \mathbf{y}_{jst} \leq \overline{Y_{t}} \qquad \text{maximum liability per period}$$
$$t = 1, \ldots, T.$$

NB: Boldface denotes random entities. All constraints hold almost surely, i.e. with probability 1.

References

I. Adler, N.K. Karmarkar, M.G.C. Resende and G. Veiga (1989). An implementation of Karmarkar's algorithm for linear programming. *Mathematical Programming* **44**, 297–335.

F. Benders (1962). Partitioning Procedures for solving mixed-variables programming problems. *Numerische Mathematik* **4**, 238–252.

J. Berger, J.M. Mulvey, E. Rothberg and R. Vanderbei (1995). Solving multistage stochastic programs using tree dissection. Statistics and Operations Research Research Report, Princeton University, Princeton, NJ.

D.P. Bertsekas (1987). *Dynamic Programming*. Prentice-Hall.

P. Billingsley (1980). *Probability and Measure*, 2nd ed. Wiley.

J.R. Birge (1985). Decomposition and partitioning methods for multistage stochastic linear programs. *Operations Research* **33**, 989–1007.

J.R. Birge, M.A.H. Dempster, H.I. Gassmann, E.A. Gunn, A.J. King and S. Wallace (1987). A standard input format for multiperiod stochastic linear programs. Mathematical Programming Society, *Committee on Algorithms Newsletter* **17**, 1–20.

S.P. Bradley and D.B. Crane (1972). A dynamic model for bond portfolio management. *Management Science* **19**, 139–151.

M.J. Brennan, E.S. Schwartz and R. Lagnado (1995). Strategic Asset Allocation. Working Paper, University of California, Los Angeles.

D.R. Cariño, T. Kent, D.H. Myers, C. Stacy, M. Sylvanus, A.L. Turner, K. Watanabe and W.T. Ziemba (1994). The Russell–Yasuda Kasai model: An asset/liability model for a Japanese insurance company using multistage stochastic programming. *Interfaces* **24**, 29–49. Reprinted in this volume, 609–633.

D.R. Cariño and W.T. Ziemba (1995). Formulation of the Russell–Yasuda Kasai financial planning model. Research Report, Frank Russell Company, Tacoma, Washington.

D.R. Cariño, D.H. Myers and W.T. Ziemba (1995). Concepts, technical issues, and uses of the Russell–Yasuda Kasai financial planning model. Research Report, Frank Russell Company, Tacoma, Washington.

G. Consigli and M.A.H. Dempster (1996). Solving dynamic portfolio problems using stochastic programming. *Zeitschrift für Angewandte Mathematik und Mechanik.* Proceedings of the GAMM96 Conference, Charles University, Prague.

X. Corvera Poiré (1995a). STOCHGEN User's Manual. Department of Mathematics, University of Essex.

X. Corvera Poiré (1995b). Model Generation and Sampling Algorithms for Dynamic Stochastic Programming. PhD Thesis, Department of Mathematics, University of Essex.

CPLEX Optimization, Inc. *Using the Cplex Callable Library, Version* 3.0 Incline Village, NE, USA (1994).

G.B. Dantzig, M.A.H. Dempster and M.J. Kallio, eds. (1981). *Large-Scale Linear programming.* Vol.1, IIASA CP-81-S1, Laxenburg.

G.B. Dantzig and A. Madansky (1961). On the solution of two-stage linear programs under uncertainty. *Proceedings of the Fourth Symposium on Mathematical Statistics and Probability* Vol.I. Univ.of California, Berkeley, 165–176.

G.B. Dantzig and G. Infanger (1993). Multi-stage stochastic linear programs for portfolio optimization. *Annals of Operations Research* **45**, 59–76.

M.A.H. Dempster, ed. (1980a). *Stochastic Programming.* Academic Press.

M.A.H. Dempster (1980b). Stochastic programming: An introduction. In Dempster (1980a), 3–59.

M.A.H. Dempster (1988). On stochastic programming: II. Dynamic problems under risk. *Stochastics* **25**, 15–42.

M.A.H. Dempster (1993). CALM: A stochastic MIP model. Finance Research Group Internal Report, Department of Mathematics, University of Essex.

M.A.H. Dempster and H.I. Gassmann (1991). Stochastic programming: Using the expected value of perfect information to simplify the decision tree. *Proceedings 15th IFIP Conference on System Modelling and Optimization*. Zurich: IFIP, 301–303.

M.A.H. Dempster and A. Ireland (1991). Object oriented model integration in a financial decision support system. *Decision Support Systems* **7**, 329–340.

M.A.H. Dempster and R.T. Thompson (1996a). Parallelization and aggregation of nested Benders decomposition. Proceedings APMOD95 Conference, Brunel University. To appear in *Annals of Operations Research*.

M.A.H. Dempster and R.T. Thompson (1996b). EVPI-based importance sampling solution procedures for multistage stochastic linear programmes on parallel MIMD architectures. Proceedings of the POC96 Conference, Versailles. To appear in *Annals of Operations Research*.

I.S. Duff, A. Erisman and J. Reid (1986). *Direct Methods for Sparse Matrices*. Clarendon Press.

J. Dupačová (1991). Stochastic Programming Models in Banking. Working Paper, International Institute for Applied Systems Analysis, Laxenburg, Austria.

J. Dupačová (1995). Multistage stochastic programs: The state-of-the-art and selected bibliography. *Kybernetika* **31**, 151–174.

Yu. Ermoliev and R.J.B. Wets, eds. (1988a). *Numerical Techniques for Stochastic Optimization*. Springer-Verlag.

Yu. Ermoliev and R.J.B. Wets (1988b). Stochastic programming, an introduction. In Ermoliev and Wets (1988a), 1–32.

H.I. Gassmann (1987). An Algorithm for the Multistage Stochastic Linear Programming Problem. PhD Thesis, Faculty of Commerce, University of British Columbia, Vancouver.

H.I. Gassmann (1990). MSLiP: A computer code for the multi-stage stochastic linear programming problem. *Mathematical Programming* **47**, 407–423.

H.I. Gassmann (1992). MSLiP82 User's Guide. School of Business Administration, Dalhousie University, Halifax, Canada.

D. Goldfarb and J.K. Reid (1977). A practical steepest edge simplex algorithm. *Mathematical Programming* **12**, 361–371.

C.G. Gonzaga (1992). Path following methods for linear programming. *SIAM Review* **34**, 167–227.

H.J. Greenberg (1994). A Computer-Assisted Analysis System for Mathematical Programming Models and Solutions: A User's Guide for ANALYZE. Mathematics Department, University of Colorado at Denver, USA.

H.J. Greenberg (1992a). A Primer for MODLER: Modelling by Object-Driven Linear Elemental Relations. Mathematics Department, University of Colorado at Denver, USA.

H.J. Greenberg (1992b). A Primer for RANDMOD: A System for Randomizing Modifications to an Instance of a Linear Program. Mathematics Department, University of Colorado at Denver, USA.

P.M. Harris (1973). Pivot selection method of the DEVEX LP Code. *Mathematical Programming* **5**, 1–28.

G. Infanger (1994). *Planning Under Uncertainty: Solving Large-Scale Stochastic Linear Programs.* Boyd and Fraser.

International Business Machines Co. *Optimization Subroutine Library Guide and Reference*, Release 2. Document SC23-0519-03, Armonk, NY (1992).

J.G. Kallberg, R.W. White and W.T. Ziemba (1982). Short term financial planning under uncertainty. *Management Science* **28**, 670–682.

N.K. Karmarkar (1984). A new polynomial-time algorithm for linear programming. *Combinatorica* **4**, 373–395.

A.J. King (1988). Stochastic programming problems: Examples from the literature. In Ermoliev and Wet (1988a), 543–567.

A.J. King (1994). Strategic asset allocation with stochastic programming. Presented at UNICOM Conference: Applications of Novel/Newly Emerging Techniques to Financial Systems, USA Embassy, London.

M.I. Kusy and W.T. Ziemba (1986). A Bank Asset and Liability Management Model. *Operations Research* **34**, 356–376.

M. Lane and P. Hutchinson (1980). A model for managing a certificate of deposit portfolio under uncertainty. In Dempster (1980a), 473–493.

R. Levkovitz and G. Mitra (1993). Solution of large-scale linear programs: A review of hardware, software and algorithmic issues. In *Optimization in Industry*, T.A.Ciriani and R.C.Leachman, (eds.), Wiley, 139–171.

J. Liu (1985). Modification of the minimum-degree algorithm by multiple elimination. *ACM Transactions on Mathematical Software* **11**, 141–153.

D.G. Luenberger (1984). *Linear and Nonlinear Programming.* Addison-Wesley.

I.J. Lustig, R.E. Marsten and D. Shanno (1992). On implementing Mehrotra's predictor-corrector interior point algorithm for linear programming. *SIAM Journal on Optimization* **2**, 735–749.

I.J. Lustig, J.M. Mulvey and T.J. Carpenter (1991). Formulating two-stage stochastic programs for interior point methods. *Operations Research* **39**, 757–770.

N. Megiddo (1989). Pathways to the optimal set in linear programming. In *Progress in Mathematical Programming: Interior Points and Related Methods* Springer-Verlag, 131–158.

S. Mehrotra (1992). On the implementation of a primal-dual interior point method. *SIAM Journal on Optimization* **2**, 575–601.

J.M. Mulvey (1992). Integrative asset-liability planning using large-scale stochastic optimization. Technical Report, Statistics and Operations Research Series, School of Engineering and Applied Science, Princeton University.

J.M. Mulvey and E. Thorlacius (1998). The Towers Perrin global capital market scenario generation system: CAP:Link. This volume.

J.M. Mulvey and H. Vladimirou (1989). Stochastic network optimization models for investment planning. *Annals of Operations Research* **20**, 187–217.

C.E. Pfefferkorn and T.A. Tomlin (1976). Design of a linear programming system for ILLIAC–V. Technical Report SOL 76–8, Department of Operations Research, Stanford University and Technical Report 5487, NASA–Ames Institute for Advanced Computation, Sunnyvale, CA.

R.T. Rockafellar and R.J.B. Wets (1991). Scenarios and policy aggregation in optimization under uncertainty. *Maths of OR* **16**, 119–147.

A. Ruszczynski (1992). Augmented lagrangian decomposition for sparse convex optimization. Working Paper, IIASA, Laxenburg, Austria.

R.T. Thompson (1997). MSLiP-OSL User's Guide. Judge Institute of Management Studies, University of Cambridge.

R. Van Slyke, R.J.B. Wets (1969). L-shaped linear programs with application to optimal control and stochastic programming. *SIAM Journal of Appl. Math* **17**, 638–663.

R.J.B. Wets (1988). Large scale linear programming techniques. In Ermoliev and Wets (1988a), 65–93.

A.D. Wilkie (1995a). More on a stochastic asset model for actuarial use. Presented to the Institute of Actuaries, London, 24 April 1995.

A.D. Wilkie (1995b). Stochastic Investment Models. Lecture Notes, Summer School on Financial Mathematics for Actuaries, Oxford 21–23 March 1995.

R.J. Wittrock (1985). Dual nested decomposition of staircase linear programs. *Mathematical Programming Study* **24**, 65–86.

S. Zenios (1992). Asset-liability management under uncertainty: The case of mortgage-backed securities. Research Report, Hermes Lab for Financial Modeling and Simulation, The Wharton School, University of Pennsylvania.

S.A. Zenios, ed. (1993) *Financial Optimization.* Cambridge University Press.

W.T. Ziemba and R.G. Vickson, eds. (1975). *Stochastic optimization models in finance.* Academic press.

A Dynamic Model for Asset Liability Management for Defined Benefit Pension Funds*

Cees L. Dert

Abstract

This paper presents a scenario-based optimisation model for analyzing the investment policy and funding policy of pension funds, taking into account the development of the liabilities in conjunction with the economic environment. Such a policy will be referred to as an asset liability management (ALM) policy.

1 Problem Description

Here, we shall consider pension funds that have been set the task of making benefit payments to people that have ended their active income earning career. The payments must be in accordance with the benefit formulae that prescribe the flow of payments to which each participant in the fund is entitled.

In general, the pension fund has two sources for funding its liabilities: revenues from its asset portfolio (investment income and appreciation of the value of the portfolio) and contributions to the fund. Contributions are, by definition, made by the sponsor of the fund. Thus, at given points in time, the value of the assets of the fund is increased by receiving contributions and by appreciation of the value of invested assets, and it is decreased by making benefit payments. It is the responsibility of the pension fund to balance this process in such a way that the fund meets the solvency standards in force, and that all benefit payments, now and in the future, can be made timely.

We propose a model that enables one to determine an ALM policy that minimises the cost of funding while safeguarding the pension funds's ability to make all benefit payments timely, without becoming underfunded.

---

*This paper is based on part of the author's dissertation, Dert (1995). The author is indebted to Alexander Rinnooy Kan and Guus Boender at the Erasmus University, Rotterdam, for many discussions about the topic of this paper. It would have been difficult to obtain the reported computational results without the support of ORTEC Consultants, Paragon Decision Technology and the Pension Fund for the Metalworking, Pipe, Mechanical and Automotive Trades.

A starting point for the analysis is the present state of a pension fund, defined by its actuarial and financial situation (asset value, premium reserve, level of benefit payments etc.), the benefit formulae and/or contribution formulae and the characteristics of the participants. Our model has been developed so as to assist in determining what dynamic ALM policy:

- guarantees that all benefit payments can be made timely;

- guarantees an acceptably small chance of underfunding;

- guarantees sufficiently stable future contributions;

- all the above against a minimal present value of expected future contributions by the plan sponsors.

The resulting policy takes into account three exogenous sources of uncertainty that can have a major effect on the development of the financial situation of the pension fund:

- the development of the liabilities due to benefit formulae which may depend on the future value of macroeconomic variables, e.g. price inflation and wage inflation;

- the development of the characteristics of the participants;

- the future returns on investment classes in which the fund is willing to invest.

The models and illustrations in this paper assume a defined benefit pension plan; however, the approach that we present is also suited for determining investment policies for defined contribution plans.

2   Characteristics of our Approach towards ALM

We propose a mixed integer stochastic programming model which can be employed to determine dynamic ALM policies that are based on scenarios able to reflect any set of assumptions one chooses to make on future circumstances. The ALM model includes binary variables that enable one to count the number of times a certain event happens. This feature has been used to formulate chance constraints that are based on the probability distribution of states of the world that follows from the scenarios. For ALM, this property is used to model and restrict the probability of underfunding, both at the planning horizon, as well as at intermediate points in time.

The model can be employed to determine a dynamic ALM strategy, consisting of an investment strategy and a contribution policy, which accounts for the development of liabilities. Decisions to be made at any point in time presume we will be able to make state-dependent recourse decisions in the future. They are the result of a trade-off between short and long term effects. Risk is reflected by the probability of underfunding and the magnitude of deficits when they occur. The model accommodates the employment of realistic probability distributions of exogenous random variables.

3   A Scenario Generator for Asset Liability Management

3.1   Modelling an Uncertain Future by Scenarios

One of the central issues in ALM modelling is the way in which uncertainty is handled. It is modelled by a large number of scenarios, each of which reflects a plausible development of the exogenous environment. Future environments are reflected by states of the world, defined by the level of benefit payments, the value of the remaining liabilities, the level of costs of wages and the return on each of the asset classes over the previous period. These states of the world are independent of the decisions to be made with respect to asset mix and contribution level. They are defined completely by factors that are exogenous to the decision model. A path through consecutive states of the world is a scenario.

A large set of scenarios is assumed to be a reasonable representation of the uncertain future: the model assumption is made that one of these paths will materialize. Uncertainty is still preserved in that the decision maker does not know yet which scenario describes the true future states of the world.

3.1.1   Scenario Structure

To model a multistage decision process with recourse, the states must be structured so that they reflect the notion of time and the principle of information being revealed as time goes by. The desired information structure and the notion of time are ensured by imposing the tree shape scenario structure; see Figure 1. At time $t = 0$, there is only one state of the world: the state that can currently be observed. Given this state there are many others that could emerge by the end of period 1. Which one of them actually materializes will be known only at time 1. In general, given state of the world s at time t, there are many states at time $t + 1$ which succeed (t, s) with positive probability. This reflects the uncertainty regarding the future environment. At any point in time, the history by which the prevailing state of the world

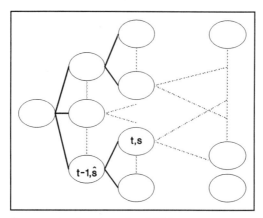

Figure 1. A scenario structure for ALM

was reached is known: the scenarios are structured so that each node has a unique predecessor.

Statistics of endogenous and exogenous state variables, such as the probability of underfunding and the expected surplus, play an important role in the ALM model. In order to compute these statistics, the scenarios are equipped with a probability structure on which the statistics can be defined. This structure specifies the probability of each state of the world occurring; unconditional, as well as conditional on the state of the world that has prevailed at the preceding point in time.

3.1.2 Consistency and Variety

The scenarios are generated so that future states of the world are consistent, i.e., the available information on stochastic and deterministic relationships between state variables at each point in time- and intertemporal relationships between state variables are reflected correctly. The variety of the states of the world suffices to capture all future circumstances with which one would want to reckon.

Figure 2 shows the scenario generator that has been used to obtain the computational results that are reported in Section 5. A time series model is employed to generate future developments of price inflation, wage inflation and returns on stocks, bonds, cash and real estate in such a way that means, standard deviations, autocorrelations and cross correlations between state variables are consistent with historical patterns. This model is discussed more extensively in Section 5.

Given the benefit formulae and relevant data on the participants (e.g. civil

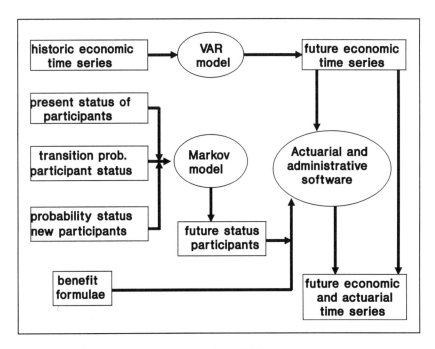

Figure 2. A scenario generator for ALM

status, age, gender, salary, earned pension rights, medical status, social status), a Markov model is employed to determine the future development of each individual currently participating in the pension fund. For an employee, for example, it is determined whether he remains alive, retires, resigns, gets disabled and/or is promoted to another job category on an annual basis. These transitions are determined by probabilities which depend on characteristics of the individuals, such as age, gender and employee category. Additional promotions and the recruitment of new employees are determined in line with the intended personnel policy.

Given the development of wage inflation, the career of each employee in each future state of the world and the current reward system, the cost of salaries, the level of benefit payments and the actuarial value of the liabilities can be computed for each state of the world.

The scenario generator has been structured in such a way, that consistency between state variables within a state, as well as consistency between states of the world is preserved.

Once the scenarios have been generated, the following information is available for each state of the world:

- the level of benefit payments;

- the level of the actuarial reserve;

- the level of the costs of wages;

- the return on each of the asset classes over the preceding period of time.

Furthermore, the scenario structure implies that for each state it is also known:

- what the preceding state of the world has been;

- which states of the world are possible successors, and what the probability is that they will emerge, given the current state.

The scenario generator used to obtain the computational results reported here will be discussed in Section 5. The ALM model can also be used in conjunction with other scenario generators. For example, one could employ a model that is based on economic theory, instead of the time series model that has been included in our scenario generator.

All the information that is contained in the scenarios is independent of ALM decisions. They are the subject of the next section.

4   A Dynamic Optimisation Model for Asset Liability Management

One of the important instruments by which the pension fund can pursue its goals is the contribution policy. However, when setting contribution levels, the fund will have to take into account the sponsor's ability and willingness to pay these contributions. It is questionable whether contributions can always be collected if they show unexpectedly large annual increases. Therefore, the rise in annual contributions, measured in percentage points of the costs of wages, will be required not to exceed a prespecified level.

Another important instrument at the disposal of the pension fund is its investment policy. We shall concentrate on the sequence of decisions regarding the allocation of assets over a limited number of asset classes, namely the asset mix.

4.1   Risk and Reward

Reward is quantified as the cost of funding; the lower, the better. These costs consist of the current value of the assets, the present value of expected regular contributions to the fund and the present value of expected remedial contributions.

Risk is associated with adverse development of the surplus (the value of assets less the value of the liabilities), i.e. with situations in which the surplus erodes to such an extent that the desired level of funding cannot be sustained any more; or worse, that the ability of the fund to meet its liabilities is at issue. Both the probability of becoming underfunded and, when it occurs, the magnitude of deficit will be taken into account.

4.2 A Mathematical Description of the ALM Process

The optimisation model for ALM allows for the determinination of sequences of time- and state-dependent decisions. We use the following notation: time is reflected by points in time t, $t = 0, \ldots, T$. By period t, $t = 1, \ldots, T$ we mean the span of time from $t - 1$ to t. A scenario or a path through time is a consecutive sequence of states of the world. Each state has exactly one predecessor and may have many successors. The predecessor of (t, s) will be indexed by $(t - 1, \hat{s})$.

Superscripts

l	lower bound on decision variables
u	upper bound on decision variables

Parameters

α	Demanded funding level
β_t	Maximal raise in contribution per period as a fraction of the cost of wages at (point in) time t
γ_{ts}	Discount factor for a cash flow at time t in state s.
l_{ts}	Benefit payments and costs to the fund at time t in state s
L_{ts}	Actuarial reserve at time t in state s
λ	Penalty parameter to penalize remedial contributions
r_{its}	Continuous return on investment of asset class i during period t in state s
M_{ts}	Large constant at time t in state s
W_{ts}	Cost of wages during period t in state s

Decision variables

A_{ts}	Total asset value before receiving regular contributions and making benefit payments at time t in state s
f_{ts}	Binary variable for remedial contributions at time t in state s
Ψ_{ts}	Probability of underfunding at time $t + 1$, given state of the world s at time t
V	Objective function value

X_{its} Amount of money invested in asset class i at time t in state s

x_{its} Fraction of asset value invested in asset class i at point in time t in state s

Y_{ts} Regular contribution during period t in state s

y_{ts} Regular contribution as a fraction of the cost of wages during period t in state s

Z_{ts} Remedial contribution at time t in state s

We develop the ALM model by using the chain of events that constitutes the ALM process.

At time 0, the contribution level for period 1 should be set. It should be between given minimum and maximum levels in dollar terms as well as relative to the costs of wages:

$$Y_{01}^l \leq Y_{01} \leq Y_{01}^u, \tag{4.1}$$

$$y_{01}^l \leq \frac{Y_{01}}{W_{01}} \leq y_{01}^u. \tag{4.2}$$

Benefit payments for period 1 should be made and the remaining wealth should be allocated over the asset classes, taking into account upper and lower bounds on the proportion to be invested in each class:

$$A_{01} + Y_{01} - l_{01} = \sum_{i=1}^{N} X_{i01}, \tag{4.3}$$

$$x_{i01}^l \left(A_{01} + Y_{01} - l_{01} \right) \leq X_{i01} \leq x_{i01}^u \left(A_{01} + Y_{01} - l_{01} \right). \tag{4.4}$$

At time t, $t = 1, \ldots, T$ the asset value has been increased by the return on investments. The asset value at time t is:

$$A_{ts} = \sum_{i=1}^{N} e^{r_{its}} X_{i,t-1,\hat{s}}. \tag{4.5}$$

The solvency requirement dictates that this asset level should at least equal the product of the required funding level with the value of the remaining liabilities:

$$A_{ts} \geq \alpha L_{ts}. \tag{4.6}$$

Including this constraint may be too restrictive: worst case realizations of L_{ts} can be arbitrarily large, requiring arbitrarily large asset values, which call for infinitely high investment returns or infinitely high contributions. The stochastic nature of the model makes constraint (4.6) unrealistic. The solvency requirement can be included in a natural and consistent way by formulating it as the chance constraint:

$$\Pr\left[A_{ts} < \alpha L_{ts} \mid (t-1,\hat{s}\,)\right] \le \Psi_{t-1,\hat{s}}^{u}. \tag{4.7}$$

Constraint (4.7) requires the probability of becoming underfunded at point in time t, given the situation at point in time $t-1$, to be less than or equal to a prespecified constant $\Psi_{t-1,\hat{s}}^{u}$. The probability of underfunding has now been modelled in such a way that:

- the model can account for any probability distribution that can be reflected by the scenarios. That includes distributions that are specified implicitly, such as the distribution of liabilities which may be given by benefit formulae in the form of computer programs;

- probabilities of underfunding are endogenous to the model;

- probabilities of underfunding are taken into account explicitly, at intermediate points in time, as well as at the planning horizon.

Assume that in case of underfunding a remedial payment is made which is precisely sufficient to restore the required funding level[1]. To reflect this assumption, constraint (4.5) should be replaced by

$$A_{ts} = Z_{ts} + \sum_{i=1}^{N} e^{r_{its}} X_{i,t-1,\hat{s}}. \tag{4.8}$$

The chance constraint (4.7) on the asset value dropping below the required level can now be replaced by a probabilistic constraint on Z_{ts} taking on a positive value:

$$\Pr\left[Z_{ts} > 0 \mid (t-1,\hat{s}\,)\right] \le \Psi_{t-1,\hat{s}}^{up}. \tag{4.9}$$

When solving the model, the chance constraint (4.9) cannot be included in this generic form. To formulate the chance constraint explicitly, we employ binary variables that register states in which remedial contributions are made:

$$Z_{ts} \le f_{ts} M_{ts}, \tag{4.10}$$

$$f_{ts} \in \{0,1\}. \tag{4.11}$$

---

[1] It is not clear what would happen in practice when a situation of underfunding occurs. The benefit formulae usually contain small print which enables the pension fund to reduce the liabilities by adjusting the benefit formulae. For instance, the decision could be made not to offer a full indexation of benefits. The fund could also call on the sponsor to make a remedial contribution to the fund. Yet another way to cope with the problem would be to accept the lower funding level for the time being and to initiate an ALM policy which is geared toward restoring the desired funding level at some point in the future. Of course, the latter solution would be possible only if there is no liquidity problem and if this is acceptable to the regulating authorities. Tax issues can be of importance with respect to the timing of remedial contributions.

If constraints (4.10) and (4.11) are included, then for any feasible solution, $Z_{ts} > 0$ implies $f_{ts} = 1$. It follows that $\Psi_{t-1,\hat{s}} \leq \sum_{s=1}^{S_t} \Pr\left[(t,s) \mid (t-1,\hat{s})\right] f_{t,s}$ and thus, any solution that satisfies constraints (4.10), (4.11) and (4.12):

$$\sum_{s=1}^{S_t} \Pr\left[(t,s) \mid (t-1,\hat{s})\right] f_{t,s} \leq \Psi_{t-1,\hat{s}}^u, \tag{4.12}$$

satisfies (4.9). Moreover, for any solution satisfying (4.9) there exists a set of values for f_{ts} such that (4.10), (4.11) and (4.12) are met. Hence (4.9) is equivalent to (4.10), (4.11) and (4.12).

Now, still at time t, it should be decided to what percentage of the costs of wages the contribution for period t should be set, observing upper bounds on the (increase of) contributions, in dollars as well as relative to the costs of wages:

$$Y_{ts}^l \leq Y_{ts} \leq Y_{ts}^u, \tag{4.13}$$

$$y_{ts}^l \leq \frac{Y_{ts}}{W_{ts}} \leq y_{ts}^u, \tag{4.14}$$

$$\frac{Y_{ts}}{W_{ts}} - \frac{Y_{t-1,\hat{s}}}{W_{t-1,\hat{s}}} \leq \beta_t \tag{4.15}$$

After receiving regular contributions and making benefit payments for period t, one must reallocate the assets,

$$A_{ts} + Y_{ts} - l_{ts} = \sum_{i=1}^{N} X_{its}, \tag{4.16}$$

subject to upper and lower bounds on the asset mix:

$$x_{its}^l \left(A_{ts} + Y_{ts} - l_{ts}\right) \leq X_{its} \leq x_{its}^u \left(A_{ts} + Y_{ts} - l_{ts}\right). \tag{4.17}$$

The equations describing the ALM process have now been specified.

Dert (1995) contains a variation on this model which allows for specifying upper bounds on trading volumes, as well as for including proportional transaction costs.

4.2.1   Selecting an Objective Function for ALM

The objective is to minimise the costs of providing a given pension insurance,

$$\text{Minimise} \quad V = A_{01} + \sum_{t=0}^{T-1} \sum_{s=1}^{S_t} \Pr\left[(t,s)\right] \gamma_{ts} Y_{ts} + \lambda \sum_{t=1}^{T} \sum_{s=1}^{S_t} \Pr\left[(t,s)\right] \gamma_{ts} Z_{ts}. \tag{4.18}$$

More precisely, we minimise the present value of the expected costs of funding, subject to constraints which ensure acceptable levels of risk in terms of solvency of the pension fund and stability of contributions by the sponsor. The costs consist of the current value of the assets, the expected present value of regular contributions to the fund and the expected present value of remedial contributions. In many instances current assets and the initial level of contributions will be given, in which case A_{01} and Y_{01} would be constants instead of decision variables. The parameter λ is a penalty reflecting the preference of asking regular contributions over remedial contributions. Its value will be chosen so that an optimal solution will not permit remedial contributions in excess of the minimal amount required to restore solvency. The discount factors γ serve to compute the present value of future cash flows. They reflect the sponsor's preferences with respect to the timing of contribution payments. The choice of the discount factor in the objective function is in principle independent of the discount rate that the actuary chooses to use to compute the actuarial value of the remaining liabilities. The actuarial principles that have been selected to value the liabilities are reflected by the values of L_{ts}. The optimization model determines an ALM policy which suits the sponsor best, given the prevailing actuarial method of valuing liabilities. See Dert (1995) for more discussion of the appropriate choice of discount factors.

The ALM model can be used to compute ALM strategies which specify investment decisions and contribution levels to be set under a wide range of future circumstances. The decisions are made so that: the present value of expected contributions to the fund is minimal, subject to raising sufficiently stable annual contributions; and the probability of underfunding at the end of each year being acceptably small when starting from the current situation, as well as from all future states of the world that the policy makers of the pension fund choose to take into account.

In summary, the ALM model is:

$$\text{Minimise} \quad V = A_{01} + \sum_{t=0}^{T-1} \sum_{s=1}^{S_t} \Pr\left[(t,s)\right] \gamma_{ts} Y_{ts} + \lambda \sum_{t=1}^{T} \sum_{s=1}^{S_t} \Pr\left[(t,s)\right] \gamma_{ts} Z_{ts}. \tag{4.19}$$

such that:

$$Y_{ts}^l \le Y_{ts} \le Y_{ts}^u, \tag{4.20}$$

$$y_{ts}^l \le \frac{Y_{ts}}{W_{ts}} \le y_{ts}^u, \tag{4.21}$$

$$A_{ts} + Y_{ts} - l_{ts} = \sum_{i=1}^{N} X_{its}, \tag{4.22}$$

$$x_{its}^l \left(A_{ts} + Y_{ts} - l_{ts}\right) \le X_{its} \le x_{its}^u \left(A_{ts} + Y_{ts} - l_{ts}\right). \tag{4.23}$$
$$(\text{For } t = 0, \ldots, T-1, \ s = 1, \ldots, S_t)$$

$$A_{ts} = Z_{ts} + \sum_{i=1}^{N} e^{r_{its}} X_{i,t-1,\hat{s}}. \qquad (4.24)$$

$$A_{ts} \geq \alpha L_{ts}. \qquad (4.25)$$

$$Z_{ts} \leq f_{ts} M_{ts}, \qquad (4.26)$$

$$\frac{Y_{ts}}{W_{ts}} - \frac{Y_{t-1,\hat{s}}}{W_{t-1,\hat{s}}} \leq \beta_t \qquad (4.27)$$

$$\sum_{s=1}^{S_t} \Pr\left[(t,s) \mid (t-1,\hat{s})\right] f_{t,s} \leq \Psi^u_{t-1,\hat{s}}, \qquad (4.28)$$

$$f_{ts} \in \{0,1\}. \qquad (4.29)$$

$$(\text{For } t = 0, \ldots, T-1, \ s = 1, \ldots, S_t) \qquad (4.30)$$

4.3 Input to the ALM Model

The bulk of input required by the ALM model consists of scenarios which reflect the discrete probability distribution that has been selected to describe future states of the world. The input which is required to formulate the ALM model consists of three groups of exogenous parameters:

- the exogenous state variables l, L, r and W;

- parameters to reflect the user's objective and requirements to the ALM policy, $\gamma, Y^l, Y^u, y^l, y^u, \Psi^u, \alpha$ and β;

- parameters for which the values are determined on computational grounds, M and λ.

4.4 Computational Complexity

The ALM model is a mixed integer linear problem, the size of which increases exponentially with the number decision moments. It is very difficult to find optimal solutions to the model for realistic problem sizes. Therefore, a heuristic has been developed by which a good, but not necessarily optimal, solution to the ALM model can be obtained.

For a T-stage problem, the heuristic uses the scenario structure with $S_t = S_2$, $t = 2, \ldots, T$, which makes the number of nodes a linear function of the number of decision moments. The heuristic consists of a backward procedure and a forward procedure. In the backward procedure, a sequence of two-stage problems is solved; one for each point in time at which state-dependent decisions can be made. The solutions to these problems serve to specify desirable situations of the pension fund in each state of the world. However, the two-stage problems have not been formulated in such a way that it is always feasible to determine an ALM strategy that results in attaining the desirable situations in all states of the world. Therefore, the backward procedure is followed by a forward one. The latter consists of solving a one-period model for each state world. Given decisions at preceding points in time, it minimises deviations from the desired situations that have been obtained from the backward procedure, subject to the constraints that the ALM policy should satisfy. The computational effort to solve the ALM model by means of the heuristic is proportional to the number of decision moments.

Dert (1995) contains a full description of the computational procedure that has been devised to solve the ALM model.

5 Computational Experiments

To obtain insight on the model on realistic problems, it has been applied to the data of a Dutch pension fund with an actuarial reserve in excess of 16 billion Dutch Guilders with approximately 1,020,000 participants of which 240,000 are premium payers.

5.1 The Exogenous Environment

The exogenous environment within which the ALM policies have to be determined is specified by the coefficients of the scenario generator, the initial state of the world and the requirements on the ALM policies. These are now discussed.

5.1.1 Historical Time Series

Consider the statistics in Table 1. They have been computed on the basis of annual observations of the following time series: Dutch inflation of wages; Dutch inflation of prices; the return on short term deposits denominated in Dutch guilders; total returns on an internationally diversified stock portfolio (Robeco); growth of Dutch gross national product; total returns on an internationally diversified property portfolio (Rodamco); and total returns on an internationally diversified bond portfolio (Rorento). Sorting the investment categories from high to low, according to average return, gives the following order: (1) stocks; (2) property; (3) bonds; (4) cash. The average return on bonds only just exceeds that on cash. Sorting the asset classes according to standard deviation of return yields the same sequence, with the exception of property and bonds. The relatively low standard deviation of property may be partly due to valuation issues. Though the return on cash is only slightly lower than that on bonds, the standard deviation of the return on bonds is more than three times as high as the standard deviation of the return on

Table 1. Statistics based on annual observations from 1956 to
1994. Note that although the time series of returns on short term
deposits has a positive standard deviation, the return on annual
deposits over the next year is known with certainty before the
asset allocation decision has been made.

	wages	prices	cash	stocks	GNP	property	bonds
Arith- metic average	5.68%	4.40%	6.02%	10.24%	3.27%	7.67%	6.38%
standard deviations and correlations							
wages	3.98%						
prices	0.52	2.81%					
cash	−0.17	0.28	2.45%				
stocks	−0.21	−0.24	−0.16	16.77%			
GNP	0.33	0.14	−0.31	−0.34	2.34%		
property	0.17	0.16	−0.06	0.33	−0.25	8.12%	
bonds	−0.07	−0.05	0.19	0.43	−0.49	0.52	8.16%

cash. Moreover, when the growth of liabilities is (partially) determined by
price inflation, then the high correlation of the return on cash with price
inflation makes cash, as compared with bonds, a very attractive asset class
since liabilities are not discounted at market rates.

Table 2 presents autocorrelations on annual data. Using the econometric
rule of thumb that autocorrelation coefficients are statistically significant if
their absolute value exceeds $2/\sqrt{n}$, where n denotes the number of observations, one should be particularly interested in figures that exceed 0.33 in
absolute value. Most figures may not come as a surprise. However, there is
an exception: the correlation between the return on stocks and the lagged
return on cash is rather high. If this statistical relationship tells us anything
about the future, then the return on Dutch deposits can be used to predict
the return on internationally diversified stock portfolios. Despite this interesting observation, we have chosen not to include this relationship in the set
of a priori expected relationships.

5.1.2 Historical Time Series and ALM

To appreciate the trade-off that has to be made in ALM for benefit-defined
pension funds, we go one step further. Suppose that liabilities are fully indexed with price inflation and that annual contributions to the fund are
precisely sufficient to fund newly acquired pension rights, exclusive of the

Table 2. Autocorrelations based on annual observations from 1956
to 1994

	wages	prices	cash	stocks	GNP	property	bonds
wages	0.33	0.52	−0.21	−0.34	0.47	−0.04	−0.28
prices	0.36	0.73	0.17	−0.19	0.22	0.07	−0.23
cash	−0.20	0.24	0.66	−0.06	−0.17	−0.03	−0.04
stocks	0.09	−0.03	0.33	−0.09	−0.32	−0.20	0.11
GNP	0.13	−0.16	−0.63	0.26	0.40	0.05	−0.22
property	0.12	0.30	0.12	−0.25	−0.03	−0.35	0.04
bonds	−0.14	0.21	0.59	−0.08	−0.27	−0.36	0.17

increase of liabilities due to indexation promises. To maintain the existing funding level, the average return on investments may not be less than
(4% + average price inflation) = 8.4%. To compose an asset mix with such
a high average return, one has to invest substantially in equities: property
and stocks are the only asset classes with an average return that exceeds 8%.
Such an asset mix, however, is also characterised by a high standard deviation
of return, which implies that it entails an unacceptably high probability of
underfunding in the shorter term, unless the sponsor is willing and able to
accept large fluctuations of annual contributions, or the pension fund has a
high funding level to start with.

5.1.3   A Time Series Model for Economic Scenarios

Following Sims (1980), we model the economic time series using a Vector
Auto Regressive (VAR) model. In addition to the arguments of Sims, when
practising our ALM approach, the simple structure of a VAR model of order
1 allows for a straightforward interpretation of the model parameters. This
provides for the opportunity to discuss results with policy makers who did not
receive a quantitatively oriented education. When called for, coefficients of
the VAR model can be adjusted in order to reflect subjective beliefs regarding
future economic developments.

It is not uncommon in financial theory to assume that asset prices follow
a lognormal distribution[2] even though real asset prices have fatter tails. In
conformity with this assumption, we shall assume that the disturbances of the
VAR model are distributed normally. The VAR model used to describe the
relationships between the continuous returns on asset classes, price inflation

---

[2]Even though empirical evidence to the contrary has been reported (see e.g. Guimaraes,
Kingsman and Taylor (1989) and Dert (1989)).

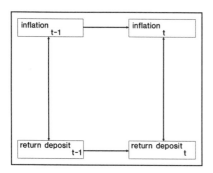

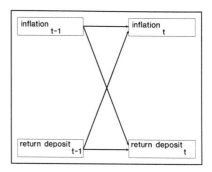

Figure 3. Cross correlation and   Figure 4. VAR correlations of
first order serial correlation   the first order

and wage inflation is:

$$R_t = \mu + \Omega \left(R_{t-1} - \mu \right) + u_t \qquad u_t \sim N(0, \Sigma), \qquad (5.1)$$

where

$$R_t = \begin{bmatrix} r_{1t} \\ \vdots \\ r_{Nt} \\ g_t \\ w_t \end{bmatrix} \qquad \mu = \begin{bmatrix} \mu[r_1] \\ \vdots \\ \mu[r_N] \\ \mu[g] \\ \mu[w] \end{bmatrix} \qquad u_t = \begin{bmatrix} u[r_{1t}] \\ \vdots \\ u[r_{Nt}] \\ u[g_t] \\ u[w_t] \end{bmatrix}, \qquad (5.2)$$

and where

$N(\mu, \Sigma)$	Gaussian distribution with mean-vector μ and covariance matrix Σ. Continuous price inflation during period t in state s
$r(L)_{ts}$	Continuous growth of liabilities during period t in state s
R_{ts}	Vector of continuous returns on asset classes, price inflation, wage inflation and increase in gross national product during period t in state s
Ω	First order autocorrelation matrix
u_t	Residual
w_{ts}	Continuous wage inflation during period t in state s
g_t	Continuous growth of gross national product during period t in state s

It allows for autocorrelation and auto crosscorrelation of the first order. Assuming that all eigenvalues of Ω are smaller than 1 in absolute value, $E\left[R_t\right]$ converges to μ for $t \to \infty$.

We have used a stepwise Maximum Likelihood method to estimate (5.1)[3]. Starting with an unrestricted ML estimation, an iterative procedure is carried

---

[3]For a discussion of the pros and cons of different estimation techniques for this type of model see Ooms (1993) or Judge $et\ al.$ (1985, 1988).

out in which one non-significant parameter per iteration is removed from the model and new ML estimates are computed for the remaining coefficients until all insignificant coefficients have been removed. In case of more than one statistically insignificant parameter, the one to be fixed at zero is selected on statistical and economic grounds. The procedure is as follows:

Step 1. Specify for each of the coefficients whether its value, based on economic theory, is expected to be negative, zero or positive.

Step 2. Compute Maximum Likelihood estimates by applying GLS iteratively.

Step 3. If all coefficients are statistically significant at the 1% level, STOP. Otherwise, select one of the non-significant coefficients to be fixed at zero, applying the following pick order:

   1. The least significant coefficient which was expected to be zero.

   2. The least significant coefficient which has a sign opposite to the expected sign.

   3. The least significant coefficient with a sign equal to its expected sign.

Go to Step 2.

5.1.4 A Vector Auto Regressive Model for Generating Economic Time Series

Given the estimated values of the parameters μ, Ω and Σ, (5.3) is applied iteratively to generate many scenarios of future time series, structured as illustrated in Figure 1: at time 0 there is only one economic state of the world, R_{01}, the current values of the economic state variables. For $t = 1, \ldots, T$, $s = 1, \ldots, S_t$ the economic scenarios can now be generated by applying (5.3) repetitively.

$$R_{ts} = \mu + \Omega \left(R_{t-1,\hat{s}} - \mu \right) + \epsilon_{ts} \tag{5.3}$$

where

$$\epsilon_{ts} = \begin{bmatrix} \epsilon[r_{1ts}] \\ \vdots \\ \epsilon[r_{Nts}] \\ \epsilon[g_{ts}] \\ \epsilon[w_{ts}] \end{bmatrix}. \tag{5.4}$$

The values ϵ_{ts} are obtained by random sampling from $N(0, \Sigma)$. Thus, given state of the world (t, s'), all states $(t+1, s)$ with $\hat{s} = s'$ have positive proba-

Table 3. Estimated coefficients and standard errors of the VAR model

dependent variable	independent variables			arith-metic mean	adjusted R^2
	intercept	price inflation	cash		
wage inflation	0.026929 (0.165032)	0.654292 (0.00828)		0.055069	0.24
price inflation	0.014001 (0.005152)	0.653854 (0.095871)		0.042115	0.49
cash	0.019525 (0.006053)		0.679611 (0.092472)	0.059059	0.42
stocks	0.084692 (0.024570)			0.085635	0.00
property	0.071748 (0.017342)			0.072410	0.00
bonds	−0.035571 (0.022893)		1.634033 (0.349911)	0.060482	0.27
GNP	0.062338 (0.006944)		−0.525310 (0.107974)	0.031594	0.36

bility of succeeding $(t-1, s')$ and they are equally likely to do so[4]. States of the world (t, s) with $\hat{s} \neq s'$ have zero probability of succeeding $(t-1, s')$. Formula (5.5) reflects the probability structure in a more formal way, assuming that there are $n_{ts'}$ states that succeed state (t, s') with positive probability.

$$\Pr\left[(t, s) \mid (t-1, s')\right] = \begin{cases} 0 & \hat{s} \neq s' \\ 1/n_{t,s'} & \hat{s} = s' \end{cases} . \tag{5.5}$$

Using the property of the scenario structure that each state has precisely one predecessor, $\Pr[(t, s)]$ for $t = 1, \dots, T$ can be computed by repeatedly applying the recursive formula $\Pr[(t, s)] = \Pr[(t, s) \mid (t-1, \hat{s})]\Pr[(t-1, \hat{s})]$.

Given a state of the world, the succeeding states of the world are defined by a deterministic component that reflects the expected development, conditional on the current state of the world, and a random component that is sampled from a distribution with expectation 0 and a covariance matrix that is determined by the correlations and standard deviations that are presented in Table 4.

---

[4]If desired, variance reduction techniques can be employed to reduce sampling errors in smaller samples (e.g. Kleijnen (1971, 1974, 1975) and Dert (1995)).

Table 4. Residual correlations and standard deviations

	wages	prices	cash	stocks	GNP	property	bonds
wages	0.03						
prices	0.28	0.02					
cash	−0.12	0.32	0.02				
stocks	−0.23	−0.31	−0.53	0.16			
GNP	0.34	0.40	0.20	−0.18	0.02		
property	0.04	−0.02	−0.19	0.33	−0.22	0.11	
bonds	−0.01	−0.27	−0.33	0.35	−0.20	0.55	0.07

Table 5. The pension fund as at the beginning of 1995.

	active participants	inactive participants	total
number	240,000	780,000	1,020,000
benefit payments 1995 in M Dfl.		300	300
contribution 1995 in M Dfl	700		700
pensionable earnings in M Dfl		4,100	4,100
indexation promise	wage inflation	price inflation	
actuarial reserve	7,600	8,800	16,400

5.1.5 The Initial State of the World

Table 5 contains data on the pension fund as at the beginning of 1995, after benefits and contributions for 1995 have been paid. The actuarial reserve amounts to 4 times the pensionable earnings and more than 20 times the annual contribution to the fund. This implies that the contribution policy will have only a limited impact on the development of the fund in comparison with the investment results.

The initial state of the world is also of importance for the scenario generator: it contains the starting values of the economic variables. These values are given in Table 6.

5.1.6 The Future According to the Scenario Generator

To facilitate the interpretation of the ALM strategies, consider Table 7. A comparison of the average investment returns with the average growth figures of liabilities and benefit payments shows that, on average, liabilities will be

Table 6. Values of the economic variables during 1994 in %.

wage	price	return on				
inflation	inflation	cash	stocks	GNP	property	bonds
1.6	2.6	5.12	−7.10	2.4	−20.75	−15.52

Table 7. Sample statistics 1995–2004 in %

	annual growth			annual investment returns			
Arith-metic	liabilities	benefits	pens. earnings	cash	stocks	property	bonds
average	10.6	5.1	3.8	6.0	10.3	8.0	6.2
Standard deviation				2.0	15.8	10.6	7.0

increased more rapidly than the value of invested assets.

The scenarios reflect an expected growth of liabilities that is rather high in comparison with the expected return on the asset classes. Therefore, feasible ALM strategies will have to maintain high funding levels in order to be able to bear the risk that comes with the selection of asset mixes with a high expected return. If the appropriate funding levels cannot be attained, then solvency requirements can be met only by dramatic rises in annual contribution. The latter becomes worse with the extent to which the fund has matured.

5.2   Requirements on the ALM Policy

ALM policies will be determined under the constraints and parameter values given in Table 8.

5.2.1   Different Settings, Different Policies

One would expect ALM decisions for a wealthy pension fund to be different from those for a thinly funded pension fund. Therefore, three settings have been selected, which differ in the initial funding level and in the amount by which annual contributions may be raised from one year to the next:

Setting 1: a low initial funding level and a low maximum increase of contributions;

Setting 2: a high initial funding level and a high maximum increase of contributions;

Table 8. Model parameters and constraints

contribution level for 1995	16% of the pensionable earnings
minimally required funding level	100%
maximum probability of underfunding	0.05
upper bound on proportional asset allocation	1
lower bound on proportional asset allocation	0 (no short selling)
horizon	10 years
number of states of the world at the end of the first year	100
number of states of the world at the end of year 2, . . . , 10	10,000
discount rate	15% per annum

Table 9. Setting-specific data

	Setting 1	Setting 2	Setting 3
Maximum rise in annual contributions if new level > 16% if new level ≤ 16%	2% points unrestricted	5% points unrestricted	3% points unrestricted
Asset value primo 1995	17,900M	32,8000M	To be determined
Funding level primo 1995	109%	200 %	by the ALM model

Setting 3: the initial funding level to be determined by the ALM model in such a way that the costs of funding are minimised subject to satisfying the solvency constraints without exceeding moderate maximum increases of contributions.

In addition to the data that have been specified in Table 8, Table 9 presents setting-specific constraints:

The settings have been selected with the intention of analyzing the ALM model on problems for which it may be expected that the driving force behind their solutions will be different for each problem.

A quick analysis of the expected development of the actuarial reserve, benefit payments and the pensionable earnings, in combination with the maximum growth of contributions reveals that a feasible ALM strategy for Setting 1 may not exist. In this case, the solution will be dominated by the search for a solution that minimises the probabilities of underfunding, while building up a sufficiently high funding level.

In Setting 2, on the other hand, a high initial funding level has been specified, in combination with a relatively high maximum on annual increases in contribution. In this setting, one would not expect any problems with respect to the solvency requirements. Here, the issue is the trade-off between reducing short-term costs (by making large restitutions in the early years and maintaining a high funding level in order to preserve the ability to meet solvency constraints) and reducing longer term costs.

If the fund has insufficient financial elbow room in Setting 1, and excessive wealth in Setting 2, then the question arises what the optimal initial funding level and the corresponding ALM strategy should be. In other words, how much should be contributed to the fund, or may be withdrawn from the fund, such that it can meet the solvency requirements with only moderate hikes of contribution, at minimal expected costs of funding. That problem is reflected by Setting 3.

In practice, one may also want to impose other constraints, such as upper and lower bounds on the proportional allocation of assets. Although the ALM model has been developed to accommodate other constraints as well, we have chosen to present computational results which are driven by policy constraints as little as possible in order not to veil the main forces that drive the ALM model.

5.2.2 Static Approaches

To compare our results with the results from static models, a static asset mix has been determined, in combination with a maximum funding level and a minimally desired funding level. These funding levels serve to determine state dependent contribution levels by the decision rules specified in Table 10.

The optimal asset mix and the optimal values of the funding levels below which regular contributions are increased and above which restitutions are made, have been determined by a brute force random search method. The static models in the literature, including mean variance approaches, do not allow for more flexible decision rules than the ones that have been used here. Therefore, the solutions that have been obtained from this static approach should be at least as good as solutions that can be obtained from static models from the literature.

Table 10. Static contribution policy

	actual funding level	contribution level
1.	> maximum funding level	restitution such that the maximum funding level remains 16% of the
2.	> minimum funding level and < maximum funding level	pensionable earnings
3.	≥ 100% and < minimum funding level	minimum of the contribution level that is required to restore the minimum funding level and the maximum contribution that follows from the constraint on annual rises of contribution
4.	< 100 %	remedial contribution to restore 100% funding level and regular contribution as in 3.

5.2.3 The Contribution of Dynamic Decisions to the Results from the ALM Model

To assess to what extent the solutions of the ALM model are due to its multi-stage character, the model has also been solved by a myopic algorithm. The myopic procedure minimises expected costs of funding in each state of the world, subject to satisfying solvency constraints that follow directly from the level of actuarial reserves at the succeeding states of the world. This method simulates a policy where optimal time- and state-dependent decisions are made, taking into account a one year horizon only. The differences between the results from the myopic procedure and the results from the proposed ALM approach can be attributed to the multistage character of the proposed approach towards ALM.

5.3 Numerical Results

5.3.1 Which Policy is the Better One?

The ALM policies are judged on three criteria: stability of regular contributions; the extent to which solvency requirements are met; and the present value of the costs of funding.

The stability requirements with respect to regular contributions are always met, by each policy, in each setting: the level of regular contributions is set at the beginning of each year so that it satisfies the upper bound that follows from the maximum rises of contribution and the regular contribution level that has been set in the preceding year. If, by the end of the year, previous decisions have led to underfunding, then a remedial payment is made to restore the minimally required funding level.

The solvency requirements are reflected by upper bounds on the annual probability that underfunding occurs. In each scenario, at the end of each year the funding level is computed before remedial contributions are made. If it is less than 100%, one has underfunding. A policy is feasible if the solvency requirements are met. That is, it is feasible if the probability of underfunding is less than or equal to 5% in each year. The extent to which this requirement is met is reflected by the average excess probability of underfunding. Feasible policies have an average excess probability of underfunding equal to 0. The extent to which a policy leads to situations of underfunding can also be measured by the magnitude of deficits when they occur. This is reflected by the present value of expected remedial payments.

The requirements on the ALM policies and the objective of minimising the present value of expected costs of funding imply that, given two sets of results, the following stepwise analysis should be used to determine which of the associated policies is the better:

Step 1 Feasibility

- If both policies are feasible go to Step 2.
- If both policies are infeasible, then the one with the lower average excess probability of underfunding is the better policy.
- If one of the policies is feasible and the other is not, then the feasible policy is the better one.

Step 2 Costs of funding

- The policy with the lower present value of total costs is preferred.

5.3.2 Results From the Myopic Procedure

The main results from the myopic procedure are in Figure 6. In Setting 1, the low initial funding level leads to the selection of an asset mix with a low volatility of investment returns. Riskier asset mixes cannot be chosen because it would imply too high a probability of underfunding by the end of 1995. The latter argument holds for years to come as well: regular annual contributions may be raised by only 2 percent points of the pensionable earnings which is insufficient to increase the funding level substantially. The average return on the 'safe' asset mixes, however, is lower than the growth of liabilities. These effects result in unacceptably high probabilities of underfunding.

Initial asset mix

setting	Cash	Stocks	Property	Bonds
1	58	17	25	0
2	0	100	0	0
3	0	100	0	0

Average total contr.(% of pens. earnings)

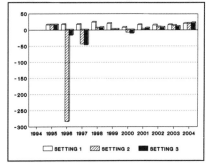

Feasibility of solutions

average excess probability of underfunding (violation of chance constraints)		
setting 1	setting 2	setting 3
10%	5%	4%

Average funding level

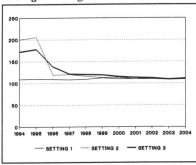

Probability of underfunding

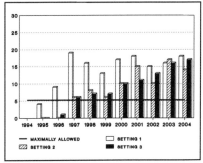

Present value (PV) of the costs of funding in mln. Dfl.

setting	Initial asset level	PV regular contributions	PV remedial contributions	PV total contributions surplus	PV terminal	PV total costs
1	18,075	4652	747	5399	1019	22,455
2	22,705	−10,337	385	−9952	1312	22,441
3	29,239	−5808	415	−5393	1063	22,783

Figure 6. Results from the myopic procedure.

The initial asset mixes in Settings 2 and 3, 100% stocks, reflect the attractiveness of a high expected investment return, combined with a surplus that is sufficiently high to bear the risk of unfavourable investment returns. More interesting, however, is the greedy character of the myopic procedure that becomes manifest in the results in Settings 2 and 3. Instead of making decisions that preserve the high initial funding levels to ensure solvency in the longer term, large restitutions are made during the early years of the planning period. As a consequence, it takes only a few years before the fund arrives at a situation that is comparable to the starting point in Setting 1: a low funding level from which it is difficult to recover. The results of the myopic procedure are a clear illustration of the risk of following a policy that is driven by short-term results only: the price for short-term reductions in the costs of funding is paid by arriving at an arduous situation from which it is difficult to recover.

In summary, due to the fact that decisions are made on a short term basis only, the forward procedure leads to infeasible policies, independent of the starting situation.

5.3.3   Results From Optimal Static Decisions

The main results from the static model follow in Figure 7. The static model differs from the myopic procedure in two important respects. The myopic procedure determines optimal state- and time-dependent decisions for a 1-year horizon, whereas the static model makes all decisions at the beginning of the planning period, independently of the scenario that will materialise, taking into account the entire 10-year planning period. Thus, for the static model, the initial asset mix will be held in all future states of the world, irrespective of the prevailing funding level.

The results of the static model are to a large extent determined by the inability to make state-dependent decisions. This is reflected by the choice of asset mixes, as well as by the development of the probability of underfunding. This effect is illustrated particularly well by the results in Setting 1. The optimal asset mix is too risky to meet short term solvency requirements: by the end of 1995, the probability of underfunding equals 15% and it rises to 28% in 1999. Thereafter, it stays high but starts to decline. This is due to the fact that, by then, the increased level of contributions in combination with relatively high average investment returns suffices to ensure a gradual rise of the average funding level. The high probabilities of underfunding that occur, despite the fact that the average funding level exceeds 120% towards the end of the planning period, reflect the limitations of the static model: the asset mix is too risky in states of the world with a low funding level and too conservative in states of the world with high funding levels. The myopic model performed much better in this setting.

In Settings 2 and 3, the optimal asset mixes are on the conservative side given the high funding levels during the early years. At the end of the planning period however, the probabilities of underfunding rise steadily which result in slight violations of the solvency constraint for 2003 and 2004 in Setting 2; an asset mix which may have been too conservative in earlier years, proves

Initial asset mix

setting	Cash	Stocks	Property	Bonds
1	49	18	33	0
2	35	28	31	6
3	14	52	32	2

Feasibility of solutions

average excess probability of underfunding (violation of chance constraints)		
setting 1	setting 2	setting 3
17%	0%	0%

Average total contr.(% of pens. earnings)

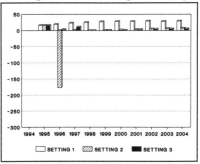

Average funding level

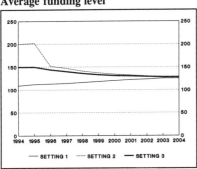

Probability of underfunding

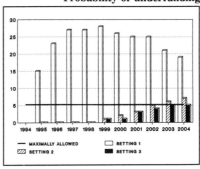

Present value (PV) of the costs of funding in mln. Dfl.

setting	Initial asset level	PV regular contributions	PV remedial contributions	PV total contributions surplus	PV terminal	PV total costs
1	18,075	−16,798	24,340	7542	2711	22,906
2	33,705	−1888	1276	−612	3030	30.063
3	24,874	4484	415	827	5311	27,099

Figure 7. Results from optimal static decisions.

to be too risky to be selected in all states of the world towards the end of the planning period.

Static policies remind one of a man who stands in the blazing sun in ice cold water up to his hips: the average temperature is fine, but the situation is far from ideal.

In summary, the inability to react to situations that have emerged at the time of decision making, which is inherent to static models, leads to poor solutions. The optimal static decisions result in dramatically high probabilities of underfunding in Setting 1. In Setting 2 and Setting 3, this shortcoming is demonstrated as clearly as in Setting 1: the high initial funding levels and the fact that the static decisions do take the entire planning period into account, result in an almost feasible strategy for Setting 2 and a feasible strategy for Setting 3.

5.3.4   Results from the Dynamic ALM Model

In relation to the myopic and static models, the ALM model should offer the best of both worlds: decisions make use of all information that is available at the time of decision making and the entire planning period is taken into account. The numerical results are in Figure 8.

The choice of initial asset mix for Setting 1 has been driven by the solvency requirement at the end of the first year: given the low initial funding level, the asset mixes with higher expected returns are too risky. Particularly interesting is the choice of asset mix for Setting 2. One would expect an initial funding level of 200% to be sufficiently high to bear the risk of the asset mix with the highest expected return, 100% stocks. This observation is correct. However, this decision is not so much driven by the short term probability of underfunding. Rather, it is driven by the implicit target of maintaining a funding level that, according to the heuristic, is optimal in view of the objective of minimising the present value of expected costs of funding. This also explains why the initial funding level selected in Setting 3 amounts to 251%: it is the minimum funding level for which all assets can be invested in stocks while satisfying chance constraints on arriving at optimal funding levels at the end of 1995.

At first sight, it may seem strange that it is optimal to maintain an unnecessarily high funding level when the expected return on investments is at most 10.3% whereas the discount rate for future contributions and restitutions is 15% per annum. This is caused by the fact that a relatively small additional investment enables one to select an asset mix which enhances the average return on the entire investment portfolio. Thus, the incremental revenues from an additional unit of investments is high in comparison with the cost of capital of this unit, the latter being reflected by the discount factor. As soon as the expected return on investments fails to increase when the initial asset level is increased (i.e., when 100% is invested in stocks), it is no longer attractive to increase the funding level: then, the incremental revenues are only 10.3% of the additional investment whereas future restitutions will be discounted by 15%.

With respect to the dependence of the optimal funding level of the remaining length of the planning period, it is interesting to compare the development of the average funding level in Setting 3 with that in Setting 2. The higher initial funding level in Setting 3 has been reduced rapidly by making large restitutions from 1996 until 2001. During the early years of the planning period, the ALM policy in Setting 2 is geared more towards maintaining the given funding level, which results in substantially higher contributions until 1998. As of 1998, the average funding levels, as well as average contributions and probabilities of underfunding in Setting 2, closely follow the development of the corresponding statistics of Setting 3. The slightly lower average funding level that is maintained in Setting 2, may reflect the greater flexibility in Setting 2 with respect to rises in regular contributions (5% of the pensionable earnings in Setting 2, compared with 3% in Setting 3).

In summary, the results from the ALM model do indeed reflect a trade-off between long and short term effects.

5.3.5 Results from the ALM Model in Comparison with Results from the Myopic Procedure

When we compare the outcomes from the myopic procedure with the results of the ALM model, it is clear that the results have been substantially affected. The graphs of the development of the average funding level and the probabilities of underfunding reflect the main difference between the two solutions. The myopic procedure apparently leads to decisions that reduce regular contributions to the maximum extent. And indeed, the present value of the contributions is lower than that of the ALM model. However, this is achieved at the cost of rendering infeasible strategies in all settings: the probabilities of underfunding are much higher than the maximum of 5% per annum. The ALM model does much better: for Setting 2 and for Setting 3, feasible solutions have been presented. Because no feasible solution could be found for Setting 1, the objective has been changed into minimising violations of the solvency constraints. The policy that has been determined by the ALM model resulted in an average probability of underfunding equal to 11% whereas the average probability of underfunding under the myopic policy amounted to 15%. Given these observations, it can be stated that the results of the policy from the ALM model are superior to those that have been obtained from the myopic model. This is due to the employment of the backward procedure in conjunction with the forward procedure.

5.3.6 Results from the ALM Model in Comparison with Results from Optimal Static Decisions

Neither the ALM model nor the static model resulted in a feasible policy for Setting 1. The extent to which solvency constraints are violated, however, differs greatly. Whereas the policy from the ALM model results in an average probability of underfunding of 11%, that of the static policy is 22%. Even more dramatic is the difference between present values of the expected costs of remedial contributions: 24,340M Dfl for the static model, compared with 699M Dfl for the ALM model. The difference between the present values of the expected total costs is only marginal. Again, the results from the static model suffer from the fact that the static decisions reflect a trade-off of their consequences in all states of the world.

The ALM model has determined a feasible policy for Setting 2 as well as for Setting 3. Let us ignore the small infeasibilities of the static policy in Setting 2. Instead, compare the present values of the expected costs of funding of the two policies in Settings 2 and 3. In both settings, the present value of the expected costs of funding associated with the policy from the ALM model is the smaller one, which shows the superiority of the ALM model over the static model.

However, there is another noteworthy observation to be made. Consider the composition of the costs of total contributions. In all settings, the remedial contributions of the static model are a multiple order of magnitude higher than those of the ALM model: they are respectively 30 times, 20 times and 60 times as high in Settings 1, 2 and 3. As a consequence, the level of total contributions to the fund according to the static model will vary enormously from one year to the next.

Finally, notice that the initial asset mixes and, in Setting 3, the initial asset level from the ALM model, are rather different from those of the static model. Thus, one cannot expect to achieve results that are comparable to those from the ALM model by following a static policy and making recourse decisions as time goes by.

6 Summary

The ALM model presented in this paper can be used to compute ALM strategies which specify investment decisions and contribution levels to be set under a wide range of future circumstances. The decisions are made so that the present value of expected contributions to the fund is minimal, subject to raising sufficiently stable annual contributions and the probability of underfunding at the end of each year being acceptably small when starting from the current situation, as well as from all future states of the world that the policy makers of the pension fund choose to take into account.

Computational results obtained from applying the model to data from one of the larger Dutch pension funds provide the following insights with respect to the proposed approach:

1. Dynamic ALM strategies lead to current decisions that are different

Initial asset mix

setting	Cash	Stocks	Property	Bonds
1	66	21	13	0
2	52	44	4	0
3	0	100	0	0

Feasibility of solutions

average excess probability of underfunding (violation of chance constraints)		
setting 1	setting 2	setting 3
7%	0%	0%

Average total contr.(% of pens.earnings)

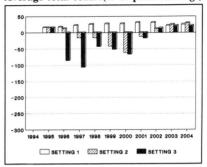

Average funding level

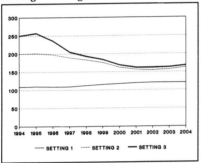

Probability of underfunding

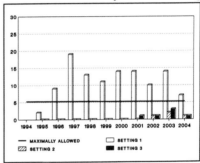

Present value (PV) of the costs of funding in mln. Dfl.

setting	Initial asset level	PV regular contributions	PV remedial contributions	PV total contributions	PV terminal surplus	PV total costs
1	18,075	6494	699	7194	2266	23,003
2	33,705	-1580	19	-1561	6788	25,356
3	42,083	-10,069	42	-10,027	7374	24,682

Figure 8. Results from the ALM model

from static policies. This applies to state-dependent as well as to time-dependent static policies.

2. In comparison to the static models, the employment of the ALM model results in strategies with lower funding costs, the probabilities of underfunding are substantially smaller and the magnitude of deficits, reflected by the costs of remedial contributions, has been reduced dramatically.

3. The favourable outcome of the comparison of policies determined by the ALM model with policies determined by optimal static decisions are to a major extent due to:

- the fact that probabilities of underfunding at intermediate points in time as well as at the planning horizon are endogenous to the model and have been modelled explicitly, and

- the dynamic character of the ALM model which enables the policies to react to situations that have emerged at the time of decision making and to reflect a correct trade-off between their longer term consequences and their short-term effects.

References

K. Aarssen (1992), *A Stochastic Programming Approach to Asset Liability* Management, Masters Thesis, Erasmus University Rotterdam.

P. Albrecht (1993), Normal and lognormal shortfall risk. In *Transactions of the 3rd International AFIR Colloquium*, 417–430.

P. Albrecht (1994), Shortfall returns and shortfall risk. In *Transactions of the 4th international AFIR Colloquium*, 87–110.

J.H.J. Almering, H. Bavinck and R.W. Goldbach (1988), *Analyse*, Delftse Uitgeversmaatschappij b.v., The Netherlands.

R.D. Arnott and P.L. Bernstein (1988), The right way to manage your pension fund, Harvard Business Review, January, p. 98.

L.J. Bain and M. Engelhardt (1987), *Introduction to Probability and Mathematical Statistics*, PWS Publishers.

C.G.E. Boender (1994), Mean variance models for asset liability management, submitted.

C.G.E Boender (1995), A hybrid simulation/optimization scenario model for asset liability management, submitted.

C.G.E. Boender and H.E. Romeijn (1991), The multidimensional Markov chain with specified asymptotic means and (auto-)covariances, *Communications in Statistics: Theory and Methods* **20** (1), 345–359.

P. Booth and A. Ong (1994), A simulation-based approach to asset allocation decisions. In *Transactions of the 4rd International AFIR Colloquium*, 217–239.

L.P. Bratley, B.L. Fox and L.E. Schrage (1987), *A Guide To Simulation*, 2nd edition, Springer Verlag.

P.L. Brocket, A. Charnes and Li Sun (1993), A chance constrained programming approach to pension plan management, *Contents of Actuarial Research Clearing House* **3** 324–357.

D.W. Bunn and A.A. Salo (1993), Forecasting with scenarios, *European Journal of Operational Research*, 291–303.

D.R. Cariño, T. Kent, D.H. Myers, C. Stacy, M. Sylvanus, A.C. Turner, K. Watanabe and W.T. Ziemba (1994), The Russell–Yasuda Kasai model: an asset liability model for a Japanese insurance company using multistage stochastic programming, *Interfaces* **24** January–February, 29–49. Reprinted in this volume, 609–633.

V. Chvátal (1983), *Linear Programming*, Freeman and Company.

A. Charnes and W.W. Cooper (1959), Chance-constrained programming, *Management Science* **6** 3–80.

M.A.H. Dempster and X. Corvera Poiré (1995), *Stochastic Programming: A New Approach to Asset/Liability Management*, Summer school Institute of Actuaries.

C.L. Dert (1989), Irrationeel Beleggersgedrag en Outperformance Strategieën, in Portefeuille-Strategisch beleggen in het Verre Oosten, Pacific Investments Research Institute, Amsterdam, The Netherlands.

C.L. Dert (1995), *Asset Liability Management for Pension Funds, A Multistage Chance Constrained Programming Approach*, Ph.D. thesis, Erasmus University Rotterdam, The Netherlands.

C.L. Dert and A.H. Rinnooy Kan (1991), Fixed income asset liability management. In *Transactions of the 2nd International AFIR Colloquium* **3** 17–20 April 1991, 285–300

F.X. Diebold and G. Rudebusch (1989), Long memory and persistence in aggregate output, *Journal of Monetary Economics* **24** 189–209.

M.V. Dothan (1990), *Prices in Financial Markets*, Oxford University Press.

D. Duffie (1988), *Security Markets: Stochastic Models*, Academic Press.

R. Entriken and G. Infanger (1990), *Decomposition and Importance Sampling For Stochastic Linear Models*, IBM Research Report RC–15545.

P. Ford (1991), Cashflow matching using modified linear programming. In *Transactions of the 2nd international AFIR Colloquium* **3** 301–322.

J.M.G. Frijns, and J.H.W. Goslings (1989), Matching voor het pensioen bedrijf, *Economische statistische berichten*, September 868–871.

R.M.C. Guimaraes, B.G. Kingsman and S.J. Taylor (1989) (eds.), *A Reappraisal of the Efficiency of Financial Markets*, Springer Verlag.

P.W. Glynn and D.L. Iglehart (1989), Importance sampling for stochastic simulations, *Management Science* **35** (11), November, 1367–1391.

B. Golub, M. Holmer, R. McKendall, L. Pohlman and S.A. Zenios (1993), A stochastic programming model for money management, to appear in *European Journal of Operations Research*.

M. Griffin (1993), A new rationale for the different asset allocation of Dutch and UK pension funds. In *Transactions of the 3rd International AFIR Colloquium*, 575–588.

M. Haigigi and B. Kluger (1987), Safety first: an alternative performance measure, *Journal of Portfolio Management*, Summer.

M.R. Holmer, R. McKendall, C. Vassiadou-Zeniou and S.A. Zenios (1993), *Dynamic models for fixed-income portfolio management under uncertainty*, Report 93–03, University of Cyprus.

M.R. Hardy (1993), Stochastic simulation in life office solvency assessment, *Journal of the Institute of Actuaries* **120** (1), 131–152.

W.V. Harlow (1991), Asset allocation in a downside-risk framework, *Financial Analysts Journal*, September-October, 28–40.

R.V. Hogg and A.T. Craig (1978), *Introduction to Mathematical Statistics*, 4th edition, Macmillan.

G. Infanger (1992), Monte Carlo importance sampling within a Benders decomposition algorithm for stochastic linear programs, *Annals of Operations Research* **39** 69–95.

J.E. Ingersoll Jr. (1987), *Theory of Financial Decision Making*, Studies in Financial Economics, Rowan & Littlefield.

J. Janssen and K. Manca (1994), Semi-Markov modelization for the financial management of pension funds. In *Transactions of the 4th International AFIR Colloquium*, 1369–1387.

G.G. Judge, R.C. Hill, W.E. Griffiths, H. Lütkepohl and T.C. Lee (1985), *Introduction to the Theory and Practice of Econometrics*, Wiley.

G.G. Judge, R.C. Hill, W.E. Griffiths, H. Lütkepohl and T.C. Lee (1988), *The Theory and Practice of Econometrics*, Wiley.

P. Kall (1976), *Stochastic Linear Programming, Econometrics and Operations Research*, Springer.

P. Klaassen (1994), *Stochastic Programming Models for Interest-Rate Risk Management*, Ph.D. thesis, M.I.T., Sloan School of Management.

J.P.C. Kleijnen (1971), *Variance Reduction Techniques in Simulation*, Ph.D. thesis, Katholieke Hogeschool Tilburg, Tilburg, The Netherlands.

J.P.C. Kleijnen (1974), *Statistical Techniques in Simulation, Part 1*, Marcel Dekker.

J.P.C. Kleijnen (1975), *Statistical Techniques in Simulation, Part 2*, Marcel Dekker.

M.I. Kusy, and W.T. Ziemba (1986), A bank asset and liability model, *Operations Research* **34** (3), 356–376.

M.L. Leibowitz (1986), *Total Portfolio Duration: a New Perspective on Asset Allocation*, Salomon Brothers, Inc., February.

M.L. Leibowitz, and R.D. Henriksson (1987), *Portfolio Optimization within a Surplus Framework, a New Perspective on Asset Allocation*, Salomon Brothers, Inc., April.

P. Ludvik (1994), Investment strategy for defined constribution plans. In *Transactions of the 4th International AFIR Colloquium*, 1389–1400.

I.J. Lustig, J.M. Mulvey and T.J. Carpenter (1990), *Operations Research* **39** (5), 757–772.

H.M. Markowitz (1952), Portfolio selection, *The Journal of Finance*, **7** (1), 77–91.

H.M. Markowitz (1959), *Portfolio Selection: Efficient Diversification of Investments*, Cowles Foundation for Research in Economics at Yale University, Monograph 16, Yale University Press.

M. Minoux (1986), *Mathematical Programming, Theory and Algorithms*, Wiley.

J.M. Mulvey (1994), An asset liability investment system, *Interfaces* **24** May-June, 22–33.

J.M. Mulvey, and H. Vladimirou (1988), Solving multistage stochastic networks: an application of scenario aggregation, Report SOR–88–1, Dept. of Civil Engineering and Operations Research, School of Engineering and Applied Science, Princeton University.

J.M. Mulvey and H. Vladimirou (1991), Applying the progressive hedging algorithm to stochastic generalized networks, *Annals of Operations Research*, **31** 399–424.

J.F.J. de Munnik (1992), *The Valuation of Interest Rate Derivative Securities*, Ph.D. thesis Erasmus University, Rotterdam.

T.H. Naylor, J.L. Balintfy, D.S. Burdick and K. Chu (1967), *Computer Simulation Techniques*, Wiley.

M. Ooms (1993), *Empirical Vector Autoregressive Modeling*, Ph.D. thesis, Erasmus University Rotterdam.

C. Petersen (1990a). Pensioenen in West-Europa, *Economische Statistische Berichten*, 3 January, 10–13.

C. Petersen (1990b), *Pensioenen: uitkeringen, financiering en beleg gingen*, Stenfert Kroese, Leiden–Antwerpen

R.T. Rockafellar and R.J.-B. Wets (1991), Scenarios and policy aggregation in optimization under uncertainty, *Mathematics of Operations Research, February* **16** (1), 119–147.

W.F. Sharpe, and L.G. Tint (1990), Liabilities – a new approach, *Journal of Portfolio Management*, Winter, 5–10.

M. Sherri, (1992), Portfolio selection and matching: a synthesis, *Journal of the Institute of Actuaries* **119** (I), 87–105.

M. Sherris (1993), 221 Portfolio selection models for life insurance and pension fundors. In *Proceedings of the 3rd AFIR International Colloquium*, 915–930.

C.A. Sims (1980), Macroeconomics and reality, *Econometrica* **48** (1), 1-48.

A. Smink (1994), Numerical examination of asset liability management strategies. In *Transactions of the 4th International AFIR Colloquium*, 969–997.

F.A. Sortino and R. van der Meer (1991), Downside risk, *Journal Portfolio Management*, Summer, 27–31.

S.M. Ross (1985), *Introduction to Probability Models*, Academic Press.

L.D. Taylor (1974), *Probability and Mathematical Statistics Statistics*, Harper & Row,.

R.J.-B. Wets (1987), Large-scale linear programming techniques in stochastic programming. In *Numerical Methods for Stochastic Optimization*, Y. Ermoliev and R. Wets, (eds.) Springer Verlag.

R.J.-B. Wets (1989a), Stochastic programming. In *Handbooks in operations research and management science*, G.L. Nemhauser, A.H.G. Rinnooy Kan and M.J. Todd (eds.). Vol 1: Optimization, Elsevier.

R.J.-B. Wets (1989b), The aggregation principle in scenario analysis and stochastic optimization. In *Algorithms and Model Formulations in Mathematical Programming*, S.W. Wallace, (ed.) Springer Verlag, 91–113.

A.D. Wilkie (1985), Portfolio selection in the presence of fixed liabilities: a comment on the matching of assets to liabilities, *Journal of the Institute of Actuaries* **112** 229–277.

A.D. Wilkie (1986a), A stochastic investment model for actuarial use, *Transactions of the Faculty of Actuaries* **39** 341–403.

A.D. Wilkie (1986b), Some applications of stochastic investment models, *JIASS* **29** 25–51.

A.D. Wilkie, J.A. Tilley, T.G. Arthur and R.S. Clarkson (1993), This house believes that the contribution of actuaries to investment could be enhanced by the work of financial economists, *Journal of the Institute of Actuaries* **120** (III), 393–414.

A.D. Wilkie (1995), More on a stochastic asset model for actuarial use. Presented to the Institute of Actuaries, 24 April 1995.

A.J. Wise (1984a), A theoretical analysis of the matching of assets to liabilities, *Journal of the Institute of Actuaries* **111** 375–402.

A.J. Wise (1984b), The matching of assets to liabilities, *Journal of the Institute of Actuaries* **111** 445–501.

A.J. Wise (1987a), Matching and portfolio selection: Part 1, *Journal of the Institute of Actuaries* **114** 113–133.

A.J. Wise (1987b), Matching and portfolio selection: Part 2, *Journal of the Institute of Actuaries* **114** 551–568.

A. Zellner (1962), An efficient method of estimating seemingly unrelated regressions and tests of aggregation bias, *Journal of the American Statistical Association* **57** 348–368.

W.T. Ziemba and R.G. Vickson (eds.) (1975) *Stochastic Optimization Models in Finance*, Academic Press.

Asset and Liability Management under Uncertainty for Fixed Income Securities[*]

Stavros A. Zenios

HERMES Laboratory for Financial Modeling and Simulation,
Decision Sciences Department, The Wharton School,
University of Pennsylvania, Philadelphia, PA 19104, USA

Short-sighted asset/liability strategies of the seventies left financial intermediaries – banks, insurance and pension fund companies, and government agencies – facing a severe mismatch between the two sides of their balance sheet. A more holistic view was introduced with a generation of *portfolio immunization* techniques. These techniques have served the financial services community well over the last decade. However, increased interest rate volatilities, and the introduction of complex interest rate contingencies and asset-backed securities during the same period, brought to light the shortcomings of the immunization approach. This paper describes a series of (optimization) models that take a global view of the asset/liability management problem using interest rate contingencies. Portfolios containing *mortgage-backed securities* provide the typical example of the complexities faced by asset/liability managers in a volatile financial world. We use this class of instruments as examples for introducing the models. Empirical results are used to illustrate the effectiveness of the models, which become increasingly more complex but also afford the manager increasing flexibility.

1. Introduction: The problem and its applications

Government agencies, such as the Federal National Mortgage Association (Fannie Mae), suffered severe losses during the early 1980's: the agencies had issued short-term non-callable bonds in the early 1970's to finance the purchase of long-term assets. However, the maturity mismatch of assets and liabilities exposed the agencies to substantial interest rate risk. As rates rose substantially in the latter part of the decade, Fannie Mae found itself with maturing liabilities, while their assets had a long remaining time to maturity and were priced at a fraction of their original value. Similar problems were faced by insurance carriers: these companies used short-term liabilities (like their Guaranteed Investment Contracts) to fund long-term assets. In the upward sloping yield curve of the time, this strategy was deemed reasonable and substantial gains were expected. However, as rates rose they found themselves in the same predicament as the government agencies.

---

[*]Reprinted, with permission, from *Annals of Operations Research* 1995. Copyright 1995, J.C. Baltzer AG, Science Publishers

While it is easy to criticize these mistakes after the fact, we should not forget that, at the time, interest rates were regulated and had remained quite stable for a long period of time. In the aftermath of these shocks, however, financial intermediaries took a more holistic view of the asset/liability management problem. A policy of *duration matching* the assets and liabilities was instituted among insurance and pension fund companies, banks and government agencies. See, for example, Christensen and Fabozzi [5], Holmer [13] or Platt [21].

Portfolio immunization – i.e. selecting portfolios with duration matched assets and liabilities – was instrumental in reducing the gap between the two sides of the balance sheet created by the short-sighted policies of the 1970's. However, as these techniques rose to widespread use throughout the 1980's, their shortcomings started to become apparent. A whole new generation of models started to emerge that could more fully cope with the volatile interest rate environment of the 1990's while dealing with the spectrum of complex financial instruments that were introduced throughout this period.

The objective of this paper is to describe a series of three such models. We use as a base case the problem of managing portfolios containing mortgage-backed securities (abbreviated: MBS). These securities – being some of the most complex instruments available in the financial markets – provide the framework for discussing a complete spectrum of issues that relate to the management of fixed-income portfolios under uncertainty. In particular, they are not only sensitive to changes in interest rates, but they are also volatile due to the embedded call option: A homeowner has the option to prepay the outstanding balance of her mortgage, with no penalty, and hence call the mortgage security. Other fixed-income securities exhibit similar characteristics. We mention, for example, callable bonds issued by corporations, single premium deferred annuities offered by insurance companies, options on bonds, etc.

The remainder of this section characterizes the complexities of the problem and discusses specific applications. Section 2 classifies and specifies three asset/ liability portfolio optimization models. Calculations of the input data required in order to operationalize the models can be obtained using extensions of standard pricing models that are grounded on prevailing financial theories. This is the topic of section 3. Empirical evidence on the performance of the models, and some discussion on further validation, are considered in section 4. The final section, section 5, provides some critical analysis of the models. The models developed here have been applied in several corporate settings for managing portfolios of mortgage-backed securities and callable bonds. These experiences are published elsewhere [10,23,7].

1.1.   THE PROBLEM

We loosely define the problem addressed in this paper as follows:

Construct a portfolio of fixed-income securities whose performance measures will remain invariant under a wide range of uncertain scenarios.

For now, we leave unspecified what we mean by *performance measures* and the precise nature of uncertainty. The key idea is to decide what goals we want our portfolio to achieve, specify measures that indicate that the goals are achieved, and make sure that these goals are still met when the economic environment changes. The three models we introduce in the next section allow the portfolio managers to specify increasingly more complex goals, and ensure that these goals are met for increasingly more complex scenarios.

1.2. APPLICATIONS

The precise goal of the portfolio manager depends on the underlying application. We describe here three practical applications where one needs to deal with fixed-income securities and their inherent uncertainties:

Indexation: Passive portfolio managers would like to build a portfolio of fixed-income securities that will track a prespecified index. For example, Shearson–Lehman and Salomon Brothers publish a monthly mortgage index that is (presumably) indicative of the overall state of this segment of fixed-income markets. Investors who wish to invest in mortgages may be satisfied if their portfolio closely tracks the index. The performance measure of such a portfolio is the difference in return between the portfolio and the index. This difference has to be very small, for all changes in the index caused by interest rate movements and by variations in prepayment activity.

Liability payback: Insurance and pension fund companies are typically heavily exposed to MBS. These instruments are considered as an investment for paying back a variety of the liabilities held by these institutions. The goal of the portfolio manager is to construct a portfolio of MBS that will pay the future stream of liabilities. Uncertainty here appears once more in the form of interest rate changes and changes in the timing of payments from the MBS. Furthermore, the timing of the liability stream may also be subject to uncertain variations: For example, the timing of payments to holders of single premium deferred annuities (SPDA) may change as annuitants exercise the option to lapse.

Debt issuance: Government agencies, such as Fannie Mae and Freddie Mac, fund the purchase of fixed income assets (typically mortgages) by issuing debt. The problem of a portfolio manager is to decide which type of debt – maturity, yield, call-option – to issue in order to fund the purchase of a specific set of assets. Of course, there is no reason to assume that the assets have been prespecified: The model may choose an appropriate asset mix from a large universe of fixed income securities. The timings of both assets and liabilities may be uncertain in this application. The goal of the portfolio manager is to ensure that the payments against the issued debt will be met from the available assets, irrespective of the timing of cash flows and fluctuations in interest rates.

2.   Structured asset/liability management models

We classify the asset/liability management models into: (1) static, (3) single-period, stochastic, and (3) multiperiod, dynamic and stochastic. It is important to understand how the models address increasingly more complex aspects of the asset/liability management problem. Only then can the portfolio manager decide which model may be more appropriate for the application at hand. Of course, this decision has to be weighted against the increasing complexity – both conceptual and computational – of the models.

Static models: Such models hedge against small changes from the current state of the world. For example, a term structure is input to the model which matches assets and liabilities under this structure. Conditions are then imposed to guarantee that if the term structure deviates somewhat from the assumed value, the assets and liabilities will move in the same direction and by equal amounts. This is the fundamental principle behind *portfolio immunization*. See, for example, Christensen and Fabozzi [5] for a discussion of the finance-theoretic principles behind immunization, and Dahl et al. [8] for operational models.

Single-period, stochastic models: A static model does not permit the specification of a stochastic process that describes changes of the economic environment from its current status. However, modern finance abounds with theories that describe interest rates, and other volatile factors, using stochastic processes; see, e.g., Ingersoll [15]. Stochastic differential calculus is often used to price interest rate contingencies. For complex instruments, analysts resort to Monte Carlo simulations, an idea pioneered by Boyle [3] for options pricing. See, for example, Hutchinson and Zenios [14] for its application to the pricing of mortgage securities. A stochastic asset/liability model describes the *distribution* of returns of both assets and liabilities in the volatile environment, and ensures that movements of both sides of the balance sheet are highly correlated. This idea is not new: Markowitz pioneered the notion of risk management for equities via the use of correlations in his seminal papers [18, 17]. However, for the fixed-income world this approach has only recently received attention. It has been formalized by Mulvey and Zenios [20], and was applied at Fannie Mae by Holmer [13].

Multiperiod, dynamic and stochastic models: A stochastic model, as outlined above, is *myopic*. That is, it builds a portfolio that will have a well behaved distribution of error (error = asset return – liability return) under the specified stochastic process. However, it does not account for the fact that the portfolio manager is likely to rebalance the portfolio once some surplus is realized. Furthermore, as the stochastic process evolves across time, different portfolios may be more appropriate for capturing the correlations of assets and liabilities. The single-period model may recommend a conservative strategy, while a more aggressive approach would be justified once we explicitly recognize the manager's ability to rebalance the portfolio.

What is needed is a model that explicitly captures both the stochastic nature of the problem, but also the fact that the portfolio is managed in a dynamic, multiperiod context. Mathematical models under the general term of *stochastic programming with recourse* provide the framework for dealing with this broad problem. Stochastic programming has a history almost as long as *linear programming* – Dantzig [9], Wets [22]. However, it was not until the early seventies – Bradley and Crane [4] – that its significance for portfolio management was realized. With the recent advances in high-performance computing, this approach has been receiving renewed interest from the academic literature – Mulvey and Vladimirou [19], Hiller and Eckstein [12], Zenios [24], and Golub et al. [10]. We are also aware of research in several industrial settings for the deployment of such models in practice.

We now continue with a mathematical description of a model from each class. The formulations are general. Our goal is to describe the key components of the models, and then discuss – in section 3 – the computation of the data requirements for each. In order to operationalize each model for the application mentioned earlier, additional specifications are needed. We do not completely specify the details, since those will only distract from the general principles we want to convey.

2.1. PROBLEM FORMULATION

We are given a universe of fixed-income securities, indexed by a set J, with market prices $\{P_{0j}\}$, and a stream of liabilities $\{L_t\}$, where t denotes a time index drawn from a discrete set T. Given is also a term structure, specified by a vector of forward rates $\{r_t\}$, $t \in T$. The problem of the portfolio manager is to decide the holdings of each security x_j in a portfolio that will match the assets with the liability stream.

For the stochastic models, we also need to specify a set of scenarios S. We assume discrete and equiprobable scenarios. The scenarios can be very general: they can represent a series of term structures drawn from some stochastic process of interest rates, or they can represent levels of prepayment activity for the mortgage securities, or they can represent levels of the liability stream, and so on. Whenever a model parameter is super-scripted by an index $s \in S$, it is understood that the value of the parameter is scenario dependent. In this respect, we will use C_{jt}^s to denote the cash flow generated by security $j \in J$ (per unit face value), and r_t^s to denote the discount rate at period $t \in T$, under scenario $s \in S$.

The interest rate scenarios can be calculated using a variety of term structure models, such as the diffusion process of Cox et al. [6], or binomial lattice models such as the one proposed by Black et al. [2]. These models are designed to generate term structure scenarios that are consistent (i.e. arbitrage free) with the treasuries' yield curve and its volatility.

Most fixed-income securities, however, cannot be priced using the same discount rates implied by the treasuries' curve. In particular, the price of the security has to reflect the credit, liquidity, default and prepayment risks associated with this instrument. In order to value the risks associated with MBS, we compute an *option adjusted premium* (OAP). The OAP methodology estimates the multiplicative adjustment factor for the treasuries' rates that will equate today's (observed) market price with the "fair" price obtained by applying the expectations hypothesis, see, e.g., Babbel and Zenios [1]. The discrepancy between the market price and the theoretical price is due to the various risks that are present in most fixed-income securities, but are not present in the treasuries' market. Hence, this analysis will price the risks.

The OAP for a given security is estimated based on the current market price P_{0j}. In particular, it is the solution of the following nonlinear equation in ρ_j:

$$P_{0j} = \frac{1}{|S|} \sum_{s=1}^{|S|} \sum_{t=0}^{T} \frac{C_{jt}^s}{\prod_{i=1}^{t}(1 + \rho_j \cdot r_i^s)}. \tag{1}$$

The computed risk premium appears in several of the models in the following sections.

2.2.   A STATIC APPROACH: DURATION MATCHING

Given the term structure, a stream of projected cash flows for the fixed-income security and a stream of liabilities, we can build a *dedicated* portfolio. That is, a portfolio of least cost – or maximum yield – of fixed-income assets that will match the stream of liabilities. Let C_{jt} denote the cash flow generated by security j at period t. This stream is projected, conditional on the current term structure. We can write the following optimization model:

$$\underset{x}{\text{Minimize}} \quad \sum_{j \in J} P_{0j} x_j \tag{2}$$

$$\text{subject to} \quad \sum_{j \in J} \left(\sum_{t \in T} \frac{C_{jt}}{\prod_{i=1}^{t}(1 + r_i)} \right) x_j \geq \sum_{t \in T} \frac{L_t}{\prod_{i=1}^{t}(1 + r_i)}, \tag{3}$$

$$x_j \geq 0. \tag{4}$$

This model will choose the least-cost portfolio, with the property that the present value of the portfolio cash flows will be at least equal to the present value of the liabilities. If the timing and magnitude of assets and liabilities do not change, nor the discount factors, then it is easy to see that the portfolio will ensure timely payments against the liabilities. (It is assumed in this model that unlimited borrowing is allowed, at all time periods, at the prevailing discount rate r_i. The model can easily be modified to eliminate borrowing, or permit limited borrowing at a discount rate greater than r_i.)

To account for the stochasticity of the security cash flow, and the volatility of the term structure, the model can be extended to match the sensitivities of both assets and liabilities to such stochasticity. For example, the *duration* of a security measures the sensitivity of its price to small parallel shifts to the term structure. Hence, we extend the model to match the duration of assets and liabilities. In order to capture the complex dependency of the cash flows of a fixed-income security to changes in the term structure, we use a Monte Carlo simulation procedure:

MONTE CARLO SIMULATION FOR OPTION ADJUSTED DURATION CALCULATIONS

Step 0: Initialize the stochastic process of interest rates, based on the current term structure, and use it to compute the option adjusted premia p_j for all securities, implied by the current market prices P_{0j} (cf. equation (1)).

Step 1: Shift the term structure by -50 basis points, and recalibrate the stochastic process of interest rates.

Step 2: Sample interest rate paths $\{r_i^{-s}\}$ from the stochastic process calibrated in step 1, and use the security cash flow projection model to compute option adjusted prices:

$$P_j^- = \frac{1}{|S|} \sum_{s=1}^{|S|} \sum_{t=0}^{T} \frac{C_{jt}^s}{\prod_{i=1}^{t}(1 + p_j \cdot r_i^{-s})}.$$ (5)

Step 3: Shift the term structure by $+50$ basis points, and recalibrate the stochastic process of interest rates.

Step 4: Sample interest rate paths $\{r_i^{+s}\}$ from the stochastic process calibrated in step 3, and use the security cash flow projection model to compute option adjusted prices:

$$P_j^+ = \frac{1}{|S|} \sum_{s=1}^{|S|} \sum_{t=0}^{T} \frac{C_{jt}^s}{\prod_{i=1}^{t}(1 + p_j \cdot r_i^{+s})}.$$ (6)

Step 5: The *option adjusted duration* of the security is given by

$$\Delta_j = \frac{P_j^+ - P_j^-}{100}$$ (7)

and the *option adjusted convexity* by

$$\Gamma_j = \frac{P_j^+ - 2P_{0j} + P_j^-}{50^2}.$$ (8)

An immunized portfolio will match the present values and durations of both assets and liabilities. If Δ_l is the liability duration, we have the following linear program:

$$\text{Minimize}_{x} \quad \sum_{j \in \mathcal{J}} P_{0j} x_j \tag{9}$$

$$\text{subject to} \quad \sum_{j \in \mathcal{J}} \left(\sum_{t \in T} \frac{C_{jt}}{\prod_{r=1}^{t}(1 + r_t)} \right) x_j \geq \sum_{t \in T} \frac{L_t}{\prod_{r=1}^{t}(1 + r_t)}, \tag{10}$$

$$\sum_{j \in \mathcal{J}} \Delta_j x_j = \Delta_l, \tag{11}$$

$$x_j \geq 0. \tag{12}$$

The model could be extended further to match asset/liability convexities. Matching higher derivatives is also possible. At the limit, the derivative matched portfolio will be identical to a cash flow matched portfolio.

2.3.   A STOCHASTIC APPROACH: CAPTURING CORRELATIONS

Portfolios of fixed-income securities have traditionally been managed using the concepts of duration and convexity matching of the previous section. With the increased volatility of the term structure, following monetary deregulation in the late 70's, this approach appears overly simplistic. The difficulty is further exacerbated with the constant stream of innovations in this market. In a recent report, Mulvey and Zenios [20] observed that returns from corporate bonds have outperformed those of equities, but at the same time exhibited similar or higher volatility. That paper went further to argue that such volatile instruments should be managed in the framework of risk-return tradeoff. In this respect, the tradition of models started with Markowitz's seminal work [18] has much to offer to the managers of fixed-income portfolios. This approach has already received attention from practitioners (Holmer [13]).

In this section, we introduce a stochastic model for managing portfolios of MBS. The model explicitly recognizes the volatility of MBS prices, and the correlation of prices in a portfolio, and develops the tradeoffs between increased return and increased volatility. The optimization model we adopt is based on the *mean-absolute deviation* (MAD) framework of Konno and Yamazaki [16]. An MAD model is suitable for the fixed-income securities with embedded options that exhibit highly asymmetric distributions of return.

One of the challenges in applying the MAD model – or any other risk-return model for that matter – to fixed-income securities is that these instruments are vanishing, with a fixed term to maturity. Furthermore, the payout function of several kinds of fixed-income securities is path dependent. Hence, at any point in time we have only one observation of price variations. Therefore, we can not resort to the statistical analysis of historical data in order to capture the volatility and correlation of returns. Furthermore, the "dividends" obtained from the security – in the form of principal and interest payments, as well as prepayment, lapse or call of outstanding balance – can not be reinvested in the same security. Hence, we need to resort to Monte Carlo simulation of the short-term risk-free rates in order to obtain holding period returns of the fixed-income security during the target holding period. The Monte Carlo simulation procedure is explained in section 3. For now, we assume that a random vector of holding period returns has been generated for each security. Let $\{R_j\}$ denote this vector random variable and let $\overline{R}_j = \mathcal{E}(R_j)$ denote its expected value. Also, let $x = \{x_j\}$ denote the composition of the portfolio, which contains a deterministic liability with return ρ. The return of the portfolio is $R = \sum_{j \in J} R_j x_j + \rho$. The *mean-absolute deviation* of return of this portfolio is defined by

$$w(R) = \mathcal{E}\{|R - \mathcal{E}(R)|\}, \tag{13}$$

where $\mathcal{E}(\cdot)$ denotes expectation. Assume now that a sample of the random variables R_j is available. That is, R_j takes the value $\{R_j^s\}$ for some scenario $s \in S$, and we assume for simplicity that all scenarios in S are equiprobable. Then, an unbiased estimate for the mean-absolute deviation of return of the portfolio is

$$w(R) = \mathcal{E}\{|R - \mathcal{E}(R)|\} \tag{14}$$

$$= \mathcal{E}\left\{\left|\sum_{j \in J}(R_j - \overline{R}_j)x_j\right|\right\} \tag{15}$$

$$= \frac{1}{|S| + |J|} \sum_{s \in S}\left|\sum_{j \in J}(R_j^s - \overline{R}_j)x_j\right|. \tag{16}$$

The mean-absolute deviation (MAD) model is written as

$$\text{Minimize} \quad w(R) \tag{17}$$

$$\text{subject to} \quad \sum_{j \in J}\overline{R}_j x_j \geq \rho, \tag{18}$$

$$\sum_{j \in J} x_j = 1, \tag{19}$$

$$0 \leq x_j \leq u_j, \quad \text{for all } j \in J. \tag{20}$$

This model can be reformulated into a linear programming problem. (This is a standard reformulation for minimizing absolute values. A minimand $|x|$ is replaced by y, where y is constrained as $y \geq x$ and $y \geq -x$.) In doing so, it is also possible to differentially penalize the upside from the downside deviation of the portfolio return from its mean. Let μ_d and μ_u denote penalty parameters for the downside and upside errors, respectively. Then the MAD model can be written as the following linear program:

$$\text{Minimize} \quad \frac{1}{|S|+|J|} \sum_{s \in S} y^s \tag{21}$$

$$\text{subject to} \quad y^s + \mu_d \sum_{j \in J}(R_j^s - \overline{R}_j)x_j \geq 0 \qquad \text{for all } s \in S, \tag{22}$$

$$y^s - \mu_u \sum_{j \in J}(R_j^s - \overline{R}_j)x_j \geq 0, \qquad \text{for all } s \in S. \tag{23}$$

$$\sum_{j \in J} \overline{R}_j x_j \geq \rho, \tag{24}$$

$$\sum_{j \in J} x_j = 1, \tag{25}$$

$$0 \leq x_j \leq u_j, \qquad \qquad \text{for all } j \in J. \tag{26}$$

2.4.   A MULTIPERIOD, DYNAMIC APPROACH: STOCHASTIC OPTIMIZATION

The multiperiod, stochastic model captures the dynamics of the following situation:

> A portfolio manager must make investment decisions facing an uncertain future. After these *first-stage* decisions are made, a realization of the uncertain future is observed, and the manager determines an optimal *second-stage* (or recourse) decision. The objective is to maximize expected utility of final wealth.

The first-stage decision deals with the purchase of a portfolio of fixed-income securities. The uncertain future is the level of interest rates and the cash flows that will be obtained from the portfolio. The second-stage decision deals with borrowing (resp. lending) decisions when the fixed-income cash flows lag (resp. lead) the target liabilities. Decisions to rebalance the portfolio at some future time period(s), by purchasing or selling securities, are also included.

2.4.1. Establishing notation

We will be using the following notation. First, parameters of the model:

T : discretization of the planning horizon, $T = \{1, 2, 3, \ldots, \overline{T}\}$ and $T_0 = \{0, 1, 2, 3, \ldots, \overline{T}\}$. $\overline{T}$ denotes the end of the planning horizon.

b_j : initial holdings (in face value) of instrument $j \in J$ and b_0 initial holdings in a riskless asset (e.g., cash).

r_t^s : one-year forward interest rate at time period $t \in T_0$ under scenario $s \in S$.

spr : spread between the lending and borrowing rates.

pf_{jt}^s : cash flow generated from instrument $j \in J$ at time period $t \in T$ under scenario $s \in S$, expressed as a percentage of face value. It includes principal and interest payments of the fixed-income security, as well as cash flows generated due to defaults, prepayments, lapse, exercise of the embedded call option, etc.

ξ_{jt}^s : price per unit of face value of security $j \in J$ sold at period t under scenario s. The cost of the transaction is subtracted from the actual price to obtain this coefficient. The price at $t = 0$ is independent of the scenario and is denoted by ξ_{j0}.

ζ_{jt}^s : price per unit of face value of security $j \in J$ purchased at period t under scenario s. The cost of the transaction is added to the actual price to obtain this coefficient. The price at $t = 0$ is independent of the scenario and is denoted by ζ_{j0}.

L_t : liability due at time $t \in T$. It is assumed to be independent of the realized scenario, although this assumption can easily be relaxed.

Define now the model variables:

x_j : first-stage variable, denoting the face value of instrument $j \in J$ purchased at the beginning of the planning horizon (i.e. at $t = 0$).

x_{jt}^s : second-stage variable, denoting the face value of instrument $j \in J$ purchased at time period t under scenario s.

y_j : first-stage variable, denoting the face value of instrument $j \in J$ sold at the beginning of the planning horizon (i.e. $t = 0$).

y_{jt}^s : second-stage variable, denoting the face value of instrument $j \in J$ sold at time period t under scenario s (i.e. at $t = 0$).

z_{jt}^s : second-stage variable, denoting the face value of instrument $j \in J$ in the portfolio at time period $t \in T$ under scenario s. z_{j0} denotes the starting composition of the portfolio, after first-stage decisions have been made, and is independent of the scenarios.

w_{jt}^s : second-stage accounting variable indicating the cash flow generated by security j at time period t under scenario s.

y_t^{-s} : second-stage recourse variable indicating the amount owed at time period $t + 1$ due to borrowing decisions made at time period t under scenario s.

y_t^{+s} : second-stage recourse variable indicating the surplus invested in the riskless asset at time period t.

$U(WT^s)$: denotes the utility of terminal wealth realized under scenario s. Appropriate choices of utility functions are, for example, the isoelastic utility $-1/\gamma(WT^s)^\gamma$ – used in the models of Grauer and Hakansson [11].

2.4.2. Defining the model

The model can now be formulated as follows:

Maximize $\quad \dfrac{1}{|S|}\sum_{s\in S}U(WT^s)$

subject to $\quad z_{j0}+y_j-\dfrac{x_j}{\zeta_{j0}}=b_j,$ $\qquad\qquad\qquad\qquad\qquad\qquad j\in\mathcal{J},$

$$y_0^{+s}+\sum_{j\in\mathcal{J}}x_j-\sum_{j\in\mathcal{J}}(1-\xi_j)y_j-\frac{1}{(1+r_0^s+spr)}y_0^{-s}=b_0,\quad s\in S,$$

$$z_{jt-1}^s+x_{jt}^s-w_{jt}^s-z_{jt}^s-y_{jt}^s=0,\qquad\qquad t\in T, j\in\mathcal{J}, s\in S,$$

$$w_{jt}^s-pf_{jt}^s z_{jt-1}^s=0,\qquad\qquad\qquad\qquad t\in T, j\in\mathcal{J}, s\in S,$$

$$\xi_{jt}^s y_{jt}^s+\sum_{j\in\mathcal{J}}w_{jt}^s-\frac{x_{jt}^s}{\zeta_{jt}^s}+(1+r_{t-1}^s)y_{t-1}^{+s}-y_{t-1}^{-s}$$

$$\qquad\qquad+\frac{1}{(1+r_t^s+spr)}y_t^{-s}-y_t^{+s}=L_t,\qquad\qquad t\in T, s\in S,$$

$$WT^s=\sum_{j\in\mathcal{J}}z_{j\overline{T}}^s\xi_{j\overline{T}}^s-y_{\overline{T}-1}^{-s}.$$

The first constraint of this mathematical program reflects first-stage decision and is deterministic. Subsequent constraints depend on the realized scenario, as well as the first-stage decision. The terminal wealth WT^s is computed by accumulating the total surplus net any outstanding debt at the end of the planning horizon, and liquidating any securities that remain in the portfolio. The complete model has the dual block-angular structure of two-stage stochastic programs with recourse. See figure 1.

3.   Model data: Holding-period returns for fixed-income securities

We now turn our attention to the data required in order to implement the models. The portfolio immunization models need preset value, duration and convexity calculations. Those are fairly standard. The stochastic models, however, need a set of scenarios of holding period returns. The calculation of these data is illustrated next.

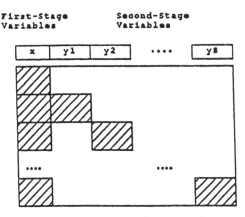

Figure 1. The constraint matrix structure of a two-stage, dynamic stochastic portfolio optimization model.

3.1. PRELIMINARIES

The rate of return of a security j during the holding period is determined by the price of the security at the end of the holding period, and the accrued value of any cash flows generated by the security. For MBS, for example, we need to estimate the accrued value of principal, interest and prepayments during the holding period, and price the unpaid balance of the security at the end of the holding period. To this end, we need a procedure for generating scenarios of the term structure, and a model that predicts the prepayment activity for each scenario. For a given interest rate scenario s, the rate of return of security j is given by

$$R_{js} = \frac{F_{js} + V_{js}}{P_{0j}}. \tag{27}$$

F_{js} is the accrued value of the cash flows generated by the security, reinvested at the short-term rates. The cash flow calculation procedures of fixed-income securities, although maybe complex as is the case in MBS, are standard.

V_{js} is the value of unpaid balance at the end of the holding period, conditioned on scenario s. This is given by $V_{js} = B_{js}P_{js}$, where B_{js} is the unpaid balance of the security and P_{js} is the price, per unit face value, of the security. Both quantities are computed at the end of the holding period, and are conditioned on scenario s. The estimation of security prices at the end of the holding period is the topic of the next section.

P_{0j} denotes the current market price of the security.

3.2. PRICING THE UNPAID BALANCE

The pricing models are based on Monte Carlo simulation of the term structure. In particular, following Cox et al. [6], we obtain the equilibrium value of the fixed-income security as the expected discounted value of its cash flow, with discounting done at the risk-free rate. We are particularly interested in pricing the security at some future time period τ. (This would be the planning horizon for the portfolio management problem.) Possible states σ of the economy at time period τ are obtained using the term structure model. From each state of the economy at instance τ, we can observe the possible evolution of interest rates further into the future, until the end of the horizon T. The price of the fixed-income security is the expected discounted value of its cash flow, with expectation computed over the interest rate paths that are *emanating* from that particular state.

In our work, we use the binomial lattice model of Black et al. [2]. A binomial lattice of the term structure can be described as a series of *base* rates $\{r_{0t}, t = 0, 1, ..., T\}$, and *volatilities* $\{k_t, t = 0, 1, ..., T\}$. The short-term rate at any state of σ of the binomial lattice at some point t is given by

$$r_t^\sigma = r_t^0 (k_t)^\sigma.$$

The base rate and volatility parameters are estimated according to the procedure developed by Black et al.

To make the idea of the pricing model precise, let S_σ denote a set of interest rate scenarios that emanate from state σ of the binomial lattice at some future time period τ. Also, let r_t^s be the short-term discount rate at time period t ($\tau \leq t \leq T$) associated with scenario $s \in S_\sigma$, and C_{jt}^s be the cash flow generated by security j at period t. A fair price for the security at period τ, and conditioned on the state σ, is

$$P_{j\tau}^\sigma = \frac{1}{|S_\sigma|} \sum_{s=1}^{|S_\sigma|} \sum_{t=\tau}^{T} \frac{C_{jt}^s}{\prod_{i=\tau}^{t}(1 + r_i^s)}. \tag{28}$$

This procedure is illustrated in figure 2.

Most fixed-income securities, however, cannot be priced using the same discount rates implied by the treasuries' curve. In particular, the price of the security has to reflect the credit, liquidity, default and prepayment risks associated with this instrument. In order to value the risks associated with a fixed-income instrument, we compute its *option adjusted premium* (OAP), cf. equation (1).

Once we have priced the various risks associated with the security, we can proceed to price the security at some future time period. The *option adjusted price* of the security $P_{j\tau}^\sigma$ can be calculated from

$$P_{j\tau}^\sigma = \frac{1}{|S_\sigma|} \sum_{s=1}^{|S_\sigma|} \sum_{t=\tau}^{T} \frac{C_{jt}^s}{\prod_{i=\tau}^{t}(1 + \rho_j \cdot r_i^s)}. \tag{29}$$

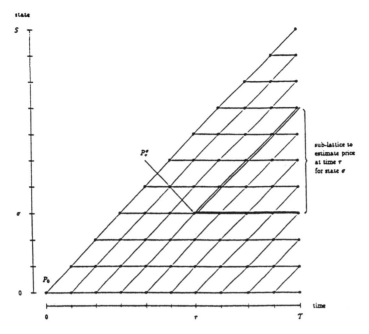

Figure 2. Estimating state-dependent prices of fixed-income securities from a binomial lattice model of the term structure.

We point out that the price $P_{j\tau}^\sigma$ may depend not only on the state σ but also on the history of interest rates from $t = 0$ to $t = \tau$ that pass through this state.[1] This difficulty can be easily resolved by sampling paths from $t = 0$ that pass through state σ at $t = \tau$. Let $S^{0,\sigma}$ denote the set of such paths, and let $P_\tau^{s(\sigma)}$, $s(\sigma) \in S^{0,\sigma}$, be the price of the security at state σ obtained by applying equation (29), conditioned on the fact that interest rate scenarios s in S_σ originate from scenarios $s(\sigma)$ in $S^{0,\sigma}$. Then the expected price of the security at σ is given by

$$P_{j\tau}^\sigma = \frac{1}{|S^{0,\sigma}|} \sum_{s(\sigma) \in S^{0,\sigma}} P_{j\tau}^{s(\sigma)}. \tag{30}$$

---

[1] In particular, the cash flows generated by some fixed-income securities at periods after τ will depend on the economic environment experienced prior to τ. For example, if the security has experienced periods of high repayments, lapse or defaults, then subsequent changes of interest rates will have less impact on the generated cash flows. Although the short-term rates prior to τ do not appear explicitly in the pricing equations, the economic activity prior to τ is used in the estimation of the cash flows C_{jt}^s for $t \geq \tau$.

4.   Do the models work? Some empirical evidence

In this section, we provide some empirical evidence that supports the validity of the models and highlights their advantages. First, we illustrate the performance of an MAD model in the context of an indexation problem. Then we use both portfolio immunization and the MAD model in structuring the funding of an insurance liability claim. In the former case, the model is shown to be very effective in tracking the index. In the latter case, we demonstrate the superiority of stochastic models (even single period ones, such as MAD) over traditional portfolio immunization techniques.

4.1.   AN EXAMPLE OF INDEXATION

Worzel et al. [23] built a mean-absolute deviation model to track the Salomon index of mortgage-backed securities. The index consists of a representative of all traded fixed-rate, pass-through securities, issued by FNMA, GNMA and FHLMC. The index is a sanitized image of the mortgage market: For example, cash flows generated by the mortgage pools are assumed to be reinvested in the index itself. There are also holdings in very small pools, but actual investments in such polls may be impossible due to liquidity difficulties. Finally, the composition of the index is changing from month to month without incurring any transaction costs. The Salomon index realized an annual return of 13.96% over the period January 1989–December 1991. Hence, creating a *tradeable* portfolio that closely tracks the index is of great interest to investors.

The model estimates holding period returns of all securities in the index, and builds a portfolio that minimizes the mean-absolute deviation of the returns of the portfolio from the expected return of the index. Upside and downside risks are penalized differentially, with no penalty on upside risk and infinite penalty on downside risk. (Downside risk is realized when the portfolio under-performs the index by a small margin, set to be −5bp in monthly terms.)

The indexation model was tested over the period January 1989–December 1991. During this period, the index realized an annual return of 13.96%. In back-testing the model, we used the following methodology: At the beginning of each month, a binomial lattice was calibrated based on the term structure of that day. The lattice was used to estimate holding period returns, and the MAD model was used to select a portfolio. The performance of the portfolio was recorded at the end of the month, based on observed market prices, and the process was repeated. Transaction costs of 2/32bp were charged, and cash flows from the mortgage pools were reinvested at the 1-month treasury rate. The very first portfolio (January 1989) was selected using the method outlined above, but no transaction costs were paid.

During the testing period, the portfolio realized an annual return of 14.18%, +22bp over the index return. The portfolio never under-performed the index by more than −3.6bp in monthly return, while the over-performance was more substantial; see figure 3. The standard deviation of the index return over the test period was

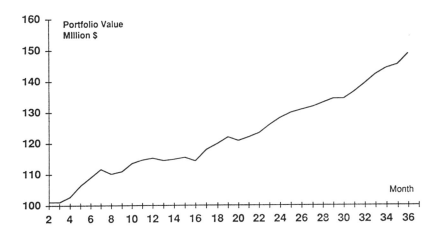

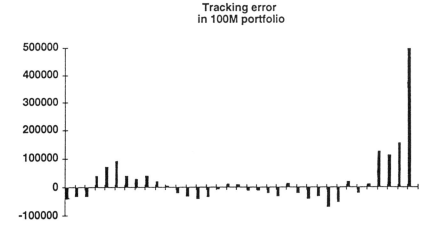

Figure 3. Return of a $100M investment in the *Salomon Brothers Mortgage Index* over the period January 1989–December 1991, and tracking error of the indexed portfolio.

0.155, while the portfolio return had a standard deviation of 0.158. The portfolio would typically consist of approximately 25 securities, none of which accounted for more than 12% of the total portfolio.

4.2.   AN EXAMPLE OF LIABILITY FUNDING

We apply the MAD model for funding a liability stream obtained from a major insurance corporation. Using the term structure of April 26, 1991, we calculated the following descriptors of the liability:

Term	100 months
Present value	$166,163,900.00
Modified duration	4.1792 years

4.2.1. An immunized portfolio

Using the portfolio immunization models, we built a portfolio that was duration and convexity matched against the liabilities. The portfolio was built from a universe of both MBS and US Treasury securities. Different levels of exposure to the mortgage market were imposed on an ad hoc basis. We list below the percentage savings realized when the liability is funded using a portfolio of MBS and treasuries, over the cost of funding the liabilities using only the risk-free rate.

Cost of portfolio using Treasuries only	$166,163,861.00
(Savings)	0.00%
Cost of portfolio using up to 25% MBS	$152,993,690.99
(Savings)	7.92%
Cost of portfolio using up to 50% MBS	$142,529,529.00
(Savings)	16.58%
Cost of portfolio using up to 100% MBS	$137,489,656.00
(Savings)	21.07%
Cost of mixed US Treasury–MBS portfolio	$136,124,130.99
(Savings)	22.07%

While it is clear from this example that using MBS in an integrated asset/liability management system produces substantial gains, the savings summarized above will not be necessarily realized in practice. They will be realized only if the term structure shifts in parallel and in small levels from that of April 16, 1991. To

Table 1

Performance of immunized and MAD portolios.

Model	100% MBS portfolio		Mixed portfolio	
	Exp. return	Std. dev.	Exp. return	Std. dev.
Immunized	10.469	0.406	10.448	0.293
MAD (equal risk)	10.783	0.405	10.692	0.293
MAD (equal return)	10.469	0.234	10.448	0.206

assess the return of the portfolio in different environments, we conducted a Monte Carlo simulation of the returns of the immunized portfolio over the holding period; see table 1. The 100% MBS portfolio produced expected returns of 10.469%, with a standard deviation of these returns of 0.406. The mixed portfolio of US Treasuries and MBS has a slightly reduced expected return of 10.448%, but a substantially reduced standard deviation of 0.293.

4.2.2. An MAD portfolio

We also developed an MAD portfolio for funding the insurance liability. The efficient frontier is shown in figure 4. In the same figure, we show the results of the simulation of the immunized portfolio. First, we observe from this figure that substantial rates of return can be realized with relatively little risk. Second, we observe that the portfolio obtained using standard immunization techniques lies below the efficient

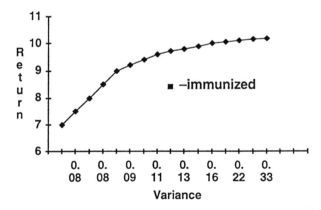

Figure 4. Efficient frontier of the mean-absolute deviation portfolio of MBS and the risk-return profile of an immunized portfolio.

frontier. These results illustrate the superiority of the MAD model in managing portfolios of complex instruments, such as MBS, as opposed to traditional fixed-income management tools. Table 1 summarizes the expected return and standard deviation of both immunized and MAD portfolios.

5. Critique and conclusions

We have presented a sequence of asset/liability management models that capture the increasing complexities of the fixed-income markets. Empirical testing has illustrated the validity of the models. Two observations are particularly interesting. First, it is possible to use optimization models, such as MAD, to solve complex problems in a very volatile environment. The performance of the index-tracking portfolio has been particularly encouraging. Second, the use of stochastic models is well justified given their superior performance over traditional immunization techniques. While a bond portfolio manager may find immunization techniques adequate – and we concur with this approach – the complexities of other fixed-income instruments leaves little choice. The volatility and co-movements of these securities must be explicitly modeled. The direct comparison of an immunized with an MAD portfolio has clearly demonstrated this point.

It is still an open question whether the multiperiod, dynamic stochastic programs offer any real advantage to portfolio managers. The state-of-the-art has reached the point where large-scale models of this form can be built and solved efficiently. However, their real advantage over either traditional immunization techniques, or the single-period stochastic models, remains to be established. We are currently completing a large-scale experimentation that pits the MAD model against several variants of the stochastic programming models in managing portfolios of MBS and callable bonds. The outcome of these studies are reported elsewhere [10, 7].

Acknowledgements

The research leading to this paper was funded by NSF University/Industry Research Collaboration Grant No. SES-91-00216. The contributions of B. Golub and L. Pohlman from Blackstone Financial Management, and M. Holmer from the Federal National Mortgage Association are gratefully acknowledged. Partial funding was also provided by NSF Grant No. CCR-9104042, AFOSR Grant No. 91-0168, and a research contract from the Metropolitan Life Insurance Company. Computing resources were made available by the Army High-Performance Computing Research Center at the University of Minnesota, and the North East Parallel Architectures Center of Syracuse University.

References

[1] D.F. Babbel and S.A. Zenios, Pitfalls in the analysis of option-adjusted spreads, Fin. Anal. J. (July/August, 1992) 65–69.

[2] F. Black, E. Derman and W. Troy, A one-factor model of interest rates and its application to treasury bond options, Fin. Anal. J. (Jan./Feb. 1990) 33–39.

[3] P. Boyle, Options: A Monte Carlo approach, J. Fin. Econ. 4(1977)323–338.

[4] S.P. Bradley and D.B. Crane, A dynamic model for bond portfolio management, Manag. Sci. 19(1972)139–151.

[5] P.E. Christensen and F.J. Fabozzi, Bond immunization: An asset liability optimization strategy, in: *The Handbook of Fixed Income Securities*, ed. F.J. Fabozzi and I.M. Pollack (Dow Jones Irwin, 1987).

[6] J.C. Cox, Jr., E. Ingersoll and S.A. Ross, A theory of the term structure of interest rates, Econometrica 53(1985)385–407.

[7] C. Vassiadou-Zeniou and S.A. Zenios, Robust optimization models for managing callable bond portfolios, Euro. J. Oper. Res. (1995).

[8] H. Dahl, A. Meeraus and S.A. Zenios, Some financial optimization models: I. Risk management, in: *Financial Optimization*, ed. S.A. Zenios (Cambridge University Press, 1993) pp. 3–36.

[9] G.B. Dantzig, Linear programming under uncertainty, Manag. Sci. 1(1955)197–206.

[10] B. Golub, M. Holmer, R. McKendall, L. Pohlman and S.A. Zenios, Stochastic programming models for money management, Euro. J. Oper. Res. (1995).

[11] R.R. Grauer and N.H. Hakansson, Returns on levered actively managed long-run portfolios of stocks, bonds and bills, Fin. Anal. J. (Sept. 1985) 24–43.

[12] R.S. Hiller and J. Eckstein, Stochastic dedication: Designing fixed income portfolios using massively parallel Benders decomposition, Manag. Sci. 39(1994)1422–1438.

[13] M.R. Holmer, The asset/liability management system at Fannie Mae, Interfaces 24(1994)3–21.

[14] J.M. Hutchinson and S.A. Zenios, Financial simulations on a massively parallel Connection Machine, Int. J. Supercomp. Appl. 5(1991)27–45.

[15] J.E. Ingersoll, Jr., *Theory of Financial Decision Making*, Studies in Financial Economics (Rowman and Littlefield, Totowa, NJ, 1987).

[16] H. Konno and H. Yamazaki, Mean-absolute deviation portfolio optimization model and its applications to Tokyo stock market, Manag. Sci. 37(1991)519–531.

[17] H. Markowitz, Portfolio selection, J. Fin. 7(1952)77–91.

[18] H. Markowitz, *Mean-Variance Analysis in Portfolio Choice and Capital Markets* (Basil Blackwell, Oxford, 1987).

[19] J.M. Mulvey and H. Vladimirou, Stochastic network optimization models for investment planning, Ann. Oper. Res. 20(1989)187–217.

[20] J.M. Mulvey and S.A. Zenios, Capturing the correlations of fixed-income instruments, Manag. Sci. 40(1994)1329–1342.

[21] R.B. Platt (ed.), *Controlling Interest Rate Risk*, Wiley Professional Series in Banking and Finance (Wiley, New York, 1986).

[22] R.J-B Wets, Stochastic programs with fixed resources: The equivalent deterministic problem, SIAM Rev. 16(1974)309–339.

[23] K.J. Worzel, C. Vassiadou-Zeniou and S.A. Zenios, Integrated simulation and optimization models for tracking fixed-income indices, Oper. Res. 42(1994)223–233.

[24] S.A. Zenios, Massively parallel computations for financial modeling under uncertainty, in: *Very Large Scale Computing in the 21st Century*, ed. J. Mesirov (SIAM, Philadelphia, PA, 1991) pp. 273–294.

Part VIII
Case Studies of Implemented Asset-Liability Management Models

Modelling and Management of Assets and Liabilities of Pension Plans in The Netherlands

C.G.E. Boender, P.C. van Aalst and F. Heemskerk

Abstract

Most Dutch pension plans carry out relatively expensive and inflation sensitive defined benefit final pay pension schemes, with the objective of a full compensation of the earned pension rights for the cost of living. Due to the ageing of the beneficiaries, the costs of these plans are becoming prohibitively high and volatile, while the sponsors, driven by international competition, require that the contributions are in accordance with the moderate international standards. In addition to a reconsideration of the currently prevailing pension and funding schemes, this increasingly urges pension plans to reallocate the investments to more risky and inflation hedging asset classes, while maintaining the solvency requirements of the regulating authorities.

This important practical issue, referred to as Asset/Liability Management, initiated an enormous flow of academic research, mathematical models and decision support systems, which are successfully applied to sustain the long term planning of almost every medium and large sized Dutch pension fund.

In this paper we first describe the full scope of the asset/liability problem, with particular emphasis on the point of view that the policy instruments of pension plans are not restricted to the allocation of the assets. Next we focus on the scenario analysis approach (including the application of simulation and optimisation models) which is used to sustain the asset/liability decision making process. The paper concludes with a discussion of the proven usefulness of asset/liability modelling in real-life practice.

1 Problem Definition

1.1 Introduction and terminology

With more than \$400 billion in pension assets, and the largest amount of pension assets per capita, The Netherlands are a major pension country. This is a result of an early founding of the Dutch pension funds, the generous pension schemes, the relatively small public pensions and a full capitalisation of the liabilities.

Almost all pension funds have a defined benefit pension plan in which the members are guaranteed certain nominal rights, which are frequently accrued by a final pay pension scheme. The expected present value of the nominal pension rights are the plan's nominal liabilities, which are usually determined at a fixed discount rate of $r = 4\%$. The extent in which the assets are sufficient to cover these liabilities is referred to as the funding level. This funding level (comparable to the ABO liabilities in the FASB rules) is the main variable which Dutch pension funds have to manage and control, where they in particular have to take into account that 'significant' underfundings with respect to this funding level (also referred to as deficits) are not allowed by the regulating authorities.

Although the definition of the funding level frequently does not take into account any costs of future indexations of the nominal obligations, it is a crucial objective of Dutch pension funds that the pension rights of the inactive participants, for which the sponsor does not donate contributions, are each year fully indexed with a rate i of inflation of prices or wages. Hence, Dutch pension funds require a portfolio return of $r = 4\%$ on the pension assets which meet the nominal obligations, whereas on the pension assets which meet the liabilities of the inactive participants an excess return of $i\%$ is needed for indexation. Shortfalls with respect to this target level may either lead to underfundings, additional contributions or the renouncement from full indexation. Surpluses are allocated to strengthen the funding level, to make up indexation arrears, and to rebates at the sponsor.

1.2 Current practice

Figure 1 depicts the wishes and requirements of the agents which a pension fund has to take into account. Figure 1 displays that the current practice can be described as follows[1]:

- Due to the growing international competition, Dutch plan sponsors require that their contributions to the pension plan are in accordance with the moderate international ranges. However, on the other hand the contributions have a tendency to become prohibitively high and volatile. This is first of all due to the employees who wish to maintain the generous, but relatively costly and inflation-sensitive benefit defined final pay schemes. Secondly, due to the ageing of the population, pension plans are confronted with an enormous growth of the costs and risks of the

---

[1]Note that tax-issues have never been relevent up till now in The Netherlands, neither with respect to the contributions, nor with respect to large surplus buffers.

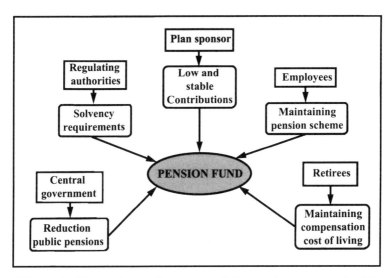

Figure 1: The wishes and requests of the parties concerned with pension plans.

pursued indexations of the earned pension rights. Thirdly, pensioners are confronted with deteriorating public pensions, which they wish to be compensated by the pension plan, with obvious negative consequences for the required contributions as well.

- The above issue could be partly solved by an appropriate amount of high return investments in the strategic asset mix. However, this is impeded by the regulating authorities who require that ultimo each year the investments are sufficient to cover the liabilities. This forces Dutch pension funds to act partly as short term investors, such that they can insufficiently profit from the higher expected (long term) returns of equities, with obvious opportunity costs for the future indexations and contributions.

1.3   Asset/Liability management

Pension fund management thus faces the issue to determine a dynamic long term strategy which balances the above interests of the plan sponsor and the plan members, while satisfying the solvency standards of the regulating authorities. To accomplish this objective, it is now well acknowledged that the available asset- and liability policy instruments have to be determined in an *integrated* fashion, which explains that this strategic planning problem is referred to as Asset/Liability Management (ALM); see Figure 2.

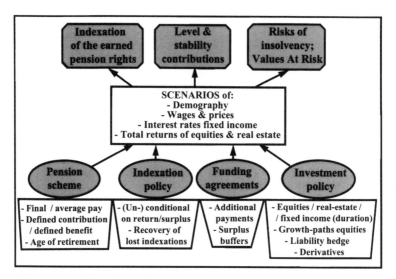

Figure 2: The Asset/Liability Management problem: Instruments, objectives and risks.

As can be verified from Figure 2 pension fund management has several policy instruments at her disposal to accomplish her objectives:

1. **Investment policy:**
   First of all, since historical and theoretical evidence indicates that equities in the long run outperform fixed income securities (cf. the research in Bernartzi and Thaler (1993) on the risk premium of risky securities), the Dutch pension levels could (partly) be maintained at more acceptable contribution rates by changing the investment portfolio, which initially contained very large amounts of loans and bonds, to a higher risk-return profile. However, if the market exposure of the investment portfolio would be increased such that the expected portfolio returns would be sufficient to maintain the pension scheme at internationally competitive costs, the funding level risk (due to the market risk of the investments and the inflation risk of the liabilities) would in most cases strongly violate the solvency standards set by the regulating authorities. Therefore, additional measures have to be considered.

2. **Funding policy:**
   An important measure is that the plan sponsor enables the pension plan to build-up or maintain a surplus buffer, and/or contributes additional funds if the plan does not satisfy the solvency standards of the regulating authorities. The trade-off of this measure is obvious. In the short

run the sponsor will frequently have to provide additional contributions to the plan. However this implies a risk reduction for the pension plan, which allows a more aggressive asset allocation, with additional returns that can be allocated to retributions to the sponsor, and to the indexation of the earned pension rights of the beneficiaries.

3. **Indexation policy and the pension scheme:**
   However, unless the previous decade with low inflations and high asset returns extrapolates into the long future, the above asset- and funding instruments will with high probability not be sufficient to maintain the generous Dutch pension system at acceptable costs. Therefore, the plan members have to participate as well in creating a feasible balance between the (future) pension levels and the required contributions. This can be accomplished by a transition to average pay schemes, or even to defined contribution systems, but also by making explicit agreements with the plan members at which funding levels the plan is allowed to decrease or postpone the indexation of the earned pension rights. Until now the pension community is not well aware that these measures are not only cost effective, but risk reducing as well. Due to these measures the inflation risk of the plan will reduce considerably, which, analogous to the funding measures, could allow a more risky asset allocation, with obvious merits for the plan sponsor and the indexations of the earned pension rights.

Thus, we quite agree with the growing community of ALM-researchers (in addition to the references in this volume we refer to Van Aalst (1995), Boender et al (1993), Boender and Heemskerk (1995), Boender (1995), Carino et al (1994), Dert (1995), Frijns and Goslings (1994), Kingsland (1982), Leibowitz et al (1994), Macbeth et al (1994), Mulvey (1996), Sharpe and Tint (1990), Winklevoss (1982)) that asset-only approaches may yield quite misleading policies. Asset policies should first of all properly take into account the (dynamics) of the liabilities, and should secondly not be evaluated in terms of risk and return of the investment portfolio, but rather in terms of several underfunding quantities (cf. Boender (1995)), the dynamics of the contributions (cf. Boender and Heemskerk (1995), Macbeth et al (1994)), or the so called funding ratio return (see Leibowitz et al (1994), Van Aalst (1995)), or by a combination of these measures. The future liabilities can not be considered as given uncontrollable quantities, but, as illustrated in Figure 2, all the possible adjustments at the liability side have to be considered as active policy instruments as well. This creates a challenge to determine a jointly optimal dynamic strategy for the asset and liability instruments which balances the interests and requirements of all agents involved.

2   Scenario analysis approach

Sharpe and Tint (1990) conclude that *'pension-fund managers show faint interest in consideration of liabilities in pension-fund asset-allocation strategies'*. Even more recently, it is argued in Macbeth *et al.* (1994) that *'most sponsors have a sketchy understanding of actuarial techniques, and actuaries do not go out of their way to make their craft understandable to sponsors'*.

In the past years we encounter in The Netherlands a completely opposite situation. Almost all large pension funds, more-and-more followed by the intermediate and smaller ones, actively apply asset/liability models to analyse, evaluate and determine coherent asset/liability strategies. (The application of these models and systems to other (European) countries has yet to occur.)

In our opinion this fortunate situation is mainly due to:

- **Practical relevance:**
  As we described above, the relevance is mainly due to the the urge to reduce the pension costs on the one hand, and to maintain the royal pension system in an ageing population on the other;

- **Integration of research and practice:**
  Enthusiastic practice-driven ALM-research, modelling, and consultancy, which is carried out in close co-operation between researchers and policy-makers at universities, consultancy firms, and pension funds;

- **Level of ambition:**
  The acceptance by top management of the scenario analysis approach, which aims at applying mathematical models and systems to enable management to *learn* about their business, rather than at using mathematical modelling to prescribe optimal strategies.

It can be verified from the literature (cf. the review paper of Bunn and Salo (1993)) that scenario analysis is in a growing extent used for long term planning and strategic analyses. With respect to definitions of scenarios in the literature, asset/liability modelling closely affiliates with Brauers and Weber (1988) who describe a scenario as *'A description of a possible state of an organisations future environment, considering possible developments of relevant interdependent factors of the environment'*. From our (ALM) point of view, the scope of scenario analysis is appropriately described in the scenario literature by Huss (1988) who argues that *'At their best, scenarios reveal new strategic options and threats, and because they record explicit assumptions about the future and provide a common framework for discussion, they also contribute to a better understanding between managers'*.

We use scenarios in an *iterative managerial learning process* of evaluating and improving asset/liability strategies on a range of scenarios of possible economic and demographic developments (rather than using poor long term forecasts, or limited intuitive judgement), sustained by simulation and optimisation. This process can be distinguished in two phases. In the first *diagnostic* phase the asset/liability playing field is explored to reveal how potential asset/liability strategies behave in various economic environments with respect to costs and risks. In the second phase of *judgement and decision making* this process of successively testing and improving strategies is repeatedly carried-out until a strategy emerges which is agreed upon by all who carry responsibility for the pension fund and her sponsors and beneficiaries.

3   Management Flight Simulator

In asset/liability projects the scenario analysis approach is sustained by a (partly tailor-made) Asset/Liability Scenario model ALS. The model contains three main submodels (cf. Figure 3). The first submodel ($\rightarrow$ '*economic scenarios*') is used to generate ranges of scenarios of the future economic environment. For each economic scenario the second submodel ($\rightarrow$ '*liability scenarios*') is applied to determine the corresponding development of the pension rights and payments which the beneficiaries will acquire whatever the strategy-dependent development will be of the plans funding level. This provides a range of combined scenarios of economic circumstances and the corresponding development of the guaranteed earned pension rights and payments (see the square in Figure 3), which provides the input of the asset/liability model ($\rightarrow$ '*ALM*'). This model applies simulation techniques to evaluate strategies with respect to the resulting future developments of the contributions, indexations and insolvencies, as well as optimisation models to determine (parts of) strategies which optimally satisfy preset constraints. The submodels will now be described in more detail.

3.1   Economic scenarios

Consider the statistics of the economic time series in Table 1 (note that we work with log-transformations of the raw data, and that the figures of the asset categories concern total rates of return in Dutch guilders). In addition to the figures such as the risk premium and volatility of equities, the correlations of cash and real-estate with the inflation of prices are of particular importance to construct portfolios whose returns match the returns of the liabilities (cf. Van Aalst (1995) and Sharpe and Tint (1990)). Evidently, following the scenario definition mentioned in Section 2, these statistics have to be taken into account in the construction of economic environments to analyse, evaluate and optimise asset/liability strategies.

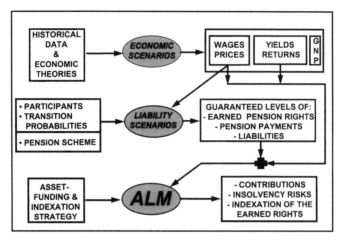

Figure 3: Architecture of the decision support system.

	Geometric average	Standard deviation	Auto-correlation	Correlation with inflations	Correlation with wage growth
Inflation	4.4%	2.8%	0.69	-	0.74
Wage growth	6.4%	4.2%	0.60	0.74	-
Bonds	7.1%	3.6%	0.14	0.20	-0.15
Cash	6.0%	2.5%	0.65	0.27	-0.02
Equities	10.7%	17.2%	-0.04	-0.28	-0.30
Real estate	5.5%	4.7%	0.64	0.67	0.58
Nominal GNP growth	3.3%	2.4%	0.38	0.14	0.35

Table 1: Statistics based on annual Dutch data over the period 1956 up to 1993.

3.1.a.   One-year dynamics

Following Sims (1980) in the paper '*Macroeconomics and reality*' the dynamics of the above time series is modelled by a Vector Auto Regressive lognormal econometric time series process (VAR). Diagnostic tests indicated that the order of the process should chosen to be equal to 1, such that the model reads:

$$y_t \sim N\left(\mu + \Omega * \{y_{t-1} - \mu\}, \Sigma\right) \tag{3.1}$$

where $N(\mu, \Sigma)$ denotes a Gaussian distribution with mean μ and covariance matrix Σ, and y_t is the vector containing the (lognormal transformations of) the above time series in year t. Thus, the underlying assumption of the model is that the realisation of *any* time series in year t is an explanatory variable for the distribution of *any* time series in year $t + 1$, i.e., in this most simple VAR-model all the time series serve both as independent and as dependent variable in the regression (see Section 3.1.b. for more realistic versions).

We estimate the model by row wise Ordinary Least Squares (OLS). Statistically insignificant parameters of the model may be stepwise removed, followed by a Seemingly Unrelated Regression step to re-estimate all the rows of Ω simultaneously, using the covariances of the errors which result from the previous OLS-procedure (cf. Judge *et al.* (1988)). It can be shown that OLS and SUR yield identical estimates if the matrix of the covariances of the errors Σ is diagonal, or if each element of Ω is used as explanatory variable in the regression.

3.1.b. Scenario generation

The estimated values of the parameters μ, Ω and Σ, and the values y_0 of the time series in the initial period, completely specify the probability distribution for period $t = 1$. Drawing a sample point from this distribution generates a value for the vector y_1, which in turn specifies the distribution for y_2. Iteratively repeating this process generates a time path of vectors $\{y_t | t = 0, \ldots, T\}$, which is used as a scenario of the development of macroeconomic circumstances. Successively restarting in $t = 1$, additional scenarios are generated until the sample variance of the set of scenarios is small enough for purposes of analysis, which is typically accomplished at a sample size of (a few) thousand(s).

We make a few observations about this procedure of scenario generation:

- **Scenario requirements:**
  In case of sufficiently many historical data it can be shown that a sample of generated scenarios displays identical long-term expected values, standard deviations, correlations and autocorrelations as observed in the data which are used to estimate the parameters of the VAR-model by which the scenarios are generated (cf. Boender and Romeijn (1991)). Thus, application of the scenario generation procedure guarantees that important scenario requirements such as '*considering relevant interdependent factors*' are satisified.

- **Historical periods:**
  The 'statistical equivalence' of a sample of generated scenarios and the
  properties of the time series in the historical period which is used to estimate the VAR-model by which the scenarios are generated, is also important from a consultancy point of view. That is, rather than viewing
  a scenario set as the best long term prediction, this feature of the scenario generator enables management to evaluate potential asset/liability
  policies against different historical periods, which of course improves the
  understanding on the perils and merits of studied strategies.

- **Arbitrage free scenarios:**
  The approach can be extended to generate a *scenario tree* (i.e., in any
  year of any scenario there are more than one subsequent states, rather
  than 1), which is arbitrage free. This is of crucial importance for a
  proper implementation of dynamic optimisation models which in each
  year of each scenario compute an optimal solution, assuming that an
  optimal strategy is followed thereafter (cf. Dert (1995)).

- **Economic regimes and long term equilibria:**
  The VAR-model can be extended to a Vector Error Correction Model
  (VECM) which additionally takes into account economic regime changes
  and long term equilibria. Examples include the regime of high inflation
  in the seventies, followed by the successful attempts to control inflation
  in the eighties and the nineties, and the long term relationship between
  the accumulated values of the inflations of prices and wages. Applying
  the error correction extension of the original VAR model gives the ALM-
  analyst the possibility to evaluate potential strategies with respect to
  scenarios which explicitly take into account these economic phenomena.

  Before introducing the VECM, we first reformulate the VAR-model by
  denoting the inflations and returns y_t as first order differences of the
  corresponding indices x_t. Then the above VAR-model is:

  $$x_t = \underbrace{x_{t-1} + \mu} + y_t, \qquad (3.2)$$

  where the first terms form the deterministic component, and y_t is the
  stochastic time-series component:

  $$y_t \sim N(\Omega * y_{t-1}, \Sigma). \qquad (3.3)$$

  The extended error correction version of this model is:

  $$x_t = \underbrace{x_{t-1} + \mu_1 \times I_{\{T_1\}} + \mu_2 \times I_{\{T_2\}}} + y_t \qquad (3.4)$$

in which the first terms are the regime-dependent determinstic component, and y_t is again the stochastic time series component:

$$y_t \sim N\left(\underbrace{\Omega_1 y_{t-1}} + \underbrace{\Omega_2.\mathbf{C}^T \left(x_{t-1} - \mu_1 I_{\{T_1\}} - \mu_2 I_{\{T_2\}} \right)}, \Sigma \right), \qquad (3.5)$$

where the Ω_1 term corresponds to the short term dynamics, and the Ω_2 term is the long term correction.

The extended model distinguishes itself with respect to the following two aspects from its predecessor:

1. **Economic regimes:**
   From the first equation of the correction version it can be verified that the average growth of the time series is not required to be identical in all time periods. In the above specification of the model there is a first economic regime in the years of the index set T_1 with growth vector μ_1, and a second regime in the years of the set T_2 with growth vector μ_2.

2. **Long term equilibria:**
   Secondly, consider the long term equilibria $\alpha_1 x_{1t} + \alpha_2 x_{2t} + \alpha_3 x_{3t} = 0$, and $\beta_1 x_1 + \beta_2 x_2 = 0$. It can be simply verified from the second equation of the model that the additional term

   $$C^T \{ x_{t-1} - \mu_1 I_{\{T_1\}} - \mu_2 \cdot I_{\{T_2\}} \} \qquad (3.6)$$

   generates an impulse to these equilibria if the first two columns of C are chosen as $[\alpha_1, \alpha_2, \alpha_3, 0, 0, 0, 0]$ and $[\beta_1, \beta_2, 0, 0, 0, 0, 0]$ respectively. The matrix Ω_2, which is estimated from the data, determines the rate in which these impulses restore violations of the equilibria.

With respect to economic regimes, the main results of stability research on the data which are summarised in Table 1, consist of the following regimes which are statistically significant with t-values $\gg 10$:

- A period of high wages in 1964–1977 with mean value 10.7%, and a low average value of 2.5% in the remaining years of 1956–1993;

- A regime of high inflation of prices with mean value 7.2% in 1969–1980, and a low inflation regime with mean value 2.8% in the other years of 1956–1993.

The main long term equilibria which are identified in these data are:

$$
\begin{aligned}
x_i &= 1.2 * x_j & (t\text{-value } -4.4) & \qquad (3.7) \\
x_k &= -4.8 * x_j - 8.2 * x_l & (t\text{-values } -3.2 \text{ and } -3.6) & \qquad (3.8)
\end{aligned}
$$

where the subscripts i, j, k, and l respectively denote wages, prices, equities, and GNP.

Incorporation of these results in the error correction version of the VAR-model improves the average adjusted R^2 from 26% to 42%, which implies a considerable improvement of the quality of the generated scenarios. Moreover, the extended model enables the asset/liability policy-maker to evaluate potential strategies at sets of scenarios which include the distinguished economic regimes in the extent in which these are considered plausible or representative for the unknown future. Thus, in addition to the improved quality, the scenarios sets also have a more clear economic interpretation, which obviously improves the process of ALM-scenario analysis described in Section 2.

3.1.c. Scenario risk measures

Once the consequences of an asset/liability strategy have been simulated on a range of combined scenarios of the economic environment and the corresponding development of the plans guaranteed pension payments and earned rights, the funding level F_{ts} is known in each year t of each scenario s $(t = 1, .., T; s = 1, .., S)$. This enables us to define various risk measures:

- **The probability of underfunding in year t:**
  This risk measure is defined as the percentage of funding levels in year t which are smaller than 100%.

- **The probability of ruin in year t:**
  This risk measure is defined as the percentage of funding levels in t which are smaller than 100%, whereas the value in $(t-1, s)$ and $(t-2, s)$ are also smaller than 100%, with one of these funding levels being less than a prespecified value of say, 80%.

- **The conditional expected underfunding in year t:**
  The conditional expected underfunding in year t is defined as the expected value of the funding levels which are smaller than 100%.

- **The downside deviation of surplus in year t:**
  Finally, analogously to the definition of downside deviation of portfolio return (cf. Sortino and van der Meer (1991)), we define the downside deviation of surplus in year t as the standard deviation of the negative surpluses in year t, i.e.

$$D_{ts} = \left[\int_{\{-\infty, 0\}} \varphi^2 f(\varphi) \mathrm{d}\varphi \right]^{\frac{1}{2}} \tag{3.9}$$

where φ denotes a (negative) surplus, and $f(\varphi)$ denotes the probability that a surplus occurs of the size φ.

In practice, more than one (if not all) of the risk measures are applied to improve the understanding how the funding risks evolve, where usually the risk measures in subsequent years are averaged (sometimes with a discount rate) to arrive at an overall risk over the planning horizon. Analogous scenario measures can be determined of the level and volatility of the required contributions and the realisations of the indexations of the earned rights.

3.2 Liability scenarios

For each macroeconomic scenario the liability module of the system is used to determine the corresponding values of the guaranteed payments and earned rights. This is accomplished in two steps.

1. In the first step a Push Markov model is applied to determine the future status of each individual plan member. For an employee this implies that in each year it is determined if he/she remains alive, retires, terminates, gets disabled and/or gets a promotion to a higher function level. These transitions are determined by probabilities which depend on characteristics of the individuals such as age, gender and employee-category. Disabled, resigned and retired participants are treated analogously. Next, given the status of each current employee in each future year, the Pull-part of the model determines additional promotions and hires new employees such that the number of employees in each category in each future year is as much as possible in accordance with prespecified values.

2. In the second step the pension rules (which may require considerable tailoring activities, and which may comprise thousands of lines of software) are applied to compute the guaranteed pension payments and the earned pension rights, which in combination with the corresponding economic environment comprise the input of the ALM-models which are described in the next section.

3.3 Asset/liability management models

The asset/liability models employ both simulation and optimisation concepts to evaluate and construct asset/liability strategies.

3.3.a. Simulation

In addition to the specifications which can be made with respect to the development of the insured population (cf. Section 3.2), the strategies whose

consequences can be simulated against the generated scenarios are defined as
follows:

1. **Strategic asset-mix:**
   The most simple (but most frequently applied) procedure is that a rebalancing policy is carried out (fixed mix). That is, in each year t of
   each scenario s the simulation process rebalances the assets to a fixed
   prespecified allocation, whatever the values are of relevant factors such
   as the funding level and the (real) interest rates in (t, s). This implies
   a buy low, and sell high policy.

   The model allows the possibility that the asset allocation is adjusted
   according to prescribed strategies (with the obvious consequence that
   the development of the asset allocation is stage dependent, rather than
   fixed). Three instances concern:

   - **Dynamic stage dependent growth-paths of equities:**
     This part of the model is used to sustain the frequently occurring
     situation that based on the results of a long term ALM-project,
     pension management has decided to expand the amount of equi-
     ties and real estate in the asset mix, while the current funding level
     is not yet sufficient to do so. This is handled by prespecifying a
     minimally required funding level (satisfying the standards of the
     regulating authorities) F_q for each amount q of equities and real
     estate. In the simulation process the amount of equities and real-
     estate in (t, s) is enlarged to q only if the funding level in (t, s) has
     reached the required level F_q. These 'safe' growth-paths are some-
     times accelerated by providing additional funding if the funding
     level grows too slowly to the level which is required to justify the
     asset allocation which pension management pursues.

   - **Duration:**
     Because the liabilities in The Netherlands frequently are not mar-
     ket valued, the duration of the fixed income portfolio plays a minor
     role in the Dutch ALM-practice in relation to e.g. the USA (cf.
     Leibowitz et al (1994)). Nevertheless, a duration-policy can be
     carried out. Then, given the interest rate and/or funding level
     in (t, s), the duration d_{ts} required by the policymakers is accom-
     plished using an optimisation model which minimises the volume
     of the required transactions.

   - **Derivatives strategies:**
     Only very recently have Dutch pension plans gained interest in
     applying long dated puts to insure the funding level against drops
     which are not permitted by the regulating authorities. Initial ex-

periments with the simulation model have shown that the application of one year puts primo year t of scenario s to insure the funding level ultimo year t of s does not lead to any efficiency gain. That is, such a policy turns out to be equivalent to carrying a portfolio which contains more fixed income than the current portfolio. However, it turns out that spectacular efficiency gains can be accomplished if primo (t, s) long puts are applied to insure the funding level at, say, ultimo $(t + 5, s)$. We will further report on this recent work in a later research paper.

2. **Funding:**

- **Permitted fluctuations of the contribution-rates:**
  A policymaker can set a maximum allowable fluctuation-rate from one year to the next, where the permitted fluctuation may depend on the level of the contributions and the funding level.

- **Minimal and maximal funding levels:**
  In addition to the the permitted fluctuations of the contribution-rates a crucial decision of the ALM-policymakers is the minimum and maximum value of the funding level. We distinguish between soft and hard extrema:

  - If the funding level at (t, s) lies outside the soft range, the contribution is reduced or increased with the objective to achieve a funding level within this specified range. However, in carrying out these adjustments, the restrictions on the fluctuation of the contribution-rates are taken into account, which may imply funding levels outside the soft range.

  - If the funding level moves outside the hard range, these extrema are restored, whether or not this violates the agreed maximal contribution-fluctuations. A hard minimum value of the funding rate implies that the sponsor carries full responsibility to restore this level, if necessary.

3. **Indexation of the guaranteed earned pension rights for the cost of living:**
   Dutch pension plans are now attempting to reduce inflation risk by making agreements with the plan members at which circumstances the indexation can be reduced or postponed. Typical examples are that the indexation in (t, s) is set dependent on portfolio return in (t, s), the funding level in (t, s), or both, where sometimes it is agreed that arrears are made up when excess portfolio returns (over the discount rate plus inflation) are made, or when the funding level has restored to safe levels.

3.3.b. Optimisation

The asset/liability decision making process is supported by several optimisation models. The most sophisticated version is a multistage stochastic programming model which determines an optimal ALM-policy for each year t and each scenario s (cf. Boender *et al.* (1993), Dert (1995)). However, these models frequently appear difficult to grasp by management, which may lead to a hesitation to gain insight from the results, at least initially. Therefore, there is ample room for a hybrid simulation/optimisation method. Applying concepts from the field of nonlinear global optimisation (cf. Boender and Romeijn (1994)), this method proceeds as follows. In the first step many, say 500, asset mixes are randomly generated. Then the simulation model is run to evaluate the consequences of these mixes (given the other specifications about the adjustment of the asset allocation and the funding and indexation policy), which is practically feasible since one run of the simulation program for, say, 1000 scenarios over an horizon of 25 years only takes a few seconds CPU on a 80586PC. Next the random search is iteratively concentrated in the neighbourhood of the promising regions until some stopping criterion is satisfied.

Applying this hybrid simulation/optimisation method, *efficient frontiers* of initial asset allocations are determined which optimally trade-off the contribution rates or pension levels against the risks of insolvency, *while consistently taking into account all the assumptions, (asset) adjustment rules and strategies which are specified in the simulation model described above.* As an example we refer to Figure 4 which for several initial asset allocations (the pies) displays the scenario contribution (x-axis) and the downside deviation of surplus (y-axis). In addition to the figures (not shown here) which display the scenario-dependent development over time of the funding level, the contribution rate and the lost indexations, graphical presentations such as Figure 4 play a crucial role in the analysis and evaluation of asset/liability strategies.

The hybrid optimisation model can be extended to determine jointly optimal asset allocations and funding policies, rather than aiming for the asset allocation which optimally matches a prespecified funding policy. Furthermore, due to its objective 'only' to identify the optimal specification of the initial asset allocation of the simulation model, the procedure can be 'positioned' as a time saving mechanism for policymakers which normally would only use the simulation model, which prevents that the optimisation procedure meets with the still prevailing scepticism of top managers to mathematical optimisation.

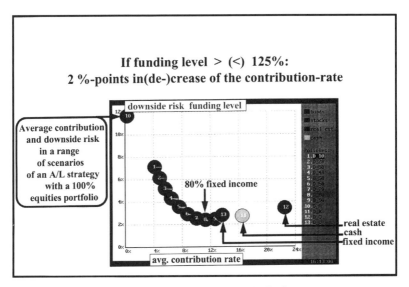

Figure 4. Contribution Rate vs. Downside Risk.

4 Practical Usefulness Asset/Liability Modelling

We distinguish two (correlated) motives to sustain asset/liability management with quantitative scenario analyses:

1. **Efficiency:**
   Obtaining more efficient and more appropriate coherent strategies which provide a better trade-off of the requirements and interests of the sponsors, the beneficiaries and the regulating authorities (cf. Figure 5).

2. **Managerial learning and consensus:**
   Improving insight of pension fund management in how potential strategies will perform in various economic and demographic environments, and establishing consensus between the officials which represent the (partly conflicting) interests of the agents concerned.

In our experience the most important practical consequences of asset/liability modelling concern:

- **Risk/return:**
  As a consequence of the improved insight, consensus, and the integral approach to ALM, many of our clients have considerably extended their

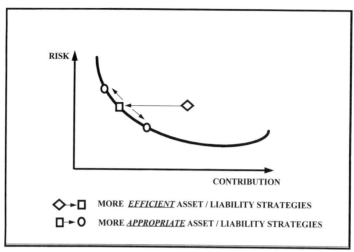

Figure 5. Usefulness ALM

investments in risky securities. At some of the clients this extension is carried out along state-dependent growth-paths, where the investments in risky securities are stepwise increased when the funding level attains preset safety levels. This phenomenon has recently attracted the attention of the Dutch press, which have published several articles in the daily financial newspapers on this 'landslide' in the investment portfolios of Dutch pension funds.

- **(In)efficient portfolios:**
  Efficient frontiers, such as depicted in Figure 4, which determine the asset allocations which for prespecified values of the scenario contributions minimise an underfunding risk, attain their minimal risk frequently, if not always, at asset allocations with considerable amounts of equities and real-estate. This will for example be the case if strong restrictions are imposed on increases of the contributions, which implies that the risk premium of equities is needed to safeguard the expected level of the funding rate from collapses to underfunding. In practice, the asset allocation which minimises the underfunding risks usually falls in the range of 20% to 50% of equities and real estate, depending on the funding rules. Evidently this implies that asset-mixes with less risky securities are inefficient (cf. Figure 4), which in our asset/liability projects frequently 'shocks' the many officials who (still) consider fixed income securities as the most 'safe' investment. This insight gives these officials more confidence to reallocate the investments to a more efficient c.q. more appropriate asset-mix.

- **Liability hedging credit:**
  Second, the asset/liability projects have made management more aware of the growing maturity, and the resulting deterioration of the inflation sensitivity of the liabilities. As a consequence the liability hedging credit of assets is getting more interest. In the absence of Duth index linked bonds, this concerns first of all real estate, in particular Dutch real estate, which due to the correlation with prices (cf. Table 1) appears in significant amounts in the mathematically optimal and practical asset allocations, although the expected return of this asset category is outperformed by its competitors. Also, this insight has initiated an ALM-focus on the duration of the fixed income portfolio: pension fund management has gained a much better understanding that the duration is not only a tool to anticipate changes in the interest rates, but should also be adjusted to the growing maturity and inflation sensitivity of the liabilities.

- **Coherent policies:**
  Third, not only ALM-researchers, but also pension fund management is now thoroughly convinced that the available policy instruments should be determined in an integrated fashion. In addition to the instances mentioned above we mention several projects where it has been agreed at which insufficient funding levels either the sponsor supplies additional contributions and/or the participants temporarily renounce from indexations. These measures of risksharing enabled the pension fund to expand the investments in high rewarding risky securities, with obvious benefits for all parties concerned.

- **Action:**
  Management has gained a much better understanding of the dynamics of their business, which has given them the confidence to actively engage in long term planning at all.

References

P.C. van Aalst (1995), *Risico analyse en matching bij pensioenfondsen*, Ph.D. thesis, Erasmus University Rotterdam.

S. Benartzi and R.H. Thaler (1993), 'Myopic loss aversion and the equity premium puzzle', Paper 4369, National Bureau Economic Research, Cambridge.

C.G.E. Boender and H.E. Romeijn (1991), 'The multi-dimensional Markov chain with prespecified means and (auto-)covariances', *Communications in Statistics: Theory and Methods* **20**, 345–360.

C.G.E. Boender, C.L. Dert, P.C. van Aalst, D. Bams and F. Heemskerk (1993), 'Scenario-approaches to asset/liability management', Inquire meeting, Bath.

C.G.E. Boender and F. Heemskerk (1995), 'A static scenario optimization model for asset/liability management of defined benefit plans', report 9512/A Econometric Institute, Erasmus University Rotterdam.

C.G.E. Boender (1995), 'A hybrid simulation/optimization scenario model for asset/liability management', report 9513/A Econometric Institute, Erasmus University Rotterdam, accepted for publication to the 20th anniversary special issue of *European Journal of Operational Research*.

C.G.E. Boender and H.E. Romeijn (1994), 'Stochastic methods'. In *Handbook of Global Optimization*, P. Horst and P.M. Pardalos (eds.), Kluwer.

J. Brauers and M. Weber (1988), 'A new method of scenario-analysis', *Journal of Forecasting* 4 772–781.

D.W. Bunn and A.A. Salo (1993), 'Forecasting with scenarios', *European Journal of Operations Research* 68 291–301.

D.R. Carino, T. Kent, D.H. Myers, C. Stacy, M. Sylvanus, A.L. Turner, K. Watanabe and W.T. Ziemba (1994), 'The Russell–Yasuda Kasai model: An asset/liability model for a Japanese insurance company using multistage stochastic programming', *Interfaces* 24 29–49. Reprinted in this volume, 609–633.

C.L. Dert (1995), *A multistage stochastic programming approach to asset/liability management*, Ph.D. thesis, Erasmus University Rotterdam.

J.M.G. Frijns and J.H.W. Goslings (1994), 'Defined benefit plans, asset-allocation and surplus management', to appear.

W.R. Huss (1988), 'A move towards scenario-analysis', *International Journal of Forecasting* 3 123–132.

G.G. Judge, R.C. Hill, W.E. Griffiths, H. Lutkepohl and T. Lee (1988), *Introduction to the Theory and Practice of Econometrics*, Wiley.

L. Kingsland (1982), 'Projecting the financial condition of a pension plan using simulation analysis', *Journal of Finance* 37 (2), 577–594.

M.L. Leibowitz, S. Kogelman and L.N. Bader (1994), 'Funding ratio return', *Journal of Portfolio Management*, Fall 1994, 3–47.

J.M. Macbeth, D.C. Emanuel and C.E. Heatter (1994), 'An investment strategy for defined-benefit plans', *Financial Analysts Journal*, May/June, 34–41.

J. Mulvey (1996), 'It always pays to look ahead', *Balance Sheet*, 4 (4), Winter 1995/1996, 23–27.

W.F. Sharpe and L.G. Tint (1990), 'Liabilities: A new approach', *Journal of Portfolio Management*, Winter, 5–10.

C.A. Sims (1980), 'Macroeconomics and reality', *Econometrica* 48 1–48.

F.A. Sortino and R. van der Meer (1991), 'Downside risk', *Journal of Portfolio Management*, Summer, 27–31.

H.E. Winklevoss (1982), 'PLASM: Pension Liability and Asset Simulation Model', *Journal of Finance*, May, 585–594.

Integrated Asset-Liability Management: An Implementation Case Study

Martin R. Holmer

This article discusses integrated asset-liability management, a new management perspective that is evolving at the more innovative financial intermediaries in response to problems caused by the older functional management
perspective. The older perspective calls for an organization to be structured
into functional units (e.g., marketing, asset management, etc.), the decisions
of which are coordinated by a corporate plan based on a macroeconomic
forecast. Growing disenchantment with the accuracy of macroeconomic forecasting in the face of a more complex and volatile capital markets environment has led some financial intermediaries to begin implementing a new management perspective. This new management perspective, termed integrated
asset-liability management (or more simply, integrated ALM) calls for an
organization structured into integrated units that include all the functional
activity related to a line of business. The integrated staff makes decisions
regarding the product(s) in the business line with the help of computer models that represent both the assets and liabilities associated with the business
line, characterize the uncertainty of the future business environment, and
produce strategies for structuring the assets and liabilities of the business
line in ways that are profitable across a range of alternative future environments. In short, the older functional management perspective calls for
functional units to make decisions using profitability calculations based on a
single-scenario planning forecast, while the newer integrated ALM perspective calls for business units to make decisions using risk-adjusted or hedged
profitability calculations based on multiple-scenario projections.

Alternatively, the article may be read as a discussion of the technical and
organizational issues raised by the use of modern financial simulation and
optimization techniques in the management of financial intermediaries. The
conceptual discussion is illustrated with an extended case study of a large
financial intermediary that has implemented elements of integrated ALM in
response to a number of problems that arose during the years a functional
management perspective was in wider use.

The article is divided into three sections. The first section introduces the
basic issues involved in managing a financial intermediary and identifies the

problems associated with the functional perspective in handling these management issues. The concepts of integrated ALM are outlined and the section concludes with a discussion of the challenges facing a financial intermediary's implementation of integrated asset-liability management.

The second section describes FNMA (the Federal National Mortgage Association, more commonly known as Fannie Mae) and its three major business activities: mortgage insurance, REMIC underwriting, and mortgage investment. The description focuses on the general features of each kind of activity that make them distinctive businesses.

The third section uses the conceptual management discussion and the outline of FNMA's major business activities as background for an account of how certain business problems associated with the functional management perspective led to the adoption of elements of the integrated asset-liability management perspective. This implementation case study focuses on issues related to managing interest rate risk in the mortgage investment business, managing default risk in the mortgage insurance business, managing marketing risk in the REMIC underwriting business, and managing the allocation of capital across different businesses.

1   Integrated ALM as Innovative Management

The evolution of management perspectives can be seen as an innovative management response to business problems. The individual concepts combined in the newer perspective have almost always been well developed and accepted techniques. The problems with the older perspective have often been understood in theory, but have not become manifest. It is usually the occurrence of a major business problem that triggers the critique of the older management perspective and the introduction of new management techniques. Unless there is an unusually high degree of internal management initiative, the newer management perspective evolves through the piecemeal implementation of new management techniques introduced to solve concrete problems arising under the older management perspective. It may take some time before there is a realization that a series of these new management techniques have combined to produce a new management perspective or system that is fundamentally different from the older perspective.

This section describes the major tasks involved in managing a financial intermediary, the older functional management perspective and its problems, and the concepts and challenges of the newer integrated asset-liability management perspective.

1.1 Managing a Financial Intermediary

Financial institutions serve as intermediaries in the capital markets between ultimate savers and ultimate investors. This intermediation involves issuing liabilities that savers or other intermediaries buy to hold as assets. These liabilities are financial securities that promise a future schedule of cashflow. A liability's scheduled cashflow may be certain or it may be contingent in the sense that it depends on the occurrence of certain future events. Liabilities with contingent cashflows are inherently risky because a potential buyer does not know which future events will occur. Other things equal, a buyer will expect to pay less for an intermediary's liability the more inherent risk it possesses.

A financial intermediary typically uses the proceeds of the liability's sale plus equity capital to buy assets, which are the liabilities of investors or other intermediaries. The cashflow of these assets are used to pay the liability's scheduled cashflow. Any asset cashflow remaining after the payment of the liability cashflow is profit for the financial intermediary. If the asset cashflow is not sufficient to pay the scheduled liability cashflow, then the losses of the financial intermediary may create doubts about the credibility of the liability's scheduled cashflow. Other things equal, potential buyers will expect to pay less for an intermediary's liability the less credible is the payment of its scheduled cashflow.

Since most financial intermediaries issue liabilities with contingent cashflow schedules and buy assets with contingent cashflow schedules, the future profitability of the intermediary is potentially quite uncertain. The basic management objective for a financial intermediary is to sell liabilities and buy assets in a way that the net cashflow or profit is both substantial relative to equity capital and consistent across the range of contingent events that affect future asset and liability cashflow.

Most financial intermediaries sell liabilities that are quite varied in their scheduled cashflow, and therefore, are bought by quite different groups of customers. Obviously, such a multi-product financial intermediary faces additional coordination tasks that make for a more complicated management problem than faced by a single-product intermediary. At the corporate level, the management objective is to allocate the intermediary's equity capital across the individual product or business lines in a way that satisfies both shareholders and regulators.

If a financial intermediary is managed so that it is relatively profitable in a wide range of event environments, then its liabilities will command a relatively high price and its profitability will be reflected in a relatively high price for its equity shares. On the other hand, a poorly managed intermediary that experiences uneven profitability and/or losses will be forced to sell its liabilities at relatively low prices and experience a relatively low price for its equity shares. If the credibility of a poorly managed intermediary erodes far enough, it can experience a run and liability buyers can experience losses resulting from a failure to pay scheduled cashflow. But more often the losses have fallen on taxpayers when insolvent intermediaries have been closed by government agencies that insure their liabilities.

This view of the management problem facing financial intermediaries has been developed in more detail by Holmer and Zenios (1995). Their analysis uses a less familiar vocabulary that describes the productivity of alternative management technologies using terms more often employed in reference to manufacturing companies. Such terminology is consistent with a flow-of-funds economic framework even though it makes financial intermediaries seem similar to manufacturing companies. The use of such terminology serves the useful purpose of suggesting parallels between recent management changes that are well understood in the manufacturing sector, but less well appreciated in the financial sector.

1.2   Functional Management Problems

In a multi-product financial intermediary organized and managed under a functional perspective, the central coordination tasks are performed by the annual corporate plan, which is based on a single-scenario macroeconomic forecast of interest rates and other business conditions. Functional units – for example, those that specialize in marketing (i.e., selling liabilities) and manufacturing (i.e., buying assets) financial products – all make decisions using the corporate plan as expectations. In addition to its business forecast element, the corporate plan usually includes guidelines for marketing (e.g., pricing rules) and manufacturing (e.g., asset allocation rules and return targets). For an intermediary with an established set of financial products operating in a relatively stable business environment, this mode of operation works reasonably well.

The problems associated with this functional management perspective become apparent when an intermediary's environment becomes more volatile or when it begins offering new financial products in response to competitive pressures.

The problem with reliance on a corporate plan in a more volatile business environment is that functional unit decisions, which are collectively profitable in the forecasted scenario, can become unprofitable if the scenario that actually does occur differs substantially from the forecast. If the macroeconomic and corporate forecasting models were quite accurate in practice, then this would be only a hypothetical problem. But, in fact, the macroeconomic models at the core of most corporate planning models, have a poor record at forecasting major shifts in the business environment. The models' repeated failures to forecast major movements in interest rates and other aspects of the business environment and the resulting disenchantment with macroeconomic modeling in both financial and non-financial businesses have been described recently by Passell (1996).

For financial intermediaries, perhaps the most stressful problems caused by making decisions based on inaccurate forecasts have been associated with major shifts in interest rate levels. The substantial increase in interest rates during the early 1980s, as well as the smaller, but sharp rise during 1994, exposed a wide range of problems among different kinds of intermediaries. Assets bought to cover contingent liability cashflow turned out to be insufficient in the altered interest rate environment. The resulting runs and regulatory closures involved large losses. There were noticeable losses in the insurance industry and local governments, but the major losses occurred in the thrift industry as described by Kane (1989) and White (1991).

In addition to problems caused by inaccurate forecasts, a financial intermediary operating under the functional management perspective is prone to introducing new financial products that are poorly designed.

The new product difficulties experienced by manufacturing businesses that are organized into separate functional units are widely appreciated. The traditional design process involving sequential activities by engineering, manufacturing, and marketing staff, has been found to be too slow and unreliable in producing new products that are demanded by customers at a price that permits profitable manufacturing.

Some established financial intermediaries that have developed new products while organized in functional manner have experienced similar difficulties. For example, several decades ago the life insurance industry introduced the guaranteed investment contract (GIC) product. While certainly attractive to customers, subsequent experience with the product suggests that insufficient attention was paid to how it could be manufactured profitably in a wide range of interest rate environments. Life insurance companies experienced substantial losses for years as a result of this poorly designed new product.

In addition to these problems at the product level, financial intermediaries have also had difficulties at the corporate level, where the management task is to allocate equity capital among the different products in a way that promotes high and steady profits. Increased regulatory pressure for capital adequate to buffer against risks has been an additional problem for intermediaries employing a functional management perspective. Use of a single-scenario plan leaves financial intermediaries ill-equipped to respond to regulatory requests for risk estimates that are based on financial performance across many scenarios representing a wide range of future business environments. In addition, the inability of single-scenario forecasts to represent risk may also be consid-

ered a problem by shareholders, who have an interest in not only high, but
also stable profits.

1.3 Concepts of Integrated ALM

As the business environment has become more volatile over the past few
decades, innovative financial intermediaries have identified elements of the
older functional management perspective as the cause of their business problems and have changed their mode of operation in ways that reflect the newer
integrated asset-liability management perspective.

A key realization was that reliance on an error-prone, single-scenario economic forecast with functional units making rule-based decisions in light of
that forecast was just too risky for the more volatile business environment.
"As a result, most big companies today are giving short shrift to economic
forecasting and focusing instead on reducing their exposure to risk," according to Passell (1996). This shift is evident in both the manufacturing and the
financial sectors, because all firms face the same pressure from shareholders,
who prefer a stable profit stream over a volatile profit stream if the average
levels of the two profit streams are the same. The pressure to shift from
forecast-based decisions to hedging-based decisions has been even stronger
for financial institutions that have been pressured by regulators who insure
their liabilities.

This new focus on risk management has had far-reaching effects on both the
technical and organizational aspects of operations at financial intermediaries.

The major technical ideas used in the integrated ALM perspective were
developed independently over the course of the past half century. Methods
for characterizing attitudes towards risk and making decisions in the face
of uncertainty were formalized beginning in the 1940s as described by von
Neumann and Morgenstern (1947). Monte Carlo simulation methods were
developed extensively during the 1940s and 1950s as described by Hammersley and Handscomb (1964). The basic techniques of portfolio optimization
by means of risk diversification were developed during the 1950s and 1960s
as described by Markowitz (1959) and by Mossin (1968). The development
of modern finance theory, which produced both analytic and computational
methods for assessing the value of contingent cashflows as well as numerous other advances, began in the 1970s as described by Black and Scholes
(1973) and by Merton (1992). Ideas from these research areas form the key
technical elements of integrated ALM. They began to be practical management techniques beginning in the 1980s because of the continuing advance in
the power of computers of all sizes ranging from supercomputers to desktop
workstations.

These technical ideas are combined in the integrated ALM perspective in specific ways that facilitate management at both the product level and the corporate level.

At the individual product level, these technical ideas are employed to develop tools that allow management to simulate the profitability of a set of product design, pricing, and funding decisions across many scenarios, each one of which represents an alternative future business environment. Monte Carlo simulation methods are used to generate the contingent cashflows for assets and liabilities and modern finance theory is used to price the assets and liabilities at both the beginning and end of the assumed horizon or holding period. The result of the simulation is a scenario distribution of return on equity (ROE) or some other profit measure. Given the management's expectations and aversion to risk, each scenario is assigned an occurrence probability and expected utility theory is used to calculate the certainty-equivalent ROE or other risk-adjusted profit measure. Since the assets, liabilities, and equity capital associated with the product or business line are considered a leveraged portfolio, the final step is to use the mapping from one set of decisions to a value for certainty-equivalent ROE to perform a portfolio optimization which identifies the particular set of design, pricing, and funding decisions that produce the highest product certainty-equivalent ROE.

One important implication of this way of combining the technical ideas is that all decisions relevant to a product or business line are interdependent. The product design decision has implications not only for the pricing of the liability, but also for the kind of asset acquisition strategies that could profitably fund the liability. The pricing decision must take into account both the inherent risk of the product relative to competing securities available in the capital markets and the sales proceeds that will be available for buying the funding assets. And the kind of assets selected to fund the liability has implications for the design and pricing of the product. In other words, it would be difficult to compartmentalize decision-making regarding the design, pricing, and funding of a financial product since decisions in all these areas affect the risk-adjusted profitability of the product.

At the corporate level, these technical ideas are employed to develop an equity capital balance sheet which can be simulated and optimized in a manner similar to each product line's balance sheet. The assets on this corporate balance sheet represent each product or business line to which equity capital is allocated. The corporate balance sheet is simulated under the same set of scenarios used in the individual product balance sheet simulations. The scenario return distribution associated with each asset on the corporate balance sheet is the scenario ROE (or other profit measure) generated by a dollar of equity capital invested in that product or business line. For each allocation of available capital across the products or business lines, the scenario ROE for total corporate capital can be calculated and converted into a corporate certainty-equivalent ROE. The optimization of this corporate balance sheet is accomplished by finding the allocation of equity capital that maximizes the corporate certainty-equivalent ROE.

Also, this kind corporate equity capital balance sheet can be used to analyze the risks facing a financial intermediary that is operating under some kind of risk-based capital regulation. It is vital to have a single framework that permits management to estimate how different decisions affect average profit and profit variability from the perspective of both shareholders' expectations and regulators' rules.

Implementation of these sorts of technical ideas produces a need for organizational changes at both the product management level and the corporate management level.

At the product level, the interdependence of design, pricing, and funding decisions calls for an integrated product team rather than separate functional units. Under integrated ALM the ongoing use of product portfolio simulation and optimization tools replaces the functional perspective's decentralized application of operational rules-of-thumb in light of the corporate plan's forecast. The nature of the technical decision-support tools calls for a tightly integrated staff responsible for all decisions related to a product or business line. This is true for existing products and even more so for new financial products where understanding the interdependencies among the design, pricing, and funding decisions is particularly important to the success of the new product.

Recent trends in the manufacturing sector – especially among firms facing frequent new product introductions – provide an organizational model for financial intermediaries implementing elements of integrated ALM. Innovative manufacturing firms have had success with the concurrent engineering approach that involves establishment of a product team that combines marketing, engineering, and manufacturing staff. This approach speeds up the new product introduction cycle largely by providing the integrated team an opportunity to consider the interdependencies of all the different decisions simultaneously rather than sequentially by different functional units.

For a financial intermediary that is implementing technical aspects of integrated ALM for management of a product or business line, the obvious organizational corollary is to establish a group that integrates staff from several functional areas and appoint a group manager that is responsible for all decisions related to that product or business line. It is the responsibility of that group and its manager to use the technical simulation and optimization tools in a way that internalizes the interdependencies of all the decisions. The group and its manager are evaluated based on the prospective certainty-equivalent ROE of the product or business line rather than on retrospective profitability, which can be influenced by good or bad luck in the realization of the business environment. The use of a prospective, risk-adjusted performance measure is essential if the potential future implications of current decisions of a product group manager are to be assessed by top management.

These organizational ideas are consistent with those of the business reengineering movement as described by Hammer and Champy (1993), for example. The objective of reengineering is to redesign the work process in a way that provides quality products produced in a cost-effective manner by taking advantage of the automation possibilities presented by modern computer systems. The computer-based technical tools of integrated ALM provide the automation opportunities for reengineering the financial product management process.

This description of the technical and organizational aspects of integrated asset-liability management is developed in more detail by Holmer and Zenios (1995).

1.4   Challenges of Integrated ALM

The adoption of elements of the integrated ALM perspective by innovative financial intermediaries indicates that some top managers believe that, all things considered, it is likely to produce better results than would the older functional management perspective. While judged better, it is not perfect in the sense that that there are no problems or challenges associated with integrated ALM. It has been chosen because its problems have been judged less difficult to live with than the problems associated with the functional management perspective. This section concludes with list of some of the challenges facing top managers who decide to implement changes inspired by the integrated asset-liability management perspective.

The technical challenges are mostly related to development and use of the portfolio simulation and optimization tools that form the technological core of integrated ALM. Depending on the nature of the financial product, such a simulation-optimization model can be quite complex conceptually and this complexity may present substantial computational demands. The conceptual complexity requires staff with substantial skills in both applied finance and software development since appropriate tools are almost never available on the market and hence need to be designed, constructed, and maintained by internal staff. The computational demands can often be met by creating tools that distribute processing across a network of workstations. This is far less expensive than buying a supercomputer, but the equipment savings requires skilled staff to develop the distributed-processing tool and to maintain a reliable network of workstations.

Once a simulation-optimization model has been developed for a product or business line, most financial intermediaries are likely to experience challenges in providing the asset and liability data necessary to support the model's calculations. More broadly, much of the data required to support the integrated ALM perspective will be new, and therefore, effort will be required to extract these new data from existing operational computer systems. These data problems are not conceptually difficult, but may require substantial effort to solve depending on the state of the intermediary's operational computer systems. This data collection challenge may be viewed as an example of the problems associated with the development of any new decision-support database.

After the development of the simulation-optimization model and implementation of the process by which its decision-support database is updated, problems associated with determining the best way to use the model in daily decision making remain to be solved. There are conceptual as well as practical problems involved in determining how the model can best be used to support management decision making. Many of these problems can be solved by staff training on the theoretical concepts and practical use of the model. Other problems are more difficult to handle because they are related to how staff are organized and how their performance is measured. These organizational challenges require strong leadership from the top managers who are initiating the shift from the functional management perspective to the integrated ALM perspective.

The organizational challenges are mostly related to the fact that the highly interdependent product decisions that once were made in separate functional units loosely coordinated by the corporate plan, are now best made in product groups. It would be difficult to overestimate the problems involved in transforming an organization consisting of functional units into one where all decisions regarding a product are the responsibility of a product group and its manager. To accomplish this transformation, there is no substitute for a clearly thought out reorganization plan and the commitment of top managers with substantial leadership skills. Such a reorganization may be implemented over a period of years, but no matter what the pace, it will face the following issues at a minimum.

Each product or business line group and its manager need to have performance measures by which the results of their work will be judged by top management. The kinds of prospective risk-adjusted performance measures that are consistent with the integrated ALM perspective are often unfamiliar to managers who have experience with retrospective measures. Beyond this acceptance problem, periodic reviews of product group managers that take place in a timely fashion will produce additional demands on the staff responsible for gathering data and operating the model.

The remaining two organizational challenges mentioned here are problems likely to be faced by the financial intermediary's top management in their coordination of the decisions made by several product groups or business lines. Unlike the challenges mentioned above, I am not aware of broadly accepted solutions to these problems. If they are unsolved problems, their description can be viewed as an attempt to formulate management research topics.

The first of these product group coordination challenges arises when the scenario ROE distribution of one group is potentially negatively correlated with the scenario ROE distribution of another product group. Consider the simplified situation in which the first group can make one of two sets of product decisions called A and B. The A decisions produce a scenario ROE distribution that is not correlated with the scenario ROE distribution of the second group and has a relatively high certainty-equivalent. The B decisions produce a scenario ROE distribution that is negatively correlated with the scenario ROE distribution of the second group, but has a certainty-equivalent that is not as high as that produced by the A decisions. Suppose the strength of the negative correlation causes the certainty-equivalent ROE for the two groups considered together to be higher when the first group makes the B decisions than when it makes the A decisions. When the first group makes the B decisions, it can be said that it is a natural hedge for the scenario ROE distribution of the second group. How can the corporate-wide benefits that would result from the first group making the B decisions, which are suboptimal from the myopic view of the first group, be made apparent to the manager of the first group? And, if a scheme were devised to send the correct signal to the first group, how should the benefits of the natural hedge be allocated across the product groups?

The second of these product group coordination challenges arises when top management allocates capital across the several product groups or business lines. From the shareholders' point of view the objective of this allocation of equity capital should be to achieve a desirable corporate-wide scenario ROE distribution. What exactly that desired distribution will look like depends on the diversification opportunities available to shareholders in the capital markets. From the regulator's point of view the objective of this allocation of equity capital should be to reduce the probability of insolvency, and therefore, the likelihood that the regulator would have to make good on its insurance of the financial intermediary's liabilities. Since the regulator has few, if any, diversification possibilities, there is little reason to believe that these two different points of view on proper capital allocation will imply similar allocations. This problem could be conceptualized as a complex dual objective optimization problem, but currently most regulated financial intermediaries grapple with this difficult problem using more *ad hoc* methods.

2   FNMA as a Financial Intermediary

This section provides a short description of FNMA (the Federal National Mortgage Association or Fannie Mae) and its three major business lines. The description includes short accounts of FNMA's mortgage insurance business, its REMIC (real estate mortgage investment conduit) underwriting business, and its mortgage investment business. These accounts are intended to provide background information for the subsequent section's discussion of how FNMA has introduced elements of integrated asset-liability management to solve past business problems.

FNMA is a United States government-sponsored enterprise that is chartered by an act of Congress, owned by shareholders, and regulated by OFHEO (the Office of Federal Housing Enterprise Oversight, which was established in 1993). FNMA was created in the 1930s with the objective of providing financial products and services in the secondary mortgage market that increase the availability and affordability of housing for low-, moderate-, and middle-income Americans.

2.1   Mortgage Insurance Business

FNMA does not directly extend mortgage loans to home buyers. Rather than operate in the primary market, its role is to maintain a secondary mortgage market by engaging in different kinds of transactions with the financial intermediaries that make mortgage loans to individual home buyers. One important kind of secondary-market transaction involves FNMA insuring a group of mortgage loans against payment delay or default by the home buyer and pooling that group of mortgage loans into a single-class mortgage-backed security (MBS). The ownership of the MBS may be shared between several investors, with the total monthly payment proportional to each investor's ownership share and the mixture of a given month's principal and interest the same for each investor.

This transaction may be thought of as creating a structured balance sheet with the liabilities consisting of the single MBS and the assets consisting of numerous home mortgage loans. All the monthly (scheduled and prepaid) principal payments and most of the monthly interest payments from the mortgage loans are passed from the home buyers, through the financial intermediary that services the loan and through FNMA, to the investors who own fractions of the MBS. The MBS agreement calls the servicer to remove a small fraction of the interest payments each month as a charge for the administrative costs incurred in servicing the loan. And FNMA removes a small fraction of each month's interest payments as an insurance premium – that is, a charge for insuring the amount and timing of the mortgage principal and interest payments to MBS investors.

A substantial fraction of home mortgage loans have been pooled into insured MBS by FNMA or its main competitor, FHLMC (the Federal Home Loan Mortgage Corporation or Freddie Mac), which is also a government-sponsored enterprise. This pooling and insuring activity has the important capital market effect of facilitating the sale of mortgage loans by the originators to investors whose liabilities are better suited to long-term mortgage assets than are the short-term liabilities of most mortgage originators. Decades ago before the current high volume of MBS creation, mortgage originators such as banks, S&Ls, and other thrift institutions, typically held fixed-rate mortgage loans as long-term assets even though they created a severe duration mismatch with their short-term deposit liabilities. The rise in interest rates during the 1970s and early 1980s exposed this problem and resulted in widespread regulatory closures of insolvent financial institutions. Now the remaining mortgage originators are more likely to convert their loans into MBS and sell the MBS to investors with longer-term liabilities such as pension funds and life insurance companies. These investors find MBS much more desirable than a group of individual loans because of the administrative simplicity of dealing with a bulk security that can be easily traded and because of the credit enhancement provided by the insurance. After the MBS creation and sale, the originator is left with the loan servicing responsibilities. But even that responsibility and the associated servicing fee can be sold to other companies that specialize in providing this kind of computer-intensive administrative service. This leaves the mortgage originator with the option of specializing exclusively in the business of underwriting loans for an origination fee.

From FNMA's perspective, these transactions comprise a mortgage insurance business. As in any insurance business, premiums constitute income and claims constitute expenses. The delay between premium receipt and claim payment permits investment of premium proceeds and the earning of addition investment income on these reserve assets and on the equity capital assigned to this business line. The key management activities involved in conducting this business are establishing and enforcing loan underwriting guidelines for the mortgage originators, developing a pricing schedule for MBS transactions that involve a variety of loan types, and determining an investment strategy for the reserve assets. The underwriting guidelines may be thought of as direct regulation of the kinds of mortgage loans that are eligible to be insured, while the pricing schedule may be viewed both as risk-based insurance premiums and as financial incentives that influence the kinds of mortgage loans that are actually converted into MBS. Since the possibility of mortgage prepayment makes premium income uncertain and the claims associated with mortgage defaults or payment delays are highly uncertain, the prospective profit and return on equity generated by an MBS transaction is sensitive to movements in interest rates in the national capital markets and to movements in prices in local housing markets. And, of course, the pattern of reserve asset investments can substantially influence some aspects of this profit sensitivity.

2.2   REMIC Underwriting Business

Despite its considerable advantages over individual mortgage loans, the MBS presents problems for investors who want assets with certain target durations. The monthly principal payments generated by a fixed-rate MBS can be spread over several decades unless a sharp fall in rates produces a large volume of prepayments as home owners refinance during a short period of time. The response to this demand for mortgage securities with more targeted durations was the collateralized mortgage obligation (CMO), the first one of which was issued in 1983 by FHLMC. The federal tax status of these multiple-class securities was clarified in the tax reform of 1986, which created a type of CMO called a real estate mortgage investment conduit (REMIC). For an account of the development of pre-REMIC CMOs, see Parseghian (1987).

The REMIC is a legal trust whose assets are MBS and whose liabilities are multiple classes of bonds, the interest and principal payments of which are determined by the timing of the MBS cashflow and by the rules of the trust. The rules are fixed, which means that the trust can be thought of as a completely passive asset-liability management scheme that passes on to investors in the different bond classes all the interest-rate risk inherent in the underlying MBS collateral and trust rules. These rules can be relatively straightforward or quite complicated depending on how the REMIC is designed to channel the MBS cashflow to interest and principal payments on each REMIC bond class. Residual cashflow is paid to owners of the REMIC's equity class.

Both the number of REMIC bond classes and the complexity of the rules that channel MBS cashflow to the classes increased considerably during the 1980s. The early issues consisted of only three or four classes and a simple sequential principal payment rule. In such a simple issue, all the MBS principal would be paid to the first bond class until it was completely retired. Then all the MBS principal would be channeled to the second bond class, which had been receiving fixed-rate interest payments while the first class was being paid down. This kind of sequential principal payment rule creates bond classes with very different durations, some of which are suitable for investors with relatively short duration liabilities (e.g., banks) and others of which are suitable for investors with relatively long duration liabilities (e.g., pension funds).

By the 1990s, a typical REMIC issue would consist of one or two dozen bond classes of great variety. Some classes would have coupon interest rates that floated according to formulas that sometimes magnified the movement in the underlying index interest rate; others would have simple fixed rates. Some classes would have their principal payments governed by trust rules that stabilized their timing and duration as long as MBS prepayments remained within a certain range; others served as prepayment shock absorbers making their duration quite sensitive to the uncertain timing of MBS prepayments. In other words, the trust rules did nothing to change the contingent nature of the MBS cashflow, but did produce REMIC bond classes with a wide variety of contingent cashflow characteristics.

The high volume of REMIC issuance by FNMA and FHLMC during the late 1980s and early 1990s produced a situation in which about two-thirds of all MBS had become REMIC collateral by 1994. This development has had important capital market effects. The more extensive tailoring of the contingent cashflow characteristics of REMIC bond classes (relative to those of the MBS collateral) provides investors with a wider range of assets with which to match their liabilities. And the high demand for MBS as REMIC collateral has caused a relative increase in their prices, enabling mortgage originators that securitize their loans to offer home buyers lower rates.

From FNMA's perspective, the issuance of REMICs comprises a large, but specialized securities underwriting business. The fees generated by this underwriting business arise from selling the REMIC bond classes and equity class for more than the cost of purchasing the MBS collateral that constitute the REMIC's assets. Other than the administrative costs associated with making the monthly principal and interest payments and providing REMIC investors with tax statements, there are no costs to REMIC issuance after the trust has been endowed with its MBS collateral. Since the MBS are relatively standardized and trade in a relatively active market, the key to high underwriting fees is structuring the bond classes in a way that they can be sold at relatively high prices to the different kinds of investors for which they have been designed. This requires extensive knowledge of the investment objectives of a wide variety of potential investors as well as maintenance of trading relationships with all the investors. These requirements mean that the two key management activities involved in conducting this business are designing the REMIC bond classes and then marketing the REMIC bonds to investors.

2.3   Mortgage Investment Business

FNMA helps maintain a secondary mortgage market by also engaging in mortgage investment transactions that are completely different from the securitization activities involved in the mortgage insurance and REMIC underwriting

businesses. The mortgage investment business involves buying mortgage securities and holding them as portfolio investments. Most of the funds used
to buy the portfolio's mortgage assets are obtained by issuing FNMA liabilities such as bonds. The remaining funds represent the portion of FNMA's
equity capital that has been allocated to this business line. While most of
the portfolio's assets are individual mortgage loans that have been purchased
from originators, some of the portfolio's assets are MBS and REMIC bonds
that have been purchased on the open market.

The purchase of mortgage securities for portfolio investment helps maintain a secondary mortgage market in at least two significant ways. First, it
provides mortgage originators with sales opportunities for small amounts of
loans that in larger volume would be appropriate for MBS pooling. And second, it provides originators a sales outlet for unusual kinds of loans for which
there is not yet an MBS market. In both these ways, the mortgage investment
business contributes to the overall objective of increasing the availability and
affordability of mortgage funding.

From FNMA's perspective, these mortgage investment and liability issuance transactions comprise a classic portfolio management business. Since
the cashflow of the assets in highly contingent on future interest rates and
prepayments and the portfolio is large and highly leveraged, there is a substantial challenge involved in the asset-liability management of the interest
rate risk potential of the portfolio. The key management activities involved
in conducting this business are deciding what mortgage assets to buy and
what kind of bonds to issue. An additional important management decision
is the mixture of debt and equity funding of the mortgage assets.

3   FNMA's Movement to Integrated ALM

This final section presents accounts of how FNMA introduced elements of
integrated asset-liability management to solve business problems that have
arisen over the course of the 1980s and 1990s. The first account focuses on
the large financial losses experienced by the mortgage investment business
during the early 1980s and on the subsequent shift to integrated management of the interest rate risk inherent in that business. The second account
discusses the creation of the mortgage insurance business in the early 1980s
in response to the problems in the mortgage investment business and the
subsequent development of integrated techniques to manage the credit risk
inherent in that business. The third account examines the development of
the REMIC underwriting business following the 1986 tax reform, the collapse
in underwriting volume following the sharp rise in interest rates during 1994,
and the management options available to deal with this collapse. The last
account focuses on early 1990s changes in the management of corporate eq-

uity capital that were introduced in response to an increase in the scope and intensity of regulation.

3.1   Investment Business Management Changes

According to OFHEO (1995, Historical Data Tables), FNMA earned $162 million in 1979 and ended the year with a balance sheet that consisted of $49.8 billion in mortgage assets, $1.5 billion in other assets, and $48.4 billion in debt outstanding. At the end of the 1970s, FNMA had issued no MBS and the CMO/REMIC had not yet been invented. The book value of its equity capital was about $1.5 billion. All profits were being generated by the mortgage investment business, which was the only line of business being pursued by FNMA.

The rise in interest rates early in the 1980s created a severe problem for FNMA since its mortgage investment business (like that of most thrifts) was not well structured at that time. According to Holmer (1994, page 5), the duration of FNMA's fixed-rate mortgage assets was estimated to be 62 months in 1980, while the financing debt had an estimated duration of 26 months, producing a duration gap of 36 months. The rise in rates left FNMA with a negative net interest income on its mortgage portfolio and overall losses of about $200 million in both 1981 and 1982. Smaller losses were experienced in 1984 and 1985. The book value of equity capital dropped by more than a third and did not surpass its 1979 value until the end of 1987.

Following the 1981 change in corporate leadership, FNMA solved this mortgage investment business problem in two stages. The first stage continued to use functional management techniques including a corporate planning model that produced a forecast of the mortgage portfolio's expected return on equity given macroeconomic forecasts of interest rates and mortgage origination volume. The first-stage innovation was the introduction as a management goal of asset and liability duration matching. The new duration matching strategy was supported by a simulation model – inherited from the 1970s – that used a number of *ad hoc* interest rate scenarios to estimate the duration of mortgage assets and financing debt. This model was also used to gauge the desirability of mortgage investments by estimating for each scenario the internal rate of return of a debt-financed mortgage investment considered on a stand-alone basis in isolation from the existing portfolio.

The mortgage portfolio's duration gap was reduced from its 1980 value of 36 months to 8 months in 1987 and to 3 months at the end of 1990 through a combination of measures aimed at decreasing the duration of mortgage assets and increasing the duration of financial liabilities. The asset duration in 1990 had decreased to 41 months (from 62 months in 1980) through the acquisition of both adjustable-rate mortgages, which first appeared in the early 1980s, and 15-year fixed-rate mortgages. The liability duration in 1990 had increased to 38 months (from 26 months in 1980) through the issuance of longer term non-callable bonds.

Once the mortgage portfolio had become roughly duration matched, the limitations of the first-stage management techniques became apparent. These techniques presented at least three key problems that prevented effective management of the trade-off between the portfolio's expected ROE and risk. First, expected return was measured in ROE percentages and risk was measured in duration gap months, which raised difficult questions about how to integrate the two into a comprehensive risk-adjusted measure of prospective portfolio performance. Second, the desirability of new transactions was analyzed on a stand-alone basis in isolation from the existing portfolio of mortgage assets and debt, making it impossible to gauge the effect of a transaction on the portfolio's performance. And third, the stand-alone analysis of new transactions operated in a what-if mode rather than using optimization methods that could systematically find the combination of transactions that most improve the portfolio's performance.

During the late 1980s, FNMA began development of a new simulation-optimization model that could overcome these problems. The ALMS (Asset-Liability Management Strategy) system, whose conceptual design and computer implementation are described in detail by Holmer (1994), uses Monte Carlo simulation, modern finance, expected utility, and optimization techniques to find the combination of assets and liabilities that maximize the expected utility of the portfolio's holding-period ROE distribution across future interest rate scenarios. After a prototype met expectations, a multi-user, distributed-processing version of the system was developed using a client-server architecture with a relational database management system running over a network of Unix workstations.

This production version of the ALMS system became operational company wide in the early 1990s and its use led to a number of important changes in the conduct of FNMA's mortgage investment business.

From the earliest prototype days, the ALMS system's portfolio analysis results played a major role in rationalizing FNMA's shift toward callable debt financing of its mortgage investments. The use of expected utility theory and the certainty-equivalent ROE concept, which combines the expected return and interest-rate risk of a portfolio in a rigorous fashion, allowed senior management to characterize portfolio risk reduction that involved issuing higher-coupon callable debt as an activity that increased risk-adjusted ROE. Use of the certainty-equivalent ROE concept became routine among FNMA financial managers and the concept was used in presentations to Wall Street equity analysts and shareholders. As a result, FNMA moved from having no callable debt financing of its mortgage assets in the 1980s to being one of the world's largest corporate issuers of callable debt in the 1990s.

As the production version of the ALMS system was being implemented across the company, FNMA undertook for other reasons a comprehensive review of its computer operations. The consultants who were assigned this task were advocates of organizational reengineering concepts and were puzzled by the mismatch between the new integrated asset-liability management tool and the old functional management structure and staff organization. The company-wide availability of the ALMS system and the reengineering perspective of the consultants focused senior management on the question of how best to use the new tool. After several years of staff reorganization, the organization of the mortgage investment business has become more integrated and focused on use of the ALMS system as its analytical tool. By the mid-1990s the older models had fallen out of use and the group that had been responsible for developing the corporate plan disappeared.

3.2 Insurance Business Management Changes

FNMA entered the mortgage insurance business in 1981 in an attempt to reduce its reliance on the mortgage investment business, which was losing money. At that time the business was dominated by FHLMC which engaged in mortgage insurance but not mortgage investment activities. The management strategy was obvious: start a new business line that could generate profits to offset losses in the existing business line and expand the new business line by turning a monopoly into a duopoly.

The main challenges facing FNMA in starting this new mortgage insurance business were establishing mortgage loan underwriting guidelines and creating procedures to determine guarantee fee quotations for mortgage originators interested in transforming loans into MBS. A functional management structure was implemented to handle these start-up challenges. The credit policy group developed underwriting guidelines for the new MBS business and the marketing group broadened its loan buying activities to include pooling mortgage loans into MBS. Guarantee fee quotes were developed informally in reference to FHLMC quotes. Representatives of these groups met regularly to coordinate actions and consider non-routine MBS transactions.

The expected challenges of starting a new business line were compounded by early problems with Texas and Alaska mortgage defaults. After the sharp drop in crude oil prices during the mid 1980s, the economies of Texas and Alaska collapsed leading to substantial declines in house prices in those areas. As mortgage interest rates dropped, many homeowners in those areas found themselves paying high rates on houses that were worth as little as half the mortgage loan amount. The ensuing wave of mortgage loan defaults placed considerable pressure on the start-up business line. Management responded by tightening underwriting guidelines and by initiating the development in the finance group of a Monte Carlo simulation model that would determine guarantee fee levels that would produce a target rate of return on MBS comprised of mortgage loans with different risk attributes.

Development of the Financial Analysis Simulation Model began in 1985 and continues to be directed by Mike Goldberg, a vice president in the finance group. The model was originally based on mortgage prepayment rate and default probability equations estimated with data on Federal Housing Administration mortgages. Using Monte Carlo methods and a short- and long-rate stochastic process from the finance literature, the model estimates the minimum guarantee fee that generates a target risk-adjusted internal rate of return on the cashflow associated with a particular pool of mortgage loans that is being proposed as an MBS security. Later during the 1980s, corporate data on the prepayment and default history of various types of mortgage loans were used to estimate the prepayment and default probability equations. At the beginning of the 1990s, the model was moved off the mainframe onto Unix workstations and work began on preparing corporate data so that they could be used to estimate the parameters of a stochastic process for house prices. By the mid 1990s, this statistical research had advanced to a stage that allowed the model to recognize regional differences in the parameters of house-price stochastic process.

Beginning in the early 1990s, the internal inefficiencies of the functional organization of the mortgage insurance business and competitive pressures from FHLMC lead to major changes in the mortgage insurance business. The reengineering consultants recommended an approach in which the central-office meetings of the functional group staff representatives would be replaced by integrated regional-office marketing staff querying an internal, on-line guarantee fee quotation system. This system provides a range of risk-based guarantee fee quotes for a very large number of different mortgage pool types, thus providing marketing staff with a quick indication of the appropriate bargaining range for a wide variety of different kind of MBS transactions.

By the mid-1990s, this internal re-engineering effort had expanded to focus on the nature of business relations between FNMA and mortgage originators. FNMA has begun to offer a variety of software products and network services that automate a wide range of mortgage origination activities including the application of underwriting guidelines, credit checks, and property appraisals. FHLMC is also beginning to offer similar software and network connections to mortgage underwriters. These initiatives are significant enough to characterize as the beginnings of a re-engineering of the mortgage origination industry in the United States. For a more detailed discussion of this effort, its effect on mortgage originators, and the possible risks for FNMA and FHLMC, see the regulatory discussion by OFHEO (1995, 1–7). Given the existence of the internal guarantee fee quotation system and this initiative to re-engineer the business relationship between FNMA and mortgage originators, the next logical step would be to consider extending some version of the on-line guarantee fee quotation system directly to mortgage originators. Whether FNMA management would consider this development to be advantageous is unclear, but it is rapidly becoming technologically feasible.

These model development and re-engineering activities draw the attention of staff in many nominally separate groups to the task of developing the automated system through which FNMA interacts with mortgage originators. The focal point of all this activity are the staff that actually develop the software components of the system. Regardless of the nominal organizational structure, the credit policy, finance, marketing, and information systems staff who develop this system have become an integrated work group. It is the technological imperatives created by the use of computer-based management and decision-support tools, which incorporate all the functional aspects of the business decisions, that drives the organization toward functional integration.

FNMA's entry into this new business line has proven successful over the past decade and a half. FNMA's share of MBS outstanding grew from 0% in 1980 to 48% in 1990 and 51% in 1994, according to figures supplied by OFHEO (1995, Historical Data Tables). And this new business volume is reasonably profitable since the mortgage insurance business contributes a substantial fraction of FNMA's earnings.

3.3 Underwriting Business Management Changes?

FNMA entered the REMIC underwriting business following the changes in tax law of 1986. The volume of REMIC bonds underwritten grew from nothing in 1985 to nearly $211 billion during 1993, then dropped according to OFEHO (1995, page 45) to an annual rate of $5 billion during the first quarter of 1995 following the rise in interest rates during 1994. During the 1986–1993 initial expansion phase of the business, FNMA essentially subcontracted the two main management activities of the business – REMIC bond design and marketing – to Wall Street investment banking firms. The underwriting profits, which are generated by selling the REMIC bonds for more than the cost of the REMIC's MBS collateral, are split between the investment banking firm and FNMA. During these years the business grew rapidly and made a non-trivial contribution to FNMA's profits.

As described above, the bond classes of a typical REMIC vary considerably in their interest rate risk. Since the basic mortgage prepayment risk is complex and the REMIC trust rules have typically become quite complicated, there is a substantial challenge involved in assessing the nature and degree of interest rate risk inherent in many REMIC bonds. It seems clear in retrospect, that a two-tier REMIC market arose during the late 1980s and early 1990s. In the upper tier of the market, REMIC bonds were purchased by investors who either bought low-risk bonds or knowledgeably bought higher-risk bonds. This latter group in the upper tier typically quantified the interest rate risk of REMIC bonds using modern finance models, which allow them to determine if a bond is priced fairly given its risk attributes and to ascertain whether the particular kind of interest rate risk exhibited by a bond is compatible with the overall risk exposure of the portfolio to which it would be added. In the lower tier of the market, REMIC bonds were purchased by less knowledgeable investors who typically bought higher-risk bonds without the advantage of a model's assessment of their fair price or risk attributes. Buyers in the lower tier of the REMIC market fall into two classes: smaller institutional investors (e.g., small pension funds or local governments) and individual investors. REMIC bonds were sometimes designated as "retail classes" by the investment banking firms if the marketing strategy was to sell these bonds to individual investors. To the extent that REMIC bonds were able to be sold in the lower tier for higher prices than in the upper tier of the market, this REMIC bond design and marketing strategy lead to higher underwriting profits than would a strategy of marketing solely to upper-tier investors.

One way to interpret the collapse in REMIC underwriting volume following the 1994 interest rate rise is that the substantial losses incurred by lower-tier investors drove up the REMIC bond prices in that tier enough to eliminate the profit on most REMIC underwritings. Given this interpretation of the development of a two-tier REMIC market, FNMA is currently facing a difficult business problem for which the solution is not obvious. The loss in underwriting profits is non-trivial, but it is not clear whether a change in the way the REMIC underwriting business is managed would solve the business problem.

One management strategy is to make no changes in the current practice of delegating REMIC bond design and marketing decisions to the Wall Street investment firms. This could be the best strategy if one thinks that other factors (e.g., a relatively flat yield curve) are the main causes of the collapse in REMIC underwriting volume. The no-change strategy could also be best if one thinks that the memories of 1994 losses will fade and REMIC bond prices will rise in the lower tier of the market enough to make most REMIC underwritings profitable. Even if these developments do not materialize, low REMIC underwriting volumes have a silver lining: the lack of REMIC collateral demand lowers the prices of MBS making them more profitable from the point of view of the mortgage investment business. Whether the no-change strategy is best depends on the alternative management strategies available to FNMA.

The strategy of not delegating REMIC bond design and marketing to Wall

Street investment firms is clearly not viable. To handle these tasks effectively, FNMA would have to develop extensive advisory and trading relationships with investors who consider REMIC bonds as just one of many kinds of portfolio investments. This would require enormous effort in an area where FNMA has no substantial experience. Even if these difficulties could be some-how overcome, this no-subcontracting strategy still leaves open the question about how to deal with the two-tier REMIC market, which the analysis above suggests is at the heart of the business problem facing FNMA.

One alternative strategy is to continue to delegate REMIC bond design and marketing activities, but to impose some structure on the kinds of bonds that can be designed and marketed. This strategy would require each REMIC bond to be an example of one of several dozen generic types of REMIC bond. The working hypothesis behind this strategy is that lower-tier investors over reacted to the 1994 losses because the risk they face in buying REMIC bonds is unknown since they have no access to sophisticated pricing and risk models. By ensuring that all REMIC bonds were of know generic types, it becomes possible to produce and sell the kinds of financial models not now available to lower-tier investors. The main reason why such models are not available on the market is that the constant "innovation" of new kinds of REMIC bonds means that models become out-dated quickly and the cost of keeping them up to date is able to be borne only by larger upper-tier investors. Whether such models were sold by third-party vendors or by FNMA itself, the availability of such models could reduce to fear of currently unknown interest rate risk in the lower tier and actually lead to a rise in REMIC bond prices in that tier.

It is too early to know how FNMA will try to solve this business problem. FHLMC also faces the same problem. The competition between the two may generate pressure for innovative solutions to the problem. If either firm gives serious consideration to this regulated-subcontracting strategy, many of the same kinds of technical and organization issues associated with earlier introductions of modern quantitative finance modeling techniques will have to be dealt with.

3.4 Capital Allocation Among Businesses

Since the early 1990s a number of developments have lead to changes in how FNMA decides to allocate equity capital among each of its major business lines. Several forces have lead to the increasing use of modern finance tech-niques to support these management decisions – the intellectual momentum generated by the earlier adoption of an integrated asset-liability management approach in the mortgage investment and mortgage insurance businesses, the arrival of new set of top managers, an increase in regulatory pressure to be adequately capitalized, an increase in political pressure to expand business activities that serve lower-income home buyers, and the continuing pressure to increase profits and raise the firm's stock price.

By the mid 1990s, FNMA had identified a manager for each business line or activity and developed a quarterly assessment procedure employing prospective measures of the profitability and risk associated with capital invested in each business line. The ALMS system, which was originally developed to support integrated asset-liability management of the mortgage investment business, has been extended to help support this assessment procedure. The logical momentum of earlier integrated asset-liability management efforts and the desire by the new top management team to organize things differently both contributed to the adoption of this kind of capital allocation procedure.

Another powerful force that continues to force changes in the way FNMA allocates equity capital is the development of capital regulations by OFHEO, the new regulator of FNMA and FHLMC. The advent of explicit capital standards has lead to a number of changes in the quantitative models used to support decision-making in the mortgage insurance and mortgage investment businesses. The regulations are not final and as they evolve, the simulation and optimization models used by FNMA will continue to the revised so they can support decision making in the more regulated environment.

While the impetus to adopt or enhance integrated asset-liability management techniques has often come from unanticipated business problems or from increased regulatory and political pressure, the use of these modern finance techniques has enabled FNMA not only to deal effectively with the problems and pressures, but also to provide its shareholders with a more compelling case about why they should invest.

References

Black, Fischer and Scholes, Myron (1973), 'The Pricing of Options and Corporate Liabilities,' *Journal of Political Economy* **81** 637–659.

Hammer, Michael and Champy, James (1993), *Reengineering the Corporation: A Manifesto for Business Revolution*, Harper Business.

Hammersley, J.M., and Handscomb, D.C. (1964), *Monte Carlo Methods*, Chapman and Hall.

Holmer, Martin R. (1994), 'The Asset-Liability Management System at Fannie Mae,' *Interfaces* **24** 3–21.

Holmer, Martin R. and Zenios, Stavros A. (1995), 'The Productivity of Financial Intermediation and the Technology of Financial Product Management,' *Operations Research* **43** 970–982.

Kane, Edward J. (1989), *The S&L Insurance Crisis: How Did It Happen?*, Urban Institute Press.

Markowitz, Harry M. (1959), *Portfolio Selection: Efficient Diversification of Investments*, Wiley.

Merton, Robert C. (1992), *Continuous-Time Finance, Revised Edition*, Basil Blackwell.

Mossin, Jan (1968), "Optimal multiperiod portfolio policies,' *Journal of Business* **41** 215–229.

OFHEO (1995), *Annual Report to Congress 1995*, Office of Federal Housing Enterprise Oversight.

Parseghian, Gregory J. (1987), 'Collateralized Mortgage Obligations'. In *The Handbook of Fixed Income Securities*, 2nd Edition, Fabozzi, Frank J. and Pollack, Irving M. (eds.), Dow Jones-Irwin.

Passell, Peter (1996), 'The Model Was Too Rough: Why Economic Forecasting Became a Sideshow,' *The New York Times*, page D1, February 1.

von Neumann, John and Morgenstern, Oscar (1947), *Theory of Games and Economic Behavior*, 2nd Edition, Princeton University Press.

White, Lawrence J. (1991), *The S&L Debacle: Policy Lessons for Bank and Thrift Regulation*, Oxford University Press.

Part IX
Total Integrative Risk Management Models

The Russell–Yasuda Kasai Model: An Asset/ Liability Model for a Japanese Insurance Company Using Multistage Stochastic Programming *

David R. Cariño, Terry Kent,
David H. Myers, Celine Stacy,
Mike Sylvanus, Andrew L. Turner,
Kanji Watanabe and William T. Ziemba

Abstract

Frank Russell Company and The Yasuda Fire and Marine Insurance Co., Ltd., developed an asset/liability management model using multistage stochastic programming. It determines an optimal investment strategy that incorporates a multiperiod approach and enables the decision makers to define risks in tangible operational terms. It also handles the complex regulations imposed by Japanese insurance laws and practices. The most important goal is to produce a high-income return to pay annual interest on savings-type insurance policies without sacrificing the goal of maximizing the long-term wealth of the firm. During the first two years of use, fiscal 1991 and 1992, the investment strategy devised by the model yielded extra income of 42 basis points (¥8.7 billion or U5$79 million).

The Yasuda Fire and Marine Insurance Co., Ltd. (Yasuda *Kasai*, or 'fire,' to distinguish it from the life insurer), has been a major company in the Japanese insurance business for over 100 years. Yasuda Kasai is the second largest Japanese property and casualty insurer and is the seventh largest property and casualty insurance company in the world based on revenue. At the end of fiscal year 1991, Yasuda Kasai had ¥3.47 trillion (or US$26.2 billion) in assets.

In recent years, Yasuda, like others in the insurance industry worldwide, has been issuing a growing number of savings oriented policies with an increasing range of policy maturities. As a result, its liabilities are of a different character and more complex than in the past. Also, insurance laws and other

*First published in *Interfaces* **24** (1), 29–49, (1994). Copyright 1994, The Institute of Management Sciences and the Operations Research Society of America (currently IN-FORMS), 2, Charles St, Suite 300, Providence RI 02904, USA. Reprinted, with permission.

regulatory guidelines impose many restrictions on insurance company practices. Further, competitive pressures give rise to the potentially conflicting goals of obtaining a competitive current yield and a long-run high total return. The basic question that Yasuda asked Frank Russell Company was how it should manage its investment assets in the face of this changing and complex environment. It clearly needed better tools than the static mean-variance analysis that it had used in the past.

Yasuda needed a comprehensive asset/liability management model because of the changing environment in the Japanese insurance and financial industries. In the 1980s, Japanese financial markets were deregulated and realigned. As the Ministry of Finance relaxed the borders between financial institutions with respect to savingstype instruments, banks, securities firms, and insurance companies began to compete in earnest for the large pool of individual investors' savings. Insurance companies like Yasuda began offering policies that included savings features.

A savings policy is an insurance policy of a fixed term (typically three to five years) that, in addition to providing property and casualty insurance protection, pays a refund at the maturity of the policy. The maturity refund consists of a portion of the premiums paid plus interest at a rate guaranteed in the policy. At the end of each fiscal year, the insurance companies declare and credit policy bonuses in addition to the guaranteed rate; insurance companies compete on these bonuses. The savings portion of the policy premium is often more than 90 per cent of the premium. In addition to the changing nature of the insurance policies, the insurance bureau within the Ministry of Finance allowed companies to set up separate accounts (special savings accounts) for the management of the assets of these savings policies. Previously, all insurance assets were held in a general account.

As the popularity of these new savings policies grew, Yasuda's business and assets began to reflect the increased emphasis on the savings portion of the policies. With this change in business, Yasuda was becoming more of a financial services firm built on the foundation of a Japanese property and casualty company. These changes necessitated Yasuda's search for a better way to manage the business in order to address the increased dependency of both assets and liabilities on the financial markets.

An appropriate asset/liability management model had to include Yasuda's multiple and conflicting objectives combined with numerous regulatory restrictions. One type of regulation restricts the amounts invested in certain asset classes both within the special savings accounts and on a firm-wide basis. Another type of regulation restricts the sources and uses of funds. For example, in one type of savings account, the regulations require that interest accruing to policies be credited only from income earned (for example, coupon yield from bonds or dividend yield from equities) and not from capital gains.

The regulators prefer the income return because high-yielding assets are typically fixed income instruments, which provide more stable, albeit lower, total returns over time. This regulation encourages investors to seek high current-yield instruments instead of high total-return assets (deep-discount bonds rather than equities, for example). Such a regulation creates conflicting goals, as exemplified by one its first consequences: During the mid-1980s Japanese insurance companies bought large amounts of foreign bonds with very high coupon rates from countries such as Australia, Canada, and the US for the income return, while sacrificing total return through the loss of principal and currency depreciation. A useful asset/liability model would have to be able to balance the dual objectives of high income return to maintain and attract policyholders with the firm's desire to maximize its underlying capital.

Yasuda's goals for the project were to develop an easily understood working model that:

(1) Adequately represented the book and market value goals of its business;

(2) Incorporated the regulations imposed on it by insurance laws and practices;

(3) Reflected its multiple and conflicting goals, including its desire to maximize the long-run value of the company and to provide high-quality service to its customers;

(4) Reflected the multiperiod nature of those goals and various constraints;

(5) Reflected the inherent uncertainty of the investment process and financial markets;

(6) Could be solved in less than three hours on a computer work station; and

(7) Was believable and understandable by managers with various backgrounds and levels of technical expertise.

Prior to the Russell–Yasuda Kasai (RY) model, Yasuda's asset/liability work employed Markowitz [1952, 1959] mean-variance analysis and simulation. Using a static mean-variance asset allocation model, Yasuda selected a policy on the efficient frontier. Then it simulated the policy allocation's return to check income goals and solvency or reserve requirements. It tested the allocation's performance against the goals by comparing the probability of missing these goals with a minimum allowable probability of failure. If a particular asset allocation policy did not meet a minimum-probability-of-failure level, Yasuda tested another policy on the efficient frontier until it found a policy with a satisfactory probability of meeting the goals. There was no guarantee that a policy chosen in this manner was truly the optimal policy or even a desirable choice. It was simply a policy that met an acceptable level of expected performance. For Yasuda's problem, the drawbacks of this method are clear. First, as a static model, it cannot synchronize the cash flows of assets with liabilities over time, since there is a dynamic element in liability payments over time that must be faced. Second, volatility of total return as a measure of risk bears little resemblance to the risks that the decision makers felt they actually faced.

Since income return was very important, a model that distinguished the characteristics of income from total return was crucial. We saw mean-variance models, which measure risk as volatility of return, as less appropriate for Yasuda's planning activities than shortfall models, which measure risk as the cost of falling below given return targets. Shortfall models specifically address the desire for positively skewed distributions and more accurately reflect the aims of the investor [Cariño and Fan 1993].

These problems are not unique to Yasuda. Insurance companies and other financial institutions have traditionally used simulations on the liability side from their actuarial firms and mean-variance techniques on their asset side. Rarely have the two sides met as they have in the Russell–Yasuda Kasai model.

Previous Models

With the above background of simulation and mean-variance techniques, the project team searched for a methodology that overcame their shortcomings. We found a number of financial planning models that have a character similar to Yasuda's problem in the field of stochastic programming. While previous models inspired us to believe that a stochastic programming model tailored to Yasuda might be possible, there are important aspects of Yasuda's problem that differ from and expand on these previous models.

First, the problem is not one that basically involves fixed-income-like securities. The BONDS model of Bradley & Crane [1972], the Lane & Hutchinson [1980] model, the MIDAS model of Dempster & Ireland [1988] and Gassmann & Ireland [1990], the stochastic dedication model of Shapiro [1988], Hiller & Shapiro [1989], and Hiller & Eckstein [1993] all have a fixed-income focus. Yasuda's investments include all classes of investment assets. Second, there are general asset class restrictions that destroy any network structure in the problem. The models of Mulvey & Vladimirou [1989, 1992], Vladimirou [1990], and Zenios [1991] all exploit network structure in certain financial-planning problems. Third, liabilities are treated differently than in most of these previous models. In most of these models, where liabilities are addressed, the liability cash flows are emphasized. The RY model also considers liability balances in a manner similar to their treatment by Kallberg, White & Ziemba [1982]; Kusy [1978]; and Kusy & Ziemba [1986]. Although all of these are close relatives, perhaps Kusy & Ziemba [1986] is the nearest predecessor to the RY model, albeit with many differences in details.

Our search led us to choose a multistage stochastic linear programming approach for Yasuda's asset/liability management problem. Such a dynamic optimization model would allow Yasuda to make asset allocation and liability management decisions with regard for the uncertain outcomes of events relevant to the company's business environment. The environment includes regulation, multiple accounts, multiple horizons for different goals, provisions for end effects, and the uncertainties of future assets and liabilities.

Overview of the Model Formulation

Yasuda's total assets are divided among several accounts: a general account and several types of savings product line accounts. The fundamental decision variables of the model are the allocations of the total market value of each account among eligible allocation classes. Allocation classes include both direct investment in asset classes and indirect investment in asset classes through tokkin funds or through foreign subsidiaries (described later). The various accounts are interrelated because of regulatory restrictions on allocation classes that extend across accounts and because of certain funds transfers among accounts that occur at the end of a fiscal year.

Linear equations describing the accumulation of asset value over time relate the allocations at one stage to the account value to be allocated at the next while accounting for cash flows into and out of the account. Rates of return (price and income) appear as random coefficients in the accumulation relations. Altogether, the model equations describe the flows of funds among accounts over time. An abbreviated statement of the model equations is provided in the appendix.

Associated with each account is a liability model. The liability model projects the future cash flows, both into and out of the account, and projects the future liability balances of the account at dates extending to the planning horizon. Cash flows and balances reflect both existing policies and new policies Yasuda expects to sell in the future. The liability model builds up the cash flows and balances for an account from disaggregated blocks of policies with similar characteristics so that the aggregate cash flows and balances are internally consistent.

The objective of the model is to maximize the expected market value of the total firm at a horizon date less accumulated penalties for shortfalls of various types, while taking into account end effects. The use of shortfall costs to characterize risk allows more tangible expression of the risks the decision

makers feel they actually face than does the traditional risk measure, variance
of total return.

The Shortfall Model

A mean-variance model equally penalizes both high and low returns relative to
their mean, but it is unclear that excessive returns are undesirable. They are
certainly more desirable than low returns. In a shortfall model, the relevant
measure of risk for the firm is the expected amount (if any) by which goals
are not achieved [Cariño & Fan 1993]. Expected shortfalls are penalized in
the RY model's objective function.

The penalty costs of shortfalls may be based either on the expected financial impact to the firm for missing a goal or on psychological costs. The
financial consequences might include higher borrowing costs if there is a down-
grade in rating or the loss of policyholder confidence. Psychological costs may
be tied more to management's beliefs on how the firm should be run – how
conservatively. Regardless of the source of the penalty cost levels, the important advantage of such relative rankings is that the trade-offs of balancing
many objectives over multiple periods can be viewed directly. The optimization clearly brings into view how changing one goal affects others. Reports are
generated that can be placed in management's hands reflecting the expected
impacts of these decisions.

One of the principal areas for shortfall concerns in the RY model is related
to the regulatory restrictions on reserve levels of the assets backing policies.
There is a minimum reserve level that constitutes a legal operating restriction, but corporate policy and general prudence dictate that reserve levels be
maintained at a level significantly above the minimum reserve. Shortfalls are
measured according to the amount by which the prudent and desired reserve
levels are missed in order to reflect corporate policies and goals (Figure 1).

Because the interpretation of the penalty costs includes possible psychological aspects of preferences, it is difficult to specify precise cost coeffficients
in advance of actually running the model. It is possible, however, to identify
critical breakpoints, such as the level of reserves from which shortfalls should
begin to be measured and the level at which costs would be nearly infinite.
More important than the exact shape of the cost functions are the relative
weightings, or at least rankings, of one type of shortfall versus another. One
can set cost coeffficients by experimenting with various settings and studying
the resulting shortfall probability distributions until one achieves acceptable
probabilities of shortfalls.

The relative importance of these goals can also be expressed and examined
across time periods. Meeting a desired reserve level may be more important
over the long run than it is in a short period. The multiperiod capability of

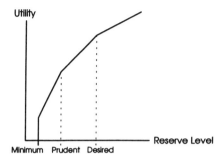

Figure 1: The graph of reserves minus shortfall penalties plotted against reserve level gives a utility curve for reserves. The shortfall cost coefficients in the objective function translate into the slopes of segments of the utility curve.

the model allows for multiperiod goals and for multiperiod decision making.

It was also valuable to Yasuda to develop a model that would determine an optimal solution based on sound stochastic optimization methods. Simulated solutions may lead to satisfactory allocations, but simulation provides no guarantee that a much better solution is not waiting to be found. The optimization framework guarantees that the solution determined is the best possible course of action given the model's assumptions. Simulations typically handle the changing of policies through time based on a system of rules: for example, if risk-based capital reserve levels are low, then the investor would be wise to decrease equity exposure. In an optimization approach, the model chooses the optimal policy through time in hopes of avoiding reserve levels falling so low that there must be a drastic and more expensive policy switch.

Liabilities and Shortfalls

To understand the model, one must understand the role of liabilities. Yasuda's savings-type policies work essentially like certificates of deposit. Each new policy sold represents a deposit, or inflow, of funds. Interest is periodically credited to the policy until maturity, typically three to five years, at which time the principal amount plus credited interest is refunded to the policyholder. The crediting rate is declared year by year and is determined competitively by the industry. It is usually related to a market index like the prime rate.

In a deterministic world, this rate would be known for all time in the future; therefore the final interest payment and the interim liability balance could be computed for every policy. Also in a deterministic world, sales of new policies would be known for all time in the future; therefore by aggregating over all policies, both those extant and those to be sold in the future, one could compute the aggregate net cash flows and aggregate liability balances in advance of their actual occurrence.

In a stochastic world, crediting rates are uncertain since they depend on market conditions; new policy sales can also be uncertain. We cannot say with certainty what future cash flows or balances might be. We can, however, calculate projected liability cash flows and balances under various scenarios; liability variables are scenario dependent. One major module of the system computes these scenario-dependent cash flows and balances, aggregated over all policies.

Insurance business regulations stipulate that interest credited to some policies be earned from investment income, not capital gains. Realized capital gains must be credited to a liability account called 'Reserve under Article 86 – Insurance Business Law,' from which subsequent capital losses may be deducted. In contrast, investment income may be credited to policies or to shareholders' equity. Yasuda's practice is to transfer the excess of income earned over interest credited from the product line account to the general account at the end of the fiscal year, thus contributing directly to reported profit. If there is a shortfall in income, that amount is transferred from the general account to the product line account, directly reducing reported profit.

Such a restriction could be formulated as a hard constraint, in which income earned in a year must be greater than or equal to the interest credited. However, this requirement for income is more like a goal, or a preference, since it ultimately relates to a desire for accounting profit. If Yasuda does not achieve adequate income, the decision makers will not like it, but they have other recourse actions available. Hence, we model this requirement as a soft constraint, creating an income shortfall variable to which we apply penalty costs in the objective function. Kusy & Ziemba [1986] have shown, in the context of their model, that using soft constraints such as these leads to better decisions than hard constraints.

Income shortfalls are only one of several types of shortfalls in the model. For example, a given asset portfolio might produce a steady income return but a loss of principal due to fluctuations in its market value. To discourage that sort of outcome, the model penalizes reserve shortfalls. Again, we defined soft constraints to measure the amount by which total asset value falls below total liability value.

Allocation of Assets

Yasuda's basic role is to invest these deposited funds to earn return from which to pay the maturity refunds. In addition to ensuring that the maturity cash flows are met, the firm must seek to minimize interim shortfalls in income earned versus interest credited. Such shortfalls can have regulatory as well as market impacts on the company. In fact, it is the risk of not earning adequate income quarter by quarter that the decision makers view as the primary component of risk. Of course, Yasuda wishes to realize an even higher return in order to make a profit. But loosely speaking, if Yasuda's decision makers were assured that, with certainty, adequate income could be achieved, then the decision makers would invest without regard to volatility of total return. Thus, they would behave as risk-neutral, expected-wealth maximizers.

The problem is to determine the optimal allocation of the deposited funds. Yasuda allocates among typical asset categories that include cash, loans (fixed rate and floating rate), bonds, equities, real estate, and other assets. It further segregates bond and equity categories into individual countries or groups of countries. If the funds are all placed in cash-equivalent assets, asset growth will be steady and predictable, but the return on cash will most likely be less than the crediting rate declared on the underlying policies. There will be shortfalls at maturity as well as shortfalls in each interim period.

Suppose all funds are placed in fixed-income securities. In a deterministic world, one might find a bond with a current yield greater than the crediting rate, and in such a world, it would clearly be desirable to hold such a security. In the real world, however, interest rates are stochastic, so although a fixed-income security may deliver steady income, its market value fluctuates with movements in interest rates and there is no guarantee that it can be sold at a favorable price. If the funds are all held in equities, there is no certainty of income or market value, so there is obviously the potential for shortfalls. Clearly, we should invest funds in a portfolio of assets rather than all in a single asset. Since we can revise the portfolio allocations over time, the decision we make is not just among allocations today but among allocation strategies over time.

Indirect Investments and Regulatory Constraints

Regulations require that credited interest for some accounts must be earned from investment income rather than from capital gain. There are vehicles, however, by which the firm can convert capital gains to income. These vehicles are tokkin funds and foreign subsidiaries. Both of these are companies set up to invest funds and pay dividends back to the parent company. The subsidiary can pay dividends from capital gains, while the parent company can consider the dividends as income. Because the parent may direct the activities of these subsidiaries, the firm can invest in an asset class directly, receiving as income the income component only, or can invest indirectly through a subsidiary, receiving as income the total return of the asset class.

Because of the possibility of circumventing the Article 86 regulation by means of these indirect investments, regulators impose limits on the fraction of the firm's assets that are invested indirectly. For example, total tokkin investments must not exceed seven percent of total assets currently. Yasuda follows regulatory guidance regarding indirect investments.

Multiple Accounts

Another major feature of the formulation is that deposits from different product lines are held in separate accounts. As part of changing regulations in the Japanese insurance industry, the regulators allow a special type of savings policy to which different restrictions apply than those that apply to conventional savings policies. However, regulations stipulate that special savings-type policy assets must be held in an account separate from conventional policy assets. Consequently, we formulated the model to include separate accounts so that we could apply some restrictions within accounts and others across the firm as a whole. To reflect Yasuda's practice mentioned earlier, we modeled fiscal yearend transfers of income to the general account.

Loans and Taxes

The RY model also accommodates the illiquidity of the loan asset class. Loans represent a substantial portion of the typical insurance portfolio. Business and personal loans are common. Unlike market-traded assets, whose prices and returns can be measured by market indexes, loan assets must be appraised to obtain market values and returns. A module of the system calculates the market values of Yasuda's loan portfolio at each stage under the given scenarios. Further, since loans cannot be bought and sold with the same ease as market-traded assets, we placed restrictions on the amounts by which the loan allocation can change from one stage to the next.

Another formulation element addresses the cash flow for payment of taxes, which are a percentage of taxable income, not of total return. A realistic dynamic asset-accumulation model must account for the payment of taxes. Since the model distinguishes income return from price return and computes changes in liability balances, the model also computes taxable income and appropriately accounts for the payment of taxes.

End-Effects Methodology

The stochastic linear programming model (the base problem) has uncertainty in many coefficients; this uncertainty is modeled through scenarios. Given that each scenario has a discrete probability of occurrence for any finite horizon, the stochastic linear program is equivalently represented by a large deterministic linear program in extensive form called the grand linear program. Yasuda has an ongoing business, so its true situation might be modeled as an infinite horizon problem.

Given the size of the problem induced by the many variables and scenarios, it is not practical to solve the infinite horizon problem directly; a modeler must use an approximation. One approach is to simply truncate the problem at a finite horizon. If both the horizon and the discount rate are sufficiently large, this approximation will not have much impact on the early period decisions. Specific calculations showing this error for given horizons and discount rates appear in Grinold [1980]. The errors are considerable. A better approach is to augment the base problem extending to a finite horizon with a steady-state term in the objective function. This term and the constraints associated with it yield one additional stage in the base problem. Hence, if the number of stages to the horizon is not too large, the problem is manageable. We implemented such a steady-state stage beyond the horizon, following a methodology described by Grinold [1983] called the dual-equilibrium procedure. The essence of the methodology is that dual variables are assumed to grow steadily past the horizon.

Scenario Generation

A stochastic programming model requires scenarios of the possible paths of stochastic elements. The random elements of the RY model include price and income returns for all of the asset classes as well as policy crediting rates. Creation of scenario inputs is analogous to creation of means, variances, and correlations for a meanvariance model; they are ultimately an expression of the decision maker's probability beliefs. The problems of creating good scenario inputs are not unique to the stochastic programming approach; they are essentially the same problems as those of forecasting asset returns for any asset allocation method.

In principle, one may sample from an asset return model to create scenarios. We developed asset models that create asset returns that are independent over time as well as models that include intertemporal dependence. Our tests show that the power of an asset allocation model depends heavily on the quality of the forecasts it relies on to generate scenarios.

Different scenarios may share a common path up to some stage and then diverge; a representation of all possible paths of the stochastic elements takes the form of an event tree (Figure 2). The stages of the decision problem have nodes or branching points of the tree. Each branching point represents an opportunity to make a decision and each branch represents a possible

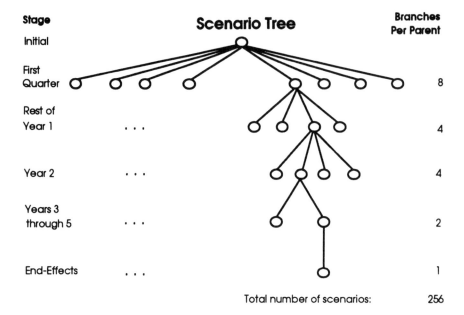

Stage		Branches Per Parent
Initial		
First Quarter		8
Rest of Year 1	⋅ ⋅ ⋅	4
Year 2	⋅ ⋅ ⋅	4
Years 3 through 5	⋅ ⋅ ⋅	2
End-Effects	⋅ ⋅ ⋅	1
	Total number of scenarios:	256

Figure 2: Modeled outcomes of random asset returns form an event tree. Each node represents a joint outcome of all the random variables at that stage. Each path through the tree is a scenario. The numbers of branches per parent node by stage summarize the scenario-tree structure.

outcome of the uncertain elements for that stage. Each path through the tree is a scenario representing one possible sequence of outcomes of the stochastic elements throughout the time horizon of the problem.

The total number of nodes in the tree is a determinant of the size of the stochastic program. To keep the model tractable, we specify the number of branches per parent in later stages to be a smaller number than in earlier stages. To describe the structure of a tree, we refer to the number of branches per parent in each stage. For example, a 1–8–4–4–2–1 tree has 256 scenarios.

The scenario generator builds the tree according to a tree-structure specification. A tree-structure specification contains a list of stage names, the period length for each period, and the number of outcomes (branches) from the parent period. While the model offers the user the flexibility to specify all of these parameters, the baseline model typically used has stages identified as Initial, End of Quarter 1, End of Year 1, End of Year 2, End of Year 5 (the planning horizon), and End-Effects, with a 256–scenario tree.

The scenario generator module creates random asset returns for the scenarios in one of three possible ways. The first method creates returns that are independent between periods. The second method creates returns from

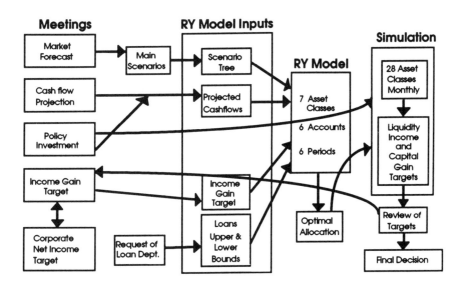

Figure 3: The RY model is central to Yasuda Kasai's asset/liability decision process. The process begins with a series of meetings to prepare the market forecasts, scenarios, cash-flow projections, targets, loan bounds, and other inputs. Planners then run the model, simulate the allocations, review the targets, and report a final decision.

a factor model that incorporates dependence between periods. The third alternative allows the user to specify each random asset return at each node. The first two methods use the computer to generate scenarios; hence, they are very easy for the user to operate. The third method requires more work, but the user has complete control over the scenarios and tree structure.

In the first method of scenario generation, where stochastic outcomes are independent between time periods, the user provides a pool of joint return outcomes for each stage. One can generate these returns by random sampling from an asset return model; one can simply use historical data; or, as Yasuda does, one can carefully construct the scenario pool from a combination of forecasting models and judgment. Figure 3 shows where scenario creation fits in Yasuda's asset/liability decision process. Given the pool, the model reduces the number of stage outcomes down to the number specified by the tree structure for the time period. The reduction method, described below, preserves the important attributes – the means and standard deviations – of the outcome pool while reducing the number of outcomes to a tractably small number.

The method begins by pairing up outcomes in the base set, determining a new, consolidated outcome from each outcome pair. The consolidated scenario has a value equal to the sum of its two precedent outcomes weighted by their probabilities. The probability of the new outcome is simply the sum of the probabilities of the original outcome pair. In this manner the method reduces the number of outcomes to half the number of the original set. It combines consecutive pairs of the resulting set if further reduction is necessary. Because of this rule, the base set of outcomes must contain a power-of-two multiple of the desired number of outcomes.

One of the features of this reduction method is the guarantee that the means of the random variables will remain equal to the desired means. However, the resulting standard deviations may not equal the desired standard deviations. Therefore, the model then applies a variance adjustment that simultaneously moves all outcomes toward or away from the mean of the random variable until the desired variance is achieved. It moves outcomes by amounts proportional to their distances from the mean. By adjusting all outcomes, the method preserves the overall general shape of the distribution; it simply expands from or contracts toward the mean by the desired amount.

The model applies this variance-adjustment procedure to each random variable individually. In reducing the number of outcomes, the method inevitably changes some characteristics of the base distribution. For example, since the number of outcomes is smaller, the method unintentionally increases the magnitudes of correlations among the random variables. However, the procedure does preserve the signs of the correlations of the base distribution.

The second method for generating scenario outcomes is one in which outcomes of the stochastic events are dependent between time periods. Using statistical factor analysis, we found that three factors captured the dependence among asset returns adequately. The three factors strongly relate to interest rates, equity returns, and currency exchange rates. Therefore, we chose the long-term government bond yield to represent the interest factor, the return to the TOPIX index (the capitalization-weighted average of all stocks on the first section of the Tokyo Stock Exchange) to represent the equity factor, and the yen versus US dollar currency return to represent the currency factor. Domestic bond, loan, and cash returns depend on the interest factor, domestic equity price returns depend on the equity factor, foreign equity return depends on the currency factor, and foreign bond return depends on both the interest and the currency factors. We modeled the movements of the factors over time by a time-series process that captures intertemporal dependence.

The model also accommodates a third method of scenario generation in which the user specifies the outcomes of the stochastic elements for each node of the scenario tree. The method allows the user complete flexibility in describing the structure and outcomes in the scenario tree; the user may even specify an arbitrary tree structure with differing numbers of branches per parent within a given period. The model then takes the specified scenario tree as given and makes no adjustments to outcomes.

Implementation

We began to develop the model in September 1989 and completed it in time for installation at Yasuda prior to its use for the fiscal year 1991, which began April 1, 1991. Russell staff and consultants and Yasuda's staff developed the model and its input in close collaboration. Yasuda used the model on a trial basis for the 1991 fiscal year (April 1, 1991 to March 31, 1992). The trial was successful, and Yasuda fully integrated the model into its financial planning process (Figure 3) for fiscal 1992 and beyond. For fiscal 1993, we added two additional new accounts to the model.

We originally conceived of a model involving 17 asset classes, 10 periods, and 2,048 scenarios. The base linear program (LP) for this problem consisted of 324 rows, 637 columns, and 3,731 nonzero coefficients. The extensive form of the stochastic problem, or 'grand LP,' consisted of 249,909 rows, 348,401 columns, and 1,556,456 non-zero coefficients. Recognizing that the size of the full stochastic problem might be too large to solve in a reasonable amount of time, we experimented by scaling back each of the LP attributes in order to determine which were most important to approximating the optimal solution. In the course of this experimentation, we developed a model generation mechanism that allowed us to create a wide variety of models reliably and automatically. The resulting model involved seven asset classes, six stages, and 256 scenarios. The base LP for this problem had 263 rows, 431 columns, and 2,950 nonzero coefficients. The grand LP for this model consisted of 31,062 rows, 44,004 columns, and 318,121 nonzero coefficients. Further details of the model appear in Carirno *et al.* [1993].

In creating and solving stochastic problems, we adopted a modular structure composed of six modules – a matrix generator, a scenario generator, a liability generator, a coefficient generator, a solver, and a report generator.

The matrix generator creates a base LP and identifies random coefficients in terms of the underlying random variables. The matrix generator builds in the flexibility to allow the user to easily adjust the LP attributes that affect the problem size. The user can, for example, select from any number of asset classes and allow or disallow particular assets in particular time periods. Shortfall costs can be specified for each period, account, and shortfall type. The penalty costs can be expressed simply as linear functions of shortfalls or as piecewise–linear convex cost functions, which are more realistic. The user can change the names and the number of accounts, the names and number of investment types, and the names and number of asset classes. Additional features allow the user to include or exclude transaction costs, set bounds on specific variables, change these bounds, or fix the allocations of assets at specified percentage values.

The scenario generator builds data structures that represent the scenario tree and calculates the random returns for each asset class in each scenario. The liability generator uses the tree structure and the random interest rates from the scenario generator to project the random liability cash flows and balances for each node in the tree. The liability cash flows and balance variables all appear on the right-hand sides of these constraints. Although stochastic, they may be computed off-line, prior to the optimization, without impact on the solution time. Hence, we programmed a very detailed model to compute what are essentially data inputs to the optimization model. We designed the input format for the insurance policy cash flows so that the user may change the number of periods or period lengths in the tree-structure specification without changing the cash-flow data inputs.

The coefficient generator calculates the value for each random coefficient from its underlying random variables for each node in the tree. For example, $(l + \text{RILF} + \text{RPLF})$ is an expression that describes a random coefficient. The terms RILF and RPLF are the names of random variables that represent the income and price returns to a particular asset class. A random coefficient specification consists of the row name, column name, and an expression for the value of a random coefficient in terms of the underlying random variables. The coefficient generator parses the coefficient specification, replaces the names of the random variables with the values of the random return or random liability that was generated earlier, and calculates the value of the random coefficient.

The solver first assembles the information from the base LP, the tree structure, and the values for the random liabilities and random coefficients into a form suitable for solution. The solver then calls an appropriate optimization algorithm (for example, Wets [1988]), which searches for a solution to the formulated problem. After the problem is solved, the report generator produces a set of reports detailing various aspects of the solution.

Depending on the number of scenarios considered, the system runs in one to three hours. It can run even faster if it is provided with a previous solution called a hot start. Computing a solution from a hot start often takes less than 20 minutes. Currently, we use an IBM R5/6000 Model 530 to solve the model. We coded the computer modules using C and FORTRAN and called routines from IBM's Optimization Subroutine Library (OSL). Since we started using the system over two years ago, cheaper, faster, and more powerful work stations have become available, meaning larger and more complex problems can now be run as fast or faster than our original problem.

While the technology has improved enough that we can generate and solve multiperiod models, there are still practical limits on the size of the problem. Therefore, we believe it important to model those periods that have the most impact on the business. We find it useful to consider quarterly decisions early in the planning period, reflecting immediate quarterly reporting periods and the transfer of information to the public, and to switch to annual decision periods after the first year. The modeling process helps to determine which periods are most important and how many to include. Ultimately, there is a trade-off among computing power, solution time, and model sensitivity to the periods.

Output

Reports that the system produces include:

(1) Allocations for the initial period;

(2) Expected allocations for each modeled period;

(3) Expected income statements for each modeled period;

(4) Expected balance sheets for each modeled period;

(5) Probability distributions of shortfalls; and

(6) The status of each regulatory constraint row activity.

Comparison to Mean Variance

Yasuda's previous methodology was to calculate portfolio weights using a mean–variance model. In this section, we compare and contrast the multi-period stochastic programming model with a static mean–variance model over one– and five–year horizons. Table 1 illustrates the main differences between the two models in terms of objective function, decision stages, types of allocations, accounts that are handled, and risk measures used in the objective function.

The mean–variance model is a quadratic programming model that determines a portfolio with minimum risk (measured as standard deviation of returns) for a given level of expected return or, equivalently, determines a portfolio with maximum expected return for a fixed level of risk. The optimal solutions for different risk–return trade-offs, or risk–tolerance levels, lie on an efficient frontier of solutions. In multiperiod applications the model is rerun on new data with a one-period horizon each period.

For mean-variance calculations we used the Russell Asset Allocation Model (RAAM), a quadratic programming optimizer based on the Markowitz formulation. To make the two models comparable in terms of structure, we made certain assumptions for the mean-variance model. To make the risk and

	Russell-Yasuda Kasai	Mean Variance
Objective	Piecewise linear	Quadratic
Stages	Multiple with end-effects period	Single
Allocations	Dynamic	Static
Accounts	Multiple	Single
Risk Measures	Multiple shortfalls	Variance

Table 1: The RY model differs from a mean-variance model in several respects.

Model used for Allocations	Model used for Objective Function	
	RY	MV
RY	I	II
MV	II	IV

Figure 4: We compared the two models, RY and mean variance (MV), by computing the values of the two models' objective functions given each of the two models' allocations. This gives four solutions, denoted I–IV. We compared I with II and III with IV.

return assumptions comparable in the two models, we used the forecasted scenarios from the RY model to generate the means, standard deviations, and correlations for the meanvariance (MV) model.

There are various ways to compare models. One may build a larger meta model and then determine which of the simple models most accurately approximates the larger model. This is a useful approach with econometric models but impractical for our project. Another approach is to formulate models of a given situation and then, using statistical tests, compare their actual or simulated performance (Kusy and Ziemba [1986] use such an approach). In our situation, we evaluated each of the allocations produced by the RY model and by the MV model using each of the objective functions. This produced four types of solutions, termed I to IV, as shown in Figure 4. We then compared I with II and III with IV.

When comparing the RY model solutions of types I and II, we used the shortfalls and expected terminal wealth as measures of success. When com-

Case	INI	QO1	YO1	Y02	Y05	YFF	Scenarios
1	1	16	4	2	2	1	256
2	1	16	2	2	2	2	256
3	1	–	16	–	–	–	16
4	1	8	4	4	2	1	256
5	1	8	4	2	2	2	256
6	1	–	8	–	–	–	8

Table 2: Test cases for the model comparisons had the scenario–tree structures shown. For each case, the table shows the numbers of branches per parent node by stage and the total number of scenarios. Column headings are the stage names.

paring the mean-variance solutions III and IV, we used portfolio means and variances as the relevant measures. In choosing MV portfolios for comparison with the RY model in tests II and IV, we selected portfolios on the efficient frontier with risk and expected return similar to those of the RY model.

We generated six test cases to compare the two models. Two of the cases had only a single period of uncertainty with a horizon of one year. The other four cases had four uncertain periods extending to a five-year horizon plus the end-effects period. The six cases differed in their numbers of scenarios; Table 2 shows the tree-structure specifications for the six cases.

In comparing solutions of types III and IV, we found that when the RY model allocations are input to RAAM as current allocations, the current allocation portfolio in the majority of cases lies close to the efficient frontier. In the six cases tested, the RY model allocation was within 0.2 per cent of the effficient frontier on a return basis and within 0.3 units on a risk basis.

In comparing solutions of types I and II we found that when the MV allocations were input to the RY model as a set of fixed allocations applied to all accounts, the MV fixed allocations resulted in dramatically lower (worse) objective value in each of the six test cases. In the best case, shortfalls from the MV allocation were merely 60 per cent greater than the RY model portfolio. In four of the six cases, they were from 3,500 to 6,000 per cent greater. These results held for both the MV portfolio with risk identical to the RY model portfolio and for the MV portfolio with identical expected return. The differences in objective values in the one-year cases (cases 3 and 6) were not as dramatic as the five-year cases (cases 1, 2, 4, and 5), but the objective values obtained using MV allocations are still worse than the optimal RY objective values.

Both comparisons of types I and II and comparisons of types III and IV support the conclusion that the solutions determined by the RY model are superior to those determined in a mean-variance framework. The RY model solutions always lie near the MV-efficient frontier; sometimes they are on the frontier. MV solutions have lower expected market values and higher penalty costs in all cases of type I and II comparisons.

In addition, the RY model enjoys other advantages over the MV model. While the MV allocations of similar-risk and similarreturn portfolios are anchored by the current portfolio choice from the RY model solutions, there are substantial differences. The RY model optimizes across all accounts, while MV is restricted to optimize each account separately. The flexibility in the RY model is evident in cases in which certain assets are found primarily in a single account.

Also, the MV approximation becomes less accurate as the planning horizon length increases. We computed RY model solutions for cases in which the allocations were constrained to equal MV allocations up to a given period and were unconstrained beyond that point. When constrained to MV allocations up to periods beyond one year, the model gives extremely poor solutions. The results reflect the impact of the multiperiod nature of the RY model. The RY model incorporates the decision makers' ability to change portfolios throughout the planning horizon to react to changing circumstances. For example, different allocation decisions would be made two years hence depending on whether the firm was in a strong or a weak position. By fixing allocations, we lose this flexibility; the severity of this loss increases as more periods are fixed. The more periods that are fixed to the MV allocations, the less flexibility the RY model has, incurring greater costs (lower objective values) as a consequence.

Conclusions

Yasuda derived much of the benefit of the project from the comprehensive modeling of its business (Figure 3). Instead of a black-box technology, the flexibility of the technology actually allowed for a more understandable system. In the past, for computational reasons, Yasuda had to make simplifying assumptions. Today, the power of work stations and the mathematics of stochastic programming make it possible to create a more realistic model of the firm's business. A few of the more salient features of the model include a multiple-stage planning horizon, realistic business goals, multiple investment and liability accounts, and the inclusion of government regulations and book- and market-value information.

Another aspect of the RY model is the capability to follow the economic value of the firm while meeting the book-value rules of accounting and of the regulatory restrictions. While this may seem a trivial aspect of the model, it is crucial to Yasuda's business. Many of the goals, such as actuarial valuation of the liabilities, and restrictions that are imposed are done on an accounting- or book-value basis, while the assets are subject to market swings. Thus, for the model to capture a necessary level of realism of the business, it had to handle both methods of tracking the assets and liabilities [Henriques 1991]. Another realistic aspect of the business is the complex structure of multiple lines of business and multiple investment accounts.

While simulation techniques can often handle much of the complex structure, the extra layers of difficulty are significant in solving an optimization model. Because of these layers of complexity, it is only now possible with the latest computer hardware and software to attempt to tackle problems of this size in a reasonable time frame and cost.

Yasuda compared the RY model's actual performance with the performance that Yasuda would have obtained using its old methodology, mean-variance analysis. In the first fiscal year (fiscal 1991, which ended March 31, 1992), the RY model's allocations produced a 15–basis-point (or US$25 million) increase in income yield over what Yasuda's old methodology would have produced. For the 1991 and 1992 fiscal years combined, the RY model delivered an extra 42 basis points (or U5$79 million) in income return over a constant-mix strategy's income return and an extra five basis points (or US$9 million) in total return (income plus capital gain). Income yield is important to Yasuda since increased income allows it to pay larger bonuses on its savings policies to its policyholders. This helps it both to maintain policyholders and to attract new customers as its competitive advantage increases. Thus, the economic benefit of the model was much greater than the US$79 million increase in income yield.

The model has provided Yasuda's asset/liability department with a state-of-the-art decision-making and risk-management tool that gives insight into the complex choices and restrictions to which the business is exposed. Yasuda's use of the model has become central to its asset/liability management system (Figure 3). The company values the fact that risk is not measured by an abstract measure like standard deviation of total return. It does not have to deal with the ambiguous notion of 'risk tolerance' and instead is able to express its preferences in terms of tangible shortfalls. Further, Yasuda feels the model fairly represents the complexity of the business it is in. The solution is directly implementable since the model considers all of its particular liabilities and constraints. Yasuda decision makers' knowledge of the model and its software implementation allows them to use the model for their own tests and analysis.

In summary, Yasuda Kasai's management recognized the need for more sophisticated financial planning tools. We developed a stochastic programming model tailored for Yasuda's liabilities and constraints, and it has been successfully integrated into the firm's ongoing planning process.

Acknowledgments

We are very grateful to K. Sasamoto, director and deputy president of Yasuda Kasai for his consistent support and to the members of the Yasuda Kasai team who assisted with this project: Y. Tayama, Y. Yazawa, Y. Ohtani, T. Amaki, I. Harada, M. Harima, T. Morozumi, and N. Ueda. For their help with many details, we thank T. Eguchi and S. Murray of Frank Russell Company. We also thank K. Sawaki of Nanzan University, Japan, R. Wets of the University of California, Davis, and C. Edirisinghe of The University of Tennessee for advice during the project. For assistance with OSL, we thank R. Clark, D. Jensen, and A. King of IBM. Some basic research related to this work by the last author was partially supported by the Natural Sciences and Engineering Research Council of Canada and the US National Science Foundation .

APPENDIX: Abbreviated Russell–Yasuda Kasai Model Formulation

Stages are indexed by $t = 0, 1, \ldots, T$. Decision variables of the stochastic program are

$$
\begin{aligned}
V_t &= \text{total fund market value at } t, \\
X_{nt} &= \text{market value in asset } n \text{ at } t, \\
w_{t+1} &= \text{income shortfall at } t + 1, \text{ and} \\
v_{t+l} &= \text{income surplus at } t + 1.
\end{aligned}
$$

Random variables appearing in coefficients of the stochastic program are

$$
\begin{aligned}
\text{RP}_{nt+l} &= \text{price return of asset } n \text{ from end of } t \text{ to end of } t + l, \text{ and} \\
\text{RI}_{nt+l} &= \text{income return of asset } n \text{ from end of } t \text{ to end of } t + 1.
\end{aligned}
$$

Random variables appearing in the right-hand side are

$$
\begin{aligned}
F_{t+l} &= \text{deposit inflow from end of } t \text{ to end of } t + 1, \\
P_{t+l} &= \text{principal payout from end of } t \text{ to end of } t + 1, \\
I_{t+l} &= \text{interest payout from end of } t \text{ to end of } t + 1, \\
g_{t+l} &= \text{rate at which interest is credited to policies} \\
&\qquad \text{from end of } t \text{ to end of } t + 1, \\
L_t &= \text{liability valuation at } t.
\end{aligned}
$$

A parametized function appearing in the objective is

$$c_t(\cdot) = \text{piecewise linear convex cost function.}$$

The objective of the model is to allocate fund value among available assets to maximize expected wealth at the end of the planning horizon T less expected penalized shortfalls accumulated throughout the planning horizon.

$$\text{Maximize} \quad E[V_T - \sum_{t=1}^{T} c_t(w_t)]$$

subject to:

budget constraints

$$\sum_{n} X_{nt} - V_t = 0;$$

asset accumulation relations

$$V_{t+1} - \sum_{n}(l + \text{RP}_{nt+1} + \text{RI}_{nt+l})X_{nt} = F_{t+1} - P_{t+1} - I_{t+1},$$

income shortfall constraints

$$\sum_{n} \text{RI}_{nt+l}X_{nt} + w_{t+l} - v_{t+l} = g_{t+1}L_t$$

and

nonnegativity constraints

$$X_{nt} \geq 0, \quad v_{t+1} \geq 0, \quad w_{t+1} \geq 0,$$

for $t = 0, 1, 2, \ldots, T-1$. Liability balances and cash flows are computed so as to satisfy the liability accumulation relations

$$L_{t+1} = (1 + g_{t+1}L_t + F_{t+1} - P_{t+1} - I_{t=1}, \quad t = 0, 1, 2, \ldots, T-1$$

For simplicity, this abbreviated formulation does not include some model elements: additional types of shortfalls, indirect investments (tokkin funds and foreign subsidiaries), regulatory restrictions, multiple accounts, loan assets, the effects of taxes, and end effects.

Full details of the model formulation appear in Cariño *et al.* [1993].

References

Bradley, S.P. and Crane, D.B. 1972), 'A dynamic model for bond portfolio management,' *Management Science* **19** (2) 139–151.

Cariño, D.R. and Fan, Y. 1993, 'Alternative risk measures for asset allocation,' *Gestion Collective internationale*, (2) (July/August), 47–51.

Cariño, D.R., Kent, T., Myers, D.H., Stacy, C., Sylvanus, M., Turner, A., Watanabe, K. and Ziemba, W.T. 1993, 'The Russell-Yasuda Kasai financial planning model,' working paper, Frank Russell Company.

Dempster, M A H. and Ireland, A.M. 1988, 'A financial expert decision support system.' In *Mathematical Models for Decision Support*, B. Mitra (ed.) NATO ASI Series, Vol. F48, Springer-Verlag, 415–440.

Gassmann, H.I. and Ireland, A.M. 1990, 'Input/output standards for the MIDAS optimization model,' working paper WP–90–2, School of Business Administration, Dalhousie University, Halifax, Canada.

Grinold, R.C. 1980, 'Time horizons in energy planning models.' In *Energy Policy Modeling: United States and Canadian Experiences*, W.T. Ziemba and S.L. Schwartz (eds.) Vol. 2, Martinus Nijhoff, 216–237.

Grinold, R.C. 1983, 'Model building techniques for the correction of end effects in multistage convex programs,' *Operations Research* **31** (3) 407–431.

Henriques, D.B. 1991, 'A better way to track your assets,' *The New York Times*, **CXL**, No. 48,556 (Sunday, March 31), Section 3 (Business), p. 11.

Hiller, R.S. and Eckstein, J. 1993, 'Stochastic dedication: Designing fixed income portfolios using massively parallel Benders decomposition,' *Management Science* **39**, (11) (November), 1422–1438.

Hiller, R.S. and Shapiro, J.F. 1989, 'Stochastic programming models for asset/liability management problems,' IFSRC Discussion Paper No. 105–89, Sloan School of Management, MIT, Cambridge, Massachusetts.

Kallberg, J.G., White, R.W. and Ziemba, W.T. 1982, 'Short term financial planning under uncertainty,' *Management Science* **28** (6) 670–682.

Kusy, M.I. 1978, *An Asset and Liability Management Model*, PhD thesis, Faculty of Commerce, University of British Columbia, Vancouver, British Columbia, Canada.

Kusy, M.I. and Ziemba, W.T. 1986, 'A bank asset and liability management model,' *Operations Research* **34** (3) 356–376.

Lane, M. and Hutchinson, P. 1980, 'A model for managing a certificate of deposit portfolio under uncertainty.' In *Stochastic Programming*, M.A.H. Dempster (ed.), Academic Press, 473–496.

Markowitz, H.M. 1952, 'Portfolio selection,' *Journal of Finance* **7** (1) (March), 77–91.

Markowitz, H.M. 1959, *Portfolio Selection: Efficient Diversification of Investments*, Cowles Foundation Monograph 16, Yale University Press.

Mulvey, J.M. and Vladimirou, H. 1989, 'Stochastic network optimization models for investment planning,' *Annals of Operations Research* **20** 187–217.

Mulvey, J.M. and Vladimirou, H. 1992, 'Stochastic network programming for financial planning problems,' *Management Science* **38** (11) 1642–1664.

Shapiro, J.F. 1988, 'Stochastic programming models for dedicated portfolio selection.' In *Mathematical Models for Decision Support*, B. Mitra (ed.) NATO ASI Series, Vol. F48, Springer-Verlag,587–611.

Vladimirou, H. 1990, *Stochastic Networks: Solution Methods and Applications in Financial Planning*, PhD thesis, Department of Civil Engineering and Operations Research, Princeton University, Princeton, New Jersey.

Wets, R.J-B. 1988, 'Large scale linear programming techniques in stochastic programming.' In *Numerical Techniques for Stochastic Optimization*, Y. Ermoliev and R.J-B Wets (eds.), Springer-Verlag, 65–93.

Zenios, S.A. 1991, 'Massively parallel computations for financial planning under uncertainty.' In *Very Large Scale Computing in the 21st Century*, J. Mesirov (ed.), SIAM, Philadelphia, Pennsylvania, 273–294.

Postscript

According to Kunihiko Sasamoto, director and deputy president of Yasuda Kasai,

> The liability structure of the property and casualty insurance business has become very complex, and the insurance industry has various restrictions in terms of asset management. We concluded that existing models, such as Markowitz mean variance, would not function well and that we needed to develop a new asset/liability management model.
>
> The Russell-Yasuda Kasai model is now at the core of all asset/liability work for the firm. We can define our risks in concrete terms, rather than through an abstract, in business terms, measure like standard deviation. The model has provided an important side benefit by pushing the technology and efficiency of other models in Yasuda forward to complement it. The model has assisted Yasuda in determining when and how human judgment is best used in the asset/liability process.

The Home Account Advisor™, Asset and Liability Management for Individual Investors

Adam J. Berger* and John M. Mulvey

Abstract

Asset and liability management systems designed for large financial institutions are extended for individual investors. We describe the Home Account Advisor™, a multi-stage model for optimizing an investor's financial objectives in conjunction with investment, savings, and borrowing strategies. A system of stochastic differential equations generates alternative economic scenarios. The stochastic model identifies robust strategies for maximizing a financial goal index across these scenarios. A case study illustrates the concepts.

Keywords

asset-liability managment, stochastic optimization, financial planning

1   Introduction

Innovative institutional investors apply asset and liability management (ALM) systems to maximize their shareholders' wealth over time. Prominent examples from a number of organizations are presented in this book, including Frank Russell [7], Towers Perrin [24], Falcon Asset Management, Inc. [28], ORTEC [4], and FANNIE MAE [16]. These long term risk management systems give confidence that the firm will be viable across a range of economic scenarios. In many cases, therefore, the organization can be more aggressive with its investment strategy, thereby achieving a higher long-term growth path.

Today, with the Home Account Advisor™, integrated risk management is practical for the individual investor. In this chapter, we show that ALM system employed by institutions can address the special needs of individuals. For instance, most people should consider tax deferred investment vehicles (e.g. IRA, KEOGH, and 401–K accounts in the US) in addition to taxable investment choices. An important feature is planning for future expenditures not usually considered liabilities, such as college tuition or retirement. In this

---

*This paper was written when the first author was at Home Account Network, Inc.

regard, individual planning presents a case of Total Integrated Risk Management [25].

Achieving long-term financial goals requires discipline to carry out a consistent plan of action. Recently, a woman from New York City who retired in the 1950s with $5000 gave Yeshiva University a gift of $22 million. The lady grew her small nest egg into a sizable fortune during her long retirement years by doggedly following a 'buy and hold' investment strategy. For planning, individuals require information regarding the impact of current decisions on the future. For example, we know that $5000 will grow to about $5,000,000 in 50 years at 15% compounding per year (ex taxes).

Unfortunately, most financial decisions involve a substantial degree of uncertainty and complexity. Not only are returns on assets such as stocks and real-estate random, but also individuals possess liabilities and goals whose cashflows depend upon stochastic events. In the case of a floating rate mortgage, for example, the actual payments are a function of interest rates at specified points in time. Another example is an investor who is saving to buy a house in 5 years, where the price of the house depends upon several factors – including inflation and supply/demand considerations. These issues are clearly stochastic.

The proprietary and patented Home Account Advisor™ developed with the assistance of the authors is based on an investment model with decision rules. There are four major elements: (1) a stochastic system for generating economic scenarios across a long-term planning horizon (20+ years); (2) a multi-objective preference function for addressing the intertemporal tradeoffs among savings, consumption, and wealth accumulation; (3) a set of well-defined decision rules and strategies for investing, borrowing, and income management; and (4) optimization of the decision rules to recommend strategies for the individual.

The ALM system must distinguish a decision making process from a standalone decision, especially for adjusting goals. In life, individuals must be flexible. In order to portray the long term consequences of current decisions, we simulate the investor's financial position with regard to wealth accumulation, savings and borrowing strategies, income, and expenses. At each time stage we rebalance the investment portfolio, change our borrowings, increase or decrease savings, and so on. The range of uncertainties is addressed via a stochastic scenario generator. Typically, a carefully selected set of scenarios form the basis for evaluating the recommended decision strategy.

Solving a multistage stochastic program requires vast numbers of conditional decisions. Often we must provide recommendations within several minutes so that the individual can evaluate alternative strategies and interact with the planning system. Accordingly, we incorporate a set of well-defined rules and strategies for the key decisions – investing, borrowing, savings, and

income management. The result is a multistage stochastic program with decision rules. These decision rules are readily understood by the investor, and
the search time to discover a recommended solution is greatly reduced.

As with institutional ALM systems, risk is calculated as it relates to the
projected uses for the funds – investing for a purpose. Individual financial
planning can be rather elaborate: we must pay taxes on portions of income;
we possess mechanisms for deferring taxes (such as retirement accounts); our
life spans are uncertain (leading to difficulties in determining the planning
horizon as well as end effects); we possess multiple and conflicting objectives
involving consumption and savings, etc.

This chapter is organized as follows. The next section describes the financial planning model, including governing equations, objective function,
scenario generation, and extensions of the base model. Section 3 summarizes
the solution algorithms. Since the stochastic program is non-convex with the
Home Account Advisor^{TM}, we employ search methods that will not stall at
local optimal points. Scatter search and adaptive memory programming (an
advanced form of Tabu search) have proven effective. A test case, illustrat-
ing the features of individual asset liability management, and some of the
capabilities of the system as implemented, is presented in Section 4. Last, we
describe promising areas for future research.

2   Problem Description

2.1   Model overview

This section describes the Home Account Advisor^{TM} as a dynamic investment
model for long-term financial planning. We discuss the modeling topics at a
general level, but reference the specific issues as they pertain to our application. Our model derives from applications of asset/liability management to
institutional investors ([20], [27]).

The system helps individuals make significant financial decisions and establish their priorities in a consistent fashion over a long-term horizon. Typical
decisions include: where to invest, how much to save, and when and how to
borrow. The planning system evaluates a consumer's position on a period
by period basis over the planning horizon. For ease of exposition, we assume
that review points occur annually, although they could be shorter, longer,
or of unequal length (see Figure 1). The system renders changes to the asset and liability positions at the start of each period, evaluating the results
over the coming period, conditional on stochastic scenarios; rebalancing assets or liabilities at times between the review points is not allowed. We start
with the customer's current balance sheet, including major asset positions
(investments, homes, etc.) and liability positions (mortgages, car loans, other

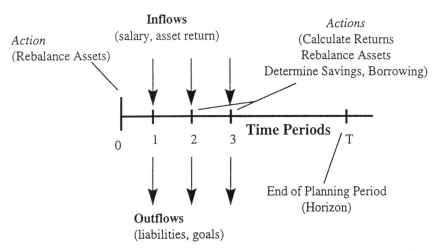

Figure 1: Event timeline. Rebalancings and policy changes occur at specified, discrete points in time over the planning period.

outstanding debts). We calculate an income statement and a cash flow statement over the period, including salary and investment income, tax and loan payments, and other sources and uses of funds. This modifies the balance sheet at the start of the next period; the process repeats for each period in the planning horizon.

Decision variables divide naturally into two sets. On the asset side of the balance sheet, decisions indicate the percentage allocation of assets to each investment category (traditional asset allocation). On the liability side, decisions involve borrowing and leverage. For the description below, we simplify the liability decision to choosing the best mortgage for each of the customer's homes from a prespecified list of mortgage types.

The model evaluates decision variables and financial strategies across a pre-defined set of *scenarios*. A scenario consists of returns on asset categories (e.g. European stocks and US corporate bonds) and economic factors (e.g. interest rates) over time. Each scenario is internally consistent: Bond prices are determined by interest rates, not by a simple covariance matrix with other asset categories. Taken as a group, the scenarios represent a range of possible economic outcomes. Performance, as determined by a multi-objective preference function, is found by evaluating the decision variables on each scenario in turn and summarizing the performance across all scenarios. Section 2.6 contains more details.

We include financial goals in addition to traditional liabilities. In contrast to a liability, a goal depicts a large planned expenditure that is not legally binding. Typical goals include retirement, a new home purchase, and tuition for children's college. Unlike many liabilities, the timing of goals can be flexible: An individual may wish to retire at 60, but may be willing to work another five years if necessary. Other goals are more time sensitive, such as college tuition.

The individual's preference function consists of four primary components: wealth at the end of the planning period, fluctuation of wealth over time, timeliness of meeting goals, and consumption over time. We capture the information in a time-dependent preference function called the Financial Goal Index (FGI)^{TM} as detailed in Section 2.5.

We extend the static, single-period investment model to include a variety of decision rules for investing and savings. Dynamic strategies serve several purposes. First the number of decision variables is greatly reduced as compared with the traditional MSP; conditional decisions in second and subsequent stages are implicit in the decision rules – thus improving computational efficiency. Unfortunately, the resulting model becomes a nonconvex program, complicating the search for the global optimal solution. See [21] for global optimization algorithms that generate a performance bound.

Second, decision rules are conducive to 'out of sample' analyses. Precision tests generate valid confidence limits on the risk measures – wealth at a point in the planning horizon, preference function value, etc. Also, we gain confidence in the recommendations after performing sensitivity analyses on the critical modeling parameters, such as the coefficients of the stochastic differential equations within the scenario generation procedure.

Third, decision rules can be readily interpreted by individual investors. Investors appreciate strategic rules; for example, professional portfolio managers often employ strategies for their own investments. In addition, decision rules can be based on sound theoretical principles. For example, variants of the dynamically balanced rule are optimal for long-term investing in the face of linear transaction costs ([8] and [23]).

For most multiperiod decision rules, the ALM model yields a nonconvex optimization problem (Figure 2). In addition, the constraint set may not be continuous due to tax rules and mortgage choice decisions. To address nonconvexity, we employ a global optimization procedure based in part on Tabu search over a fixed set of scenarios (Section 3).

For selected individuals, the decision rules can be largely eliminated and the model solved as an unconstrained multistage stochastic program. However, computational and informational costs are greatly increased; financial recommendations are generated in the absence of an underlying investment strategy. Nevertheless, a stochastic program may lead to improved recommendations for individuals who have confidence in the process and are able

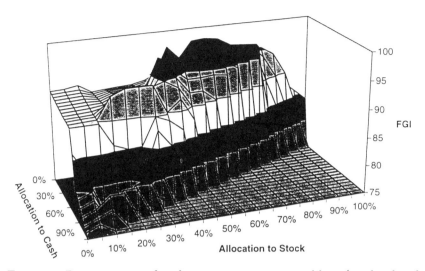

Figure 2: Decision space for three asset category problem (stocks, bonds, cash). Bond position is calculated implicitly as (1-stocks-cash). The objective function is explained in Section 2.5.

to accept the additional computational and informational costs.

2.2 Governing equations

This section defines the constraints of theinvestment model. We illustrate an asset mix decision rule called dynamic rebalancing, in which asset positions are reset to specified fixed percentages at every period, such as 60% stock, 30% bonds, 10% cash. The fixed mix strategy possesses several desirable theoretical properties, including optimality for long term investors under certain restricted conditions such as linear transaction costs ([8]). Also, the amount of turnover is small relative to other investment strategies that take sizeable positions (see [5] in this volume). As such, the approach can be a relatively conservative method which may be employed as a benchmark for other more complex investment strategies.

Other decision rules are easily incorporated (see Section 2.4). We define the relevant sets, decision and accounting variables, inputs, and the governing equations for the system. Additional decision rules and other ancillary matters, such as dynamic leveraging and splitting out taxable and tax deferred investments, are discussed in Section 2.4. Key functions such as the mortgage and tax calculators are discussed in the next subsection.

Define the following sets:

Discrete times at which the portfolio will be rebalanced: $\{1, \ldots, T\}$.

Scenarios which contain asset returns and economic variables: $\{1, \ldots, S\}$.

Asset categories: $\{1, \ldots, I\}$.

Mortgages: $\{1, \ldots, M\}$.

Mortgage categories: $\{1, \ldots, N\}$

Define the following decision variables:

λ_i: Fraction of assets invested in asset category i. $\sum_{i=1}^{I} \lambda_i = 1$, $\lambda_i \geq 0$.

m_j: Mortgage type for mortgage number j.

Define the following accounting variables:

$\tilde{x}_{i,t}^s$: Amount of money (in dollars) in asset category i, time period t, under scenario s *before* rebalancing.

$x_{i,t}^s$: Amount of money (in dollars) in asset category i, time period t, under scenario s *after* rebalancing.

sal_t^s: Income (salary) at time t in scenario s.

maxsav_t^s: Maximum possible savings at time t in scenario s.

sav_t^s: Actual savings at time t in scenario s.

w_t^s: Wealth at time t in scenario s.

tax_t^s: Tax on asset gains from time $(t-1)$ to t, scenario s.

trc_t^s: Total transaction costs at time t, scenario s.

gain_t^s: Appreciation in asset value from time $(t-1)$ to t, scenario s.

$\text{bal}_{m,t}^s$: Outstanding balance on mortgage m, time t, scenario s.

$\text{prin}_{m,t}^s$: Portion of mortgage payment for repayment of principal, scenario s.

$\text{mortint}_{m,t}^s$: Portion of mortgage payment for interest on balance, scenario s.

Define the following coefficients (customer input):

w_0: initial wealth (assets minus liabilities) $x_{i,0}$: Initial allocation of assets in each category sal$_0$: Initial income (salary)

salrate: rate of salary increase with respect to inflation.

l_t^s: liability payment at time t in scenario s.

goal$_t^s$: goal payment at time t, scenario s.

bal$_{m,0}$: initial balance on mortgage m.

mortrate$_m$: interest rate on mortgage (for fixed rate mortgages).

length$_m$: length of mortgage from initial period.

core$_0$: initial core living expenses: food, clothing, utilities, property taxes, etc.

corerate: rate of core expense increase with respect to inflation.

savrate: fraction of discretionary income allocated to savings

tr$_i$: percent transaction cost on asset category i.

k: tax rate - can also be a function of current wealth.

Define the following coefficients (derived from the scenario generation system):

$r_{i,t}^s$: percentage return for asset i, scenario s, from time t to time $(t+1)$

int$_t^s$: short interest rate at time t in scenario s

inf$_t^s$: inflation rate from time t to time $(t+1)$ in scenario s

The following steps execute at each time period t in the specified order. They depict the governing equations of the investment system. Unless otherwise indicated, indices cover the following three sets: $t \in \{1, \ldots, T\}$, $s \in \{1, \ldots, S\}$, $i \in \{1, \ldots, I\}$.

1. Calculate asset position at start of period:

$$\tilde{x}_{i,t}^s = x_{i,t-1}^s * r_{i,t-1}^s$$

$$\text{gain}_t^s = \sum_{i=1}^{I} (\tilde{x}_{i,t}^s - x_{i,t-1})$$

2. Increase salary and expenses at the rate of inflation:

$$\text{sal}_t^s = \text{sal}_{t-1}^s * \left(\inf_{t-1}^s + \text{salrate}\right)$$

$$\text{core}_t^s = \text{core}_t^s * \left(\inf_{t-1}^s + \text{corerate}\right)$$

3. Calculate mortgage balance, and principal and interest payments:

$$\text{bal}_{m,t}^s = \text{bal}_{m,t-1}^s - \text{prin}_{m,t-1}^s$$

$$(\text{prin}_{m,t}^s, \text{mortint}_{m,t}^s) = \text{Mortgage}_{\text{Calc}}(m_j, \text{bal}_{m,t}^s,$$

4. Calculate taxes based on asset gains, salary, mortgage interest, and transaction costs:

$$\text{tax}_t^s = \text{Tax}_{\text{Calc}}(\text{gain}_t^s, \text{sal}_t^s, l_t^s, \text{trc}_{t-1}^s)$$

5. Determine maximum savings - remaining funds available for either savings or consumption:

$$\text{maxsav}_t^s = \text{gain}_t^s + \text{sal}_t^s - \text{prin}_{m,t}^s - \text{mortint}_{m,t}^s - \text{tax}_t^s - \text{trc}_{t-1}^s - \text{core}_t^s - \text{goal}_t^s$$

6. Divide maximum savings between savings and consumption:

$$\text{sav}_t^s = \max(0, \text{maxsav}_t^s * \text{savrate})$$

$$\text{con}_t^s = \max(0, \text{maxsav}_t^s * (1 - \text{savrate}))$$

7. Calculate current wealth:

$$w_t^s = w_{t-1}^s + \text{sav}_t^s$$

8. Rebalance to target fixed mix percentages:

$$x_{i,t}^s = w_t^s * \lambda_i$$

9. Calculate total transaction costs on rebalancing (paid next period):

$$\text{trc}_t^s = \sum_{i=1}^{I} \text{tr}_i * |\tilde{x}_{i,t}^s - x_{i,t}^s|$$

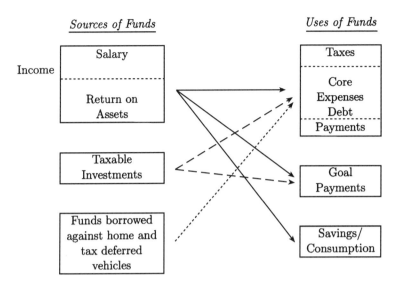

Figure 3: Fund use diagram.

These nine steps fall within the traditional mathematical domain. Two functions in Steps 3 and 4, described in the next section, and the maximization calculation in Step 6, require clarification. One approach to handling Step 6 (i.e. converting a maximum of two quantities into a mathematical programming formulation) is to create a variable for the positive portion of maxsav_t^s. The new variable, $\text{maxsav}+$ requires a negative coefficient in the objective function and four additional constraints:

$$
\begin{aligned}
\text{maxsav}+_t^s &\geq \text{maxsav}_t^s \\
\text{maxsav}+_t^s &\geq 0 \\
\text{sav}_t^s &= \text{maxsav}_t^s * \text{savrate} \\
\text{con}_t^s &= \text{maxsav}_t^s * (1 - \text{savrate})
\end{aligned}
\tag{2.1}
$$

Difficulties arise when insufficient funds exist to meet scheduled goals or liabilities. In this case, funds must be transferred from other sources in order to pay expenses. This situation leads to possible difficulties in providing a continuous mathematical model due to logical conditions.

As an example, suppose a middle-aged couple, Mr. and Mrs. Garrett, are preparing to send their son to college in the current year. After paying taxes, living expenses, and mortgage, there is insufficient money from income to pay the tuition. The Garretts decide to borrow against their home equity to pay the tuition. If the Garretts still did not have adequate funds, they might sell assets in order to pay the tuition.

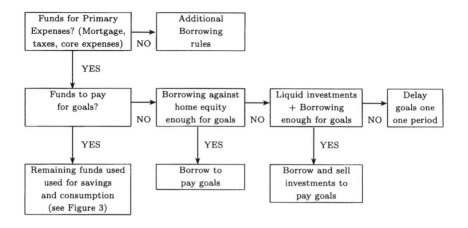

Figure 4: An example income management rule.

The procedure can be formalized into sequential decision steps. First, allocate funds for taxes, core living expenses such as food and clothing, and outstanding debt payments (Steps 5 and 6). If insufficient income exists to meet scheduled goals or core living expenses, borrow against home equity. If there are still not enough funds to meet core expenses, sell investments (stocks, bonds, etc.) or borrow against other non-liquid assets, such as a retirement account. After paying core expenses, if there are not enough funds available from borrowing against home equity or selling assets, postpone goals until the next period. A maximum leverage amount restricts the total available credit. See Figure 3 for a funds usage diagram and Figure 4 for a flow-chart of the selling/borrowing process.

The above procedure provides an example of an *income management rule*. There are many income management rules. For example, the borrowing sequence may be changed, or borrowing to pay for goals may be prohibited.

Income management rules are difficult, but not impossible, to represent in a mathematical program. We illustrate the first part of the above rule, in which funds are assured for core expenses by either selling assets or borrowing. Define the following variables:

invcore_t^s: investments sold to cover core expenses

invtot_t^s: total investments available at time t

invrem_t^s: investments remaining after core expenses

The next equation assures that adequate investments are sold to cover expenses:

$$\text{invcore}_t^s \geq \sum_{m=1}^{M} (\text{prin}_{m,t}^s + \text{mortint}_{m,t}^s) + \text{tax}_t^s + \text{trc}_{t-1}^s + \text{core}_t^s - \text{gain}_t^s - \text{sal}_t^s$$

$$\text{invcore}_t^s \geq 0$$

Herein, the coefficient for invcore_t^s in the objective function must be negative. If all investments are sold and expenses remain, funds must be borrowed:

$$\text{borrow1}_t^s \geq \text{invcore}_t^s - \text{invtot}_t^s$$

$$\text{invrem}_t^s \geq \text{invtot}_t^s - \text{invcore}_t^s$$

$$\text{borrow1}_t^s, \text{invrem}_t^s \geq 0$$

Again, there must be a negative coefficient on borrow1_t^s and invrem_t^s in the objective function. In addition, we can constrain borrow1_t^s based on total wealth w_t^s or other measures such as home value.

The rule–based formulation has the advantage of flexibility. Any income management rule detailing a sequence for fund usage and borrowing, tailored to the desires of the individual, may be employed. In addition, it presents a more accurate picture of future events. For example, the rule–based system gives a distribution of times to achieve goals (Figure 10); if a goal is delayed, the system may try to meet the goal in subsequent periods. In a mathematical program with objective function penalties, information is limited to the unattained goals. A disadvantage of the rule–based system is that the formulation yields a non-convex, integer program which can be extremely difficult to solve, although the number of variables is greatly reduced. Our solution method which utilizes tabu search over a fixed set of scenarios is described in Section 3.

2.3   Model functions

The two functions in Steps 3 and 4 calculate mortgage and tax payments. The mortgage function returns required principal and interest payments. For traditional mortgages, a standard amortization schedule over the remaining years of the mortgage determines the payments. For adjustable mortgages, the principal payment is calculated initially for the entire term of the mortgage and the interest payment is based on the current outstanding balance and governing interest rate. We also include the Home Account mortgage, a specialized and patented mortgage, in which no principal payments are made, only interest payments on the outstanding balance.

When the mortgage choice is exogenous, the mortgage function can be represented as follows:

$$\text{prin}^s_{m,t} = \text{bal}^s_{m,t} * i * \frac{i^{(\bar{t}-t)}}{1 - i^{(\bar{t}-t)}} \qquad (2.2)$$

$$\text{mortint}^s_{m,t} = \text{bal}^s_{m,t} * i$$

where i equals the initial mortgage rate for fixed rate mortgages, $i = \text{int}^s_t$ for adjustable rate mortgages, and $\bar{t}$ is the termination date of the mortgage (the superscripts on i denote exponentiation).

Although the mortgage principal calculation involves nonlinear operations, if optimization does not occur over the mortgage decision, the calculations are performed prior to the model run. If choice of mortgage is required, we generate M binary variables representing each possible mortgage choice, and add the restriction that exactly one of these variables is nonzero. Given difficulties that arise when solving this program, an investment model decision rules approach appears warranted.

The tax calculator separates ordinary income from capital gains, both having separate tax brackets and rates. Mortgage interest is deducted from salary and wages (within certain limits) before income tax is calculated.

2.4   Model extensions

We extend the basic Home Account Advisor™ ALM model along several dimensions. A primary extension splits the aforementioned asset categories into taxable and tax deferred asset classes. Tax deferred assets include investments in IRAs, KEOGHs, 401-K, and other defined contribution retirement plans. As a rule, we assume that funds remain within the tax deferred categories until retirement, but may be borrowed against in certain situations (see Figure 3). By doubling the number of asset decision variables, we allow separate mix percentages for taxable and tax deferred investments. Separate mixes are desirable from a financial planning standpoint, since tax deferred investments should target long term goals, whereas taxable investments may be cashed in for near-term goal payments.

A second extension involves determining the savings rate parameter (savrate). Recall that savrate equals the percentage of discretionary income allocated to savings as opposed to bonus consumption (consumption over and above core living expenses). Instead of fixing a percentage across all time periods and scenarios, we determine savrate based on the likelihood of meeting the specified goals at a particular time in a specific scenario. If a customer has adequate funds to meet all goals, savrate will be close to 0%, but if goals will be hard to achieve, savrate will be close to 100%. The decision is based on a net present value of future goals, compared with current asset position. This approach corresponds to corporate pension plan contributions: If funding status is adequate, the yearly pension plan contribution is minimal. Define two new customer inputs and one accounting variable:

minsav: minimum savings percentage of discretionary income

maxsav: maximum savings percentage of discretionary income

$NPVgoals_t^s$: net present value of goals

Then:
$$savrate_t^s = \max(maxsav, f(wealth_t^s, NPVgoals_t^s))$$

where f is a function indicating willingness to reduce savings when meeting goals appears likely.

A third extension involves an alternative asset allocation decision rule. The extension is the life cycle mix rule, in which there are two classes of asset categories: core and tactical. The tactical asset class consists of asset categories that produce greater risks and possibly greater long-term returns (such as small cap stocks). These categories are generally more appropriate for younger investors with longer planning horizons. Core assets are less volatile and generate income (such as government bonds and money market funds). For young investors, a large percentage of investments should be placed in tactical assets, but for older investors core assets make up a relatively larget portion of the portfolio. The life cycle mix allows the percentage in core and tactical assets to shift over the investment horizon. The mix percentages for asset categories within the tactical asset class (and core asset class) remain fixed over time.

The life cycle mix rule is a generalization of the fixed mix rule. If all the asset classes are put in the core asset category and the percentage of assets in the core category is fixed at 100%, then the life cycle mix problem is equivalent to the fixed mix problem. Although the life cycle mix allows a wider range of solutions, it is harder to solve since there are extra decision variables. For other dynamic rules it is not clear *a priori* which will offer a better solution.

The following additional sets and variables are required for life cycle mix:

Set of core assets: $\{1, \ldots, I_1\}$

Set of tactical assets: $\{1, \ldots, I_2\}$.

λ_i^c: Fraction of wealth in asset class $c \in \{1, 2\}$ invested in asset category i.

$$\sum_{i=1}^{I_1} \lambda_i^1 = 1, \sum_{i=1}^{I_2} \lambda_i^2 = 1, \lambda_i^c \geq 0$$

γ_t: percent of wealth in core assets at time t.

Instead of Step 8, the rebalancing equations are now:

$$x_{i,t}^s = w_t^s * \gamma_t * \lambda_i^1 \quad i = 1, \ldots, I_1 \tag{2.3}$$
$$x_{i,t}^s = w_t^s * (1 - \gamma_t) * \lambda_i^2 \quad i = 1, \ldots, I_2$$

More generally, we can expand the decision-rule framework by allowing the decision variables to depend on time period and scenario. Thus, we replace λ_i with $\lambda_{i,t}^s$, the percentage of assets allocated to category i at time t under scenario s. If we solve this problem without additional constraints to achieve the best allocation in each scenario at every time, we would be assuming perfect foresight with respect to market and economic conditions. The dynamically balanced rule prevents this anticipation and is equivalent to:

$$\lambda_{i,t1}^{s1} = \lambda_{i,t2}^{s2} \quad \forall s1, s2 \in \{1, \ldots, S\} \quad \forall t1, t2 \in \{1, \ldots, T\} \tag{2.4}$$
$$i = 1, \ldots, I \tag{2.5}$$

Other decision rules should be considered, in addition to the life cycle mix described above. For example, allowing the rebalancing percentages to change at different life-stages, such as before and after children are grown:

$$\lambda_{i,t}^{s1} = \lambda_{i,t}^{s2} \quad \forall s1, s2 \in \{1, \ldots, S\}$$
$$i = 1, \ldots, I \qquad t = 1, \ldots, \tilde{t} \tag{2.6}$$

$$\lambda_{i,t}^{s1} = \lambda_{i,t}^{s2} \quad \forall s1, s2 \in \{1, \ldots, S\}$$
$$i = 1, \ldots, I \qquad t = \tilde{t} + 1, \ldots, T \tag{2.7}$$

where a life change occurs between time $\tilde{t}$ and time $\tilde{t} + 1$. This model has two stages with distinct rebalancing mixes for each stage.

Model parameters can be generalized in a number of ways. For example, inputs may be dependent on scenarios: goal payments depending on the inflation rate, and salary growing at, above, or below the inflation rate.

2.5   Objectives

Constructing a meta-objective function for investor preferences over an extended time horizon is a poorly understood task. First, there are temporal considerations, including the tradeoff of short-term increased saving versus achieving long-term goals such as retiring with adequate income. Many investors are unable to save a sufficient amount, despite the belief that they will eventually regret their poor planning. Actions and preferences are intertwined. Second, including uncertainty within an extended time horizon complicates the decision process – partially by creating confusion and potential regret, and partially by preventing the establishment of any hard and fast rules. For example, most financial experts agree that equities hold the greatest potential for investors with long range goals since equities have historically outperformed inflation by a wide margin. Yet, individuals often view stocks as more risky than cash – even investors who have a long term horizon such as 25-year-olds with retirement accounts.

We must define a specific objective function, of course, in order to search for a recommended investment strategy. Our approach specifies a figure-of-merit for each scenario path, and then combines the figures into an aggregate number, the *Financial Goal Index* (FGI)™. The index represents the likelihood of meeting one's goals given a set of plausible outcomes under a fixed set of scenarios and given some risk aversion characteristics. The approach is based on the goal programming.

The index evaluates each wealth path over the planning horizon. A path corresponds to a set of asset decisions, a debt management program (such as choosing a type of mortgage), and a savings/consumption strategy, given a single economic scenario. Four items contribute to the calculation. First, if the investor is unable to meet his planned obligations – not only his legal liabilities, but also his planned goals (e.g. retirement), a penalty is received. The degree of the penalty, p_d is a function of the investor's answers to some preliminary preference questions. The goal index is:

$$\text{INDEXgoal} = 100 - \sum_{t=1}^{T} \frac{1}{1+\rho_{\text{goal}}} * p_d * \text{delay}_t$$

where delay_t equals 1 if a period t goal can not be met in period t and equals 0 otherwise and ρ_{goal} is a discount factor. As an example, suppose an investor has a retirement goal and has chosen the penalty $p_d = 5$ and discount factor $\rho_{\text{goal}} = 0$. A INDEXgoal = 90 indicates that, on average, the retirement goal will be delayed two years.

The next element focuses on volatility of the wealth path. First, an ideal wealth path is determined whereby wealth grows in a smooth manner over the planning horizon. Then, for each scenario, the projected paths are compared with the ideal path so that a correspondence can be established between the idealized path and the actual outcomes. Penalties occur when the outcome turns out to be inconsistent with the investor's preferences. For instance, if the investor displays aversion to volatility, even if the final wealth is relatively large, a penalty of p_p is issued when the wealth path displays greater volatility than the idealized path (Figure 5).

$$\text{INDEXpath} = 100 - \sum_{t=1}^{T} \frac{1}{1+\rho_{\text{path}}} * p_p * \text{short}_t$$

where short$_t$ equals 1 if wealth in period t falls below the 'ideal' wealth path and equals 0 otherwise and ρ_{path} is a discount factor. This example captures downside risk without penalizing upside potential (see [19] and [22]).

The third element of the Financial Goal Index™ measures the bonus consumption over time. Recall that discretionary income (income remaining after all core expenses are met for the year) is divided between savings and bonus consumption. A general approach seeks a utility function of consumption across time, which we simplify by assuming separability and a constant discount rate (ρ_{con}):

$$
\begin{aligned}
\text{INDEXcon} &= U(\text{con}_1, \text{con}_2, \ldots, \text{con}_T) \\
&= \textstyle\sum_{t=1}^{T} \rho_{\text{con}}^t U(\text{con}_t)
\end{aligned}
\tag{2.8}
$$

The fourth element of the index involves the *surplus* position at the end of the planning horizon. Surplus extends the usual definition of wealth to include net present value of future goal payments:

$$
\text{surplus}_t = \text{wealth}_t - \text{NPV}(\text{goals})
$$

Surplus provides a more accurate picture of the investor's current financial position than legal wealth since it includes goals in addition to liabilities. By comparing wealth with surplus, the customer can receive immediate feedback on the impact of goals on his financial outlook.

For the surplus component, the utility function is evaluated at the end of the planning horizon:

$$
\text{INDEXhor} = U(\text{surplus}_T)
$$

In finance, we often maximize criteria involving the horizon wealth and consumption over time. For example, maximizing the discounted utility of consumption is the objective in [8], and [18] combines both the consumption and horizon wealth objectives. For comparisons, the ALM system incorporates these objectives via the weighting on the four criteria and setting the discount parameter appropriately. As an example, the discounted expected utility of consumption model is achieved by setting the weights on INDEXgoal, INDEXpath, and INDEXhor to zero.

Determining a recommended solution given the four objectives just described fits within the realm of multi-criteria optimization, which can be solved with nonlinear goal programming. This framework extends to handle risk characteristics over long time horizons. See Jia and Dyer [17] and Bell [2] for practical methods for dealing with asymmetric risks in a static setting.

We define a utility function over the indices and maximize expected utility:

$$
\text{INDEX} = \sum_{s=1}^{S} p_s U_s(\text{INDEXgoal}, \text{INDEXpath}, \text{INDEXcon}, \text{INDEXhor}) \tag{2.9}
$$

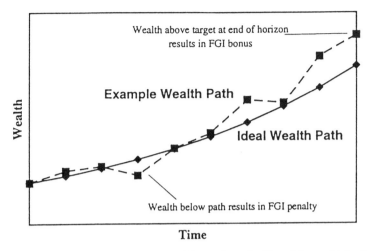

Figure 5: Index calculations for wealth path volatility (downside). Wealth path falling below target path results in a penalty.

Determining a multi-attribute utility function on a rapid basis for individuals is a challenging task. As an initial step, we might simplify equation 2.9 by taking a weighted average of the four indices:

$$U_s = w_1 \text{INDEXgoal}_s + w_2 \text{INDEXpath}_s + w_3 \text{INDEXcon}_s + w_4 \text{INDEXhor}_s$$

where $w_i \geq 0$ and $\sum_i w_i = 1$. Values of these weights can be refined through an iterative process between the individual and an interactive system.

Figure 6 shows an example INDEX for an individual investor. Given that the value lies above 100, this investor will likely achieve his plans based on the system's assumptions, such as savings in accordance with the recommended prescriptions and the chosen scenarios. If the index lies in a lower range, say below eighty, it is less likely that the investor will be able to meet his goals – given the plans as presented. Either the investor will have to increase savings, reduce expenses, or modify future goals or preferences, in which case an explanation of alternatives is warranted.

2.6 Scenario generation process

Modeling the stochastic parameters for the ALM system requires a set of stochastic equations. With the Home Account Advisor™, a sampling procedure identifies a group of representative scenarios. Recall that the optimization proceeds in a discrete time and discrete scenario context. An alternative

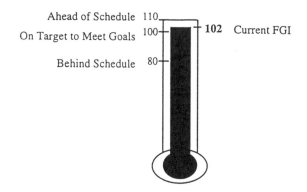

FINANCIAL GOAL INDEX

Figure 6: An example of a goal index. The investor is likely to meet goals in a timely fashion since the current index value exceeds 100.

approach employs dynamic stochastic control in which the state space and the planning periods are discretized, but not the underlying probability distributions ([5, 6]).

The process is diagrammed in Figure 7. First, we generate a time series for economic factors, including short and long government interest rates. Several economic factors display mean reversion, a key ingredient for interest rates and related series. Given these factors, we link the returns for the major asset categories (e.g. stocks, bonds) along with liabilities, borrowing costs, and all other monetary data. As an example, college tuition has risen at an annual rate 3% to 5% above inflation over the past forty years.

The remaining asset returns (secondary categories) depend upon the major asset categories through a conditional variance/covariance structure. Secondary categories include small US stocks, international stocks, and corporate bonds.

Calibrating the economic projection system requires careful consideration of historical relationships and current market conditions. It is important to present a variety of plausible paths of the economic factors over the planning horizaon. Thus, we might generate scenarios in which interest rates and inflation might rise to double digit levels (as in the 1970s), as well as scenarios in which long interest rates stay below 5% for many years (as in the 1950s). To the extent possible, we attempt to meet specified summary statistics for the scenario set S, such as autocorrelations, cross-correlations, and expected values. We employ a model for conducting the calibration task, by extending the simulated moment estimation (SME) procedure proposed by Duffie and

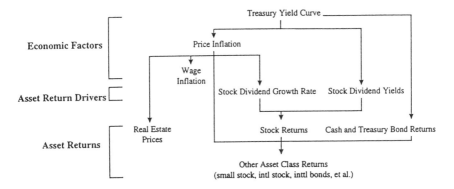

Figure 7: Structure of the scenario generation procedure.

Singleton [9]. Unfortunately, the model results in a non-convex optimization problem, hence global optimization on general search procedures are required.

3   Solution Strategies

Both the dynamically balanced and life cycle mix models give rise to non-convex stochastic optimization models (see also [15] and [21]). We therefore incorporate several efficient metaheuristic search techniques. The first step is to develop a list of high quality starting points, called *reference solutions*, which are based on previous experience, user preferences, or scatter search techniques. Once reference solutions are identified, local searches called *intensifications* are performed around a subset of the reference solutions. Finally, a *reduction* among local optima identifies and combines the best features of these solutions. In essence, we coordinate a number of strategies to search for the global optimum point based on variants of tabu search and scatter search techniques over a fixed set of scenarios.

As an alternative solution strategy, we could employ stochastic programming algorithms for solving the investment planning problem [3]. Due to the non-convexities in the decision rules, however, we must rerun the algorithm from a variety of starting points since the programming algorithms will converge only to local optimal points, i.e. those which satisfy the Karush-Kuhn-Tucker conditions. With this approach, there is no guarantee that a local solution is equivalent to the global solution. See [21] for a global method for solving the fixed mix problem with performance guarantees over a fixed set of scenarios.

3.1   Reference solutions

In a search algorithm, the identification of high quality starting points is important. When an ALM system is implemented, many customers possess similar investment profiles. Solutions previously found optimal for a customer with a like situation are ideal candidates for the reference solution list. In addition, the customer may have a mix that he or she wants tested by the model. An efficient approach to generating reference solutions is to deploy a scatter search on a restricted class of assets. A small set of asset categories, called *key categories* are chosen to represent most or all of the categories in the entire analysis. As an example, we employ Cash (for money market and t-bills), US Bonds (for corporate and international bonds) and Large Capitalized US Stocks (for all other equities). These assets depict aggregate categories (the program operates on several aggregate and disaggregate levels). The mix percentage outside of the key categories is set equal to zero. The key category solution space is divided into a number of equally sized regions, depending on the number of reference solutions desired. A solution is then chosen from each region and added to the reference list.

Scatter search generates a number of plausible solutions in different regions of the solution space by restricting the number of variables analyzed initially. This procedure assumes, for example, that a solution with 60% US government bonds will have similar properties and objective function value as a solution with 60% US corporate bonds.

3.2   Intensification

Intensifications occur around points chosen from a reference solution list. First, we rank all the solutions on the reference list. The best solution is picked and an intensification performed, which may be a strict local optimization or a tabu search. In a strictly local optimization, all neighbor solutions are checked and the algorithm moves to the best neighbor. A neighbor consists of the current mix with two asset category percentages adjusted: one category percentage increased by the stepsize, the second category percentage decreased by the stepsize. More formally, if the current mix vector is λ^0, then λ' is a neighbor if and only if

$$\exists i, j, i \neq j, s.t.$$
$$\lambda'_i = \lambda^0_i + \delta$$

$$\lambda'_j = \lambda^0_j - \delta$$
$$0.0 < \lambda'_i < 1.0$$
$$0.0 < \lambda'_j < 1.0$$

where $\delta > 0$ is the stepsize. If an improved neighbor can not be found, the stepsize is reduced and closer neighbors are rechecked (stepsize may also be increased when substantial improvements occur). If there are no better neighbors via the minimum stepsize, a local optimal solution is found.

3.3 Adaptive memory

The local optimization meshes tabu search. Tabu search prevents certain moves to often-visited neighbors. For example, after finding a local optimum, the tabu procedure continues to new neighbors, thus prohibiting a return to the local optimum for a specified number of moves. The goal is to move away from the current local optimum.

Another strategy prevents moves based on levels of certain *attributes* of the solution. In this case, an attribute might be allocation percentage to a particular asset category or choice of mortgage for the liability decision. If all recent iterates contain a particular type of mortgage, for example, then solution points containing that mortgage could not be visited as part of the search path for a specified number of iterations. This tactic encourages exploration of new regions.

4 Sample Customer Case Study

To illustrate some of the ALM features, we describe a sample family who has engaged the individual ALM system for long-term financial planning. In this example, the investors choose among eight asset categories: cash, US government bonds (long term), US corporate bonds, US small stocks, US large stocks, international bonds, international stocks, and real estate. We have employed the Home Account Advisor system for solving the investor example in the next section. The HOME Account Advisor as implemented posesses specific income management rules, FGI functions, and the other elements described throughout this Chapter.

4.1 Problem description

Steve and Sally Li, a couple in their early forties with two children, are planning their children's college education as well as retirement. They currently possess $90,000 in taxable investments, divided as follows: $30,000 cash (checking/savings and money market accounts), $20,000 government bonds, $30,000 large cap stocks, and $10,000 international stock mutual fund. Their tax deferred portfolio consist of an IRA, in which they have $5,000 in a money market fund and $10,000 in a small stock mutual fund. Their other major asset is a home which is valued at $200,000.

On the liability side, the Li's possess a traditional fixed-rate mortgage: 8.25%, with a balance of $120,000 and 15 years remaining. They currently owe $3,500 on two credit cards at an interest rate of 18% per annum.

The Li's are planning for their two children's college tuition. Michael is starting this fall at an estimated cost of $20,000 per year for four years (including tuition, room, board, and other expenses). Julie will be starting in the year 2000 and the Li's would like to plan for the same expense rate, adjusted for inflation. In addition, Steve and Sally anticipate retiring when Steve reaches 60, in 16 years. Based on discussions with a financial planner, they have selected a reasonable target equal to 80% of their final salary as an annuity going forward.

To meet these goals, the Li's realize they must save a considerable portion of their income. Their pretax income is $60,000 per year, and after taxes ($15,000) and mortgage payments (about $15,000) and core living expenses ($12,000) – leaving the Li's about $18,000 for discretionary purchases (bonus consumption) and savings. They plan on devoting one-third of this amount to savings, about $6,000 per year. Two issues are under study: finding the best investment mix for their asset portfolio, and determining whether they are on track to meet their education and retirement goals.

4.2   Performance results

We solved the Li's problem via the Home Account Advisor[TM] ALM system. The object-oriented code is written in C++ and compiled under Solaris 2.4, Sun Microsystems' version of UNIX, with Sun's standard C++ compiler. We solved all the problems in this section on a SPARCstation 20 with a 75MHz Supersparc chip. The code easily ports to other environments, such as UNIX workstations and PCs.

Execution took 42 seconds (and 45 iterations) to analyze the Li's financial situation. Crucial to the success of global optimization methods is the choice of starting points. Runtimes can be greatly reduced by generating elite starting points tailored for the individual (Section 5). Table 1 presents comparable results on a suite of test problems with varying customer data but consistent starting points and standard options.

As with any global optimizer, performance guarantees are difficult to estimate. To test the algorithm's behavior, we disabled termination tests and plotted the objective function value as a function of iterations. Little improve-

Name	Asset Rule	Mortgage Choice	Dynamic Savings	Number of Iterations	Run time (seconds)
li1	DB	No	Yes	42	45
li2	DB	No	Yes	54	87
investor0	DB	Yes	No	126	56
investor1	DB	Yes	Yes	170	59
investor2	DB	No	No	97	259
investor3	LC	No	No	141	53
investor4	DB	No	No	97	280
investor5	DB	No	Yes	46	90
investor6	DB	Yes	Yes	23	69
investor7	LC	Yes	Yes	36	111
investor8	LC	Yes	No	17	65
investor9	LC	Yes	Yes	21	33
investor10	DB	No	Yes	24	50
investor11	LC	Yes	Yes	226	75
investor12	DB	No	Yes	81	87
investor13	DB	Yes	No	12	18
investor14	DB	Yes	No	40	43
mean				73.7	87.1

Table 1: Iterations and run times for Li's and other test problems. DB - dynamically balance, LC - life cycle. Mortgage Choice includes mortgage selection as part of the optimization. Dynamic Savings adjusts *savrate* parameter every review point.

ment occurs after the algorithm's normal termination criteria (Figure 8). This observation is consistent across a variety of customers.

4.3 Solution presentation and customer interaction

Based on the preliminary ALM analysis, the Li's appear unlikely to meet their goals: Their current index equals 73. By changing their investment mix to the optimal mix their index improves to 88. By examining the individual components of the index (Figure 9) we discern the critical issues. The INDEXgoal index is 65, far below the target value of 100, indicating that the

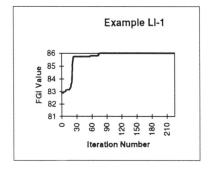 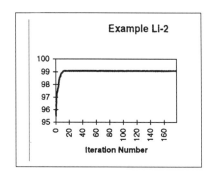

Figure 8: Improvement in solution value with increasing iterations.

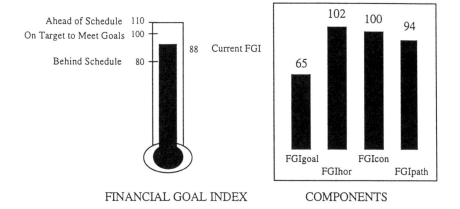

Figure 9: Financial Goal Index for Li customer example

Li's cannot meet their retirement goal by the desired date. The retirement graph (Figure 10), showing the chance of meeting their retirement goal by year, pinpoints this problem: Retirement at 80% of final salary occurs for most scenarios across years 2019-2023, instead of the desired 2013.

The Home Account Advisor™ ALM system makes suggestions for the Li's

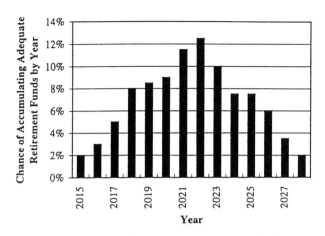

Figure 10: Chance of meeting retirement goals by year.

to improve their financial position. We have designed an interactive display by which the investor adjusts inputs and sees changes in the index. Currently, a sensitivity analysis gives changes in the index with changes in savings rate, expense rate, retirement age, and tuition costs. For the Li's, increasing savings by $900 per year yields an index improvement of 1.6 units. Alternatively, postponing retirement boosts the index by approximately one point per year, decreasing the retirement annuity percentage by 10% will increase the index by 5.2 units.

The Li's can also find key information by comparing the wealth (Figure 11) and surplus (Figure 12) graphs. The former shows assets minus liabilities at each time period over the planning horizon. The Li's appear to be performing well until they must make a large payment at retirement for their annuity. In contrast, the surplus graph provides a more accurate picture; it includes the present value of the tuition and retirement goal expenditures. The Li's will need to save more and garner additional investment returns in order to achieve a positive surplus.

Based on these observations and a reevaluation of their spending and savings patterns, the Li's rerun the financial analysis after adjusting selected inputs (runtime data is in Table 1 on row Li2). In particular, they decide to increase savings by $2,000 per year, for a total of $8,000 (44% of discretionary income) and to retire at age 65 (5 years later) on 70% of salary. The resulting index equals 98 (Figure 13). The Li's are now close to meeting their goals with a reasonable level of confidence over the defined set of scenarios.

The retirement graph in Figure 14 shows the Li's have about a 50% chance

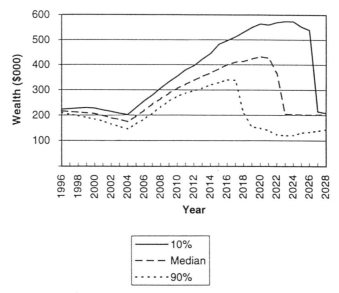

Figure 11: Wealth (assets minus liabilities) by year. Median result with (10%,90%) confidence bands.

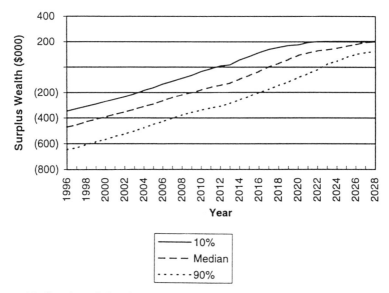

Figure 12: Surplus, defined as assets minus liabilities minus net present value of goals. Median result with (10%,90%) confidence bands.

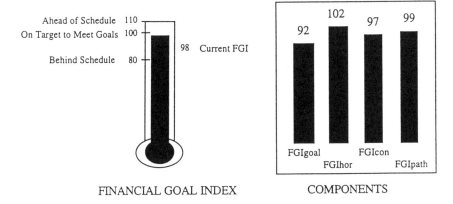

Figure 13: Goal index for the Li's, after adjustments. The improved index is primarily due to increased savings and decrease in retirement annuity.

of meeting their retirement goal by 2018, and about an 80% chance of meeting the goal within two years of their target. Since INDEXgoal lies below 100 (at 92), some scenarios occur wherein the Li's will not have the specified funds until later. Again the recommendation depends upon the chosen set of economic scenarios, and the modeling assumptions and parameters.

5   Conclusions

This paper shows that integrated risk management provides a viable approach for individuals. Both computational and information limits are fast disappearing as computers and real time databases become widely available. ALM systems developed for institutions such as insurance companies and pension plans can now be adapted for individuals.

What are some directions for future research? A primary issue involves constructing investment/savings/borrowing strategies that are tied to individual behavior. The Home Account Advisor™ utilizes two decision rules that have proven effective in practice. Undoubtedly, as greater numbers of consumers engage the system for financial planning, we will find new strategies for targeted groups. Mapping decision rules to individuals appears to be a worthwhile activity. Further, traditional multistage stochastic programming, and perhaps dynamic stochastic control ([5, 6]), promises to be appropriate

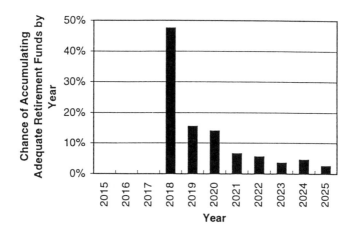

Figure 14: Chance of meeting retirement goals by year, after customer adjustments.

for individuals who are willing to pay the higher computational and informational costs.

Many economists in the US (and elsewhere) argue that the current US savings rate is far too low to improve our long-term standard of living. Part of the cause of this phenomenon, in our experience, is the lack of comprehensive *and* easy to use tools for financial planning. It should not take a 40 page form and 4–6 hours to organize one's financial affairs when meeting with a financial planner. Accordingly, it would be ideal to extract information from disparate sources electronically and without customer intervention, both over the World Wide Web and dedicated computer links with mutual fund, insurance, and other financial companies. Research on mechanizing these data collection, filtering, and maintenance tasks is needed.

A critical task relates to the dynamic and temporal aspects of risks. The standard approach discounts future consumption and possibly wealth, at a constant rate. We have found little empirical evidence for this approach ([1]), except for simplifying the computational tasks. The discussed preference function is an initial effort to approximate a multi-objective and temporal decision analysis. Further research is clearly needed.

Potential users of individual financial planning systems number in the millions of people. Given this perspective, Glover, Mulvey, and Bai [14] have put together an approach for analyzing customers in reference to an evolving underlying population. The approach, called Integrated Population Analysis, gives rise to many applications. For instance, we can improve the computational efficiency of solution algorithms by over 90% via the formation of 'computational neighborhoods' [14]. Other noteworthy applications occur in target marketing, financial engineering, and new product development.

Over the past twenty years in the financial industry, new concepts and products, such as futures, options, options on futures, index funds, mortgage backed securities and index-linked securities, are first deployed by large institutional investors – Wall Street firms, insurance companies, and pension plans. Eventually, the products find their way to individual consumers. We envision a similar path in the area of asset/liability management systems. Consumers deserve to be given the best advice and tools for planning their financial future. The Home Account Advisor's investment model with decision rules provides an integrated framework for conducting financial analysis for individuals.

Acknowledgements

This paper describes the Home Account Advisor™, part of the Home Account System for financial planning and management. The Home Account and its systems are protected by US patent 4953085 and other US and international patents pending.

References

[1] G. Ainslie and N. Haslam. Hyerbolic discounting. In G. Loewenstein and J. Elster, editors, *Choice Over Time*, pages 57–92. Russell Sage Foundation, New York, 1992.

[2] D.E. Bell. Risk, return and utility. *Management Science*, 41:23–30, 1995.

[3] A.J. Berger, J.M. Mulvey, and A. Ruszczyński. An extension of the DQA algorithm to convex stochastic programs. *SIAM Journal on Optimization*, 4:735–753, 1994.

[4] C.G.E. Boender, P.C. van Aalst and F. Heemskerk Modelling and management of assets and liabilities of pension plans in The Netherlands This volume, 561–580.

[5] M.J. Brennan and E.S. Schwartz. The use of Treasury bill futures in strategic asset allocation programs. This volume, 205-228.

[6] M.J. Brennan, E.S. Schwartz, and R. Lagnado. Strategic asset allocation. Technical report, University of California, Los Angeles, 1995 (to appear in *Journal of Economic Dynamics and Control*).

[7] D.R. Cariño, T. Kent, D.H. Myers, C. Stacy, M. Sylvanus, A. Turner, K. Watanabe, and W.T. Ziemba. The Russell–Yasuda Kasai financial planning model. *Interfaces*, 24:29–49, 1994. Reprinted in this volume, 609–633.

[8] M.H.A. Davis and A.R. Norman. Portfolio selection with transaction costs. *Mathematics of Operations Research*, 15:676–713, 1990.

[9] D. Duffie and K. Singleton. Simulated moments estimation of Markov models of asset prices. *Econometrica*, 61:929–952, 1993.

[10] J.S. Dyer. Interactive goal programming. *Management Science*, 19:62–70, 1972.

[11] F. Glover. Tabu search: A tutorial. *Interfaces*, 20:74–94, 1990.

[12] F. Glover. Tabu search fundamentals and uses. Technical report, University of Colorado, Boulder, Colorado 80309, 1995.

[13] F. Glover. Genetic algorithms and scatter search: unsuspected potentials. Technical report, University of Colorado, School of Business, Boulder, CO 80309, Aug 1993.

[14] F. Glover, J.M. Mulvey, and D. Bai. Improved approaches to optimization via integrative population analysis. Technical Report SOR-95-25, Program in Statistics and Operations Research, Princeton University, Princeton, NJ 08544, 1996.

[15] F. Glover, J.M. Mulvey, and K. Hoyland. Solving dynamic stochastic control problems in finance using Tabu search with variable scaling. In I.H. Osman and J.P. Kelly, editors, *Meta-heuristics*, pages 429–448. Kluwer Academic Publishers, Boston, 1996.

[16] M.R. Holmer. Integrated financial product management: an implementation case study. This volume, 581–605.

[17] J. Jia and J.S. Dyer. A standard measure of risk and risk-value models. Technical Report 1, Risk-Value Study Series, University of Texas at Austin, July 1994.

[18] I. Karatzas, J.P. Lehoczky, and S.E. Shreve. Optimal portfolio and consumption decisions for a 'small investor' on a finite horizon. *SIAM Journal on Control and Optimization*, 25:1557–1586, 1987.

[19] A.J. King. Asymmetric risk measures and tracking models for portfolio optimization under uncertainty. *Annals of Operations Research*, 45:165–177, 1993.

[20] M.I. Kusy and W.T. Ziemba. A bank asset and liability management model. *Operations Research*, 34:356–376, 1986.

[21] C.D. Maranas, I.P. Androulakis, C.A. Floudas, A.J. Berger, and J.M. Mulvey. Solving stochastic control problems in finance via global optimization. *Journal of Economic Dynamics and Control*, 21:1405–1425, 1997.

[22] H. Markowitz, P. Todd, G. Xu, and Y. Yamane. Computation of mean-semivariance efficient sets by the critical line algorithm. *Annals of Operations Research*, 45:307–317, 1993.

[23] R.C. Merton. Lifetime portfolio selection under uncertainty: the continuous-time case. *Review of Economic Statistics*, 51:247–257, 1969.

[24] J.M. Mulvey. Generating scenarios for the Towers Perrin investment systems. *Interfaces*, 26:1–15, 1996.

[25] J.M. Mulvey. It always pays to look ahead. *Balance Sheet*, 4:23–27, Winter 1995/96.

[26] Mulvey, J.M., S. Correnti and J. Lummis. Total integrated risk management: insurance elements. Princeton University Report, SOR–97–2, 1997.

[27] J.M. Mulvey and H. Vladimirou. Stochastic network programming for financial planning problems. *Management Science*, 38:1642–1664, 1992.

[28] J.C. Sweeney, S.M. Sonlin, S. Correnti and A.P. Williams Optimal insurance asset allocation in a multi-currency environment. This volume, 341–368.